Acta Physica Austriaca

Supplementum XXIII

Proceedings of the
XX. Internationale Universitätswochen für Kernphysik 1981
der Karl-Franzens-Universität Graz
at Schladming (Steiermark, Austria)
February 17th—26th, 1981

Sponsored by
Bundesministerium für Wissenschaft und Forschung
Steiermärkische Landesregierung
International Centre for Theoretical Physics, Trieste
Sektion Industrie der Kammer der
Gewerblichen Wirtschaft für Steiermark

1981

Springer-Verlag
Wien New York

New Developments in Mathematical Physics

Edited by
Heinrich Mitter and Ludwig Pittner, Graz

With 54 Figures

1981

Springer-Verlag
Wien New York

Organizing Committee

Chairman

Prof. Dr. H. Mitter
Institut für Theoretische Physik
der Universität Graz

Committee Members

L. Mathelitsch
L. Pittner
W. Plessas
F. Widder

Secretary

M. Krenn
E. Neuhold

Softcover reprint of the hardcover 1st edition 1981

Library of Congress Cataloging in Publication Data

Internationale Universitätswochen für Kernphysik der Karl-Franzens-Universität Graz (20th : 1981 : Schladming, Austria)
New developments in mathematical physics.

(Acta physica Austriaca. Supplementum ; 23)
"Proceedings of the XX. Internationale Universitätswochen für Kernphysik 1981 der Karl-Franzens-Universität Graz at Schladming (Steiermark, Austria) February 17th-26th, 1981. Sponsored by Bundesministerium für Wissenschaft und Forschung ... [et al.]"--
1. Mathematical physics--Congresses. I. Mitter, Heinrich. II. Pittner, Ludwig. III. Austria. Bundesministerium für Wissenschaft und Forschung.
IV. Title. V. Series.
QC19.2.I58 1981 530.1'5 81-16720

ISSN 0065-1559
ISBN-13:978-3-7091-8644-2 e-ISBN-13:978-3-7091-8642-8
DOI: 10.1007/978-3-7091-8642-8

CONTENTS

SEMINARS

FOREWORD

The papers contained in this volume are lectures and seminars presented at the 20th "Universitätswochen für Kernphysik" in Schladming in February 1981. The goal of this school was to review some rapidly developing branches in mathematical physics. Thanks to the generous support provided by the Austrian Federal Ministry of Science and Research, the Styrian Government and other sponsors, it has been possible to keep up with the - by now already traditional - standards of this school. The lecture notes have been reexamined by the authors after the school and are now published in their final form, so that a larger number of physicists may profit from them. Because of necessary limitations in space all details connected with the meeting have been omitted and only brief outlines of the seminars were included. It is a pleasure to thank all the lecturers for their efforts, which made it possible to speed up the publication. Thanks are also due to Mrs. Krenn for the careful typing of the notes.

H.Mitter L.Pittner

Acta Physica Austriaca, Suppl. XXIII, 3–28 (1981)

CLASSICAL SCATTERING THEORY+

by

W. THIRRING
Institut für Theoretische Physik
Universität Wien, Austria

1. INTRODUCTION

It was first recognized by Hunziker [1] that the notions of scattering theory play an important role in classical mechanics. It turned out [2] that it leads to non-trivial information for the global properties of the solutions of the classical trajectories. For instance it shows that in the three body problem there are large regions in phase space with 2n - 1 = 17 constants of motion and all trajectories in this region are homotopic to straight lines. Furthermore Wigner's [3] time delay has a simple geometrical meaning [4] for the trajectories. Recently Bollé and Osborn [5] succeeded in deriving even a classical analogue to Levinson's theorem. In these lectures I shall, following [6], show in detail how the phase shift corresponds to the generator of the S-transformation. For this purpose we define in the next section canonical coordinates for a one-, two- and three-dimensional configuration space such that this statement assumes a

+Lectures given at the XX.Internationale Universitätswochen für Kernphysik, Schladming,Austria,February 17-26, 1981.

simple form. This sheds some light on how trajectories with large time delays or loopings generate resonances of the quasiclassical phase shift. In the following section we give an alternative proof of the classical form of Levinson's theorem and illustrate its subtle feature by some examples. Finally we give a simple derivation of how a Dollard's [7] change in the free motion leads to the Coulomb phase shift as generator of the classical S-transformation for a 1/r-potential.

We shall employ the following

Notations:

$$\Theta(x) \equiv \begin{cases} 1 & \text{for} \quad x > 0 \\ 0 & \text{for} \quad x < 0 \end{cases} \qquad \text{(step function)}.$$

$$f \circ g(x) = f(g(x)) \qquad \text{(composition of maps)}.$$

$$\sup_x f(x) = \text{least upper bound of } f.$$

$$\vec{a} \wedge \vec{b} = \text{vector product}.$$

2. THE S-TRANSFORMATION

Scattering theory investigates the asymptotic behaviour of the trajectories in phase space. Only some observables, typically functions of the momentum, will converge for $t \to \pm\infty$. This suffices to define the scattering angle

$$\Theta = \sphericalangle (p_+, p_-) , \qquad p_\pm = \lim_{t\to\pm\infty} p(t) ,$$

but if the potentials decrease faster than 1/r for $r \to \infty$ additional information becomes available. Then time evolution Φ_t: $(x(0),p(0)) \to (x(t),p(t))$ though not tending to a limit for $t \to \pm\infty$ approaches the free time evolution Φ_t^o such that $\Phi_{-t} \circ \Phi_t^o$ tends to a limit in some regions $D_\pm$:

$$\Omega_\pm = \lim_{t\to\pm\infty} \Phi_{-t} \circ \Phi_t^o .$$

This implies the convergence of $\vec{x}(t) - \vec{p}(t).\, t$ and allows the definition of the time delay compared with the free motion. Since Φ and Φ^o are canonical transformations the "Möller-transformations" $\Omega_\pm$ will in general be a local canonical transformation mapping the domains $D_\pm$ into ranges $R_\pm$. Closed orbits will be excluded from $R_\pm$ but in reasonable [8] cases their union with R_+ or R_- will fill almost all of phase space. The scattering transformation

$$S = \Omega_+^{-1} \circ \Omega_- = \lim_{t\to\infty} \Phi^o_{-t/2} \circ \Phi_t \circ \Phi_{-t/2} \tag{1}$$

transforms D_- into D_+. If Φ_t^o is the free time evolution having straight lines as trajectories, S has a simple geometrical meaning. For negative t,$\Phi_{-t} \circ \Phi_t$ means that you follow the straight trajectory back for a time $-|t|$ and then continue with the actual time evolution for the same length of time. If the forces have a finite range and Φ and Φ^o coincide outside a certain region then $\Phi_{-t} \circ \Phi_t^o$ will become independent of t as soon as t leads you outside this region (Fig. 1). Then the limit is attained, Ω_- maps the straight line onto this trajectory of Φ_t which is asymptotically tangent to it. Similar arguments for Ω_+ show that S maps (Fig. 1) the straight lines tangent for $t \to -\infty$ onto the ones tangent for $t \to +\infty$. It follows from its definition that S commutes with the free time evolution: $S \circ \Phi_t^o = \Phi_t^o \circ S$. As a canonical transformation one should be able to exhibit its generator which actually is possible. We first study the special cases.

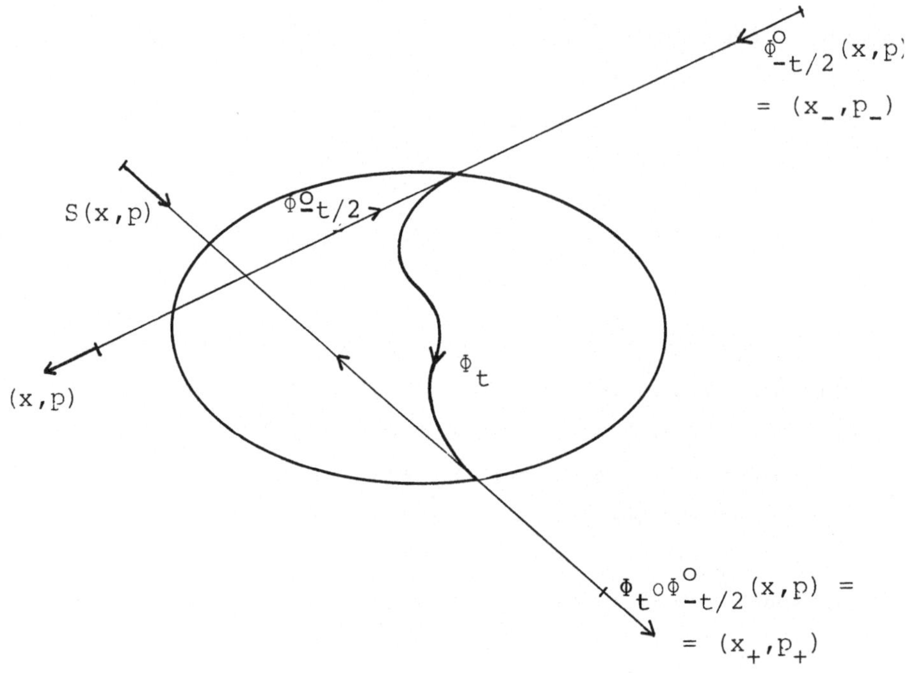

Fig. 1. Motion in configuration space.

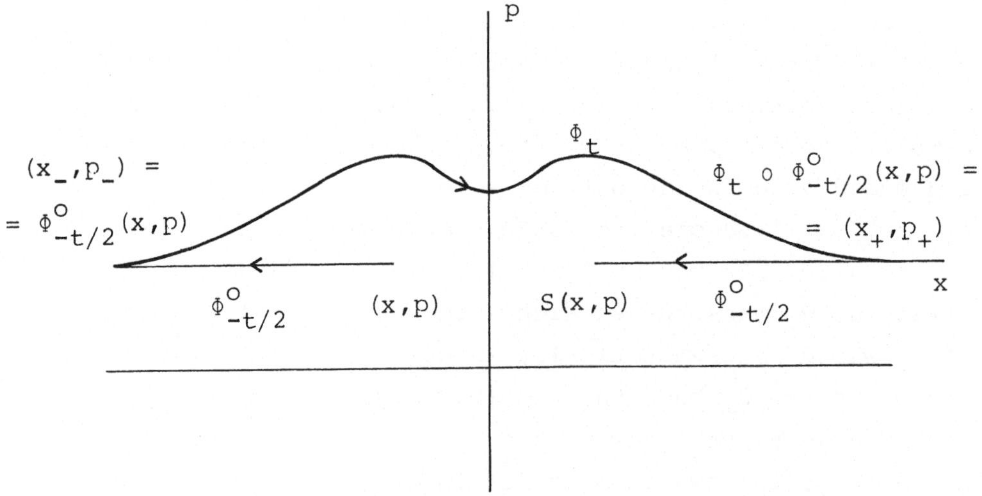

Fig. 2. One-dimensional motion in phase space.

Examples

a) One-dimensional motion

Let (x,p) be the canonical variables and consider the motions due to $H^o = p^2/2$, $H = p^2/2 + V(x)$ where V(x) has finite range or falls off sufficiently fast. The corresponding flows are

$$\Phi_t^o: (x,p) \to (x + pt, p)$$

$$\Phi_t(x,p) \to (x(t), \sqrt{p^2 + 2(V(x) - V(x(t)))} \quad . \tag{2}$$

If $p^2 > \sup_x 2V(x)$, so that there is no reflection, x(t) is determined by

$$t = \int_x^{x(t)} \frac{d\alpha}{\sqrt{p^2 + 2(V(x)-V(\alpha))}} \quad .$$

In this case the S-transformation can be easily constructed: If we call $x_- = \Phi^o_{-t/2}(x) = x - pt/2$, $x_+ = \Phi_t(x_-)$ then $\Phi^o_{-t/2} \circ \Phi_t \circ \Phi^o_{-t/2}$ acts as

$$(x,p) \xrightarrow{\Phi^o_{-t/2}} (x-pt/2,p) \xrightarrow{\Phi_t} (x-pt/2+\int_{x_-}^{x_+} d\alpha, \sqrt{p^2+2V(x_-)-2V(x_+)})$$

$$\xrightarrow{\Phi^o_{-t/2}} (x-\frac{t}{2}(p+\sqrt{p^2+2V(x_-)-2V(x_+)})+\int_{x_-}^{x_+} d\alpha, \sqrt{p^2+2V(x_-)-2V(x_+)}) \tag{3}$$

where

$$t = \int_{x_-}^{x_+} \frac{d\alpha}{\sqrt{p^2 + 2(V(x_-) - V(\alpha))}} \quad .$$

For $t \to \infty$ we have $x_- \to -\infty$, $x_+ \to +\infty$ and $V(x_-)$, $V(x_+) \to 0$. Then

$$(x,p) \overset{S}{\to} (x-p \int_{-\infty}^{\infty} d\alpha (\frac{1}{\sqrt{p^2-2V(\alpha)}} - \frac{1}{\sqrt{p^2}}),p) \equiv (x-p\tau,p) \quad , \tag{4}$$

i.e. S changes x by p times the time delay τ. The latter is the difference of the times the actual and the free time evolutions need for the trajectory from (x_-,p_-) to (x_+,p_+) in the limit $x_- \to -\infty$, $x_+ \to +\infty$ (Fig. 2). S is the canonical transformation $(x,p) \to (x - 2\partial\delta(p)/\partial p, p)$ where

$$\delta(p) = \frac{1}{2} \int_{-\infty}^{\infty} d\alpha(\sqrt{p^2 - 2V(\alpha)} - \sqrt{p^2}).$$

If there is an x_1 such that $p^2 < 2V(x_1)$ then a trajectory with $x < x_1$, $p > 0$ will be reflected to the left. In this case the action (3) of S is changed to ($V_- \equiv V(x_-)$, etc.)

$$(x,p) \xrightarrow{\Phi^0_{-t/2}} (x - pt/2, p) \xrightarrow{\Phi_t} (x_+, -\sqrt{p^2+2V_- -2V_+}) \xrightarrow{\Phi^0_{-t/2}}$$

$$\to (x_+ + \frac{t}{2}\sqrt{p^2 + 2V_- - 2V_+}, -\sqrt{p^2 + 2V_- - 2V_+}) .$$

Here

$$t = \int_{x_-}^{x_o} \frac{d\alpha}{\sqrt{p^2+2V_- -2V(\alpha)}} + \int_{x_+}^{x_o} \frac{d\alpha}{\sqrt{p^2+2V_- -2V(\alpha)}}$$

where the reflection point x_o is the smallest x with $2V(x) = p^2 + 2V_-$. If $x_o < 0$ we may write

$$x_+ = - x_- - \int_{x_+}^{x_o} d\alpha - \int_{x_o}^{-x_o} d\alpha - \int_{-x_o}^{-x_-} d\alpha ,$$

and thus for $t \to \infty$ we obtain

$$(x,p) \overset{S}{\to} (\lim_{t\to\infty}(-x+pt/2-\int_{x_+}^{x_o} d\alpha-\int_{x_o}^{-x_o} d\alpha- \int_{-x_o}^{-x_-} d\alpha+ \frac{t}{2}\sqrt{p^2+2V_- -2V_+}), -\sqrt{p^2+2V_- -2V_t}) =$$

$$= (-x + 2p \int_{-\infty}^{x_o} d\alpha\{\frac{1}{\sqrt{p^2-2V(\alpha)}} - \frac{1}{\sqrt{p^2}}\} - \int_{x_o}^{-x_o} d\alpha, -p) .$$

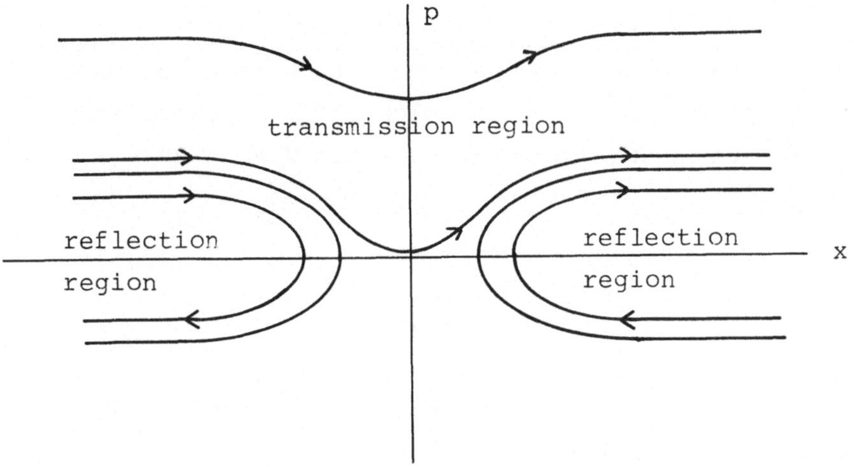

Fig. 3. Trajectories in phase space for a repulsive potential.

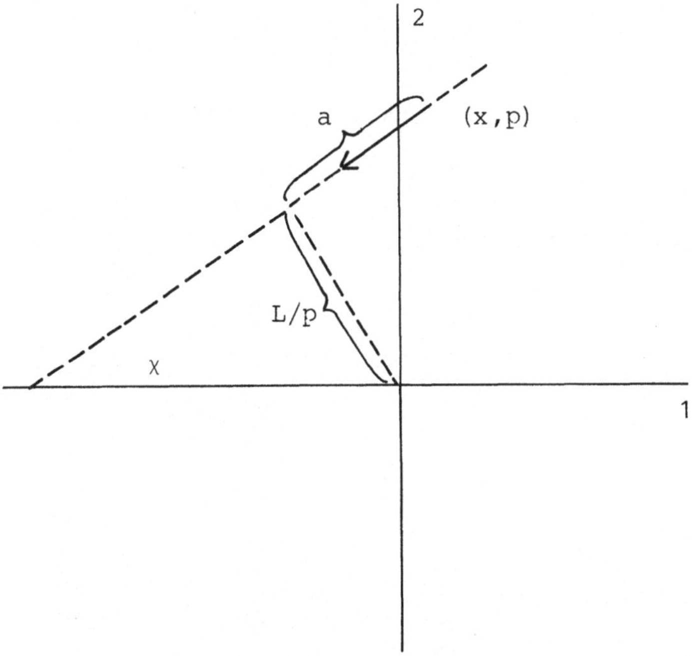

Fig. 4. The canonical coordinates in 2 dimensions.

Thus the time delay is in comparison with a free motion going up to the origin. If V is twice differentiable it becomes infinite when p^2 approaches $2 \sup_x V(x) = 2V(x_o)$ because

$$\int \frac{d\alpha}{\sqrt{V'' \cdot (x_o - \alpha)^2}} = \infty .$$

Then there is an orbit which approaches x_o for $t \to \infty$ and for this value of p S does not exist. It separates the region in phase space where S is of the form (3) and the present case (Fig. 3) where

$$(x,p) \overset{S}{\to} \left(-x + 2\frac{\partial\delta(p)}{\partial p}, -p\right) ,$$

$$\delta(p) = \int_{-\infty}^{x_o(p)} d\alpha\left(\sqrt{p^2 - 2V(x)} - \sqrt{p^2}\right) - p|x_o(p)| .$$

For example the square well potential, $V(x) = V\,\Theta(R-|x|)$, gives

$$\delta(p) = R\left(\sqrt{p^2 + 2|V|} - \sqrt{p^2}\right) \quad \text{for} \quad V < 0$$

$$= R\left(\sqrt{p^2 - 2V} - \sqrt{p^2}\right) \quad \text{if} \quad p^2 > 2V$$

for V>0 .

$$= -pR \quad \text{if} \quad p^2 < 2V$$

b) Two-dimensional motion

Consider again $H^o = \frac{\vec{p}^2}{2}$, $H = \frac{\vec{p}^2}{2} + V(x)$. It is convenient tc use the variables $p = |\vec{p}|$, $a = \frac{\vec{p}\cdot\vec{x}}{p}$, $L = |\vec{x} \wedge \vec{p}|$ and $\chi = \arccos p_x/p$, i.e. (Fig. 4)

$$p_x = p \cos\chi , \quad x = a\cos\chi - \frac{L}{p}\sin\chi ,$$

$$p_y = p \sin\chi , \quad y = a\sin\chi + \frac{L}{p}\cos\chi .$$

Since L generates a rotation and $(\vec{p}\cdot\vec{x})$ the dilation $(x,p) \to (x\, e^{\beta},\ p\, e^{-\beta})$ one sees readily that the new variables are canonical, i.e. $\{a,p\} = \{\chi,L\} = 1$, the other Poisson brackets vanishing. The free time evolution is simply $(a,\chi;\ p,L) \to (a + pt,\chi;\ p,L)$ but in general Φ_t will be complicated. S maps free trajectories into free trajectories and will be of the form

$$(a,\chi;p,L) \overset{S}{\to} (a - p\tau,\chi';p,L') \quad ,$$

χ' and L' independent of a. If V is a central potential $V(|\vec{x}|)$ or more generally of the form $V(|\vec{x}|,L)$ so that L is constant, then Φ_t can be reduced to a one-dimensional problem and S constructed explicitly. The chain (3) of maps becomes in that notation

$$(a,\chi;p,L) \xrightarrow{\Phi^{o}_{-t/2}} (a-pt/2,\chi;p,L) \xrightarrow{\Phi_t} (a-pt/2+\int_{a_-}^{a_+} d\alpha,\chi';\sqrt{p^2+2(V_- - V_+)},L) \to$$

$$\xrightarrow{\Phi^{o}_{-t/2}} (a-\tfrac{t}{2}(p + \sqrt{p^2+2(V_- - V_+)}) + \int_{a_-}^{a_+} d\alpha,\chi';\sqrt{p^2+2(V_- - V_+)},L) . \quad (5)$$

Again for $t \to \infty$ we have $|a_+|,\ |a_-| \to \infty$, $V_+,\ V_- \to 0$. The time t for the actual motion Φ_t is readily expressed as integral over $r \equiv |\vec{x}| = \sqrt{a^2+L^2/p^2}$ with $E = p^2/2 + V_-$,

$$t = \int_{r_o}^{r_-} \frac{dr}{\sqrt{2E-L^2/r^2-2V(r)}} + \int_{r_o}^{r_+} \frac{dr}{\sqrt{2E-L^2/r^2-2V(r)}} \quad ,$$

where $\sqrt{\ } = 0$ for $r = r_o$. Since $a = \sqrt{r^2 - L^2/p^2} = r + O(1/r)$ and $d\alpha = p\, dr/\sqrt{p^2 - L^2/r^2}$ we have for $r_+,\ r_- \to \infty$:

$$\int_{a_-}^{a_+} d\alpha = \left(\int_{L/p}^{r_-} + \int_{L/p}^{r_+}\right) \frac{p\, dr}{\sqrt{p^2-L^2/r^2}} .$$

Thus in the limit $t \to \infty$ we arrive at

$$(a,\chi;p,L) \overset{S}{\to} \left(a - 2\,\frac{\partial\delta(p,L)}{\partial p},\quad - 2\,\frac{\partial\delta(p,L)}{\partial L},\ p,L\right),$$

$$\delta(p,L) = \lim_{R\to\infty} \left(\int_{r_o}^{R} dr\,\sqrt{p^2 - L^2/r^2 - 2V(r)} - \int_{L/p}^{R} dr\,\sqrt{p^2-L^2/r^2}\right). \tag{6}$$

Here we have used the well-known expression for the scattering angle $\chi-\chi'$.

Let us consider the two-dimensional S-transformation for typical potentials.

$1/r^2$-potential:

If $V(r) = c/r^2$ then S exists only for $c > - L^2/2$. In the attractive case trajectories with the impact parameter and therefore L too small, spiral into the origin. For the others the r-integral is easily calculated, for instance by complex integration. Evaluating the residue at the origin we find

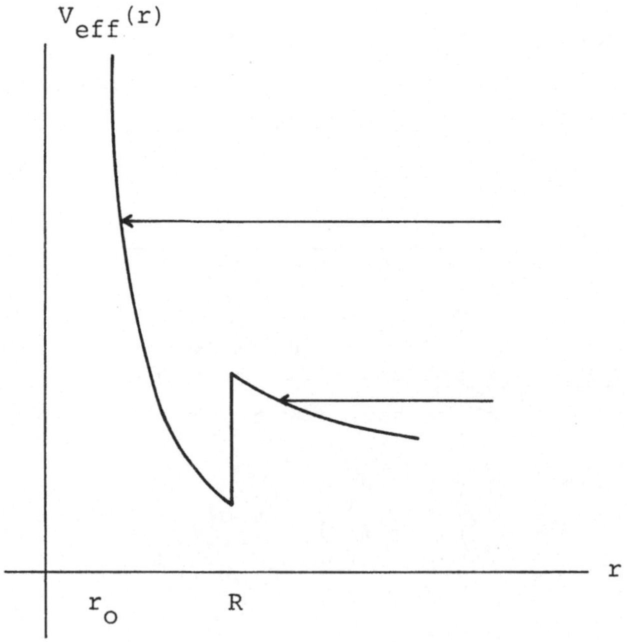

Fig. 5. The effective square well potential.

$$\delta(p,L) = \frac{\pi}{2}\,[\sqrt{L^2+2c} - L]\ .$$

Since δ is independent of p the time delay is zero. This is related to the dilation properties of the Hamiltonian (compare [2, 3.4.15.3]. The scattering angle $\pi\left(\frac{L}{\sqrt{L^2+2c}} - 1\right)$ tends to ∞ for $c<0$ for those L where spiral orbits set in.

Square well potential:

If $V(r) = c\,\Theta(R-r)$ we find

$$\delta(p,L)=\{\sqrt{(p^2-2c)R^2-L^2}+L \text{ arc sin}\frac{L}{R\sqrt{p^2-2c}}\}\Theta(p^2R^2-L^2)\Theta((p^2-2c)R^2-L^2)\cdot$$

$$-\{\sqrt{p^2R^2-L^2}+L \text{ arc sin}\frac{L}{Rp}\}\ \Theta(p^2R^2 - L^2)\quad .$$

In the attractive case δ is discontinuous for $p^2 = 2c + L^2/R^2$ because there r_o jumps from $L/\sqrt{p^2-2c}$ to L/p (see Fig. 5).

Thus δ is given in the two regions $p^2 \lessgtr 2c + L^2/R^2$ by different expressions whereas on the separating hypersurface S does not exist. For a rounded off potential there are the trajectories which asymptotically reach the maximum of the potential but never get over it. For a twice differentiable potential the time delay would become ∞ whereas for the square well

$$\tau = \frac{2}{p}\frac{\partial\delta}{\partial p} = \frac{2}{p^2}\{\sqrt{p^2R^2 - L^2 - 2cR^2} - \sqrt{p^2R^2 - L^2}\}$$

remains finite. In any case this is the closest similarity to a quantum mechanical resonance since the special value $\delta/\hbar = \pi/2$ has classically no significance.

c) Three-dimensional motion

There are many canonical coordinate systems such that the free motion just shifts one coordinate. We shall choose one where $|\vec{p}|$, L and L_z occur such that for central potentials the S-transformation is simple. A convenient choice are the coordinates used in [2, 5.3.4] with $\vec{x}$ and $\vec{p}$ exchanged: (see Fig. 6)

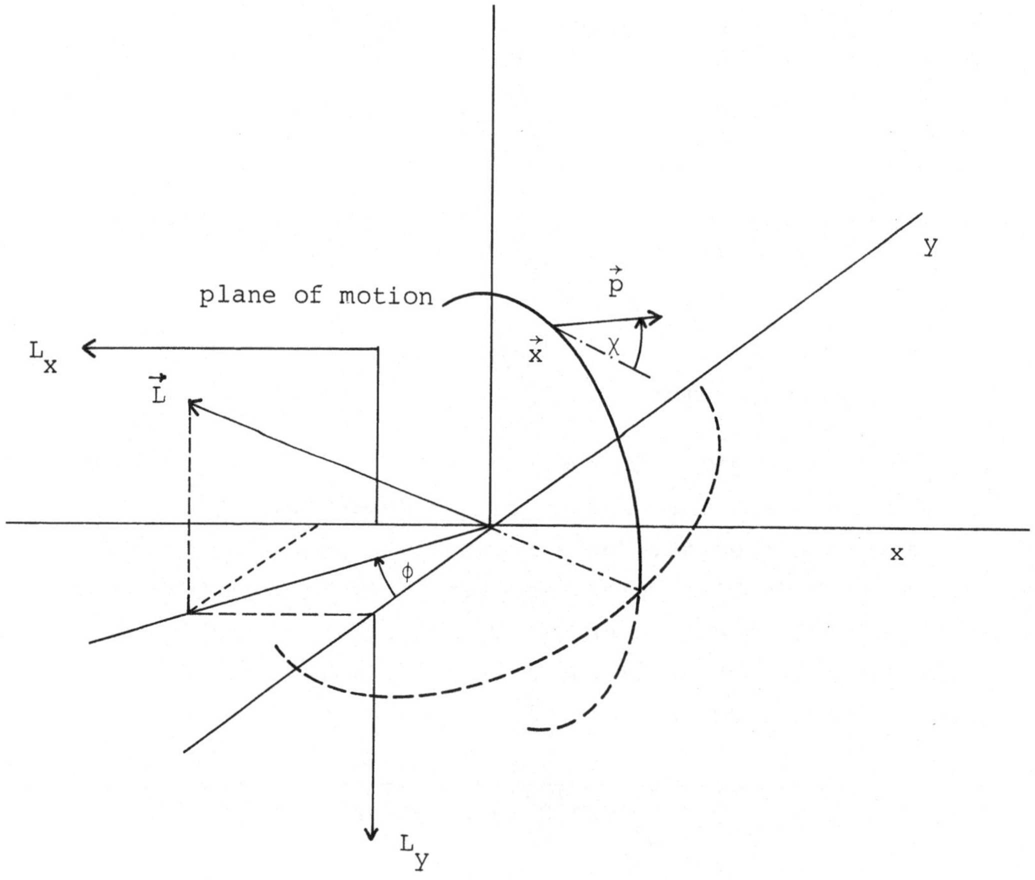

Fig. 6. The canonical coordinates in 3 dimensions.

$$p = \sqrt{p_x^2 + p_y^2 + p_z^2}\ , \quad L = |\vec{x} \wedge \vec{p}|, \quad L_z = [\, x \wedge p]_z\ ,$$

$$a = \frac{\vec{x}\cdot\vec{p}}{p}\ , \quad \chi = \text{arc cos}\ \frac{L_x p_y - L_y p_x}{p\sqrt{L_x^2 + L_y^2}}\ , \quad \phi = \text{arc cos}\ \frac{L_y}{\sqrt{L_x^2 + L_y^2}}\ .$$

χ is the angle of momentum in the plane of motion and ϕ is the angle of the projection of $\vec{L}$ in the (x,y)-plane. For the proof of the canonicity of these coordinates see (2). The free motion is $(a,\chi,\phi;p,L,L_z) \to (a + pt,\chi,\phi;p,L,L_z)$. For a central potential the trajectory remains in a plane and the S-transformation can be found as in b):

$$(a,\chi,\phi;p,L,L_z) \to \left(a - 2\,\frac{\partial\delta(p,L)}{\partial p}\ , \chi - 2\,\frac{\partial\delta(p,L)}{\partial L},\phi;p,L,L_z\right)\ ,$$

$$\delta(p,L) = \lim_{R\to\infty} \left(\int_{r_o}^{R} dr\ \sqrt{p^2 - L^2/r^2 - 2V(r)} - \int_{r_o'}^{R} dr\ \sqrt{p^2 - L^2/r^2}\right). \tag{7}$$

<u>Remarks</u>

1. We see that the generator of S is the so-called [8] quasiclassical approximation for the quantum mechanical phase shift $\delta/\hbar$. A close relation is to be expected since the quantum mechanical S-matrix $S = e^{2i\delta/\hbar}$ generates the above transformation. However the expression (6) is classically not an uncontrollable approximation but the exact result.

2. The limit $R \to \infty$ exists if V decreases for $r \to \infty$ as $r^{-1-\varepsilon}$, $\varepsilon > 0$. In this case one sees easily that also the limit $t \to \infty$ in the definition of S exists.

3. In the two- and three-dimensional case the time delay is measured by comparing the time of the trajectory with the following free motions: Follow the one tangent for $t \to -\infty$ up to r_o', i.e. the point of closest approach

to the origin, then switch over to the free trajectory tangent for $t \to +\infty$ at the same r_o'.

4. Since pa = (xp) generates dilations its change under S is given by a generalization of the virial theorem to infinite trajectories. One finds for the time delay [4]

$$\tau = \frac{1}{E} \int_{-\infty}^{\infty} dt \left(2V(x(t)) + \vec{x}(t) \cdot \vec{\nabla} V(x(t))\right) \quad .$$

5. Since $\delta(p,L)$ goes to zero for $L \to \infty$ we may write

$$\delta(p,L) = \frac{1}{2} \int_{L}^{\infty} dL' (\chi'(p,L,\chi) - \chi) \quad .$$

δ has the dimension of an action. Choosing $\hbar$ as unit of angular momentum

$$\frac{1}{\hbar} \delta = \frac{1}{2} \int \frac{dL'}{\hbar} (\text{—"—})$$

becomes dimensionless. If the deflection angle exceeds π over an interval $\hbar$ and otherwise keeps the same sign the $\delta/\hbar$ goes beyond 90°. Thus we see the following connection between resonances and looping trajectories. A looping for all angular momenta in the interval $(L', L' + \hbar)$ and $L' > L$ implies a "resonance" in the sense that the quasiclassical $\delta(p,L)$ is larger than 90°. Generally resonances occur for those L for which the sum of the deflection angles for larger L's reaches 180°. An analogous statement can be made with respect to the time delay since

$$\delta(p,L) = \int_{p}^{\infty} p' \, dp' \, \tau(p',L) \quad .$$

6. If $H = H^o + \lambda V$ we see

$$\frac{\partial \delta}{\delta \lambda} = -\frac{1}{2} \int_{-\infty}^{\infty} dt \, V(x(t)) \quad .$$

This is the classical version of an analogue to the Hellmann-Feynman formula in scattering theory [10]. Thus also classically $V \gtrless 0$ implies $\delta \lessgtr 0$. If we were to confine the system in a ball with radius R the right hand side above is $- R/p \cdot$ (time average of V). This is a classical analogue of Schwinger's relation between phase shift and energy shift of the system in ball. This furthermore shows that Kato's monotonicity [13] is classically obvious.

7. Although we don't have a general explicit expression for $\Omega_\pm$ outside the range of the potential for $a > 0$, Ω_+ is π and Ω_- therefore equals S. Similarly for $a < 0$ $\Omega_- = \pi$ and $\Omega_+ = S^{-1}$.

8. Everybody conversant with modern classical mechanics has more powerful methods available than the pedestrian ones employed sofar. He would generalize our results as follows. Since the time evolution changes the canonical 1-form by the exterior derivative of the action (see [1], (3.2.9)) we have with the previous notation

$$\vec p_- \cdot d\vec q_- = \vec p \cdot d\vec q + dw_-^o, \quad (q_+, p_-) = \Phi^o_{-t/2}(q,p), \quad w_-^o = -\int_{-t/2}^{o} dt' \Phi^o_{t'}\left(\frac{\vec p^2}{2}\right),$$

$$\vec p_+ \cdot d\vec q_+ = \vec p_- \cdot d\vec q_- + dw, \quad (q_+, p_+) = \Phi_t(q_-, p_-), \quad w = \int_o^t dt' \; \Phi^o_{t'}\left(\frac{\vec p_-^2}{2} - V(x_-)\right),$$

$$\vec p_s \cdot d\vec q_s = \vec p_+ \cdot d\vec q_+ + dw_+^o, \quad (q_s, p_s) = \Phi^o_{-t/2}(q_+, p_+), w_t^o = -\int_{-t/2}^{o} dt' \Phi^o_{t'}\left(\frac{\vec p_+^2}{2}\right).$$

Adding these equations up gives

$$\vec p_s \cdot d\vec q_s = \vec p \cdot d\vec q + d(w + w_-^o + w_+^o)$$

and thus for $t \to \infty$ the general form of the generator of $S: (q,p) \to (q_s, p_s)$. For the evaluation of the path integrals one may use

$$\int^{t} dt' \, \Phi_{t'} \left(\frac{p^2}{2} - V(q)\right) = \int_{q}^{q_t} dq' \sqrt{p^2+2V(q)-2V(q')} - Et$$

and rederive the previous expressions. Jajima [12] has shown that the difference between the action and the free action is the classical limit of the quantum phase-shift. To make it useful as generator of a classical canonical transformation is a matter of finding the appropriate canonical variables.

3. THE CLASSICAL LEVINSON THEOREM

Levinson's theorem relates the change of the phase between $E = 0$ and $E = \infty$ to the number of bound states. This seemingly wave-mechanical statement corresponds to a classical geometrical fact relating the volume in phase space of the bound orbits to the integral over the time delay. We shall now give a simple derivation of this relation using the fact that Ω and S as canonical transformation preserve the volume in phase space (Compare[14]).

Before embarking on calculations let me illustrate the intuitive background in the case of an attractive potential in one dimension. Consider the effect of Ω_+ on the rectangular region $\{(x,p): |x| < R, 0 < p < \sqrt{2E}\}$ in phase space. If R is larger than the range of the force the line $x = R$ is left invariant whereas $x = -R$ is shifted by $p\tau < 0$. $p = 0$ is mapped just above the bound orbits whereas $p = \sqrt{2E}$ is slightly shifted upward in the range of the potential but for large E the change in momentum becomes small (Fig. 7). Since Ω_+ is an area preserving transformation the region lost to the bound orbits must equal the strip on the left gained by the shift by τp. If E tends to ∞ then $\int_0^\infty dp \, p|\tau|$ should become equal to the area of the bound orbits which is the classical version of Levinson's theorem.

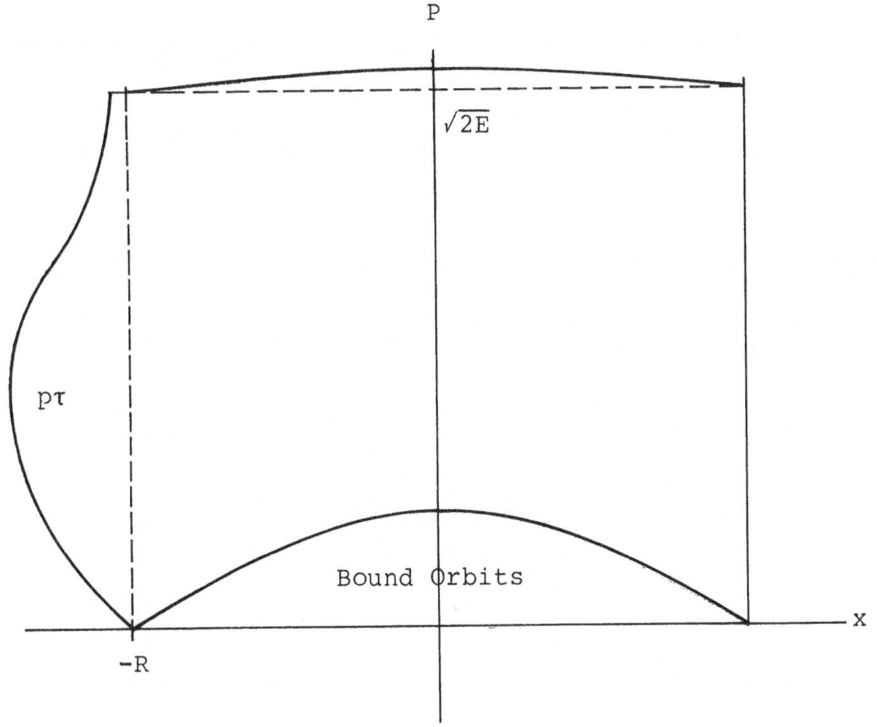

Fig. 7. Action of Ω_+ in Phase Space.

Let the phase space be decomposed by the characteristic functions χ_b and χ_s into the regions of bound orbits and scattering trajectories. χ_b is 1 in the former and 0 in the latter region and χ_s vice versa. Furthermore we first confine the integration to compact regions in phase space by the characteristic function

$$\Theta_E(x,p) = \begin{cases} 1 & \text{if } H(x,p) < E \\ 0 & \text{otherwise} \end{cases} ,$$

$$\Theta_R(x,p) = \begin{cases} 1 & \text{if } x^2 < R^2 \\ 0 & \text{otherwise} \end{cases} .$$

Now

$$\int d^\nu x\, d^\nu p\, \Theta_E\, \Theta_R = \int d^\nu x\, d^\nu p\, \Theta_E\, \Theta_R(\chi_b + \chi_s) =$$

$$= \int d^\nu x\, d^\nu p\, \Theta_E\, \Theta_R\, \chi_b + \int d^\nu x\, d^\nu p\, \Theta_E \circ \Omega_+\, \Theta_R \circ \Omega_+ \tag{8}$$

since $\chi_s \circ \Omega_+$ is one. Furthermore $\Theta_E \circ \Omega_+ = \Theta(E - p^2/2)$ such that

$$\int d^\nu x\, d^\nu p\, \Theta_E\, \Theta_R\, \chi_b = \int d^\nu x\, d^\nu p\, \Theta_R[\Theta_E - \Theta(E - p^2/2)] +$$

$$+ \int d^\nu x\, d^\nu p\, \Theta(E - p^2/2)[\Theta_R - \Theta_R \circ \Omega_+] . \tag{9}$$

We now have to consider the limits $R \to \infty$, $E \to \infty$ and assume that the potential is reasonable enough so that this can be done with impunity. For the discussion of the right hand side we distinguish between different dimensions:

1) $\underline{\nu = 1}$

Assume that $V(x)$ is uniformly bounded, $c_1 \leq V(x) \leq c_2$. Then

$$\lim_{R\to\infty} \int dx\, dp\, \Theta(R^2 - x^2)[\Theta(E - \frac{p^2}{2} - V(x)) - \Theta(E - \frac{p^2}{2})] =$$

$$= \int dx\, 2(\sqrt{2(E - V(x))} - \sqrt{2E}) \qquad \text{for} \quad E > c_2 . \tag{10}$$

This integral exists if V is integrable (which is necessary for the existence of scattering theory) and tends to zero for $E \to \infty$ as $-\frac{2}{\sqrt{2E}} \int dx\, V(x)$. For discussing the last term we assume first that the potential has finite support such that $V(x) = 0$ for $|x| > R_o$. Then according to remark 7 in § 2

$$\Theta_R \circ \Omega_+ = \Theta(xp)\Theta_R + \Theta(-xp)\, \Theta_R \circ S^{-1} . \tag{11}$$

If the potential does not have finite support but decreases faster than $1/|x|^{1+\varepsilon}$ then (11) can be replaced by

$$\lim_{R\to\infty} [\Theta_R \circ \Omega_+ - \Theta(x\ p)\ \Theta_R - \Theta(-x\ p)\ \Theta_R \circ S^{-1}] = 0 \quad . \tag{12}$$

The effect of S^{-1} on x is known to be the shift $p\tau(p)$. The last term contributes only for $xp < 0$ and then only if an $|x| < R$ is shifted by S to be $> R$ or vice versa. Thus the x integral picks up $p\tau$ from one end or the other:

$$\lim_{R\to\infty} \int dx\ dp\ \Theta(E - \frac{p^2}{2})\ [\Theta_R - \Theta_R \circ \Omega_+] = - \int dp\ |p|\tau(p)\Theta(E- \frac{p^2}{2}) =$$

$$= - 2 \int d\varepsilon\ \tau(\varepsilon)\ \Theta(E - \varepsilon) \quad .$$

Thus finally in the limit $E \to \infty$ we arrive at

$$\int dx\ dp\ \chi_b = - \int dp\ |p|\ \tau(p) \quad . \tag{13}$$

<u>Examples:</u>

1. If $V(x) < 0$ then

$$\int dx\ dp\ \chi_b = \int dx\ dp\ \Theta(2|V(x)| - p^2) = \int dx\ 2\sqrt{2|V(x)|} \quad .$$

On the other hand

$$-\int dp\ |p|\tau(p) = -\int dx \int |p|\ dp \left(\frac{1}{\sqrt{p^2-2V(x)}} - \frac{1}{\sqrt{p^2}}\right) = 2\int dx \sqrt{2|V(x)|} \, .$$

2. Repulsive square well: $\chi_b = 0$. According to § 2 we have

$$\tau = 2R \cdot \begin{cases} \left(\dfrac{1}{\sqrt{p^2-2V}} - \dfrac{1}{\sqrt{p^2}}\right) & \text{for} \quad p^2 > 2V \\[2ex] -\dfrac{1}{|p|} & \text{for} \quad p^2 < 2V \ , \end{cases}$$

and in fact

$$\int_0^\infty dp\ p\tau(p) = \int_0^{\sqrt{2V}} dp(-2R) + \int_{\sqrt{2V}}^\infty dp\ 2R\ \left(\frac{1}{\sqrt{p^2-2V}} - \frac{1}{\sqrt{p^2}}\right) = 0 .$$

2) $\underline{\nu = 2}$

For $E > c_2$ the first term in (9) becomes simple and E-independent:

$$\int d^2x\ d^2p[\Theta(E - \frac{p^2}{2} - V(x)) - \Theta(E - \frac{p^2}{2})] = -2\pi\int d^2x\ V(x) .$$

For calculating the last integral (11) remains valid for potentials with compact support. The statement (12) becomes too weak for two dimensions, because the x-integration runs over a sphere and there it has to be replaced by the condition

$$\lim_{R\to\infty} R^{1+\varepsilon}[\Theta_R \circ \Omega_+ - \Theta_R \circ S^{-1}\Theta(-xp) - \Theta_R\ \Theta(xp)] = 0 . \qquad (12')$$

This condition is met by potentials decreasing faster than $1/r^{2+\varepsilon}$.

Fixing $\vec{p}$ the integration over x runs over half of the sphere after replacing Ω by S^{-1}. The time delay depends on p and L and ϕ (which we can forget after fixing $\vec{p}$). Since we already assumed that V vanishes sufficiently at infinity it follows that

$$\lim_{L\to\infty} L^{1+\delta}\tau(p,L,\phi) = 0$$

so that τ is integrable in L. Now we turn to the coordinates introduced in § 2 for $\nu = 2$. Since $|\vec{x}| = a + O(1/a)$ the action of S for large $|\vec{x}|$ becomes $S(|\vec{x}|) = S(|\vec{x}| - p\tau)$. Thus for $R \to \infty$ $\Theta_R - \Theta_R \circ S^{-1}$ becomes $\Theta(R - |\vec{x}|) - \Theta(R - - |\vec{x}| - p\tau)$. Since the coordinates are canonical the volume element in phase space is dp da dL dϕ and the integral can be treated as for $\nu = 1$. We find

$$\lim_{R\to\infty} \int d^2x\, d^2p[\Theta_R - \Theta_R \circ S^{-1}]\Theta(-xp)\Theta(E - \frac{p^2}{2}) =$$

$$= - \int dp\, d\phi\, \Theta(E - \frac{p^2}{2})\, dL\, p\tau(p,L,\phi)$$

or in the case of spherical symmetric potential

$$= - 2\pi \int d\varepsilon\, \Theta(E - \varepsilon)\, dL\, \tau(\varepsilon,L) \quad .$$

It should be noted that for $E > c_2$ the first and second expression in (9) become E-independent, thus it follows that for $E > c_2$ we must have

$$\int dL\, d\phi\, \tau(p,L,\phi) \equiv 0 \, .$$

For spherical symmetric potentials this can be shown explicitly:

$$\tau(\varepsilon,L) = 2\int dr\, [\frac{1}{\sqrt{2(\varepsilon-V(r))-L^2/r^2)}}\Theta(r-r_o(L,p)) - \frac{1}{\sqrt{2\varepsilon-L^2/r^2}}\Theta(r-L/p)]$$

where $V(r_o) + L^2/2r_o^2 = \varepsilon$. Thus after changing the order of integration the two contributions cancel upon L-integration:

$$\int dr\, r[\int_o^{r\sqrt{2(\varepsilon-V(r))}} \frac{dL}{\sqrt{2(\varepsilon-V(r))r^2-L^2}} - \int_o^{r\sqrt{2\varepsilon}} \frac{dL}{\sqrt{2\varepsilon r^2-L^2}}] = 0 \, .$$

Thus in two dimensions Levinson's theorem reads

$$\int d^2p\, d^2x\, \chi_b = -2\pi \int d^2x\, V(x) - \int d^2p\, dL\, \tau(\vec{p},L) \, .$$

We note a correction term which has already been found in [5].

3) $\nu = 3$

The first term on the r.h.s. of (9) becomes now (again for $E > c_2$)

$$\int d^3x\, d^3p\, [\Theta(E - \frac{p^2}{2} - V(x)) - \Theta(E - \frac{p^2}{2})] =$$

$$= \frac{4\pi}{3} 2^{3/2} \int d^3x\, ([E - V(x)]^{3/2} - (E)^{3/2})$$

which in leading order in E

$$= -4\pi \int d^3x\, \sqrt{2E}\, V(x) \quad .$$

The evaluation of the last term is completely analogous as for two dimensions. The condition on the potential has to be strengthened to V(r) decreases as $1/r^{3+\varepsilon}$. Thus

$$\lim_{R\to\infty} R^{2+\varepsilon} [\Theta \circ \Omega_+ - \Theta_R \circ S^{-1}\Theta(-xp) - \Theta_R\Theta(xp)] = 0$$

suffices for replacing Ω by S. The integration over the surface of the halfball can be replaced in the limit $R \to \infty$ by that over the halfphase such that we obtain

$$\lim_{R\to\infty} \int d^3x\, d^3p\, [\Theta_R - \Theta_R \circ S^{-1}]\, \Theta(-xp)\, \Theta(E - \frac{p^2}{2}) =$$

$$= -\int dp\, d\phi\, d\chi\, \Theta(E - \frac{p^2}{2})\, dL\, dL_z\, p\tau(p,L,L_z,\phi,\chi) \quad .$$

Generally the result for $\nu = 3$ is

$$\int d^3p\, d^3x\, \chi_b = \lim_{E\to\infty} [-\sqrt{2E}\, 4\pi \int d^3x\, V(x) - \int_0^{\sqrt{2E}} dp \int d\phi\, d\chi\, dL_z\, dL\, p\tau] \; .$$

Since the left hand side of (9) becomes independent of E for sufficiently large E we obtain the relation

$$\int d\phi\, d\chi\, dL_z\, dL\tau = 4\pi \int d^3x\, [\sqrt{2(E - V(x))} - \sqrt{2E}]\ \forall E > \sup_x V(x).$$

For a spherical potential this relation can be checked explicitely since

$$\int d\phi\, d\chi\, dL_z\, dL\tau(p,L)=4\pi^2 \int_0^{L_{max}} dL^2 2\int dr\left[\frac{1}{\sqrt{2(E-V(r))-L^2/r^2}} - \frac{1}{\sqrt{2E-L^2/r^2}}\right] = 16\pi^2 \int r^2 dr\, [\sqrt{2(E-V(r))} - \sqrt{2E}]\ .$$

It should be noted that in more than 1 dimension a negative potential does not necessarily generate a negative time delay. Though the particle becomes faster it may have to cover a longer trajectory.

4. THE COULOMB POTENTIAL

For $V = e^2/r$, $\lim_{t\to\infty} \Phi_t \circ \Phi^o_{-t}$ does not exist: Φ^o_{-t} maps $\vec{x}$ into $\vec{x}-\vec{p}t$, but for the Kepler motion this quantity goes for $t \to \infty$ as $\ln t$. Following Dollard [7] one considers another Φ^o_t which has also straight trajectories but covered with a non-uniform speed. In the notation of § 2 in two and three dimensions the free motion Φ^o_{-t} is

$$a \to a - pt - \frac{e^2}{p^2} \ln(t+1)\, \frac{p}{2}\ ,$$

the other variables remaining constant. Then the chain of maps (5) changes a (for $t \to \infty$) into

$$a - pt - \frac{2e^2}{p^2} \ln tp/4 + \int_{a_-}^{a_+} d\alpha \qquad \text{where}$$

$$t/2 = \int_{r_o}^{R} \frac{dr}{\sqrt{p^2 - \frac{2e^2}{r} - \frac{L^2}{r^2}}} = \frac{1}{p}\sqrt{(R-\frac{e^2}{p^2})^2 - \frac{L^2}{p^2} - \frac{e^4}{p^4}} +$$

$$+ \frac{e^2}{p^3} \operatorname{arcosh} \frac{R-e^2/p^2}{\sqrt{\frac{L^2}{p^2} + \frac{e^2}{p^4}}} \overset{R\to\infty}{\simeq} \frac{R}{p} - \frac{e^2}{p^3} + \frac{e^2 \ln R}{p^3} -$$

$$- \frac{e^2}{2p^3} \ln(L^2 + \frac{e^4}{p^2}) + \frac{e^2}{p^3} \ln p/2$$

and $a_\pm = R + O(1/R)$. Thus for $R \to \infty$ we find

$$S(a) = a + \frac{2e^2}{p^2} + \frac{e^2}{p^2} \ln(L^2 + \frac{e^4}{p^2}) \ .$$

Using the well known expression for the Coulomb scattering angle we have in three dimensions

$$S(a,\chi,\phi,p,L,L_z) \to (a - 2 \frac{\partial\delta(p,L)}{\partial p}, - 2 \frac{\partial\delta(p,L)}{\partial L}, \Phi; p,L,L_z) ,$$

$$\delta(p,L) = \frac{1}{2i}[(L + \frac{ie^2}{p})\ln(L + \frac{ie^2}{p}) - (L - \frac{ie^2}{p}) \ln (L - \frac{ie^2}{p})].$$

Remarks

1. Φ_t^o is for fixed t a canonical transformation but Φ^o is not a one parameter group. Its choice is to a large extent arbitrary, one has only to see that the ln R term cancels. This liberty affects the time delay but not the scattering angle.
2. It is remarkable that for the smooth potentials $V = 1/r$ or $1/r^2$ the classical and quantum δ's are so similar: Essentially one has to replace L by $1/2 + \sqrt{\vec{L}^2 + 1/2}$ to obtain the quantum phase shift. In the Coulomb case we have chosen Φ^o such that there is no additional term depending only on p. Such a contribution only enters into the time delay and depends on the choice of Φ^o. Φ determines δ up to a function of p only. Also quantum-

mechanically the Coulomb phase shift can be deduced up to a function of p by studying the asymptotic properties of Φ [11].

3. The change in Φ^o does not repair Levinson's theorem for the Coulomb potential. Also the analog of Hellmann-Feynman's theorem and therefore the sign rule are not valid in this case. Note that now for $e^2 > 0$ (repulsive case) we have $\tau < 0$ whereas for $V = e^2 r^{-\nu}$, $\nu > 1$, we get according to remark 6 of § 2 that

$$\tau = \frac{e^2}{E}(2-\nu) \int_{-\infty}^{\infty} dt\, r(t)^{-\nu} .$$

References

1. W. Hunziker, Scattering in Classical Mechanics, in Scattering Theory in Mathematical Physics, J.A. Levita and J. Marchands eds, Boston, D. Reidel, 1974.
2. W. Thirring, Classical Dynamical Systems, Springer, New York (1978).
   M. Breitenecker, W. Thirring, Suppl. Nuovo Cim. 2/4 (1979) 1.
3. E. Wigner, Phys. Rev. 98 (1955) 145.
4. H. Narnhofer, Another Definition of Time Delay, to be published in Phys. Rev..
5. T.A. Osborn, R.G. Froese, S.F. Howes, Phys. Rev. A22 (1980) 101.
   D. Bollé, T.A. Osborn, Sum Rules in Chemical Scattering, Preprint KUL Leuven.
6. W. Thirring, Quantenmechanik von Atomen und Molekülen, Springer Wien (1979) 156 and
   H. Narnhofer, W. Thirring, The Canonical Scattering Transformation in Classical Mechanics, Vienna preprint 80-25, submitted to Phys. Rev..
7. J.D. Dollard, J. Math. Phys. 5 (1964) 729.

8. M. Reed, B. Simon, Scattering Theory, Academic Press, New York (1979) 11.
9. L.D. Landau, E.M. Lifschitz, Vol.III, Quantummechanics, § 126, Pergamon, Oxford (1958).
10. R. Blankenbecler, R. Sugar, Phys. Rev. 136 (1964) 472.
11. H. Grosse, H.R. Grümm, H. Narnhofer, W. Thirring, Acta Physica Austriaca 40 (1974) 97.
12. K. Jajima, Mathematical Problems in Theoretical Physics, Springer, Lecture Notes in Physics 116, p. 73.
13. T. Kato, Hadronic Journal 1,1 (1978) 134.
14. A. Martin, Il Nuovo Cimento 10 (1958) 607.

Acta Physica Austriaca, Suppl. XXIII, 29–63 (1981)

GEOMETRIC METHODS IN SCATTERING THEORY+

by

V. ENSS
Inst. f. Mathematik, Ruhr-Universität Bochum
D-4630 Bochum 1, BRD

We give an introductory outline of some of the main ideas, concepts, and techniques which are relevant in the geometric, time dependent approach to spectral and scattering theory of Hamiltonian operators in nonrelativistic quantum mechanics. For further details and references we refer to the lecture notes [7] of the Erice summer school 1980, which contain a discussion of various alternative routes, omitted here, and to [4,2,3,10,11,13,19,21].

CONTENTS

+Lecture given at the XX. Internationale Universitätswochen für Kernphysik, Schladming, Austria, February 17-26, 1981.

I. RELATION OF SPECTRAL AND SCATTERING THEORY

Consider e.g. the quantum mechanical system of a particle which is influenced by a potential V. The dynamics of the system is governed by the Schrödinger equation

$$i \frac{d}{dt} \psi(t) = H\psi(t) \tag{1.1}$$

where the state $\psi(t)$ at any time t lies in the state space $\mathcal{H}$ which is a Hilbert space, e.g. $L^2(\mathbb{R}^\nu, d^\nu x)$ if the particle moves in ν-dimensional space. The Hamiltonian H is a self-adjoint linear operator on $\mathcal{H}$. It is obtained as a perturbation of the free Hamiltonian H_o which describes the kinetic energy of the particle:

$$H_o = \frac{1}{2m} (\vec{p})^2 = - \frac{1}{2m} \Delta \tag{1.2}$$

where the latter expression applies to configuration space wave functions. Then

$$H = H_o + V \tag{1.3}$$

where typical potentials V are real multiplication operators in $\vec{x}$-space, or non local, or pseudodifferential operators describing velocity dependent forces. In general both H_o and V are unbounded operators, therefore the definition of their sum requires some care. There are plenty of methods to define such sums (see e.g. [16]) and to determine the domain $\mathcal{D}(H)$ on which H is self-adjoint. With the spectral theorem and functional calculus we can form functions of H, e.g. $\exp(-iHt)$, $t \in \mathbb{R}$. These are unitary operators, called propagators, or time-evolution operators, which map a state at some time T into that state at time T + t. The Schrödinger equation is solved by

$$\Psi(t) = \exp(-iHt)\Psi \tag{1.4}$$

with initial condition $\Psi(0) = \Psi$.

The propagators have some properties with a direct physical meaning. (i) The group property

$$\exp(-iHt)\cdot\exp(-iHT) = \exp(-iH(t+T));$$

it means that waiting one second and then three seconds is the same as waiting four seconds.

(ii) Strong continuity:

$$\lim_{t\to 0} \|\exp(-iHt)\Psi - \Psi\| = 0 , \tag{1.5}$$

i.e. a state changes arbitrarily little within a short enough time interval.

Conversely Stone's theorem asserts that any strongly continuous unitary group can be written as exp(-iHt) with a uniquely determined self-adjoint operator H, called its generator. Thus the subtle mathematical property of self-adjointness of the Hamiltonian is well motivated by physical properties which a decent time evolution of states must have. Moreover it is equivalent whether one studies the Hamiltonian H or the propagator (or other functions like the resolvent). Which one is simpler to analyse will depend on the particular question.

In the context of quantum mechanics "spectral theory" mainly is spectral theory of the Hamiltonian H. One is interested in the spectrum of H: $\sigma(H)$. It is typically bounded below:

$$\sigma(H) \subset [a,\infty), \; a \in R ,$$

and it consists of several parts: the point spectrum $\sigma_p(H)$ which is the set of eigenvalues, i.e. those λ for which

$$H\Psi = \lambda\Psi \tag{1.6}$$

has a solution $\Psi \in H$. The closed linear span of the eigenvectors, the "point spectral subspace", is denoted by H_p. Its orthogonal complement is the continuous spectral subspace

$$H_{cont} = (H_p)^{\perp} .$$

Then the <u>continuous spectrum</u> is defined as

$$\sigma_{cont}(H) = \sigma(H \restriction H_{cont}).$$

For $\lambda \in \sigma_{cont}(H)$ but $\lambda \notin \sigma_p(H)$ one can solve the partial differential equation

$$Hf = \lambda f \tag{1.7}$$

in a suitably chosen function space. The solutions f are not physical states, they are "eigenfunctions" or "improper states". However certain continuous superpositions of eigenfunctions are contained in the state space H. In what follows we will avoid completely the use of improper states because they are mathematically hard to control and their physical interpretation is rather indirect.

By the spectral theorem there corresponds to H a unique family of projections $E(\lambda)$, $\lambda \in \mathbb{R}$. The corresponding subspaces Ran $(E(\lambda)) = E(\lambda)H$ are spanned by the eigenvectors and superpositions of eigenfunctions with eigenvalues $\leq \lambda$.

The bounded function $(\Psi, E(\lambda)\Psi)$ is a monotone increasing function of λ which tends to zero for $\lambda \to -\infty$ and to 1 for $\lambda \to +\infty$ (we use normalized states $\|\Psi\| = 1$).

If $\Psi \in H_p$ (i.e. $\Psi = \Sigma\alpha_i \Psi_i$, where Ψ_i are normalized eigenvectors of H with eigenvalue λ_i) then

$$(\Psi, E(\lambda)\Psi) = \Sigma|\alpha_i|^2, \text{ i such that } \lambda_i \leq \lambda .$$

This function is the integral of the positive normalized

pure point measure

$$d(\Psi,E(\lambda)\Psi) = \Sigma|\alpha_i|^2 \delta(\lambda - \lambda_i) d\lambda \quad . \tag{1.8}$$

If on the other hand $\Psi \in H_{cont}$ then $(\Psi,E(\lambda)\Psi)$ is continuous and the same is true for the corresponding positive normalized measure $d(\Psi,E(\lambda)\Psi)$. The connection between the operator H and the measure is given by

$$D(H) = \{\Psi| \int \lambda^2 d(\Psi,E(\lambda)\Psi) < \infty\} \quad , \tag{1.9}$$

$$(\Phi,H\Psi) = \int \lambda \, d(\Phi,E(\lambda)\Psi) \tag{1.10}$$

for $\Phi \in H$, $\Psi \in D(H)$. For bounded measurable functions f one has for Φ, $\Psi \in H$

$$(\Phi,f(H)\Psi) = \int f(\lambda)d(\Phi,E(\lambda)\Psi) \quad . \tag{1.11}$$

In many cases such measures can be written as a positive density times the Lebesgue measure, i.e.

$$d(\Psi,E(\lambda)\Psi) = \rho(\lambda)d\lambda$$

where $\rho(\lambda) = d(\Psi,E(\lambda)\Psi)/d\lambda$ is integrable. In that case we say that Ψ belongs the absolutely continuous spectral subspace w.r.t. H, denoted H_{ac}.

According to general measure theory (Lebesgue decomposition theorem) there is a third type of measures besides the point measures and the absolutely continuous ones: the singular continuous measures with a corresponding spectral subspace H_{sing}. These are continuous measures which nevertheless "live" on a set of Lebesgue measure zero (like a Cantor set). We will show later that under physically reasonable assumptions on the potential the Hamiltonian does not have a singular continuous spectrum. On the other hand Pearson constructed physically understandable (although unrealistic) examples of potentials with purely singular spectrum (see Section II).

The spectrum and its type are unitarily invariant. Suppose e.g. that H_1 has absolutely continuous spectrum on $\mathcal{H}_1$, and suppose that $U : \mathcal{H}_1 \to \mathcal{H}_2$ is unitary. Then $H_2 = U H_1 U^*$ has absolutely continuous spectrum on $\mathcal{H}_2$.

(Details about the mathematics can be found in books on functional analysis or Hilbert space theory, e.g. [15].)

Let us illustrate all these notions with a simple example, the free Hamiltonian H_o. As a representation of the state $\Psi \in \mathcal{H}$ one can use equivalently its $\vec{x}$-space wave function $\psi(\vec{x})$ or the momentum space wave function $\hat{\psi}(\vec{p})$ obtained by Fourier transformation

$$\hat{\psi}(\vec{p}) = (2\pi)^{-\nu/2} \int d^\nu x \, e^{-i\vec{p}\cdot\vec{x}} \psi(\vec{x}) \quad .$$

In this representation H_o is the multiplication operator with $p^2/2m$ (where $p = |\vec{p}|$):

$$\widehat{(H_o\Psi)}(\vec{p}) = \frac{p^2}{2m} \hat{\psi}(\vec{p}) , \tag{1.12}$$

and the domain of self-adjointness consists of the states with

$$\int d^\nu p |(p^2/2m)\hat{\psi}(\vec{p})|^2 < \infty \quad . \tag{1.13}$$

The spectral projections $E_o(\lambda)$ of H_o are the multiplication operators with the characteristic functions of the balls $p^2/2m \leq \lambda$, thus

$$(\Psi, E_o(\lambda)\Psi) = \int_{p^2 \leq 2m\lambda} |\hat{\psi}(\vec{p})|^2 d^\nu p$$

$$= \int_o^\lambda d\left(\frac{p^2}{2m}\right) \left[mp^{\nu-2} \int d\Omega |\hat{\psi}(\vec{p})|^2\right] \tag{1.14}$$

is a continuous function of λ. Moreover the corresponding measure is absolutely continuous:

$$d(\Psi, E_o(\lambda)\Psi) = \rho(\lambda)d\lambda$$

with the integrable positive density

$$\rho(\lambda) = [mp^{\nu-2} \int d\Omega |\hat{\psi}(\vec{p})|^2]_{p = \sqrt{2m\lambda}} \quad . \tag{1.15}$$

Thus the free Hamiltonian has only an absolutely continuous spectrum which consists of the positive half line.

It is easy to verify (1.9) - (1.11); obviously the complex function $(\Phi, E_o(\lambda)\Psi)$ is

$$(\Phi, E_o(\lambda)\Psi) = \int_{p^2 \leq 2m\lambda} \overline{\hat{\phi}(\vec{p})}\hat{\psi}(\vec{p})d^\nu p \quad . \tag{1.16}$$

With the exception of the last example we have given so far some abstract notions of spectral theory which apply to any self-adjoint operator. It is more interesting to study the properties of all these quantities for specific Hamiltonians obtained with particular potentials or with special classes of potentials. We list some questions:

Are there eigenvalues? Where?
Are they degenerate?
Are there finitely or infinitely many eigenvalues?
Are there accumulation points of eigenvalues?
Are there eigenvalues embedded in the continuous spectrum?
Where are the eigenvectors localized?
Do they have exponential decay?
Where does the continuous spectrum lie?
What is its multiplicity?
Is there a singular continuous spectrum?

In the context of spectral theory it is not so easy to ask questions about the states in H_{cont} because the eigenfunctions are not physical states and there is no natural choice which superposition should be studied. Here another approach is better suited which relies on some physical insight and on the mathematical fact discussed above that the Hamiltonian and the propagator exp(iHt) contain exactly the same information about the physical system. Therefore we will study the propagator next.

The time evolution of an eigenstate $H\Psi = \lambda\Psi$ is extremely dull:

$$\exp(-iHt)\Psi = \exp(-i\lambda t)\Psi \quad . \tag{1.17}$$

The vector is multiplied by a phase factor, thus the state does not change, it is stationary. Moreover it is not helpful to study (1.17) instead of the usual eigenvalue equation (1.6), because in contrast to H the operator exp(-iHt) does not have a simple explicit form.

The basic ingredients of scattering theory are a few simple physical observations. Suppose the potential decays sufficiently fast towards infinity, then one observes two types of states: (i) the bound states which are stationary (ii) the states describing a particle which asymptotically moves freely. The latter alternative means that the particle has moved approximately free in the remote past, then it is influenced by the potential (scattered), and in the far future it will be approximately free again. (As a special case this includes a particle which misses the target and is free forever.) Within the limitations of the model one has not seen any other type of behaviour. If these simple observations are correct one should study the asymptotic time evolution for the states in H_{cont} which becomes free. Since we know all about the free time evolution we can conclude a lot of information about the interacting time evolution and the Hamiltonian on H_{cont}. This is scattering theory to which the main part of these lectures is devoted. In the following we concentrate on a special approach to scattering theory, the geometric, time dependent method. "Time dependent" means that we deal with physical states and that we follow their time evolution (in contrast to the time independent or stationary approach which uses eigenfunctions). And "geometric" says that the localization of states in space at various times (i.e. the propagation properties) plays a crucial role. For other important methods see e.g. [17] and the references given there.

II. EXAMPLES FOR UNEXPECTED BEHAVIOR

It was (and still is) a widespread prejudice, mainly among physicists, that any reasonable potential gives rise to a well behaved physical system. But Pearson constructed two examples of potentials which give rise to unexpected phenomena [12].

The first is a model for <u>local adsorption:</u> a particle which was an asymptotically free scattering state in the remote past has a nonzero probability to be trapped at the potential. This can happen although there is no absorptive part in the potential and the Hamiltonian is self-adjoint. The potential has a singularity which is bounded by an inverse power, and it oszillates faster and faster. The very steep slopes of the potential let the particle run back and forth near the singularity for an infintely long time. The effect is a subtle phenomenon which is rather unstable under small changes of the potential. It is hard work to prove that the Hamiltonian is self-adjoint and thus gives rise to a well defined time evolution. The Hamiltonian does not satisfy local compactness (see next section).

The second example uses an extremely nice and innocent

looking potential V(x): multiplication with a C^∞ function which is bounded (including all its derivatives). Let f(x) be a smooth positive function of compact support, and let with $\lambda_n > 0$

$$V(x) = \sum_n \lambda_n [f(x - x_n) + f(x + x_n)] \quad . \tag{2.1}$$

If the bumps are far enough separated (e.g. $x_n = \exp(n^3)$ will do) and if the strength does not decrease too fast: $\sum \lambda_n^2 = \infty$, then the spectrum of the Hamiltonian is purely singular continuous!! A particle will never be in a bound state because even low energy states can tunnel through the barriers and will eventually escape from any bounded region. On the other hand the particle will never be approximately free. If a wave packet hits a bump it will be split into a reflected and transmitted part. Due to the large separation of the bumps each encounter is approximately an independent scattering. Although there is a positive transmission probability for each finite subset of bumps, an infinite set of bumps will certainly reflect the particle (because of $\sum \lambda_n^2 = \infty$) no matter how far out the particle had travelled. Thus the particle will never become free; it might even return to a bounded region again and again. This is a typical behavior for a time evolution generated by a Hamiltonian with singular continuous spectrum (see e.g. [20]).

In both examples the forces are not unreasonable and one can understand the effects physically. But apparently nature has chosen to use potentials with less severe singularities and faster decay at infinity. We will impose suitable conditions later to exclude these examples.

III. LOCALIZATION OF BOUND STATES

As the name suggests a particle in a bound state is mainly localized in a certain region of space, like an electron which is close to the nucleus. But at the moment when a particle hits a target, it may be well localized too. Bound states are characterized by their localization for all times. Experience tells us that bound states usually are not confined to a finite region but that they have tails extending to infinity with exponential decay. Thus we adopt as our geometric characterization of bound states: $\Psi \in M_{bd}$ if

$$\lim_{R\to\infty} \sup_t \| F(|\vec{x}| > R) \exp(-iHt)\Psi \| = 0 \quad . \tag{3.1}$$

Here the operator $F(\cdot)$ is the multiplication operator in $\vec{x}$-space with the characteristic function of the indicated region, i.e. a spectral projection for the self-adjoint operator $|\vec{x}|$. (Later we will use F for spectral projections of other operators too.) In (3.1) it is the projection to that part of the state which is localized outside a ball of radius R.

In most quantum mechanics textbooks a spectral theoretic definition is given, namely that a bound state is an eigenstate of the Hamiltonian (or a superposition of eigenstates). Are the two definitions equivalent? The answer is yes under very mild conditions on the local singularities of the potential.

First it is clear that any eigenvector is localized uniformly in time since it does not change at all. The same is true for finite superpositions $\Psi = \sum^{N} \alpha_i \Psi_i$ because

$$\| F(|\vec{x}| > R) \exp(-iHt)\Psi \|$$

$$\leq \sum |\alpha_i| \, \| F(|\vec{x}| > R)\Psi_i \|$$

can be made arbitrarily small for R big enough. By a limiting argument the same applies to general superpositions, thus $H_p \subset M_{bd}$, the closed linear subspace of geometric bound states.

As we have seen in the previous section the opposite inclusion does not hold automatically. A sufficient condition on the interaction is <u>local compactness</u> which has a simple physical interpretation: A classical particle which is confined to a ball of radius R and whose energy is bounded by λ can be in classical phase space in a volume of size

$$\text{vol} = \int_{|\vec{x}|<R,\,(p^2/2m)+V(\vec{x})<\lambda} d^\nu x \, d^\nu p \ .$$

This volume is finite if the potential does not have too severe negative singularities, e.g. if $V(\vec{x}) \geq -c|\vec{x}|^{-\alpha}$, $\alpha < 2$ in $\nu = 3$ dimensions. According to the correspondence principle any quantum state requires a volume h^ν in classical phase space, thus there are only finitely many quantum states with these restrictions. In Hilbert space compact sets and compact operators are approximately finite dimensional, thus we require

$$F(|\vec{x}| < R)E(\lambda) \text{ is compact } \forall R, \lambda; \tag{3.2}$$

or equivalently

$$F(|\vec{x}| < R)\ (H + i)^{-1} \text{ is compact } \forall R\ . \tag{3.2'}$$

If H is defined as an operator sum or a form sum with an H_o-form-bounded potential V or a highly singular positive V, then $(H_o + i)^{1/2}(H + i)^{-1/2}$ is bounded. Since $F(|\vec{x}| < R)\cdot (H_o + i)^{-1/2}$ is compact (a simple exercise in Fourier analysis) the local compactness (3.2') and (3.2) follows. Thus the standard methods to construct a self-adjoint

Hamiltonian H imply local compactness. Even point potentials are allowed. In this sense the condition is extremely weak. (For further details see e.g. [1] or Section 3 of [7] and the references therein.)

To finish our discussion of bound states we give a theorem due to Ruelle [18] and later extended ([1], Section XI, 17 A of [17], Section 4 of [7]).

<u>Theorem 3.1.</u> Let local compactness (3.2) hold. Then $H_p = M_{bd}$ and for any $\Psi \in H_{cont}$, $R < \infty$,

$$\lim_{|T|\to\infty} \frac{1}{T} \int_0^T dt \, \| F(|\vec{x}| < R) \exp(-iHt)\Psi \| = 0 \,. \tag{3.3}$$

Thus we have as a first geometric characterization of the states in H_{cont} that they will leave any bounded region in space in the time mean. We won't give the proof here because we will prove a stronger result shortly under additional assumptions on V about its decay at infinity. Note that we have excluded Pearsons first example (local adsorption) by the assumption of local compactness, but we have not yet excluded the second one. This tells us that we cannot even replace the time-mean in (3.3) by a time limit $t \to \infty$ unless we impose conditions on the <u>qualified</u> decay of the potential at infinity.

IV. PROPAGATION OF SCATTERING STATES

From now on we will impose conditions on the decay of the potential at infinity to ensure that scattering states have asymptotically free motion. In this section we prove a preliminary result under weaker conditions, and for simplicity we discuss a special case. (Section 7 in [7],[8]).

First we prove an auxiliary

Lemma 4.1. Let P_{cont} denote the projection onto the continuous spectral subspace H_{cont} w.r.t. the Hamiltonian H, and let C be a compact operator. Then for $C(t) = \exp(iHt)\cdot C\exp(-iHt)$,

$$\left\|\frac{1}{T}\int_0^T dt\, C(t)P_{cont}\right\| \to 0 \text{ as } |T| \to \infty \quad . \tag{4.1}$$

Proof: Any compact operator can be approximated by a finite dimensional operator

$$\left\|C - \sum_{i=1}^{N(\varepsilon)} |\Psi_i \rangle\langle \Phi_i|\right\| < \varepsilon,\ N(\varepsilon) < \infty \quad . \tag{4.2}$$

Thus it is sufficient to estimate

$$\left\|\frac{1}{T}\int_0^T dt\, e^{iHt}|\Psi\rangle\langle P_{cont}\Phi|\, e^{-iHt}\right\|^2$$

$$\leq \frac{\|\Psi\|^2}{T^2}\int_0^T dt \int_0^T dt'|\langle P_{cont}\Phi|\, e^{-iH(t-t')}P_{cont}\Phi\rangle|$$

$$\leq \|\Psi\|^2 \{\frac{1}{T^2}\int_0^T dt \int_0^T dt'|\langle P_{cont}\Phi|\, e^{-iH(t-t')}P_{cont}\Phi\rangle|^2\}^{1/2} \quad ,$$

where we have used the Schwarz inequality in the last step. Denote by $d\mu(\lambda) = d(P_{cont}\Phi, E(\lambda)P_{cont}\Phi)$ the spectral measure; then the curly bracket is

$$\frac{1}{T^2}\int_0^T dt \int_0^T dt'\, e^{-i(\lambda-\sigma)(t-t')}d\mu(\lambda)d\mu(\sigma)$$

$$= \int\int [\frac{2}{T(\lambda-\sigma)} \sin \frac{T(\lambda-\sigma)}{2}]^2 d\mu(\lambda)d\mu(\sigma) \quad .$$

For $|\lambda-\sigma| > 2\delta$ the integrand is bounded by $(T\delta)^{-2}$ and the measures are bounded. Thus this contribution to the integral vanishes as $|T| \to \infty$ for any δ. In the remaining region $|\lambda-\sigma| < 2\delta$ the integrand is bounded by one and

$$\int_{|\lambda-\sigma|<2\delta} d\mu(\lambda)d\mu(\sigma) = \int d\mu(\lambda)[\mu(\lambda+2\delta) - \mu(\lambda-2\delta)]$$

$$\leq \sup_{\lambda}[\mu(\lambda + 2\delta) - \mu(\lambda - 2\delta)] \int d\mu(\lambda) \quad .$$

This can be made arbitrarily small by choosing δ small enough, because the continuous monotone function $\mu(\lambda)$ is bounded and thus uniformly continuous.

<u>Remark:</u> This Lemma is the crucial step to prove Theorem 3.1. Basically it is a special application of Wiener's theorem.

Let us now assume that the potential is a Kato-bounded perturbation of H_o, i.e. $D(V) \supset D(H_o)$ and for any $\Psi \in D(H_o)$

$$\|V\Psi\| \leq a\|H_o\Psi\| + b\|\Psi\| \quad , \qquad a < 1 \quad . \tag{4.3}$$

Then $D(H) = D(H_o)$ and the Schwartz space S of rapidly decaying test functions is a core for both operators. Assume moreover that V is a multiplication operator which can be split into a short-range part V_s and a remainder V_ℓ which may be of long range. They should obey

$$|\vec{x}| \; V_s(\vec{x}) \; (H_o + 1)^{-1} \quad \text{is compact} \; , \tag{4.4}$$

$$\vec{x}\cdot\vec{\nabla}V_\ell(\vec{x}) \; (H_o + 1)^{-1} \quad \text{is compact} \; . \tag{4.5}$$

Roughly this says that V_s and the gradient of V_ℓ decay a bit faster than $|\vec{x}|^{-1}$, but they could have singularities arbitrarily far out if they are weak enough. Certainly the Coulomb potential is allowed.

One way to analyse propagation properties of scattering states is to integrate the Heisenberg equations of motion for observables like $\vec{x}^2(t) = \exp(iHt)\vec{x}^2.\exp(-iHt)$. The equations are

$$(\frac{m}{2}\vec{x}^2(t))^{\bullet} = i[H,\frac{m}{2}\vec{x}^2(t)] = \frac{i}{4}[\vec{p}^2(t),\vec{x}^2(t)] = D(t) \qquad (4.6)$$

where the self-adjoint operator

$$D = \frac{1}{2}(\vec{x}\cdot\vec{p} + \vec{p}\cdot\vec{x}) \qquad (4.7)$$

is the generator of the dilation group:

$$(\exp(-i\lambda D)\psi)(\vec{x}) = (e^{-\lambda})^{\nu/2}\psi(e^{-\lambda}\vec{x}) \quad . \qquad (4.8)$$

The equation of motion for D(t) is

$$D(t)^{\bullet} = i[H,D(t)] = i[H_o(t),D(t)] + i[V(t),D(t)]$$

$$= 2H_o(t) - \vec{x}\cdot\vec{\nabla}V(\vec{x}) = :2H-I \qquad (4.9)$$

with interaction term $I = 2V + \vec{x}\cdot\vec{\nabla}V(\vec{x})$.

(4.6) and (4.9) should be understood as equations for quadratic forms on $S \times S$ which correspond to unique self-adjoint operators (all operators are essentially self-adjoint on S). Now we can integrate and obtain

$$D(t) = D(0) + \int_0^t dt'\{2H-I(t')\} =$$

$$= D(0) + 2H\cdot t - \int_0^t dt' I(t') \quad , \qquad (4.10)$$

$$\frac{m}{2}\vec{x}^2(t) = \frac{m}{2}\vec{x}^2(0) + \int_0^t dt' D(0) + 2H\int_0^t t'dt' - \int_0^t dt' \int_0^{t'} d\tau \; I(\tau)$$

$$= \frac{m}{2}\vec{x}^2(0) + D(0)\cdot t + H\cdot t^2 - \int_0^t dt' \int_0^{t'} d\tau\, I(\tau) \quad . \tag{4.11}$$

To obtain asymptotically constant quantities we divide by suitable powers of t:

$$\frac{D(t)}{t} = 2H + \frac{D(0)}{t} - \frac{1}{t}\int_0^t dt'\, I(t') \; ; \tag{4.12}$$

$$\frac{m}{2}\,\frac{\vec{x}^2(t)}{t^2} = H + \frac{D(0)}{t} + \frac{m}{2}\,\frac{\vec{x}^2(0)}{t^2} - \frac{1}{t^2}\int_0^t dt'\int_0^{t'} d\tau\, I(\tau) \quad . \tag{4.13}$$

The adequate topology to study the limits of these unbounded operators on scattering states as $|t| \to \infty$ is that of strong resolvent convergence. This means that any bounded continuous function of the operators converges strongly. Away from the point spectrum the functions may have discontinuities. A convenient sufficient condition for strong resolvent convergence is strong convergence on a core of the limiting operator.

Theorem 4.2. Let the potential satisfy (4.3) - (4.5). Then in the sense of strong resolvent convergence on H_{cont}:

$$\lim_{|t|\to\infty} \frac{D(t)}{t} = 2H \tag{4.14}$$

$$\lim_{|t|\to\infty} \frac{m}{2}\,\frac{\vec{x}^2(t)}{t^2} = H \tag{4.15}$$

$$\lim_{|t|\to\infty} H_0(t) = H \quad . \tag{4.16}$$

Proof: We give the proof for a typical but special case. The extension to the general case is an easy technical exercise which does not require new ideas. We assume that the full potential satisfies (4.5), then $I\cdot E(\lambda)$ is compact for any $\lambda < \infty$. Moreover we assume that

$$D = \{\Psi \in H_{cont} | \Psi = E(\lambda)\Psi \text{ for some } \lambda < \infty, \ \Psi \in D(\vec{x}^2)\}$$

is a core for $H \restriction H_{cont}$ (this will generally be the case). Since $D \subset D(D)$ we have for any $\Psi \in D$

$$\|t^{-1}D(0)\Psi\| \to 0 \quad \text{as} \quad |t| \to \infty ;$$

$$\|t^{-2}\,\vec{x}^2(0)\Psi\| \to 0 \quad \text{as} \quad |t| \to \infty ;$$

$$\|\frac{1}{t}\int_0^t dt'\ I(t')\Psi\| \leq \|\frac{1}{t}\int_0^t dt'[I\cdot E(\lambda)](t')P_{cont}\| \cdot \|\Psi\|,$$

and the first factor of the last expression tends to zero by Lemma 4.1. Thus (4.12) and (4.13) imply (4.14) and (4.15) resp..

To show (4.16) observe that conditions (4.3) - (4.5) imply that the difference of the free and interacting resolvent

$$(H - z)^{-1} - (H_o - z)^{-1} = -(H - z)^{-1}\ V(H_o - z)^{-1}$$

is compact. We will see shortly that as a consequence of (4.14) alone w-lim $\exp(-iHt)\Psi = 0$. Thus $(H_o(t) - z)^{-1}\Psi \to$ $\to (H - z)^{-1}\Psi$.

Because of their simple physical interpretation we single out a few consequences of Theorem 4.2. $F(\cdot)$ are the obvious spectral projections.

Corollary 4.3. Let $\Psi \in H_{cont}$; then for $E > 0$

$$\lim_{t\to+\infty} \|F(D < 2Et)\exp(-iHt)F(H > E)\Psi\| = 0, \tag{4.17}$$

$$\lim_{t\to-\infty} \|F(D > 2Et)\exp(-iHt)F(H > E)\Psi\| = 0, \tag{4.18}$$

$$\text{w-}\lim_{|t|\to\infty} \exp(-iHt)\Psi = 0 . \tag{4.19}$$

<u>Proof:</u> Since D and $H \upharpoonright H_{cont}$ have purely continuous spectra we may use discontinuous functions like the spectral projections:

$$\exp(iHt)F(D \lessgtr 2Et)\exp(-iHt) = F(\frac{D(t)}{t} < 2E) \text{ for } t \gtrless 0.$$

By (4.14) the latter converges to $F(H < E)$ which proves (4.17), (4.18). Finally observe that

$$\lim_{t\to\infty} \|F(D > 2Et)\Phi\| = 0$$

for any Φ and $E > 0$. Furthermore $E > 0$ can be chosen small enough such that $\|F(H < E)\Psi\| < \varepsilon$ for a given $\Psi \in H_{cont}$. Then

$$|(\Phi, \exp(iHt)\Psi)| \leq \|F(H \leq E)\Psi\| +$$

$$+ \|F(D > 2Et)\Phi\| + \|F(D < 2Et)\exp(-iHt)F(H > E)\Psi\| .$$

The first summand is bounded by ε and the two others tend to zero for any $E > 0$.

Similarly one concludes from (4.15)

<u>Corollary 4.4.</u> Let $\Psi \in H_{cont}$; then

$$\lim_{|t|\to\infty} \|F(|\vec{x}| \leq v|t|)\exp(-iHt)F(H \geq \tfrac{m}{2}v^2)\Psi\| = 0 , \qquad (4.20)$$

$$\lim_{|t|\to\infty} \|F(|\vec{x}| < R)\exp(-iHt)\Psi\| = 0 . \qquad (4.21)$$

This Corollary tells us that a state with strictly positive energies will leave the origin with at least the corresponding minimal velocity. In quantum mechanics a strict localization in $\vec{x}$-space and momentum space simultaneously is impossible. But asymptotically wave packets will mainly be localized where they should be according to the classical intuition. The tails into the other regions will eventually disappear. In particular a state from H_{cont} will leave any bounded region in the time limit and it will never return.

A similar classical interpretation can be given for Corollary 4.3. If for a point particle $\vec{p}\cdot\vec{x} > 0$ then the velocity is pointing away from the origin, the particle is "outgoing". Similarly we call it "incoming" if $\vec{p}\cdot\vec{x} < 0$. The quantum mechanical analog of $\vec{p}\cdot\vec{x}$ is the self-adjoint operator $D = \frac{1}{2}(\vec{p}\cdot\vec{x} + \vec{x}\cdot\vec{p})$ and we may call the vectors in its positive spectral subspace the outgoing ones: $F(D>0)H$. Similarly the incoming states span $F(D < 0)H$ [11]. Due to the quantum tails there is some arbitrariness in the definition (for several other definitions see Section 5 of [7]). But the precise cutoff is irrelevant because the "D-values" will continue to increase with time. Any state from H_{cont} will be outgoing in the far future and it was incoming in the remote past.

Note that for the special asymptotic observables which we have studied in this section the interaction does not have an effect; the results are the same for free and interacting systems.

V. FREE EVOLUTION OF OUTGOING STATES

We continue to exploit our classical intuition: Let a free classical particle at some time τ be localized outside a ball of radius R, be outgoing and let it have a minimal velocity v. Then the distance from the origin will increase with time:

$$|x(\tau + t)| \geq \sqrt{R^2 + v^2 t^2} \geq \frac{1}{\sqrt{2}} (R + vt), \tag{5.1}$$

the ball with radius $\frac{1}{\sqrt{2}}(R + vt)$ is "classically forbidden".

A similar statement is true for the free quantum time evolution. The tails of suitably chosen quantum states into the classically forbidden region decay faster than any inverse power of R and t.

Let us study states which are well localized in $\vec{x}$-space and momentum space simultaneously. Denote by $\chi(\vec{P})$ the multiplication operator in momentum space with the C_0^∞-function $\chi(\vec{p})$ which has its support near a vector $m\cdot\vec{v}$:

$$\text{supp}\ \chi(\vec{p}) = \{\vec{p}\,|\,|\vec{p} - m\vec{v}| \leq m|\vec{v}|\sin 10^\circ\} \quad . \tag{5.2}$$

A classical state with this momentum distribution starting from $\vec{a}$ at t = 0 would be localized at time t in the ball

$$B_t = \{\vec{x}\,|\,|\vec{x} - \vec{a} - \vec{v}t| \leq t\cdot|\vec{v}|\ \sin 10^\circ\} \quad . \tag{5.3}$$

In quantum physics we have to control the tails beyond this region. Consider the part of a state localized in a unit cube around $\vec{a}$ and then restricted in momentum space: $\chi(\vec{P})F_{\vec{a}}\Psi$, where

$$F_{\vec{a}} = \prod_{i=1}^{\nu} F(|x_i - a_i| < \tfrac{1}{2}) \tag{5.4}$$

is the projection to the unit cube around $\vec{a} \in Z^\nu$. Then

Lemma 5.1. For any N there is a C_N such that

$$\|F(|\vec{x} - \vec{a} - \vec{v}t| > r + |\vec{v}t|\cdot\sin 15^\circ)\exp(-iH_0 t)\chi(\vec{P})F_{\vec{a}}\| \leq$$

$$\leq C_N\ (1 + |t|)^{-N}(1 + r)^{-N} \quad . \tag{5.5}$$

Proof: By the stationary phase method (e.g. Theorem XI.14 and its Corollary in [17]) one easily gets for $|\vec{x}| > |\vec{v}t|\cdot\sin 15^\circ$:

$$(\exp(-iH_o t)\chi(\vec{P} + m\vec{v})F_{\vec{o}}\Psi)(\vec{x}) \leq C_N'\|\Psi\|(1 + |\vec{x}| + |t|)^{-N}. \quad (5.6)$$

(Note that $\chi(\vec{p} + m\vec{v})$ has its support near $\vec{p} = 0$.) A shift in momentum space by $m\vec{v}$ and in $\vec{x}$-space by $\vec{a}$ gives (5.5).

Now we can sum up those components of the state which are outgoing (and a bit more).

Lemma 5.2. Let $\chi(\vec{P})$ and $F_{\vec{a}}$ be defined as above. Denote by Σ' the sum over $\vec{a} \in Z^\nu$ such that $|\vec{a}| > R$ and the angle between $\vec{a}$ and $\vec{v}$: $\measuredangle(\vec{a},\vec{v}) < 110^\circ$. Then for $t \geq 0$

$$\|F(|\vec{x}| < \tfrac{1}{4}(R + |\vec{v}|t))\exp(-iH_o t)\chi(\vec{P})\Sigma' F_{\vec{a}}\| <$$

$$< D_N'(1 + t)^{-N}(1 + R)^{-N} \quad (5.7)$$

for any N.

Proof: For the specified angles by elementary geometry

$$|\vec{a} + \vec{v}t|^2 \geq (|\vec{a}| \sin 70^\circ)^2 + (|\vec{v}t| - |\vec{a}| \cos 70^\circ)^2$$

$$\geq \tfrac{1}{2}(|\vec{a}| \sin 70^\circ + |\vec{v}t| - |\vec{a}| \cos 70^\circ)^2 ;$$

$$|\vec{a} + \vec{v}t| \geq \frac{1}{\sqrt{2}}[|\vec{a}|(\sin 70^\circ - \cos 70^\circ) + |\vec{v}|t],$$

and therefore in the region $|\vec{x} - \vec{a} - \vec{v}t| < (r + |\vec{v}t|\sin 15^\circ)$ one has

$$|\vec{x}| > (|\vec{a}| \frac{\sin 70^\circ - \cos 70^\circ}{\sqrt{2}} - r) + |\vec{v}t|(\frac{1}{\sqrt{2}} - \sin 15^\circ)$$

$$> 0{,}4\ (|\vec{a}| + |\vec{v}t|) - r$$

$$\geq 0{,}25\ (R + |\vec{v}|t) + 0{,}4\ (|\vec{a}| - R) + 0{,}15\ R - r . \quad (5.8)$$

Thus in the region $|\vec{x}| < \frac{1}{4}(R + |\vec{v}t|)$ the estimate (5.5) holds with $r = 0{,}4\,(|\vec{a}| - R) + 0{,}15\,R$. Due to the rapid decay in r one can sum up the contributions from different $\vec{a}$'s (the number of terms grows only like $|\vec{a}|^{\nu-1}$) and still have rapid decay in R.

Let $g \in C_0^\infty(0,\infty)$; then we can decompose $g(p^2/2m) = \sum \chi_j(\vec{p})$ where the sum is finite and all χ_j are like χ discussed above with small support around $m\vec{v}_j$. Denote $v = \min_j |\vec{v}_j| > 0$. For all j the sum below is as in Lemma 5.2, but for $\vec{v}_j$. We define for $\vec{a} \in Z^\nu$

$$P_R^{out} = \sum_j \{\chi_j(\vec{P}) \sum_{\substack{|\vec{a}| > R \\ \sphericalangle(\vec{a},\vec{v}_j) < 110^\circ}} F_{\vec{a}}\} \quad . \tag{5.9}$$

This is the definition of the <u>outgoing</u> part of a state using a <u>phase space cell decomposition.</u>

Since the sum over j is finite we get immediately

<u>Lemma 5.3.</u> For any N there is a D_N such that

$$\|F(|\vec{x}| < \tfrac{1}{4}(R + vt))\exp(-iH_0 t)\, P_R^{out}\| <$$

$$< D_N\,(1 + t)^{-N}(1 + R)^{-N} \tag{5.10}$$

for $t \geq 0$.

The terms which have been omitted in the sum over $\vec{a} \in Z^\nu$ are either incoming with $\sphericalangle(\vec{a},\vec{v}_j) \geq 110^\circ$, or they are separated from the origin by less than R. We have a physically motivated decomposition

$$g(H_0) = P_R^{out} + P_R^{in} + g(H_0) \sum_{\substack{\vec{a} \in Z^\nu \\ |\vec{a}| \leq R}} F_{\vec{a}} \quad . \tag{5.11}$$

In the Appendix we prove with a stationary phase argument: for $E > 0$

$$\|F(D > 2E\tau)\ P_R^{in}\| \to 0 \text{ as } \tau \to \infty\ . \tag{5.12}$$

Intuitively this is evident because for large τ the first factor projects to clearly outgoing states. It is an estimate of the rapidly decaying tails by which the definitions differ.

For the terms with $|\vec{a}| > R$ and $70^\circ \leq \sphericalangle(\vec{a},\vec{v}_j) \leq 110^\circ$ the estimate (5.10) holds in both time directions. This part of the state "misses" a neighborhood of the origin; except for small tails it stays outside a ball of radius $(1/4)\ (R + v|t|)$. For the incoming part as defined in (5.11) with $\sphericalangle(\vec{a},\vec{v}_j) \geq 110^\circ$ the estimate (5.10) holds for negative times. The analysis of the next Section for short range interactions will show that for big R the outgoing part moves approximately free in the future, the incoming part in the past, and the part which misses the target moves approximately free for all times.

Certainly the choice of the angles 70° and 110° is arbitrary. For any pair of angles $0 < \alpha < 90^\circ < \beta < 180^\circ$ the part of the state may be called incoming (resp. outgoing or missing) if $\sphericalangle(\vec{a},\vec{v}_j) > \alpha$ (resp. $< \beta$ or between α and β) or any subset thereof. Only the support of χ_j must lie in a small enough ball around $m\vec{v}_j$.

We have shown propagation properties of the outgoing (and incoming) part which are stronger than in the previous section. We have control on the decay of the tails but only for the free time evolution. For other methods and related results see Section 6 of [7] and the references given there. Recently Perry [14] has shown similar decay properties for interacting time evolutions if the potential is dilation analytic.

VI. ASYMPTOTICALLY FREE MOTION

In Section IV we have seen that the interacting time evolution asymptotically behaves in the same way as the free time evolution does if one studies suitable observables like $\vec{x}^2$, D, etc.. Now we will compare the time evolutions themselves on scattering states. The physical intuition suggests that one has to wait a long enough time τ until the scattering is over. For the future time evolution it should not matter whether the potential is there or not. A precise formulation is

$$\lim_{\tau\to\infty} \sup_{T\geq 0} \|[\exp(-iHT)-\exp(-iH_oT)] \exp(-iH\tau)\Psi\| = 0 \quad . \tag{6.1}$$

A state Ψ which satisfies (5.1) is called an "asymptotically free scattering state" (in the future). A similar definition can be given for the past. Certainly (6.1) does not hold on bound states but we will prove it for all $\Psi \in H_{cont}$ for short-range interactions. Since the operators are bounded it is sufficient to verify (6.1) for a dense set of vectors in H_{cont}. For simplicity we assume again $D(H) = D(H_o)$ (otherwise one would need an additional technical step, see e.g. Section 8 of [7]) and our dense set consists of states with strictly positive and bounded energy:

$$D = \{\Psi \in H_{cont} | F(E_1 < H < E_2)\Psi = \Psi\} \quad . \tag{6.2}$$

Here $0 < E_1 < E_2 < \infty$ depends on Ψ. Evidently $\exp(-iH\tau)D = D$.

To obtain a bound on the difference of the time evolutions we use Cook's estimate which is valid on any $\Phi \in D(H) \cap D(H_o)$:

$$\|[\exp(-iHT)-\exp(-iH_oT)]\Phi\|$$

$$= \|[1 - \exp(iHT)\exp(-iH_oT)]\Phi\|$$

$$= \left\| - \int_0^T dt \exp(iHt)[iH - iH_o]\exp(-iH_o t)\Phi \right\|$$

$$\leq \int_0^\infty dt \, \| V \exp(-iH_o t)\Phi \| \quad . \tag{6.3}$$

This is a bound uniform in $T \geq 0$.

Now choose a particular $\Psi \in D$ and pick a $g \in C_o^\infty(0,\infty)$ such that $g(e) = 1$ for $E_1 \leq e \leq E_2$ and $g = 0$ in a neighborhood of zero. Then $g^2(H)\Psi = \Psi$ and by (4.16) of Theorem 4.2 ,

$$\lim_{\tau\to\infty} \| [1 - g^2(H_o)]\exp(-iH\tau)\Psi \| = 0 \quad . \tag{6.4}$$

With the decomposition (5.11)

$$g(H_o)\exp(-iH\tau)\Psi =$$

$$= [P_R^{out} + P_R^{in} + g(H_o) \sum_{|\vec{a}| \leq R} F_{\vec{a}}] \exp(-iH\tau)\Psi \tag{6.5}$$

we use (4.17) of Corollary 4.3 and (5.12) to show that the second summand disappears as $\tau \to \infty$. For the third term this follows from (4.20) of Corollary 4.4. Thus for any R

$$g(H_o)P_R^{out} \exp(-iH\tau)\Psi \tag{6.6}$$

is an arbitrarily good approximation of $\exp(-iH\tau)\Psi$ for large enough τ. We insert this expression into (6.1) and estimate it with (6.3) by

$$\int_0^\infty dt \, \| Vg(H_o)\exp(-iH_o t)P_R^{out} \exp(-iH\tau)\Psi \|$$

$$\leq \int_0^\infty dt \, \| Vg(H_o) \| \cdot \| F(|\vec{x}| < \tfrac{1}{4}(R + vt))\exp(-iH_o t)P_R^{out} \| \quad +$$

$$+ \int_0^\infty dt \, \| Vg(H_o) F(|\vec{x}| > \frac{1}{4} (R + vt)) \| \; \| P_R^{out} \| \quad . \tag{6.7}$$

Choose the velocity v depending on g such that Lemma 5.3 holds; then the first integral can be made arbitrarily small for big enough R.

If the potential is of <u>short range,</u> i.e.

$$\| Vg(H_o) F(|\vec{x}| > \rho) \| = : h(\rho) \in L^1 \, (R_+, d\rho) \tag{6.8}$$

for any $g \in C_o^\infty$, then also the second integral in (6.7) converges and is small for big enough R. Thus we have shown that any state $\Psi \in H_{cont}$ satisfies (6.1), it is an asymptotically free scattering state. Certainly the same analysis applies for negative times. This fact is called <u>asymptotic completeness</u> because the bound states and the asymptotically free scattering states together are complete.

Consider now the sequence

$$\exp(iH_o t) \exp(-iHt) \Psi \quad . \tag{6.9}$$

If Ψ satisfies (6.1) then (6.9) is a Cauchy sequence and the limit $t \to \infty$ exists. The analysis of Section IV applies as a special case to H_o which has all of H as its continuous spectral subspace. Therefore we get for any state $\Phi \in H$

$$\lim_{\tau \to \infty} \sup_{T \geq 0} \| [\exp(-iHT) - \exp(-iH_o T)] \exp(-iH_o \tau) \Phi \| = 0 \tag{6.10}$$

and existence of the limit

$$\lim_{t \to \infty} \exp(iHt) \exp(-iH_o t) \Phi \tag{6.11}$$

for any Φ. Therefore it is reasonable to define the outgoing wave operator

$$\Omega_- = s - \lim_{t\to\infty} \exp(iHt)\exp(-iH_o t) \tag{6.12}$$

and similarly the incoming Ω_+ (we use the physicist's sign convention). The range of Ω_- are the states Ψ such that the limit of (6.9) exists. Thus an equivalent formulation of asymptotic completeness (in its strongest form) is

$$\mathrm{Ran}\ (\Omega_{\mp}) = \mathcal{H}_{cont} \quad . \tag{6.13}$$

As strong limits of unitary operators the wave operators are unitary from $\mathcal{H}$ onto their range $\mathcal{H}_{cont}$. A simple consequence of (6.12) is the intertwining property on $D(H_o)$,

$$H\Omega_{\mp} = \Omega_{\mp}\, H_o \quad . \tag{6.14}$$

Therefore H_o and $H \restriction \mathcal{H}_{cont}$ are unitarily equivalent. In particular $H \restriction \mathcal{H}_{cont}$ has purely absolutely continuous spectrum and there is no singular continuous one because H_o has these properties as we had seen in Section I. This gives detailed spectral information about H.

Since the ranges of the wave operators are equal we can construct a <u>unitary</u> scattering operator on $\mathcal{H}$

$$S = (\Omega_-)^* \Omega_+ \quad . \tag{6.15}$$

In conclusion let us state our main result under conditions which are sufficient (and weaker than those used above). The operator $(H + i)^{-1} V (H_o + i)^{-1}$ is bounded as a difference of two resolvents. If the forces decay at infinity it is reasonable to require

$$\|(H + i)^{-1} V (H_o + i)^{-1} F(|\vec{x}| > \rho)\| \to 0 \text{ as } \rho \to \infty \quad . \tag{6.16}$$

This together with local compactness (3.2) is equivalent to

$$(H + i)^{-1} - (H_o + i)^{-1} \text{ is compact .} \tag{6.17}$$

The short range condition (6.8) can then be replaced by the weaker (w.r.t. singularities)

$$\| g(H)Vg(H_o)F(|\vec{x}| > \rho)\| \in L^1(R_+, d\rho) \tag{6.18}$$

for any $g \in C_o^\infty(R)$.

Theorem 6.1. Let the potential be such that (6.17) and (6.18) are satisfied, then asymptotic completeness holds:

$$\text{Ran}(\Omega_\mp) = H_{cont} = H_{ac} .$$

For details and other versions of the proof see Sections VIII and IX of [7].

The short-range condition says roughly that $V(\vec{x})$ decays like $|\vec{x}|^{-1-\varepsilon}$, $\varepsilon > 0$, towards infinity, but it may have singularities all the way out. Moreover it need not be a multiplication operator in $\vec{x}$-space but it may describe velocity dependent forces. The Coulomb potential in $\nu = 3$ dimensions was included in Section IV but it is not of short range.

VII. OTHER APPLICATIONS

The main ingredients of our geometric, time dependent approach are (i) the physical intuition to ask appropriate questions, (ii) patience to wait until the interaction is over, (iii) some estimates on the propagation of freely evolving states. Therefore the scheme can be easily adjusted and extended to handle the Dirac and wave equations, phonon scattering off impurities in crystals and many other simple scattering systems. See e.g. [19] and the references given in Section XII of [7].

The inclusion of the physically important Coulomb force and other long-range potentials requires modifications of the asymptotic condition (6.1). The necessary estimates were rather tedious in the pedestrian approach ([5], Section X of [7]). Recently Perry gave an elegant proof which uses the dilation analyticity of Coulomb-tails [14].

A major goal is to treat multiparticle scattering in a similarly intuitive fashion. So far only the two cluster scattering with short- and long-range forces could be treated [6]. Finally we mention that geometrical methods can also be used to study more practical problems like bounds on total cross sections [9].

VIII. APPENDIX: AN APPLICATION OF THE STATIONARY PHASE METHOD

In this appendix we give the proof of (5.12) which was omitted in Section V. As a special case of Theorem XI.14 in [17] we quote without proof:

Lemma: Let $u(q) \in C_0^\infty(R)$ and $f(q) \in C^\infty(\Theta)$ in a neighbourhood Θ of the compact support of u, and $f'(q) \neq 0$ there. Then for any m

$$\left|\int \exp(isf(q))u(q)dq\right| < \text{const.}(1 + |s|)^{-m}\|u\|_{m,\infty} \tag{8.1}$$

where

$$\|u\|_{m,\infty} = \sum_{k=0}^{m} \|u^{(k)}(q)\|_\infty \quad . \tag{8.2}$$

Moreover if f is taken from a compact subset $M \subset C^m(\Theta)$ of functions with nonvanishing derivative, then the constant can be chosen independent of f. The proof is essentially by partial integration.

We use this lemma to study how a state moves in D-space when it is shifted in $\vec{x}$-space. Observe that scaling $|\vec{p}| = p$ is a shift in $q = \ln p$. Therefore the dilation generator D can be diagonalized by a Fourier transform w.r.t. q (keeping angles fixed), which is the same as a Mellin transform (up to a sign) [13]. Define with $p = e^q$

$$\overset{\circ}{\phi}(\lambda,\vec{\omega}) = (2\pi)^{-1/2}\int dq\ \exp(i\lambda q)\, e^{q\nu/2}\hat{\phi}(e^q,\vec{\omega}) \tag{8.3}$$

which is a unitary mapping from $L^2(R_+ \times \Omega,\ p^{\nu-1}\, dp\, d\Omega)$ onto $L^2(R\times\Omega,\ d\lambda\, d\Omega)$. Then we have

$$(g(D)\Phi)^{\circ}(\lambda,\vec{\omega}) = g(\lambda)\overset{\circ}{\phi}(\lambda,\vec{\omega}) \quad . \tag{8.4}$$

If the state Φ is shifted in $\vec{x}$-space by $\vec{a} = a\vec{e} \in Z^\nu$, then its momentum space wave function is multiplied by $\exp(-i\vec{p}\cdot\vec{a}) = \exp(-ie^q a\vec{e}\cdot\vec{\omega})$. Then the D-space wave function (8.3) is proportional to

$$\int dq\ \exp\{i\lambda q - ie^q a\vec{e}\cdot\vec{\omega}\} e^{q\nu/2}\hat{\phi}(e^q,\vec{\omega}) \ . \tag{8.5}$$

This is of the form (8.1) with $s = (1 + \lambda + a)$ and the two parameter family of phase functions

$$f_{\lambda,a}(q) = (1 + \lambda + a)^{-1}[\lambda q + ae^q(-\vec{e}\cdot\vec{\omega})] \ , \tag{8.6}$$

$$f'_{\lambda,a}(q) = (1 + \lambda + a)^{-1}\lambda + (1 + \lambda + a)^{-1}ae^q(-\vec{e}\cdot\vec{\omega}) \ . \tag{8.7}$$

We are interested in $\lambda,\ a \geq 0$, q from a compact subset of R, $(-\vec{e}\cdot\vec{\omega}) \geq -\cos(100^{\circ}) > 0$. Then $f_{\lambda,a}$ is a linear combination of two C^m-functions (all m) with coefficients in a bounded interval. Thus the phase functions are contained in a compact set in all C^m. For $a \geq 1$ the derivative is bounded below by

$$\frac{1}{2}\min\ (1,\ \inf\ e^q\ (-\vec{e}\cdot\vec{\omega})) > 0 \ .$$

Therefore the estimate (8.1) can be used. For $a = 0$ we get the same result with $s = \lambda$ (usual Fourier transformation).

We want to study states of the form ($\Psi_{-\vec{a}}$ is the shifted state)

$$(\chi(\vec{P})F_{\vec{a}}\Psi)(\vec{p}) = \exp(-i\vec{p}\cdot\vec{a})\chi(\vec{p})(F_{\vec{o}}\Psi_{-\vec{a}})(\vec{p}), \tag{8.8}$$

$$u(q;\vec{\omega},\vec{a}) = e^{q\nu/2}\chi(e^{q}\cdot\vec{\omega})(F_{\vec{o}}\Psi_{-\vec{a}})(e^{q}\cdot\vec{\omega}) \quad . \tag{8.9}$$

By the support properties of χ the q-values vary over a compact set and for the incoming part the angle between $\vec{a}$ and the directions inside supp χ is at least 100°. It remains to estimate

$$\|(\frac{d}{dq})^{m} u(q;\ \vec{\omega},\ \vec{a})\|_{\infty}$$

$$\leq C_{m} \|P(|\vec{x}|)F_{\vec{o}}\|\cdot\|\Psi\| \quad . \tag{8.10}$$

Here the constant C_m is the sup over a combination of derivatives up to m-th order of $\chi(e^{q}\cdot\vec{\omega})$ multiplied with $\exp(q\cdot(m + \nu/2))$. P is a polynomial of degree m and we have used that by Schwarz' inequality the L^1-norm of a function restricted to a unit cube is bounded by the L^2-norm. Note that the bounds can be chosen uniform in $\vec{\omega}$ and $\vec{a}$.

With $m\cdot\vec{v}$ the "center" of the support of χ we have shown for any $\vec{a}$ such that

$$\measuredangle(\vec{a},\vec{v}) \geq 110^{o},\ \lambda \geq 0,\ k \in N:$$

$$|(\chi(\vec{P})F_{\vec{a}}\Psi)^{o}(\lambda,\vec{\omega})| \leq d_{k}'(1 + \lambda)^{-k}(1 + a)^{-k}\|\Psi\| \quad . \tag{8.11}$$

Now we can sum over $\vec{a} \in \mathbf{Z}^{\nu}$, $|\vec{a}| \geq R \geq 0$, $(\vec{a},\vec{v}) \geq 110^{o}$. We denote this sum by Σ''. The power increase of the number of summands is more than compensated by the rapid decay in a. We get for $d \geq 0$

$$\|F(D \geq d)\Sigma'' \chi(\vec{P}) F_{\vec{a}}\| \leq d_k (1+d)^{-k} (1+R)^{-k} \tag{8.12}$$

for any k by integrating (8.11) over $d\Omega$ and $\lambda \geq d$. The finite sum over χ's does not change this result and we obtain with P_R^{in} as defined in (5.11)

$$\|F(D \geq 0) P_R^{in}\| \leq C_k (1+R)^{-k} , \tag{8.13}$$

$$\|F(D \geq d) P_R^{in}\| \leq C_k' (1+d)^{-k} \quad \text{for any } R . \tag{8.14}$$

The latter implies (5.12).

Denote by $\hat{P}_R^{out}$ the outgoing part which is not missing the target, i.e.

$$\hat{P}_R^{out} = \sum_j \{\chi_j(P) \sum_{\substack{|\vec{a}| > R \\ \measuredangle(\vec{a},\vec{v}_j) < 70^0}} F_{\vec{a}}\} . \tag{8.15}$$

Then one gets the analogue of (8.13), (8.14) by a reflection of $\vec{x}$ or $\vec{p}$:

$$\|F(D \leq 0) \hat{P}_R^{out}\| \leq C_k (1+R)^{-k} , \tag{8.16}$$

$$\|F(D \leq -d) \hat{P}_R^{out}\| \leq C_k' (1+d)^{-k} \text{ for any } R . \tag{8.17}$$

This can be used for the treatment of scattering states in the remote past. Note that the definitions of the incoming and outgoing part of a state using the phase space cell decomposition or the dilation generator D may differ only on that part of the state which misses the target. Therefore for the study of dynamical questions the arbitrariness is irrelevant.

REFERENCES

1. W.O. Amrein, V.Georgescu, Helv. Phys. Acta 46 (1973) 635-658.
2. W.O. Amrein, D.B. Pearson, M. Wollenberg, Evanescence of States and Asymptotic Completeness, preprint Univ. of Genève UGVA-DPT 1980/05-242, 1980.
3. E.B. Davies, Duke Math. J. 47 (1980) 171-185.
4. V. Enss, Commun. Math. Phys. 61 (1978) 285-291.
5. V. Enss, Ann. Phys. (N.Y.) 119 (1979) 117-132, and Addendum, preprint Univ. Bielefeld BI-TP 79/26, 1979, unpublished.
6. V. Enss, Commun. Math. Phys. 65 (1979) 151-165.
7. V. Enss, Geometric Methods in Spectral and Scattering Theory of Schrödinger Operators, to appear in Rigorous Atomic and Molecular Physics, G. Velo and A.S. Wightman eds., Plenum, New York, ca. 1981.
8. V. Enss, Asymptotic Observables on Scattering States, in preparation.
9. V. Enss, B. Simon, Phys. Rev. Lett. 44 (1980) 319-322 and 764; Commun. Math. Phys. 76 (1980) 177-209; Total Cross Sections in Non-Relativistic Scattering Theory, to appear in Classical, Semiclassical, and Quantum Mechanical Problems in Mathematics, Chemistry, and Physics, K. Gustavson and W.P. Reinhardt eds., Plenum, New York ca. 1981.
10. J. Ginibre, La Mèthode "Dépendant du Temps" dans le Problème de la Complétude Asymptotique, preprint Univ. Paris-Sud, LP TH E 80/10, 1980.
11. E. Mourre, Commun. Math. Phys. 68 (1979) 91-94.
12. D.B. Pearson, Commun. Math. Phys. 40 (1975) 125-146; Commun. Math. Phys. 60 (1978) 13-36; Contribution in Rigorous Atomic and Molecular Physics, G. Velo and A.S. Wightman eds., Plenum, New York, ca. 1981.
13. P.A. Perry, Duke Math. J. 47 (1980) 187-193.
14. P.A. Perry, Propagation of States in Dilation Analytic Potentials and Asymptotic Completeness, preprint Caltech 1980.

15. M. Reed, B. Simon, Methods of Modern Mathematical Physics, I. Functional Analysis, Academic Press, New York 1972.

16. M. Reed, B. Simon, II. Fourier Analysis, Self-Adjointness, Academic Press, New York 1975.

17. M. Reed, B. Simon, III. Scattering Theory, Academic Press, New York 1979.

18. D. Ruelle, Nuovo Cimento 61A (1969) 655-662.

19. B. Simon, Duke Math. J. 46 (1979) 119-168.

20. K.B. Sinha, Ann. Inst. Henri Poincaré 26 (1977) 263-277.

21. D.R. Yafaev, On the Proof of Enss of Asymptotic Completeness in Potential Scattering Theory, preprint LOMI E-2-79, 1979.

Acta Physica Austriaca, Suppl. XXIII, 65–110 (1981)

THREE-BODY COULOMB SCATTERING+

by

S.P. MERKURIEV

Leningrad University, USSR

1. INTRODUCTION

In my lectures, I shall describe recent progress in mathematical justification of the three-body Coulomb scattering problem. Importance of the former for the atomic and nuclear physics is evident and I shall not repeat known arguments [1-3]. I shall pay principal attention to the mathematical aspects of this problem.

I shall consider a system of three spinless distinguishable particles. The two-body potentials $V_\alpha(x_\alpha)$ ($\alpha = 1,2,3$) will be assumed to have the form of the sum of the Coulomb part and of a "nuclear" short-range one,

$$V_\alpha(x_\alpha) = \frac{n_\alpha}{|x_\alpha|} + \mathring{V}_\alpha^{(n)}(x_\alpha), \quad x_\alpha \in R^3 \quad .$$

In the three-particle center-of-mass system I take as independent coordinates the usual pairs of three-vectors $\{x_\alpha, y_\alpha\}$ ($\alpha = 1,2,3$). I write $\{k_\alpha, p_\alpha\}$ for the independent momenta which are conjugate to the variables $\{x_\alpha, y_\alpha\}$. I shall frequently combine pairs of coordinates $\{x_\alpha, y_\alpha\}$ and momenta $\{k_\alpha, p_\alpha\}$ into six vectors $X = x_\alpha + y_\alpha$, $P = k_\alpha + p_\alpha$.

+Lectures given at XX. Internationale Universitätswochen für Kernphysik, Schladming, Austria, February 17-26, 1981.

The transition from one pair $\{x_\alpha, y_\alpha\}$ to another one $\{x_\beta, y_\beta\}$, $\alpha \neq \beta$, is given by a rotation in R^6,

$$x_\beta = c_{\beta\alpha}\, x_\alpha + s_{\beta\alpha}\, y_\alpha \ ,$$
$$y_\beta = -s_{\beta\alpha}\, x_\alpha + c_{\beta\alpha}\, y_\alpha \ , \qquad c^2_{\beta\alpha} + s^2_{\beta\alpha} = 1 \ ,$$
$$\alpha, \beta = 1,2,3 \ ,$$

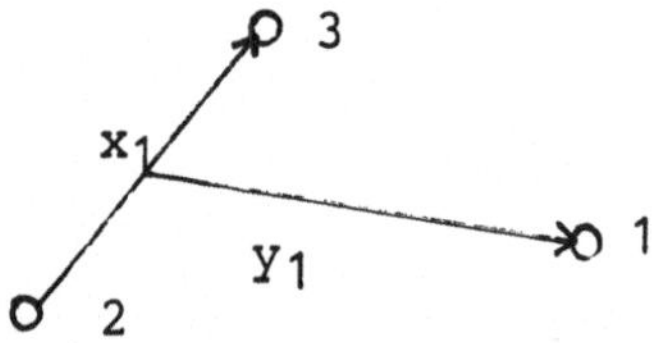

where the coefficients $c_{\beta\alpha}, s_{\beta\alpha}$ can be expressed in terms of the particle masses m_i, $i = 1,2,3$.

The three-particle Hamiltonian H is given by the expression

$$Hf = (-\Delta_{x_\alpha} - \Delta_{y_\alpha} + \sum_\alpha V_\alpha(x_\alpha))f(x) \equiv (-\Delta_x + \sum_\alpha V_\alpha(x_\alpha))f(x)$$

$$\equiv (H_k + V)f(x) \ ,$$

for $f \in C^2$. Here Δ_{x_α} (Δ_{y_α}) is the Laplace operator acting on the variables x_α (y_α) only, $H_k = -\Delta_x$ is the free three particle Hamiltonian. The Hamiltonian H is a self-adjoint operator on the Hilbert space $H = L_2(R^6)$. The Hamiltonians for the two-particle subsystems will be denoted by h_α, $\alpha = 1,2,3$,

$$(h_\alpha f)(x_\alpha) = (-\Delta_{x_\alpha} + V_\alpha(x_\alpha))f(x_\alpha) \ .$$

These Hamiltonians will be assumed to have the bound states $\psi_{\alpha,i}(x_\alpha)$ with energies $-\varepsilon_{\alpha,i}, \varepsilon_{\alpha,i} > 0$, $i = 1,2\ldots N_\alpha$. If the Coulomb force is repulsive, $n_\alpha > 0$, (attractive, $n_\alpha < 0$), the number N_α of bound states is finite (infinite). I shall denote by $H^{(\alpha)}$ a "two-body Hamiltonian" in the three-body space,

$$H^{(\alpha)}f(x) = (-\Delta_x + V_\alpha(x_\alpha))f(x) \ .$$

Let us discuss the non-stationary formulation of the scattering theory. If there is no Coulomb interaction the physical scattering is described with the aid of (non-stationary) wave operators

$$\tilde{U}_a^{(\pm)} = \underset{t\to\mp\infty}{\text{s.lim}}\, e^{itH} e^{-itH_a} P_a \ , \qquad a = 0,\{\alpha,i\} \ ,$$

$$\alpha = 1,2,3,\ i = 1,2,\ldots N_\alpha \ . \qquad (1)$$

If $a = \{\alpha,i\}$, $H_a = H^{(\alpha)}$, and the symbol P_a stands for the orthogonal projector onto the subspace of functions of the form $\psi_{\alpha,i}(x_\alpha)f(y_\alpha)$. If $a = 0$, $H_a = H_k$, $P_a \equiv I$, where I is an identity operator. Physical basis of this definition of the wave operators is the fact that the asymptotic evolution of clusters is generated by the free cluster Hamiltonians $H^{(\alpha)}$ and H_k.

Mathematical properties of the wave operators $U_a^{(\pm)}$ were studied by L.D. Faddeev [4] with the aid of integral equations in momentum space.

In the case of charged particles the limit (1) does not exist [5] and the non-stationary definition of the wave operators must be modified. The generalized Coulomb wave operators for few-body systems were found by J. Dollard [5]. To understand a physical idea of the generalization, let us consider a classical scattering on the bound pair $\alpha = 1$ in the state i. If the relative distance between center-of-mass of this pair and the third particle y_1 tends to infinity, we get the asymptotic relations

$$n_2|x_2|^{-1} \sim n_2|s_{21}|^{-1}|y_1|^{-1},\ n_3|x_3|^{-1} \sim n_3|s_{31}|^{-1}|y_1|^{-1} \ ,$$

$$m_1 = m_2 = m_3 = 1 \ .$$

Therefore an effective two-body potential describes the

long-range interaction between two fragments,

$$V^{(1)}(y_1) = \frac{n_{11}}{|y_1|} , \quad n_{11} = \sum_{\beta \neq 1} \frac{n_\beta}{|s_{\beta 1}|} . \tag{2}$$

If $t \to \pm \infty$, the relative distance between these fragments y_1 depends linearly on the time in principal order,

$$y_1 \sim 2p_1 t .$$

Hence, the asymptotic evolution of the system is directed accurately enough by the "asymptotic" non-stationary Hamiltonian

$$H_{as}^{(1)} = p_1^2 + V^{(1)}(2p_1 t) - \varepsilon_{1,i} .$$

In the quantum mechanics, we can take the corresponding Hamiltonian as

$$H_{as}^{(1)} P_{1,i} = H^{(1)} P_{1,i} + V^{(1)}(2\hat{p}_1 t) P_{1,i} ,$$

where $\hat{P}_1$ is the momentum operator.

Following to ref.[37,38], consider the differential equation for a quantum evolution operator,

$$i \frac{dU}{dt} P_{1,i} = HUP_{1,i} , \quad U|_{t=0} = I ,$$

and replace H by the asymptotic Hamiltonian $H_{as}^{(\alpha)}$. Since the operators $H_{as}^{(\alpha)}(t)$ and $H_{as}^{(\alpha)}(t')$ commute, the solution of the obtained equation is given by the formula

$$U_{as}^{(1,i)} P_{1,i} = \exp \{-i \int_0^t H_{as}^{(\alpha)}(t) dt\} P_{1,i} =$$

$$= \exp \{-iH^{(\alpha)} t + i \operatorname{sign} t \frac{n_{11}}{2|p_1|} \ln|t|\} P_{1,i} . \tag{3}$$

This relation shows that the two fragments are never asymptotically free because of the term containing $\ln|t|$. Consequently $H^{(\alpha)}$ cannot describe the asymptotic dynamics and the limit (1) does not exist by reason of the logarithmic distortion.

Comparing the relation (3) with the definition (1) of wave operators we can suggest the limit should have the expressions of the form (1) with the operator $U_{as}^{(1,i)}$ instead of the free channel evolution operator $e^{iH^{(\alpha)}t} P_{\alpha,i}$. This hypothesis was proved by J. Dollard who considered the general case of n particles [5].

Usually one uses the following normalized definition of the Dollard's wave operators:

$$\mho_{\alpha,i}^{(\pm)} = \underset{t\to\mp\infty}{\text{s.lim}} \exp\{iHt\} \exp\{-iH^{(\alpha)}t + i\phi_{\alpha,i}(t)\} P_{\alpha,i} ,$$

$$\mho_{o}^{(\pm)} = \underset{t\to\mp\infty}{\text{s.lim}} \exp\{iHt\} \exp\{-iH_k t + i\phi_o(t)\} ,$$

where the Coulomb phase operators $\phi_{\alpha,i}(t)$ and $\phi_o(t)$ are given by

$$\phi_{\alpha,i}(t) = \frac{n_{\alpha\alpha} \text{ sign } t}{2|p_\alpha|} \ln(4p_\alpha^2|t|),$$

$$n_{\alpha\alpha} = \sum_{\beta\neq\alpha} \frac{n_\beta}{|s_{\beta\alpha}|} ,$$

$$\phi_o(t) = \frac{q_o(p)\text{sign } t}{2|P|} \ln(4P^2|t|),$$

$$q_o(P) = \sum_\alpha \frac{n_\alpha|P|}{|k_\alpha|} .$$

J. Dollard proved that the modified wave operators satisfy the same properties (except the completeness) as the usual wave operators (1). Namely, the energy conservation yields the intertwining property

$$H\tilde{U}_a = \tilde{U}_a H_a P_a , \qquad a = 0 , \quad \{\alpha,i\} ,$$
$$\alpha = 1,2,3, \quad i = 1,2,\dots N_\alpha . \qquad (4)$$

These operators are partial isometries ,

$$\tilde{U}_a^{(\pm)*}\tilde{U}_a^{(\pm)} = P_a, \qquad \tilde{U}_a^{(\pm)}\tilde{U}_a^{(\pm)*} = \tilde{P}_a^{(\pm)},$$

where $\tilde{P}_a^{(\pm)}$ are orthogonal projection operators on H. The projections $\tilde{P}_a^{(\pm)}$ are mutually orthogonal,

$$\tilde{P}_a^{(\pm)}\tilde{P}_b^{(\pm)} = \delta_{ab}\, \tilde{P}_b^{(\pm)} ,$$

and its sum $P^{(\pm)} = \sum_a \tilde{P}_a^{(\pm)}$ is the orthogonal projection on H too.

However, J. Dollard did not prove the asymptotic completeness of the modified wave operators. Hence an open mathematical problem was the proof of the relation

$$\sum_a \tilde{U}_a^{(\pm)} \tilde{U}_a^{(\pm)*} = I - P_d , \qquad (5)$$

where P_d is the orthogonal projection onto three-particle bound states. Eq. (5) is a statement that three-body eigenfunctions and wavefunctions together form a complete set in Hilbert space H. This result plays a fundamental role in the scattering theory.

Different aspects of the non-stationary formulation of the scattering theory were studied in many papers (see for example ref.[6-10]).In particular, A.M. Vesselova proved that the Green's function in momentum space has the distorted pole singularities which follows from the existence of the Dollard's wave operators.

The following representation was established in ref.[6]:

$$R(P,P',z) = \frac{\tilde{R}_{oo}(P,P',z)}{(P^2-z)^{1+i\tilde{q}_o}(P'^2-z)^{1+i\tilde{q}'_o}} +$$

$$+ \sum_{\alpha,i} \frac{\tilde{R}_{o,\alpha i}(P,p'_\alpha)\hat{\psi}_\alpha(k'_\alpha)}{(P^2-z)^{1+i\tilde{q}_o}(p'^2_\alpha-z+\varepsilon_{\alpha,i})^{1+i\tilde{n}'_{\alpha\alpha}}} +$$

$$+ \sum_{\beta,j} \frac{\hat{\psi}_{\beta,j}(k_\beta)\ R_{\beta j,o}(p_\beta,P',z)}{(p^2_\beta-\varepsilon_{\beta,j}-z)^{1+i\tilde{n}_{\beta\beta}}(P'^2-z)^{i\tilde{q}'_o}} +$$

$$+ \sum_{\alpha,i,\beta,j} \frac{\hat{\psi}_{\beta,j}(k_\beta)\ R_{\beta j,\alpha i}(p_\beta,p'_\alpha,z)\psi_{\alpha,i}(k'_\alpha)}{(p^2_\beta-z+\varepsilon_{\beta,j})^{1+i\tilde{n}_{\beta\beta}}(p'^2_\alpha-z+\varepsilon_{\alpha,i})^{1+i\tilde{n}'_{\alpha\alpha}}} +$$

$$+ \tilde{R}(P,P',z) ,$$

$$\tilde{q}_o = \frac{q_o(P)}{2|P|} , \quad \tilde{n}'_{\alpha\alpha} = \frac{n_{\alpha\alpha}}{2|p'_\alpha|} , \quad \tilde{n}_{\beta\beta} = \frac{n_{\beta\beta}}{2|p_\beta|} .$$

$$\tilde{q}'_o = \frac{q_o(P')}{2|P'|} . \tag{6}$$

Here $\hat{\psi}_{\alpha,i}(k_\alpha)$ is the Fourier transform of the two-body eigenfunction $\psi_{\alpha,i}(x_\alpha)$. The functions $\tilde{R}_{oo}$, $\tilde{R}_{o,\alpha i}$, $\tilde{R}_{\beta j,o}$, $\tilde{R}_{\beta j,\alpha i}$ and $\tilde{R}$ do not contain the distorted pole singularities separating in the representation (6). The expressions for the matrix elements of the scattering operator in terms of the kernels $\tilde{R}_{oo}$, $\tilde{R}_{o,\alpha i}$, $\tilde{R}_{\beta j,o}$ and $\tilde{R}_{\beta j,\alpha i}$ were obtained in ref.[6] too.

However the method investigated by A.M. Vesselova does not permit to study the properties of the functions $\tilde{R}_{oo}$, $\tilde{R}_{o,\alpha i}$, $\tilde{R}_{\beta j,o}$, $\tilde{R}_{\beta j,\alpha i}$ and $\tilde{R}$ which are now known to have much complementary pecularities [11]. Consequently the singularities of the scattering matrix were not studied in ref.[6].

Several forms of integral equations were proposed to investigate the mathematical problems which were not solved by means of the non-stationary method. J. Noble formulated modified Faddeev equations for systems with Coulomb plus short range interactions. But the input kernels of Noble's equations involve the pure Coulomb interaction and they cannot be expressed in terms of known quantities (even if only two particles are charged). In fact, Noble's equations cannot be used to develop the stationary formulation of the Coulomb three-particle scattering theory and to solve the problems mentioned above.

In the papers of A.M. Vesselova [12,13] modified Faddeev equations were formulated for a system of three charged particles. I shall explain a principal idea of her method. Let us write the usual Faddeev equations for the wave function (23) + 1 in momentum space as one equation ,

$$\psi = \phi + A\psi ,$$

where ψ is a three-vector-function and A is a 3x3 matrix-operator. For simplicity I shall propose only particles 2 and 1 to be charged. Then, below the three-body threshold the operator A can be divided into the sum of two terms

$$A = A_o + \tilde{A} ,$$

where A_o contains the Coulomb singular term of the two-particle T-matrix t_{12} and the bound pole of the T-matrix t_{23}. The combination of these singularities **has** the form of the two-particle effective long-range interaction (2) in momentum space $n_{11}|p_1-p_1'|^{-2}$. Therefore the operator A_o-I can be inverted explicitely in terms of two-body Coulomb T-matrix. By inverting it we obtain the following modified Faddeev equations

$$\psi = \psi_o + (I-A_o)^{-1}\tilde{A}\psi, \quad \psi_o = (I-A_o)^{-1}\phi \quad .$$

A.M. Vesselova proved the resulting kernel $(I-A_o)^{-1}\tilde{A}$ to be compact below the three-particle threshold. It should be noted that the screening method was used in ref.[12] to invert the operator $(I-A_o)$ rather than the direct inversion method investigated in ref.[13].

Much more severe difficulties appear above the three-body threshold. Here the new Coulomb pecularities arise from the interplay of the pole in the free Green's function and Coulomb singularities in the two-particle T-matrix. In result, we are confronted with the problem of distinguishing an explicit inversion of the main part in the genuine three-particle integral equations. This problem is not solved up to now. Recently, E.O. Alt et al. [15-17] used the renormalization procedure proposed in ref.[12] for studying the three-body Coulomb problem to obtain modified AGS quasi-particle equations [18]. The compactness of the resulting kernels below the threshold is the consequence of the results proved in ref.[12]. It should be noted that the modified AGS equations coincide with Vesselova's equations if only two particles are charged and the short range potentials are separable. But above the three-body threshold, the two-particle renormalization procedure [12,19] is not sufficient to guarantee compactness. Moreover, if E = 0, the new singularities exist which make the resulting kernels non-compact [20,14].

We can state that the equations of ref.[12] and of ref.[15,16] are not of the Fredholm type above the threshold.

The corresponding homogeneous equations have non-trivial solutions which are the $(3 \to 3)$ wavefunctions. Our principal arguments are the following. It has been proved in [22,26] the $(3 \to 3)$ scattering amplitude to have the Coulomb singularities of the form $|p_\beta - p'_\beta|^{-3-in}$ (even if only two particles are charged). These singularities are non-integrable and may be proved to be more strong than the δ-function $\delta(p_\beta - p'_\beta)$. Evidently, they are different from the two-particle Coulomb pecularities.

One can show that the iterations of the integral equations described above do not contain such singularities [15-17]. However, it is well known that the Fredholm equation of the second type has solutions which reproduce the properties of the free terms. Therefore the singularities $|p_\beta - p'_\beta|^{-3-in}$ mentioned above should not appear if the corresponding integral kernels [12-17] should be compact.

Since the modified Faddeev' equations [12] and AGS equations [15] are not compact for positive energies, they cannot be used for mathematical justification of the scattering problem. However, compact or not, these equations should be applied for numerical calculations in some approximation. Similar situation is well known in physics. For example, the Schrödinger equation is not of the Fredholm type. But it has a unique solution satisfying the appropriate boundary conditions which can be founded numerically. It should be pointed out that the existence theorem is to be proved for AGS equations to make correct the numerical method proposed in ref.[15-17]. Namely, it is to be shown that the solution of such equations coincides in some approximation with the physical wavefunction.

A new form of modified Faddeev's equations with compact kernels was proposed in papers by author [21-25]. These equations were formulated and investigated in configuration space and the following results were obtained:

1) Smoothness and the asymptotic properties of the resolvent kernel $R(X,X',z)$ in configuration space were established for all z, $|\mathrm{Im}\, z| \geq 0$.

2) The complete set of the Coulomb wave functions was found. In particular, the asymptotic completeness (5) was proved for these functions.

3) The unitary Coulomb scattering operator was studied. The singularities of its kernels were specified.

4) The connection between stationary and non-stationary wave operators was established.

The solution of these problems completes mathematical justification of the three-body Coulomb scattering theory. I believe the theoretists can now concentrate on the physically interesting problems. In particular the correct numerical methods are to be investigated. One of such methods based on the differential formulation of the Coulomb scattering problem was proposed in ref.[39]. Below I shall describe new results for p-d scattering obtained by this method.

In my lectures, I shall cover the various attempts to formulate the principal formulas obtained in ref.[21-26] using modified Faddeev's equations in configuration space. The other approaches mentioned above were discussed in detail in many reports of E.O. Alt [17] and in the recent extended review talk of C.Chandler in Oregon [31].

2. CONFIGURATION SPACE THREE-PARTICLE COULOMB SCATTERING APPROACH

1. First, I want to explain the basic ideas of the configuration space treatment of the Coulomb problem. Let us consider the scattering of two charged particles. The most simple definition of the wavefunctions is the well-known

explicit solution of the Schrödinger equation [1,2]. The asymptotic form of this function is given by the sum of the distorted plane waves and spherical waves in almost all directions of the configuration space (except forward scattering direction). The Coulomb phase shifts $\ln(|k||x| - (k,x))$ and $\ln(2|k||x|)$ have a very simple physical meaning. They are given by eikonal formulas corresponding to the asymptotic trajectories [1]. On the contrary, one can prove the Schrödinger equation together with asymptotic boundary condition to have a unique solution. The former is the physical wave function. So, we have a simple configuration space formulation of the two-body Coulomb scattering theory.

If we investigate the three-particle Coulomb problem then we have to attack two kinds of difficulties. The first one is generated by the short-range parts of the two-body forces. The second type of difficulties is associated with long-range interaction because the asymptotic motion of particles is never free. In principle one can separate these difficulties using a following idea. We can divide two-body potentials into the sum of the short range parts and of the long-range ones. Applying Fadeev's procedure we can involve the short-range effect in mathematically correct way. However, a "non perturbated" Hamiltonian in the Faddeev's equations is to be taken an "asymptotic" one involving the "long range" part of the interactions. Then the principle problem is the study of the asymptotic dynamics. For this purpose one can use the heuristic eikonal construction.

Physically justified hypothesis is one that the asymptotic form of Coulomb wave functions should be defined by the same eikonals as in the case of neutral particles. For example, we have seen the plane eikonal (p,x) and the spherical eikonal $|x|$ are to be used for two-particle systems. The three-body case need much more eikonals which were founded in the ref.[27] of author.

2. Now, let us describe the modified Faddeev equations in the configuration space mentioned above. Let $\Omega_\alpha(\nu)$, $0 < \nu < 1/2$, be the domain in the configuration space R_c^6 where the distance $|x_\alpha|$ between the particles of the pair α is much less than the distance $|y_\alpha|$ between the third particle and the α-pair center-of-mass, when $|y_\alpha| \to \infty$,

$$\Omega_\alpha(\nu) = \{X: |x_\alpha| < a_o(1 + |y_\alpha|)^\nu\}, \quad a_o \text{ some constant.}$$

Let $\Omega_{o\alpha}(\nu)$ be the domain where the complementary inequality is true, and $\Omega_o = \bigcap_\alpha \Omega_{o\alpha}$. Let $\chi_\alpha(x,\nu)$ be a smooth finite function which equals 1 in $\Omega_\alpha(\nu)$ and 0 in $\Omega_{o\alpha}(\nu_1)$, $\nu < \nu_1 < 1/2$.

We shall separate the two-body potential V_α into two parts

$$V_\alpha(x_\alpha) = \hat{V}_\alpha(x) + V_\alpha^{(0)}(x) .$$

Here, $\hat{V}_\alpha(x)$ is the "short-range" term and $V_\alpha^{(0)}$ is the "three-particle long-range" term:

$$\hat{V}_\alpha(x) = \tilde{V}_\alpha^{(n)}(x_\alpha) + \frac{n_\alpha}{|x_\alpha|}\chi_\alpha(x) ,$$

$$V_\alpha^{(0)} = (1 - \chi_\alpha)\frac{n_\alpha}{|x_\alpha|} .$$

Let H_o be the "asymptotic" Hamiltonian generated by the "long-range" parts,

$$H_o = H_k + \sum_\alpha V_\alpha^{(0)} .$$

We shall denote by H_α the sum $H_\alpha = H_o + \hat{V}_\alpha$.

Let $R_{\beta\alpha}$ be the Faddeev-type resolvent components

$$R_{\beta\alpha} = R_o(\delta_{\alpha\beta}\hat{V}_\alpha - \hat{V}_\alpha R(z)\hat{V}_\beta)R_o \quad .$$

The components $R_{\beta\alpha}$ satisfy the modified Faddeev's equations

$$R_{\alpha\beta}(z) = (R_\alpha(z) - R_o(z))\delta_{\alpha\beta} - \sum_{\gamma\neq\alpha} R_\alpha\hat{V}_\alpha R_{\gamma\beta}(z) \quad . \tag{7}$$

These equations involve the "short-range" rescattering effects in Faddeev's form. The "long-range" specific is now concentrated in the operators R_o and R_α which have to be given. But in contrast to the case of neutral particles, there are no explicit expressions for the kernels $R_o(\mathbf{X},\mathbf{X}',z)$ and $R_\alpha(X,\mathbf{X}',z)$. These kernels are to be studied with the aid of independent equations.

To obtain such equations I have developed a special method. Its principal aspect is the construction of asymptotic solutions of the various Schrödinger equations involved. For example, I have constructed explicitely the kernel $R_{as}(X,X',z)$ satisfying

$$(-\Delta_{X'} + \sum_\alpha V_\alpha^{(O)}(x')-z)R_{as}(\mathbf{X},X',z) =$$

$$= \delta(X-X') - (R_{as}V_{as})(X,X',z) \quad ,$$

$$R_{as}(X,X',z) = \overline{R_{as}(X',X,\bar{z})} \quad ,$$

where the term $(R_{as}\hat{V}_{as})(X,X',z)$ decreases fast enough if $|X'| \to \infty$, uniformly in X and z. This kernel may be taken as a zero order approximation of the resolvent kernel $R(X,X',z)$. With the aid of Green's theorem, one can obtain an integral equation of the second kind for the kernel $R_o(z)$, namely.

$$R_o(z) = R_{as}(z) \cdot (I-V_{as}R_o(z)) \quad . \tag{8}$$

This Lippmann-Schwinger type equation may be studied by the well-known configuration space method [28]. The compactness of the operator $(R_{as}V_{as})^n$ may be proved for $n \gg 1$.

Similar equations are true for the operators R_α $(\alpha = 1,2,3)$. The smoothness and the asymptotic form of the kernels R_α may be studied.

Note that equation (8) is analogous to the modified Lippmann-Schwinger equation applying in the two-body Coulomb scattering problem [1]. The operator $R_{as}\ V_{as}$ plays the role of the product $G_c\ \tilde{V}_n$ with G_c the pure Coulomb Green's function and $\tilde{V}_n$ a nuclear short range potential.

Knowing the properties of the kernels $R_o(X,X',z)$ and $R_\alpha(X,X',z)$ one can investigate the properties of the resolvent kernel $R(X,X',z)$ by means of familiar Banach space techniques. All the technical details used for studying the equations (7,8) are described in papers of the author [21-26]. They are to numerous to repeat them here.

The known properties of the resolvent kernel permit us to study mathematical aspects of the scattering theory applying the well known functional methods [29].

3. Let us now explain the construction of the eikonal approximation which is the basis for studying the Coulomb scattering problem in configuration space.

Let L be a solution of the eikonal equation

$$|\nabla L|^2 = 1$$

in R^n. It should be noted that the number of particles may be arbitrary. Let $\hat{K} = \nabla L$ be the asymptotic direction of the motion of particles in configuration space. The n-vector X may be represented in the form of an orthogonal sum

$$X = L(X)\hat{K} + M \quad ,$$

where M is the projection of X onto the surface L(X) = = const.. Let $\gamma(t)$ be the straight line $\gamma(t) = \hat{K}t + M$, such that $\gamma(L) = X$. The asymptotic solution of the Schrödinger equation with long range interaction potential V may be represented in the following form:

$$\psi_{as} = f_L(M)A_L(X)\exp\{i\sqrt{E}L + i\tilde{W}_L\} \quad . \tag{9}$$

Here $X \in R^n$, $A_L(X)$ satisfies the continuity equation

$$2(\nabla A,\nabla L) + A\Delta L = 0 \quad ,$$

and $f_L(M)$ is an arbitrary smooth function of the "angle" variable M. The Coulomb phase shift $\tilde{W}_L$ has the form of the sum

$$\tilde{W}_L = W_L + \delta W_L$$

where the principal term W_L is given by the formula

$$W_L = -\frac{E^{1/2}}{2}\int_{\gamma(t)} dt\, V(\hat{K}t + M) \quad . \tag{9'}$$

The integration is performed along the "asymptotic" trajectory $\gamma(t)$ to the point $\gamma(L) = X$. The term δW_L decreases if $\sqrt{E}L \to \infty$ (E is the energy of scattering). This term is defined by the recurrance relations in ref.[22].

Let us describe the typical eikonal approximations in terms of which the construction of Green's function and of wavefunctions is established.

Let $Q_{\alpha,i}(y_\alpha,E_{\alpha,i})$ be the distorted spherical wave in R^3 corresponding to the two-body spherical eikonal $|y_\alpha|$,

$$Q_{\alpha,i} = |y_\alpha|^{-1} \exp\{iE_{\alpha,i}^{1/2}|y_\alpha| + iW_{\alpha,i}(y_\alpha)\},$$

$$W_{\alpha,i} = -\frac{n_{\alpha\alpha}}{2E_{\alpha,i}^{1/2}} \ln(2\, E_{\alpha,i}^{1/2}|y_\alpha|),$$

$$n_{\alpha\alpha} = \sum_{\gamma\neq\alpha} \frac{n_\gamma}{|s_{\gamma\alpha}|}, \tag{10}$$

and Q_o be the distorted spherical wave in R^6,

$$Q_o(X,E) = |X|^{-5/2} \exp\{i\sqrt{E}|X| + iW_o(X) + i\tfrac{\pi}{4}\},$$

$$W_o(X) = \sum_\alpha \frac{n_\alpha E^{-1/2}}{2|x_\alpha|}|X| \ln(2\sqrt{E}\,|X|). \tag{11}$$

The distorted spherical wave $Q_o(X,E)$ is an eikonal approximation (9), (9') with $L = |X|$. If $L = (\hat{P},X)$ is a plane eikonal, then the corresponding eikonal approximation (distorted plane wave) U_o is given by the formula (9), where

$$f_L = 1,\ A_L = 1,\ W_o = \sum_\alpha \frac{n_\alpha}{2|k_\alpha|} \ln(|k_\alpha|\ \xi^{(\alpha)}),$$

and $\xi^{(\alpha)}$ a parabolic coordinate, $\xi^{(\alpha)} = |x_\alpha| - (\hat{k}_\alpha, x_\alpha)$. The "point source" eikonal $|X-X'|$ is to be known to define the asymptotic form of the free resolvent kernel $R_k(X,X',z)$ in configuration space [30]. The corresponding eikonal approximation for three-body Coulomb problem has the form

$$\tilde{R}_{as}(X,X',z) = c_o \frac{\exp\{i\sqrt{z}|X-X'|\}}{|X-X'|^{5/2}} \tilde{G}_{as}(X,X',z), \tag{12}$$

where $c_o = e^{3i\frac{\pi}{4}} z^{3/4} (8\sqrt{2}\,\pi^{5/2})^{-1}$ and

$$\tilde{G}_{as} = \exp\{i \frac{|X-X'|}{2\sqrt{z}} \sum_\alpha \frac{n_\alpha}{|x_\alpha - x'_\alpha|} \times$$

$$\times \ln \frac{(x_\alpha - x'_\alpha, x_\alpha) + |x_\alpha - x'_\alpha||x_\alpha|}{(x_\alpha - x'_\alpha, x'_\alpha) + |x_\alpha - x'_\alpha||x'_\alpha|}\} .$$

Note that the asymptotic Green's function $R_{as}(z)$ from (8) is given by (12) almost in all directions of the configuration space.

4. Describe finally the construction of the resolvent kernel $R(X,X',z)$. Consider a set $\tilde{C}^{(2)}$ of functions $f(X)$ which are twice differentiable if $|x_\alpha| > 0$, $\alpha = 1,2,3$, and are of Hölder class if $|x_\alpha| = 0$ $(\alpha = 1,2,3)$. We denote by $\Phi_E(\hat{q})$, $q \in R^n$, $n = 2,3,\ldots$, the class of smooth bounded functions $f(X,\hat{q},E)$, $f \in \tilde{C}^{(2)}$, that may be represented as the sum

$$f(X,\hat{q},E) = \sum_{\beta=1}^{3} \sum_{j=1}^{N_\beta} \psi_{\beta,j}(x_\beta) U_{\beta,j}(y_\beta,\hat{q},E_{\beta,j}) +$$

$$+ U_0(X,\hat{q},E) , \tag{13}$$

where the terms $U_{\beta,j}$ and U_0 have a fixed asymptotic form if $|X| \to \infty$. The functions $U_{\beta,j}$ are the distorted spherical waves (10), where $E_{\beta,j} = E - \varepsilon_{\beta,j}$, and U_0 is the distorted spherical wave (11),

$$U_{\beta,j} \sim A_{\beta,j}(\hat{y}_\beta,\hat{q},E) Q_{\beta,j}(y_\beta,E_{\beta,j}) ,$$

$$U_0 \sim A_0(\hat{X},\hat{q},E) Q_0(X,E) .$$

The amplitudes $A_{\beta,j}$ and A_0 are assumed to be sufficiently smooth functions [11,24].

Let Π_ε be the complex plane C_2 with the interval $[\varepsilon,\infty)$ removed. Let ω_F be the set of points where the homogeneous equations (7,8) have a nontrivial solution and $D(\omega_F)$ be its nightbourhood . From integral equations (7) and (8) which have compact kernels in suitable Banach spaces, the following assertion may be derived:

<u>Theorem 1.</u> The resolvent kernel $R(X,X',z)$ belongs to the class $\Phi_z(X')$ $(\Phi_z(X))$ if $X'(X)$ is fixed and $|X-X'| \geq \delta > 0$, $z \in \Pi'_{-\varepsilon_0}$, $\Pi'_{-\varepsilon_0} = \Pi_{-\varepsilon_0} \setminus D(\omega_F)$, $\varepsilon_0 = \max_{\alpha,i} \varepsilon_{\alpha,i}$. The limits $\lim_{\varepsilon \downarrow 0} R(X,X',E\pm i\varepsilon)$ exist if $E \in D'$, and the asymptotic representation

$$R(X,X',E\pm i0) \sim A_0^{(\pm)}(X,P)Q_0(X,E) +$$

$$+ \sum_{\beta=1}^{3} \sum_{j=1}^{N_\beta} \psi_{\beta,j}(\mathbf{x}'_\beta) A_{\beta,j}^{(\pm)}(X,p_\beta) Q_{\beta,j}(y'_\beta, E_{\beta,j}) \tag{14}$$

is true if

$$|X| \leq C(1+|X'|)^\nu,\ \nu < 1/6,\ P = -E^{1/2}\hat{X}',$$

$$p_\alpha = -y'_\alpha\ |X'|^{-1}\ E^{1/2}\ .$$

This proposition describes the asymptotic form of the Green's function for the case when one point $X(X')$ lies much farther from the origin then the other one. The general case is considered in paper of author [24] and I do not give the corresponding formulas here. Note only we need the "spherical source" eikonal and the "rescattering" eikonals $L_\alpha^{(1)}(X,X')$, $L_{\alpha\beta}^{(2)}(X,X')$ etc. for describing the asymptotic form for arbitrary situated X and X'. The rescattering eikonals represent an "optical" distance between the points X and X' if the "ray" is n times (n = 1,2,...) in contact with the "hypersurfaces" $x_\alpha = 0$, $x_\beta = 0$ etc. An explicite expression for these eikonals may be obtained from the variational principles [24]

$$\nabla_{x_i} L^{(n)}_{\alpha_1\alpha_2\ldots\alpha_n}(X,X_1,\ldots X_n,X') = 0\ ,$$

$$x_{\alpha_i} = 0\ ,\quad i = 1,2,\ldots n\ ,$$

where

$$L^{(n)}_{\alpha_1\alpha_2\ldots\alpha_n} = \sum_{i=1}^{n-1} |X_{i+1} - X_i| + |X - X_1| + |X_n - X'| .$$

It should be pointed out the representation (14) in configuration space yields straightforward the distorted pole singularities (6) of the resolvent kernel in momentum space.

3. WAVEFUNCTIONS

1. It is easy to prove the resolvent kernel $R(X,X',z)$ satisfies a non-homogeneous Schrödinger equation

$$(-\Delta_X + \sum_\alpha V_\alpha(x_\alpha) - z)R(X,X',z) = \delta(X-X') .$$

Inserting the asymptotic representation (14) into this equation we may conclude that the amplitudes $A^{(\pm)}_{\alpha,i}(X,p_\alpha)$ and $A_o(X,P)$ satisfy the homogeneous Schrödinger equation

$$(-\Delta_X + \sum_\alpha V_\alpha(x_\alpha) - E)A^{(\pm)}(X) = 0 \tag{15}$$

with respect to X variable. As result we may define normalized wavefunctions for the Hamiltonian H by

$$\Psi_o(X,P) = -2|P|^{-3/2}(2\pi)^{5/2}A_o(X,P) ,$$

$$\Psi_{\alpha,i}(X,p_\alpha) = 2(2\pi)^{5/2}A_{\alpha,i}(X,p_\alpha) .$$

Let us now describe the asymptotic form of these wavefunction which may be proved with the aid of integral equations (7), (8). The wavefunctions Ψ_a may be represented as the sum

$$\Psi_a(X,P_a) = \tilde{\chi}^{(s)}_a(X,P_a) + \tilde{U}_a(X,P_a) , \tag{16}$$

where the term U_a belongs to the class $\Phi_{E_a}(P_a)$. Here, as usually, $P_a = P$, $E = P^2$ if $a = 0$,and

$$E_a = p_\alpha^2 - \varepsilon_{\alpha,i}, \; P_a = p_\alpha \text{ if } \quad a = \{\alpha,i\} .$$

The function $\tilde{\chi}_a^{(s)}$ contains the slowly decreasing terms in the asymptotics of the wavefunction Ψ_a, which corresponds to the initial state of the system. We shall denote by A_{ba} the amplitude of the distorted spherical wave $Q_b(\hat{X}_b, P_a)$ appearing on the right hand side of (11). Here, $X_b = X$ if $a = 0$,and $X_b = y_\beta$ if $b = \{\beta,j\}$.

The function $\tilde{\chi}_a^{(s)}$ is given by $\tilde{\chi}_{\alpha,i}^{(s)} = \psi_{\alpha,i}(x_\alpha)\cdot \phi_{\alpha,i}(y_\alpha,p_\alpha)$ for $a = \{\alpha,i\}$, and $\phi_{\alpha,i}(y_\alpha,p_\alpha)$is the two-particle wavefunction for the Hamiltonian $\tilde{h}_{\alpha,i} = -\Delta_{y_\alpha} + \tilde{V}_{\alpha,i}(y_\alpha)$ with the potential

$$\tilde{V}_{\alpha,i}(y_\alpha) = \sum_{\beta\neq\alpha} \int dx_\alpha n_\beta |\psi_{\alpha,i}(x_\alpha)|^2 |c_{\beta\alpha}x_\alpha + s_{\beta\alpha}y_\alpha|^{-1} .$$

If the quantity $n_{\alpha\alpha}$ (see (10)) is not equal to zero, the potential $\tilde{V}_{\alpha,i}(y_\alpha)$ is a long range one,

$$\tilde{V}_{\alpha,i} \sim n_{\alpha\alpha}|y_\alpha|^{-1} + O(|y_\alpha|^{-2}) \quad .$$

Note that the two-body Hamiltonians $\tilde{h}_{\alpha,i}$ have a simple physical meaning. They describe the scattering of the third particle on an spectator which is a bound pair α. The latter generate an effective two-body potential $\tilde{V}_{\alpha,i}$.

If a two-body parabolic coordinate $\tilde{\xi}^{(\alpha)}$, $\tilde{\xi}^{(\alpha)} = |y_\alpha| - (\hat{p}_\alpha, y_\alpha)$, tends to infinity, $\tilde{\xi}^{(\alpha)} \to \infty$, the asymptotics of the wavefunction may be represented by the sum

$$\phi_{\alpha,i}(y_\alpha,p_\alpha) \sim \chi_{\alpha,i}^{(0)}(y_\alpha,p_\alpha) + U_{\alpha,i}^{(0)}(y_\alpha,p_\alpha),$$

where $\chi_{\alpha,i}^{(0)}$ is the distorted two-particle plane wave

$$\chi^{(0)}_{\alpha,i} = \exp\{i(p_\alpha,y_\alpha) + iw^{(0)}_\alpha\}(1+O((\hat{\xi}^{(\alpha)})^{-1})) ,$$

and

$$w^{(0)}_\alpha = \frac{n_{\alpha\alpha}}{2|p_\alpha|}\ln(|p_\alpha|\hat{\xi}^{(\alpha)}).$$

The function $U^{(0)}_{\alpha,i}$ is the distorted spherical wave (10) with a singular amplitude $A^{(s)}_{\alpha,i}$. The principal singularities of $A^{(s)}_{\alpha,i}$ have the form of a pure Coulomb two-particle singularity. The representation

$$A^{(s)}_{\alpha,i} = A^{(0)}_\alpha + \tilde{A}^{(s)}_{\alpha,i}$$

is true, where $A^{(0)}_\alpha$ is the Coulomb amplitude

$$A^{(0)}_\alpha = -\frac{\tilde{n}_{\alpha\alpha}}{2|p_\alpha|}\frac{\exp\{-i\tilde{n}_{\alpha\alpha}\ \ln\sin^2\frac{\theta_\alpha}{2} + 2i\eta_\alpha\}}{\sin^2\frac{\theta_\alpha}{2}}$$

with $\cos\theta_\alpha = (\hat{p}_\alpha,\hat{y}_\alpha)$, $\eta_\alpha = \arg\Gamma(1 + i\tilde{n}_{\alpha\alpha})$, $\tilde{n}_{\alpha\alpha} = \frac{n_{\alpha\alpha}}{2|p_\alpha|}$. The term $\tilde{A}^{(s)}_{\alpha,i}$ is initiated by the multipole parts of the effective potential $\tilde{V}_{\alpha,i}$. This term has singularities which are less strong then the pure Coulomb one.

The sum

$$\tilde{A}_{\alpha i,\alpha i} = A^{(s)}_{\alpha,i} + A_{\alpha i,\alpha i}$$

may be interpreted as the elastic (2 → 2) scattering amplitude. Hence this amplitude has a non-integrable singularity due to the long range interaction. The rearrangement (2 → 2) amplitudes $A_{\beta j,\alpha i}(\hat{y}_\beta,p_\alpha)$, $\{\beta j\} \neq \{\alpha i\}$, are smooth bounded functions. The breakup amplitude $A_{o,\alpha i}$ is sufficiently smooth function. If the Coulomb force between the particles of the pair β is attractive, the breakup amplitudes $A_{o,\beta j}$ have singularities of the form

$\sqrt{|x_\beta|^{-1}|X|}$. The former is square integrable on the hypersphere $|X| = 1$ [11].

Next, consider the asymptotic behaviour of the wavefunction $\Psi_o(X,P)$. The asymptotics of $\Psi_o(X,P)$ may be described in terms of eikonal approximations involving four types of eikonals. I have described two of them above: plane eikonals $(\hat{P},X)$ and spherical eikonal $|X|$. Let us describe the other two, Z_α and $Z_{\alpha\beta}$, which correspond to the single and double two-particle scattering, respectively.

Let Z_α be a single scattering eikonal given by

$$Z_\alpha = |k_\alpha||x_\alpha| + (p_\alpha, y_\alpha) .$$

This eikonal corresponds to the process where the two particles of the pair α interact and the third one moves freely (see Fig.1).

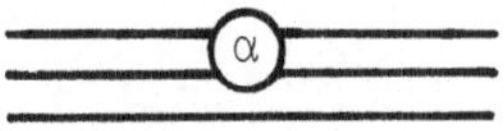

Fig. 1

To describe the double scattering eikonal we have to introduce some notation. Let $\{k_{\alpha\beta}, p_{\alpha\beta}\}$ be the Jacobi momenta of the particles after the double two-particle collisions $\beta \to \alpha$ (see Fig.2).

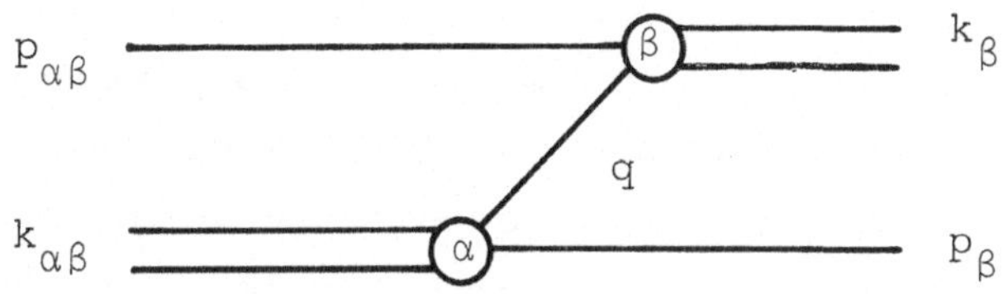

Fig. 2

The final momenta $k_{\alpha\beta}$ and $p_{\alpha\beta}$ may be expressed in terms of the initial momenta p_{β} and the intermediate momenta $q_{\beta\alpha}$ which defined from the princip of least action [11,22]. Then the double scattering eikonal is

$$Z_{\alpha\beta} = |k_{\alpha\beta}||x_{\alpha}| + (p_{\alpha\beta}, y_{\alpha}) \quad .$$

The single-scattering eikonal approximation is given by

$$\Psi_{\alpha} = C_{\alpha}(M_{\alpha})|x_{\alpha}|^{-1} \exp\{i\sqrt{E}\, Z_{\alpha} + iW_{\alpha}\} \quad . \tag{17}$$

Here, the Coulomb phase shift W_{α} has the form of the sum

$$W_{\alpha} = \sum_{\beta} W_{\alpha}^{(\beta)} \quad ,$$

where

$$W_{\alpha}^{(\alpha)} = -\frac{n_{\alpha}}{2|k_{\alpha}|} \ln 2|k_{\alpha}||x_{\alpha}|$$

if $\beta = \alpha$, and

$$W_{\alpha}^{(\beta)} = -\frac{n_{\beta}}{2|k_{\beta}^{(\alpha)}|} \ln\left(|k_{\beta}^{(\alpha)}||x_{\beta}| + (k_{\beta}^{(\alpha)}, x_{\beta})\right) \quad ,$$

$$k_{\beta}^{(\alpha)} = c_{\beta\alpha}|k_{\alpha}|\hat{x}_{\alpha} + s_{\beta\alpha}\, p_{\alpha} \quad ,$$

if $\beta \neq \alpha$. The amplitude $C_{\alpha}(M_{\alpha})$ which is a function of the "tangent" variable $M_{\alpha} = X - Z_{\alpha}\nabla Z_{\alpha}$ may be represented in the form

$$C_{\alpha}(M_{\alpha}) = f_{\alpha}(\hat{x}_{\alpha}, k_{\alpha}) \exp\{i\delta W_{\alpha}(M_{\alpha})\} \quad , \tag{18}$$

where f_{α} is the two-particle scattering amplitude for the Hamiltonian h_{α}. The function δW_{α} is an additional Coulomb phase shift which takes into account the long-range interaction before the short-range α-pair collision. The explicit expression of this function is given in ref. [11,22]. The double-scattering eikonal approximation $\Psi_{\alpha\beta}$ may be represented in the form

$$\tilde{\Psi}_{\alpha\beta} = (|x_\alpha||y_\alpha|)^{-1} A_{\alpha\beta} C_{\alpha\beta}(M_{\alpha\beta}) \exp\{iE^{1/2} Z_{\alpha\beta} + iW_{\alpha\beta}\} , \quad (19)$$

where the amplitude $A_{\alpha\beta}$ is given by

$$A_{\alpha\beta} = \left(\frac{|y_\alpha|}{\sin \tilde{\theta}_\alpha}\right)^{1/2} \left| \frac{\partial^2 Z_{\alpha\beta}}{\partial \theta^2_{\alpha\beta}} \right|^{-1/2} ,$$

$$\cos \theta_{\alpha\beta} = (\hat{p}_\beta, \hat{q}_{\beta\alpha}) , \; \cos \tilde{\theta}^2_\alpha = (\hat{p}_\alpha, \hat{y}_\alpha) .$$

The amplitude $C_{\alpha\beta}$ being a function of the tangent variables $M_{\alpha\beta} = X - \nabla Z_{\alpha\beta} Z_{\alpha\beta}$ is given by

$$C_{\alpha\beta}(M_{\alpha\beta}) = f_\alpha(\hat{x}_\alpha, k_{\alpha\beta}) f_\beta(\hat{q}_{\beta\alpha}, k_\beta) \times$$

$$\times (|s_{\alpha\beta}|^{-1} |k_\beta| \sin \theta_{\alpha\beta})^{1/2} \exp\{i\delta W_{\alpha\beta}(M_{\alpha\beta})\} . \quad (20)$$

Here f_α and f_β are the two-particle scattering amplitudes for the Hamiltonians h_α and h_β. The additional phase shift is expressed in terms of the "angle" variables $M_{\alpha\beta}$ [11,22].

Now we may describe the slowly decreasing term $\tilde{\chi}_o^{(s)}$ of the wavefunction $\psi_o(X,P)$. It may be represented as the sum

$$\tilde{\chi}_o^{(s)} = \chi_o + \sum_\alpha U_\alpha + \sum_{\alpha \neq \beta} U_{\alpha\beta} . \quad (21)$$

In the "non singular" directions, the terms χ_o, U_α and $U_{\alpha\beta}$ are given by eikonal formulas (9), (17) and (19) corresponding to the eikonals Z, Z_α and $Z_{\alpha\beta}$, respectively. These eikonal formulas are valid if all the amplitudes and phases are smooth bounded functions. However, there are many "singular directions" in configuration space where they become infinite. In these directions, the asymptotic form of $\tilde{\chi}_o^{(s)}$ is described in terms of certain special functions (see[22,23]).

As we shall see below, the "singular directions", where one of the eikonals $Z = (\hat{P},X)$, Z_α and $Z_{\alpha\beta}$ coincides with the spherical eikonal $|X|$, play an important role in the behaviour of the scattering amplitudes.

Let ξ, ξ_α and $\xi_{\alpha\beta}$ be the parabolic coordinates

$$\xi = |X|-Z, \quad \xi_\alpha = |X|-Z_\alpha, \qquad \xi_{\alpha\beta} = |X|-Z_{\alpha\beta} \quad ,$$

and define neighborhoods V_o, V_α and $V_{\alpha\beta}$ of the singular directions ,

$$V_o = \{X:\xi < (1 + |X|)^\nu\} \ ,$$

$$V_\alpha = \{X:\xi_\alpha < (1 + |X|)^\nu\} \ ,$$

$$V_{\alpha\beta} = \{X:\xi_{\alpha\beta} < (1 + |X|)^\nu\} \ ,$$

where $1 > \nu > 0$ is some constant.

The function χ_o is a smooth function of the parabolic coordinate in V_o. It has an integral representation involving the confluent hypergeometric function $\phi(a,c,\xi)$ (see [23]). If $\xi \to \infty$, the asymptotic form of χ_o is a sum of the distorted plane wave and the distorted spherical wave. Here a strong singular amplitude $A_o^{(s)}(\hat{X},P)$ is given by

$$A_o^{(s)}(\hat{X},P) = (1 - |X|^{-1}Z)^{-5/2-i\tilde{q}_o}\,\tilde{A}(\hat{X},P) \ , \tag{22}$$

where $\tilde{A}$ is a sufficiently smooth function.

In V_α, the asymptotic form of U_α is described in terms of confluent hypergeometric functions $\phi(a,c,\xi_\alpha)$ [22]. If $\xi_\alpha \to \infty$, the term U_α takes the form of the sum of the eikonal approximation (17) and of the distorted spherical wave with a singular amplitude $A_\alpha^{(s)}$ given by

$$A_\alpha^{(s)} = f_\alpha(\hat{x}_\alpha,k_\alpha)\,|\Delta_\alpha^{(s)}|^{-3-ia_\alpha}\,\tilde{A}_\alpha^{(s)}(\hat{X},P) \quad , \tag{23}$$

where $\Delta_{\alpha}^{(s)} = (1-|X|^{-1}z_{\alpha})^{1/2}$ and $\tilde{A}_{\alpha}^{(s)}$ is a smooth function (see [22]). The value a_{α} is defined by

$$a_{\alpha} = \sum_{\gamma\neq\alpha} a_{\alpha}^{(\gamma)}, \quad a_{\alpha}^{(\gamma)} = \frac{n_{\gamma}}{2|k_{\gamma}|} + \frac{n_{\gamma}}{2|k_{\gamma}^{(\alpha)}|} \quad . \tag{24}$$

The asymptotic form of the term $U_{\alpha\beta}$ depends on the direction $\hat{X}$ in configuration space. Let $V_{\alpha\beta}^{(+)}$ ($V_{\alpha\beta}^{(-)}$) be the domain where the inequality $\omega_{\alpha\beta} > 0$ ($\omega_{\alpha\beta} < 0$) is satisfied, $\omega_{\alpha\beta} = E^{-1/2}(|k_{\alpha\beta}|(\hat{p}_{\alpha\beta},y_{\alpha}) - |p_{\alpha\beta}||x_{\alpha}|)$. The quantity $\omega_{\alpha\beta}$ is equal to zero in the singular direction $V_{\alpha\beta}$. Due to the conservation of energy and momentum for classical particles undergoing double two-body collisions ($\beta \to \alpha$), $V_{\alpha\beta}^{(+)}$ is the allowed domain (asymptotically) and $V_{\alpha\beta}^{(-)}$ is the forbidden one. In $V_{\alpha\beta}^{(+)}$ the term $U_{\alpha\beta}$ has the eikonal form (19). In $V_{\alpha\beta}$, however, it changes its form, and in $V_{\alpha\beta}^{(-)}$ it becomes the distorted spherical wave with a singular amplitude $A_{\alpha\beta}^{(s)}$ given by

$$A_{\alpha\beta}^{(s)} = \frac{f_{\alpha}(\hat{x}_{\alpha},k'_{\alpha\beta})f_{\beta}(\hat{k}_{\beta\alpha},k_{\beta})}{(1-|\mathbf{X}|^{-1}z_{\alpha\beta})^{1+ia_{\alpha\beta}}} \tilde{A}_{\alpha\beta}^{(s)} \quad . \tag{25}$$

Here $\tilde{A}_{\alpha\beta}^{(s)}$ is a bounded smooth function. The quantity a is expressed in terms of kinematic variables with $a_{\alpha\beta}$ formula similar to (24) (see [22]). The vectors $k'_{\alpha\beta}$ and $k_{\beta\alpha}$ are

$$k'_{\alpha\beta} = -\frac{c_{\alpha\beta}}{s_{\alpha\beta}} p'_{\alpha} + \frac{1}{s_{\alpha\beta}} p_{\beta} \quad ,$$

$$k_{\beta\alpha} = \frac{c_{\beta\alpha}}{s_{\alpha\beta}} p_{\beta} - \frac{1}{s_{\alpha\beta}} p'_{\alpha} \quad .$$

In the intermediate region $V_{\alpha\beta}$, the exchange behaviour of the term $U_{\alpha\beta}$ is described in terms of the irregular confluent hypergeometric function $\psi(a,c,\xi_{\alpha\beta})$ [22]. Note that the analogous exchange effect also exists in the case of neutral particles, where Fresnel's integral appears instead of hypergeometric functions [27].

This completes our analysis of the asymptotic form for the wavefunctions $\Psi_a^{(+)}$. The asymptotic form for the $\Psi_a^{(-)}$ follows from

$$\Psi_a^{(-)}(X,P_a) \sim \overline{\Psi_a^{(+)}(X,-P_a)} \ .$$

2. We now formulate the boundary value problems for Coulomb wavefunctions. They may be defined as smooth solutions of the Schrödinger equation (15) having the asymptotic form (16). In that representation the slowly decreasing terms $\tilde{\chi}_a^{(s)}$ have to be fixed according to the prescription given above, while the functions of the class $\Phi_E(\hat{p}_a)$ are then specified in terms of the solution.

Alternatively, the wavefunctions $\psi_{\alpha,i}$ and ψ_o may be defined as smooth solutions of the modified Faddeev differential equations. Let

$$\chi_{o\alpha}^{(s)} = \delta_{\alpha 1}\, \chi^{(0)}(X,P) + U_\alpha(X,P) + \sum_{\gamma\neq\alpha} U_{\alpha\gamma}(X,P) \ ,$$

where the terms $\chi^{(0)}$, U_α and $U_{\alpha\gamma}$ where described above. We shall denote by $\Phi(\hat{P}_a)$, $a = 0$, $\{\alpha,i\}$, the class of smooth vector-valued functions $F = \{f_a^{(1)}, f_a^{(2)}, f_a^{(3)}\}$ whose components $f_a^{(\beta)}$ have the form

$$f_{\alpha,i}^{(\beta)} = \chi_{\alpha,i}^{(s)}(X,p_\alpha)\delta_{\alpha\beta} + U_{\alpha,i}^{(\beta)}(X,P_a) \ ,$$

$$f_o^{(\beta)} = \chi_{o\beta}^{(s)}(X,P) + U_o^{(\beta)}(X,P) \ ,$$

where the functions $U_a^{(\beta)}(X,P_a)$ belong to the class $\Phi_E^{(\beta)}(\hat{P}_a)$ of smooth bounded functions whose asymptotics (13) contain the terms $\psi_{\alpha,i}\, U_{\alpha,i}$ corresponding to one index β only.

Then we have

Theorem 2. The modified Faddeev equations in differential form

$$(-\Delta_X + \sum_{\gamma} V_{\gamma}^{(0)}(X) + \hat{V}_{\beta}(X))\Psi_a^{(\beta)} =$$

$$= - \hat{V}_{\beta}(X) \sum_{\gamma\neq\beta} \Psi_a^{(\gamma)}(X) \qquad (26)$$

have a unique solution in the class $\Phi_a(\hat{P}_a)$. The sum $\sum_{\gamma} \Psi_a^{(\gamma)}$ coincides with the wavefunction $\Psi_a(X,P_a)$. The proof of this theorem is analogous to the one for neutral particles [30].

It should be noted that the Noble's equations in differential form may be helpful for numerical calculations. These equations read

$$(-\Delta_X + \sum_{\gamma} \frac{n_{\gamma}}{|x_{\gamma}|} + \hat{\mathcal{V}}_{\beta}^{(n)}(x_{\beta}))\hat{\mathcal{U}}^{(\beta)}(X) = -\hat{\mathcal{V}}_{\beta}^{(n)}(x_{\beta}) \sum_{\gamma\neq\beta} \hat{\mathcal{U}}^{(\gamma)}, \qquad (26')$$

and may be proved to have a unique solution in the class $\hat{\Phi}_a(P_a)$ if the Coulomb forces between all particles are repulsive. I want to stress that the theorem 2 is true even if some of the Coulomb forces are attractive.

In contrast the wavefunctions $\Psi_{\alpha,i}$ and Ψ_0 may also be defined as solutions of the modified Faddeev integral equations (7) (see [11]).

It should be noted that the wavefunctions $\Psi_{\alpha,i}$ and Ψ_0 describe the physical situation completely. The wavefunctions $\Psi_{\alpha,i}(X,p_{\alpha})$ describe the scattering processes $(2 \to 2)$ and $(2 \to 3)$ with two clusters in the initial state, the pair α is in the bound state $\psi_{\alpha,i}(x_{\alpha})$. The wavefunctions $\Psi_0(X,P)$ correspond to the scattering processes $(3 \to 3)$ and $(3 \to 2)$ with three free particles in the initial state.

4. WAVE OPERATORS AND S-MATRIX

1. For defining stationary wave operators let us describe some formal constructions. Let H_o be the Hilbert space $L_2(R^6)$ and $H_{\alpha,i}$, $\alpha = 1,2,3$, $i = 1,2,\dots,N_\alpha$, be the Hilbert space $L_2(R^3)$. We shall denote by $\hat{H}$ the orthogonal sum

$$\hat{H} = H_o \oplus \sum_{\alpha,i} \oplus H_{\alpha,i} .$$

Elements of the space H_o will be denoted by $f_o(P)$ and elements of $H_{\alpha,i}$ by $f_{\alpha,i}$. The elements of $\hat{H}$ are vector-valued functions $\hat{f}$, $\hat{f} = \{f_o, f_{\alpha,i}\}$, $\alpha = 1,2,3$, $i = 1,2,\dots N_\alpha$. Let $\hat{H}_o$ be the reducible operator in $\hat{H}$ defined by

$$\hat{H}_o = \tilde{H}_o \oplus \sum_{\alpha,i} \oplus \tilde{H}_{\alpha,i} .$$

Here, $\tilde{H}_o$ and $\tilde{H}_{\alpha,i}$ are multiplication operators acting as

$$(\tilde{H}_o f_o)(P) = P^2 f_o(P) ,$$

$$(\tilde{H}_{\alpha,i} f_{\alpha,i})(p_\alpha) = (p_\alpha^2 - \varepsilon_{\alpha,i}) f_{\alpha,i}(p_\alpha) .$$

Consider the wavefunctions $(2\pi)^{-3}\Psi_o(X,P)$, $((2\pi)^{-3/2}\Psi_{\alpha,i}(X,p_\alpha))$ as the kernels of operators $U_o(U_{\alpha,i})$ from H_o $(H_{\alpha,i})$ to H. Then in the mixed X-P representation, these kernels are

$$U_o^{(\pm)}(X,P) = (2\pi)^{-3}\Psi_o^{(\pm)}(X,P), \quad U_{\alpha,i}^{(\pm)}(X,p_\alpha) = (2\pi)^{-3/2}\Psi_{\alpha,i}^{(\pm)}(X,p_\alpha) .$$

Let $U^{(\pm)}$ be an operator from $\hat{H}$ to H which acts on the elements of $\hat{H}$ as

$$U^{(\pm)}\hat{f} = U_o^{(\pm)} f_o + \sum_{\alpha,i} U_{\alpha,i}^{(\pm)} f_{\alpha,i}, \quad \hat{f} = \{f_o, f_{\alpha,i}\} ,$$

$$\alpha = 1,2,3, \quad i = 1,2,\dots N_\alpha .$$

Then we have

Theorem 3. An arbitrary function $f \in H$ may be uniquely represented as the orthogonal sum

$$f = f_d + \sum_a U_a^{(\pm)} f_a^{(\pm)} ,$$

where $f_d = P_d f \in H_d$ and

$$f_a^{(\pm)} = U_a^{(\pm)*} f \in H_a , \qquad a = 0, \{\alpha, i\} ,$$

$$\alpha = 1,2,3, \; i = 1,2,\ldots N_\alpha .$$

The representation

$$\phi(H) f = \phi(P_d H) f_d + \sum_a U_a^{(\pm)} \phi(\tilde{H}_a) f_a^{(\pm)}$$

is true for an arbitrary bounded smooth function

$$\phi(t) , \quad t \in (-\infty, \infty) .$$

This theorem may also be formulated in terms of the operators $U^{(\pm)}$, $U^{(\pm)} : \hat{H} \to H$:

Theorem 4. The wave operators $U^{(\pm)}$ satisfy the relations

$$U^* U = \hat{I}, \quad UU^* = I - P_d, \quad HU = U\hat{H}_o , \tag{27}$$

where $\hat{I}$ is an identity operator in $\hat{H}$.

The demonstrations of the completeness relations in the Theorem 3 and 4 are based on the following

Lemma 1. Let λ be a non-singular point of the integral equations (7), (8), $\lambda \neq \omega_F$. Let f, f' be smooth functions with compact support. Then the spectral function $E(\lambda)$ is differentiable with respect to λ, and

$$\frac{d}{d\lambda}(E(\lambda)f,f') = \int dP F_o^{(\pm)}(P,f)\overline{F_o^{(\pm)}(P,f')}\delta(P^2-\lambda) \quad +$$

$$+ \sum_{\alpha,i} \int dp_\alpha F_{\alpha,i}(p_\alpha,f)\overline{F_{\alpha,i}(p\ ,f')}\delta(p_\alpha^2-\varepsilon_{\alpha,i}-\lambda) .$$

Here,

$$F_o^{(\pm)}(P,f) = \int dX U_o^{(\pm)}(X,P)f(X) \ ,$$

$$F_{\alpha,i}(p_\alpha,f) = \int dX U_{\alpha,i}^{(\pm)}(X,p_\alpha)f(X) \ .$$

The proofs of this lemma may be given by usual methods basing on the known resolvent kernel behaviour. We have to use the expression of the spectral density in terms of the resolvent R(z),

$$\frac{d}{d\lambda}(E(\lambda)f,f') =$$

$$= \frac{1}{2\pi i} \lim_{\varepsilon\downarrow 0} [(R(\lambda+i\varepsilon)-R(\lambda-i\varepsilon))f,f'] \ . \qquad (28)$$

Then we apply Green's theorem to convert the right hand side (28) into a surface integral. The latter may be calculated using the known asymptotic form of the Green's function. Corresponding technical details are given in ref.[11].

To demonstrate the orthogonality of the wave operators, $U_a^{(\pm)*} U_b^{(\pm)} = \delta_{ab} I_a$, we have to use Green's theorem together with the Schrödinger equation. Then the integral involved may be calculated using known asymptotic form of the kernels $U_a(X,P_a)$ (see ref.[11]).

Finally, we shall describe the relation between stationary wave operators and non-stationary ones.

Let $I_{\alpha,i}$ and I_o be the identification operators from $P_{\alpha,i}H$ to $H_{\alpha,i}$ and from H to H_o.

Theorem 5. The non-stationary wave operators are related to the stationary ones as follows:

$$\tilde{U}_o^{(\pm)} = U_o^{(\pm)} I_o \ , \quad \tilde{U}_{\alpha,i}^{(\pm)} = U_{\alpha,i}^{(\pm)} I_{\alpha,i} P_{\alpha,i} \ , \qquad (29)$$

$$\alpha = 1,2,3, \qquad i = 1,2,\ldots N_\alpha \ .$$

The proof of this theorem is analogous to the one for neutral particles, the principal details are given in ref.[11].

This theorem plays an important role to imply the completeness of Dollard's wave operators.

2. Let us now describe the construction of the kernels of the Coulomb scattering operator. First we shall introduce some distributions in terms of which we describe such kernels.

We shall denote by t^{-1-in}, $t \in [0,1]$, a distribution defined by analytic continuation of the integral

$$\int_0^1 t^{-z} f(t)\,dt, \qquad f \in C^\infty [0,1],$$

from the domain Re $z < 1$ [32]. Note that the regularization

$$\int_0^1 t^{-1-in} f(t)\,dt = \frac{i}{n} f(0) + \int_0^1 \frac{f(t)-f(0)}{t^{1+in}}\,dt$$

may be used for this distribution.

We shall denote by $(t \mp i0)^{-1-in}$, $t \in [-1,1]$, a distribution defined by the analytic continuation of the integral

$$\int_{-1}^{+1} f(t)\,(t \mp i\varepsilon)^{-1-in} dt$$

in $\varepsilon, \varepsilon > 0$ for $\varepsilon \to 0$.

Remember that (2 → 2) elastic scattering amplitudes $A_{\alpha i,\alpha i}(\hat{p}_\alpha, p'_\alpha)$ may be represented in the form

$$A_{\alpha i,\alpha i}(\hat{p}_\alpha, p'_\alpha) = \frac{b_{\alpha,i}(\hat{p}_\alpha, p'_\alpha)}{|p_\alpha - p'_\alpha|^{2+2i\tilde{n}_\alpha}} + \tilde{A}_{\alpha i,\alpha i} \ ,$$

$$p_\alpha^2 = p_\alpha'^2 = E + \varepsilon_{\alpha,i} \ ,$$

where $b_{\alpha,i}$ and $\tilde{A}_{\alpha i,\alpha i}$ are smooth bounded functions, $\tilde{n}_\alpha = n_\alpha/(2|p_\alpha|)$. We shall associate with this amplitude a distribution which will be denoted by the same symbol $A_{\alpha i,\alpha i}(\hat{p}_\alpha, p'_\alpha)$. In the local coordinates $t = 1-\cos\theta$ with $\cos\theta = (\hat{p}_\alpha, p'_\alpha)$, this distribution becomes t^{-1-in}. Similar procedure was proposed for two-body Coulomb S-matrix by I. Herbst [33].

The (3 → 3) scattering amplitude A_{oo} has more singularities as was discussed in section 2. This amplitude may be represented as the sum

$$A_{oo} = A_o^{(s)} + \sum_\alpha A_\alpha^{(s)} + \sum_{\alpha\neq\beta} A_{\alpha\beta}^{(s)} + \tilde{A}_o$$

corresponding to the asymptotic expansion (21), where the term $\tilde{A}_o$ is a sufficiently smooth function. Other terms have non-integrable singularities generated by the three-body elastic (3 → 3) processes, as well as single scattering and double scattering processes. These terms will be considered as distributions given by

$$A_o^{(s)}(\hat{P}, P') = |P-P'|^{-5-2i\tilde{q}_o(P)} \tilde{A}_s(\hat{P}, P'), \ P^2 = P'^2 = E \ ,$$

$$A_\alpha^{(s)}(\hat{P}, P') = f_\alpha^{(s)}(\hat{k}_\alpha, k'_\alpha) \frac{\tilde{A}_\alpha^{(s)}(\hat{P}, P')}{|p_\alpha - p'_\alpha|^{3+ia_\alpha}} \ ,$$

$$A^{(s)}_{\alpha\beta}(\hat{P},P') = \frac{f^{(s)}_{\alpha}(\hat{k}_{\alpha},k'_{\alpha\beta})f^{(s)}_{\beta}(k_{\beta\alpha},k_{\beta})}{(k^2_{\beta\alpha} - k^2_{\beta} - i0)^{1+ia_{\alpha\beta}}}\tilde{A}^{(s)}_{\alpha\beta}(\hat{P},P') ,$$

where the following notations have been used: $\tilde{q}_o = q_o(P)/(2|P|)$ and $\tilde{A}_s, \tilde{A}^{(s)}_{\alpha}\tilde{A}^{(s)}_{\alpha\beta}$ are sufficiently smooth functions. The singular function $|P-P'|^{-5-2i\tilde{q}_o}$ is to be considered as a distribution. In the local coordinates, $t = 1-\cos\theta$, $\cos\theta = (\hat{P},P')$, this distribution becomes t^{-1-in}. In the local coordinates $t = 1-Z_{\alpha}|X|^{-1}$, $\hat{P} = \hat{X}$, the distribution $(p_{\alpha}-p'_{\alpha})^{-3-in_{\alpha}}$ also becomes t^{-1-in}. The singularity $(k^2_{\beta\alpha}-k^2_{\beta}-i0)^{-1-ia_{\alpha\beta}}$ is considered as a distribution $(t-i0)^{-1-in}$. Finally, the distributions $f^{(s)}_{\alpha}$ are defined by

$$f^{(s)}_{\alpha}(\hat{k}_{\alpha},k'_{\alpha}) = f_{\alpha}(k_{\alpha},k'_{\alpha})-i\exp\{2i\arg\Gamma(1+\frac{in_{\alpha}}{2|k_{\alpha}|})\}\delta(\hat{k}_{\alpha},\hat{k}'_{\alpha})$$

where $\delta(\hat{k}_{\alpha},\hat{k}'_{\alpha})$ is the δ-function given on the unit sphere in R^3. The two-body amplitude f_{α} for Hamiltonian h_{α} has the non-integrable singularities due to the Coulomb potential $n_{\alpha}|x_{\alpha}|^{-1}$. In the local coordinates $t=1-(\hat{x}_{\alpha},\hat{k}_{\alpha})$, the distribution f_{α} becomes t^{-1-in}. Note that the distributions $f^{(s)}_{\alpha}$ and $|p_{\alpha}-p'_{\alpha}|^{-3-in}$, ($f^{(s)}_{\alpha}$ and $f^{(s)}_{\beta}$, $(k^2_{\beta\alpha}-k^2_{\beta}-i0)^{-1-ia_{\alpha\beta}}$) depend on different kinematic variables.

Consider the operator in $\hat{H}$

$$S = U^{(-)*}U^{(+)}$$

which we shall call the scattering operator. If $\hat{f} \in \hat{H}$, $\hat{f} = \{f_o, f_{\alpha,i}\}$, then the vector valued function $\hat{f}'=S\hat{f}$ is given by

$$f'_a = \sum_b S_{ab}f_b, \quad f' = \{f'_a\}, \quad a = 0, \{a,i\} .$$

Theorem 6. S is a unitary operator which commutes with the asymptotic Hamiltonian $\hat{H}_o$. It is an integral operator whose kernels are given by

$$S_{ab}(P_a,P_b') = C_{ab}\,\delta(E_a-E_b)A_{ab}(\hat{P}_a,P_b')$$

where $E_a = P^2$ if $a = 0$, $E_a = p_\alpha^2 - \varepsilon_{\alpha,i}$ if $a = \{\alpha,i\}$, $C_{\alpha i,0} = C_{\alpha i,\beta j} = 2i\pi^{1/2}$, $C_{0,0} = C_{0,\alpha i} = -i\pi^{1/2}(2|P|^{-1})^{3/2}$. The singular kernels $A_{a,a}$ are distributions defined above. For the proof of Theorem 6, we have to represent the integral $S_{ab} = U_a^{(-)*}U_b^{(+)}$ as a limit of surface integrals on the spheres (R). The limit $R \to \infty$ may be performed with the aid of known asymptotic form of $U_a^{(\pm)}$ (see ref.[11]).

It should be pointed out that in contrast to the case of neutral particles the matrix elements of the Coulomb scattering operator do not contain δ-function terms. This was also shown by H.M. Vesselova who used the existence of the Dollard wave operators [6]. In the presence of long-range forces, the δ-function terms are replaced by new distributions. For example, the δ-functions $\delta(P-P')$ and $\delta(p_\alpha-p_\alpha')$ are replaced by $|P-P'|^{-5-2i\tilde{q}_0}$ and $|p_\alpha-p_\alpha'|^{-3-in_\alpha}$, respectively, and instead of the double scattering poles $(k^2-k'^2-i0)^2$ we have the distorted poles $(k^2-k'^2-i0)^{-1-in}$. It should be noted that the physical interpretation of the matrix elements S_{ab} is the same as for neutral particles. The effective cross section for the process $(b \to a)$ is equal to $C_{ab}|A_{ab}|^2$, $a,b = 0,\{\alpha,i\}$, where the coefficients C_{ab} are well known (see for example [3,30]).The non-integrable singularities of the elastic scattering amplitudes make the complete cross section infinite.

5. NUMERICAL METHODS

In the section 3, I have formulated the boundary value problems for three-particle Coulomb wavefunctions.A similar configuration space formulation is known to be very useful for numerical calculation of scattering dates for neutral particles [30]. This method permits the consideration of realistic nucleon-nucleon interaction (including a tensor contribution) [34] for n-d scattering. Its generalization for Coulomb systems given above has the same advantage as usual configuration space formulation.

Namely, in configuration space, after angular analysis, we use as input of the three-body problem only one-variable functions, i.e. the interaction potentials, the radial parts of the two-body bound-state wavefunction and pure two-body radial Coulomb wave functions instead of the three-variable two-body t-matrices in momentum space. This advantage plays a principal role in Coulomb scattering because it permits to avoid well-known difficulties associated with the singularities of the two-body Coulomb T-matrix. Remember that the momentum-space pecularities are just shifted onto the wave function asymptotic form.

The method used to solve the three-body differential equations numerically is very similar, in principle, to the one used to solve the two-body Schrödinger equation.

In order to illustrate the usefulness of the differential formulation we take the Noble-Faddeev equations (27) and consider the s-wave p-d scattering in s = 3/2 state (with the Yukawa type III potential used by Kloet and Tjon [35]). After angular momentum analysis these equations become a set of partial differential equations analogous to the ones described in ref.[30]. Neglecting the Faddeev's components with $L > 0$ (L to be the total orbital momentum) we can obtain the following equation for the s-wave component:

$$\left(-\frac{\partial^2}{\partial x^2} - \frac{\partial^2}{\partial y^2} + V_1(x) + \frac{n\mu(\theta)}{\rho} - K^2\right)U_o(x,y) \quad =$$

$$= \left(\frac{2}{\sqrt{3}y} - \frac{2}{x}\right)\psi(x)\phi_c(y)\varepsilon\left(\frac{\pi}{6} - \theta\right) +$$

$$+ \frac{1}{2} V_1(x) \int_{-1}^{+1} du \frac{xy}{x'y'} \left(U(x',y') + \chi_1(x',y')\right) \quad . \qquad (30)$$

Here

$$x' = \frac{1}{2}(x^2 - 2\sqrt{3}xyu + 3y^2)^{1/2}, \quad y' = \frac{1}{2}(3x^2 + 2\sqrt{3}xyu + y^2)^{1/2} ,$$

$$\rho = \sqrt{x^2+y^2} , \quad tg\,\theta = y/x , \quad \varepsilon(t) = \begin{matrix} 1 , & t \geq 0 , \\ 0 , & t < 0 , \end{matrix}$$

$$\mu(\theta) = \begin{cases} \dfrac{2}{\sqrt{3}\sin\theta} = \dfrac{2}{\sqrt{3}y} , & \theta \geq \dfrac{\pi}{6} , \\ \dfrac{2}{\cos\theta} = \dfrac{2}{x} , & \theta < \dfrac{\pi}{6} ; \end{cases}$$

the function $\chi_1(x,y)$ is the incident Coulomb "bound plane wave", $\chi_1 = \psi(x)\phi_c(y)$.

The function $\phi_c(y,q)$ is the solution of the two-particle Schrödinger equation

$$\phi'' - \frac{2n}{\sqrt{3}y}\phi + q^2\phi = 0 , \qquad q^2 - \kappa^2 = K^2 ,$$

$$\phi(y,q) = qye^{-\frac{n\pi}{2\sqrt{3}q}}\Gamma\left(1+i\frac{n}{\sqrt{3}q}\right)e^{iqy}\Phi\left(1+i\frac{n}{\sqrt{3}q},2,-2iqy\right) .$$

The function $\Phi(a,c,x)$ is the confluent hypergeometric function [36], $-\kappa^2$ is the bound energy. The Faddeev's components $U(x,y)$ satisfy the regularity conditions

$$U(x,0) = U(0,y) = 0$$

and the asymptotic boundary conditions

$$U(x,y) \sim a_o\psi(x)e^{iqy+iw_3} + A_o\frac{e^{ik\rho+iw_o}}{\rho^{1/2}} \quad . \tag{31}$$

The Coulomb phase shifts w_3 and w_o are given by

$$w_3 = -\frac{n}{\sqrt{3}q}\ln(2qy), w_o = -\frac{n}{2K}\mu(\theta)\ln(2K\rho) \ .$$

The function $A_o(\theta)$ is connected with the physical breakup amplitude by means of the same relations as in the case of neutral particles [30]. To obtain the elastic scattering amplitude we have to calculate the elastic coefficients a_o, $a_1 \ldots a_L$, corresponding to the different momenta L. Then the elastic amplitude f is given by the expression

$$f = f_c + f' \quad ,$$

where f_c is the pure Coulomb scattering amplitude for the charge $n\sqrt{3}/2$ and f' is expressed by a_o, $a_1 \ldots a_L$ with the aid of usual partial wave decomposition formula [1]. Let us remember a numerical method to solve the equation (30). This method was originally investigated in the ref.[30]of C. Gignoux, A. Laverne and the author. We used a finite-difference approximation of the equation similar to (30). The adoption of a grid in the (x,y) plane will transform the set of partial-differential integral equations into a set of finite difference equations for the values U(x,y) at the nodes of the grid. Easy computation of the integrals and the subsequent procedure to solve the set of finite difference equations make a polar grid the best choice. Then the problem is to determine the values of U(x,y) at the nodes of the grid defined by the circle arcs $x^2+y^2=\rho_i^2$, $i = 1,2,3,\ldots\rho_\theta$, and the radii $y/x = \tan\theta_j$, $j = 1,2,\ldots N_\theta$.

The solution of the linear system is started from the origin, where the function U is to be regular, $U(x,0) = U(0,y) = 0$. A step by step elimination deter-

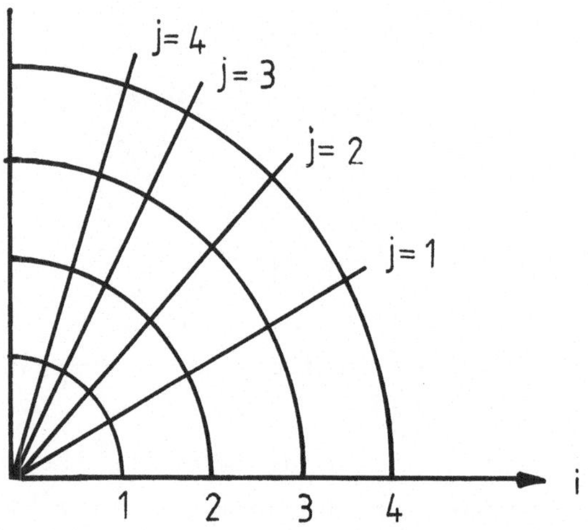

Fig. 3

mines the real matrix of an inhomogeneous linear relation between values of U on to neighbouring arcs of the circle. The rank of this matrix is only the number of discretized θ values. This process is continued until a value of the radius ρ is obtained large enough to be in a domain, where the asymptotic form (31) is true. The discretized asymptotic form (31) will lead us to another (homogeneous) linear relation between the values of U on the last two arcs of the circle once the unknowns A and a_o are eliminated. For this we first get rid of A by considering the ratio of the quantity $U(x,y) - a_o\psi(x)e^{iqy+iw_3}$ on two arcs. Then A is eliminated by taking it as the value of $U(x,y)/\psi(x)e^{iqy+iw_3}$ for x small enough so that the behaviour $\psi(x)e^{iqy+iw_3}$ is dominant. The complex coefficients (when $E > 0$) of this linear relation together with the real coefficients of the first linear relation enable us to find the values of U on one of the arcs. Finally the amplitude a_o is just given by the ratio $U(x,y)/e^{iqy+iw_3}\psi(x)$ x small, and then the amplitude A6x,y) is obtained by identifying the solution with its asymptotic form (31).

So obtained values $A(\theta)$ are presented on Fig. 4 for different energies. The grid parameters are the following

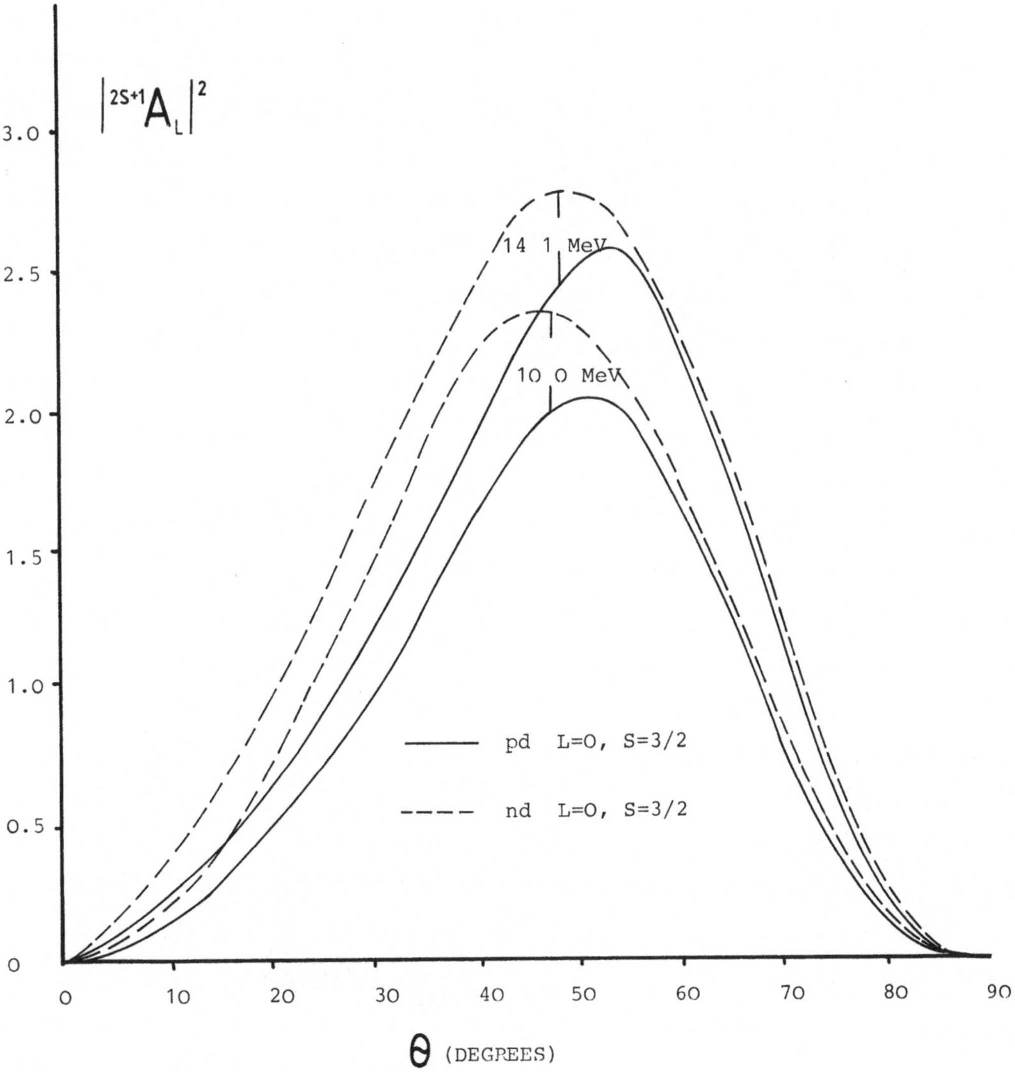

$|^{2S+1}A_L|^2$
3.0
2.5
2.0
1.5
1.0
0.5
0
14.1 MeV
10.0 MeV
pd L=0, S=3/2
nd L=0, S=3/2
0
10
20
30
40
50
60
70
80
90
θ (DEGREES)

Fig. 4

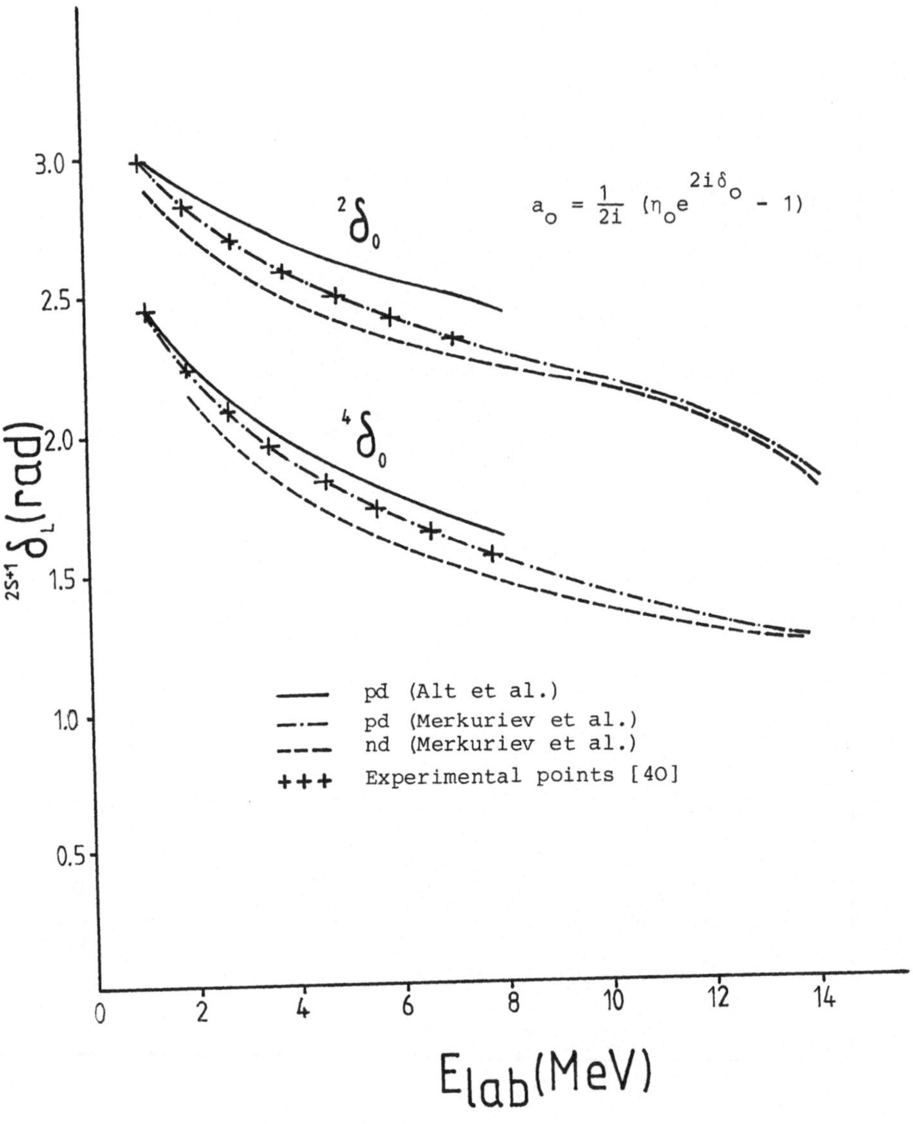

3.0
2.5
2.0
1.5
1.0
0.5
$^{2S+1}\delta_L$(rad)
$^2\delta_0$
$^4\delta_0$
$a_o = \frac{1}{2i}(\eta_o e^{2i\delta_o} - 1)$
pd (Alt et al.)
pd (Merkuriev et al.)
nd (Merkuriev et al.)
+++ Experimental points [40]
0
2
4
6
8
10
12
14
Elab(MeV)

Fig. 5

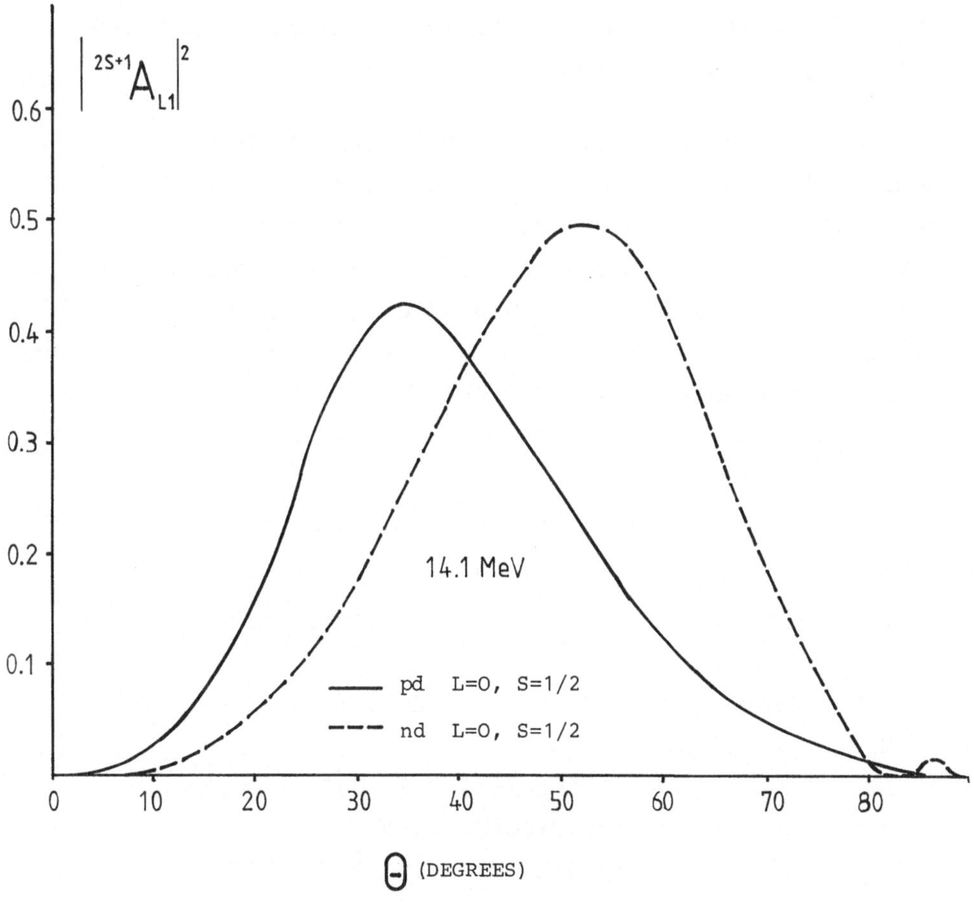

$\left|{}^{2S+1}A_{L1}\right|^2$
0.6
0.5
0.4
0.3
0.2
0.1
14.1 MeV
pd L=0, S=1/2
nd L=0, S=1/2
0
10
20
30
40
50
60
70
80
Θ (DEGREES)

Fig. 6

determines

$$N_\theta = 35, \quad N_\rho = 75, \quad \rho_{max} = 24 \text{ fm} \quad .$$

We have considered the s-wave p-d scattering in an L = 0, J = 1/2 state. The result is presented on Fig. 5,6. In this case three coupled equations of the type (30) were solved by means of the method described above.

At the end I would like to point out that the numerical task to solve the equations of the type (30) for charged particles is the same as the one for neutral particles. Only the number of the equations defines the rank of matrices involved. Hence we can consider the scattering processes with an arbitrary numerical accuracy, if we have a sufficiently powerful computer.

REFERENCES

1. R.G. Newton, Scattering Theory of Waves and Particles, Mc Graw-Hill, New York (1966).
2. M.L. Goldberger, K.M. Watson, Collision Theory, Wiley, New York (1964).
3. L.D. Landau, E.M. Lifschitz, Vol. III, Quantum Mechanics, Pergamon, Oxford (1958).
4. L.D. Faddeev, Trudy Matem. In-ta Akad. Nauk SSSR, 69 (1963).
5. J.Dollard, J. Math. Phys. 5 (1964) 729.
6. A.M. Vesselova, Teor. mat. fiz. 13 (1972) 368.
7. C. Chandler, A.G. Gibson, J. Math. Phys. 15 (1974) 1366.
8. J. Zorbas, J. Math. Phys. 17 (1976) 498.
9. G. Cattapan and V. Vanzani, Lett. Nuovo Cim. 20 (1977) 465.
10. G. Bencze et al., Lett. Nuovo Cim. 20 (1977) 248.
11. S.P. Merkuriev, On the Three-Body Coulomb Scattering Problem, Freie Universität Berlin, preprint FUB/HEP 2/80 (1980).

12. A.M. Vesselova, Teor. mat. fiz. 3 (1970) 326.
13. A.M. Vesselova, Teor. mat. fiz. 35 (1978) 180.
14. L.P. Kok and H. van Haeringen, Phys. Rev. C21 (1980) 512.
L.P. Kok, D.J. Struik and H. van Haeringen, On the Exact Solution of the Three-Particle Equations with Coulomb Interaction I (1980); II (1981), Univ. Groningen reports.
15. E.O. Alt et al., Phys. Rev. Lett. 37 (1976) 1537.
16. E.O. Alt et al., Phys. Rev. C17 (1978) 1987.
17. E.O. Alt, Invited Contributions to Few Body Conferences, Dubna (1979), Oregon, USA (1980).
18. E.O. Alt et al., Nucl. Phys. B2 (1967) 167.
19. V.G. Gorshkov, JETP 40 (1961) 1481.
20. A.M. Vesselova, Extended Contribution to the Dubna Symposium on the Few-Body Problem, Dubna (1979).
21. S.P. Merkuriev, Yad. Fiz. 24 (1976) 289.
22. S.P. Merkuriev, Teor. Mat. Fiz. 32 (1977) 187.
23. S.P. Merkuriev, Teor. Mat. Fiz. 38 (1979) 201.
24. S.P. Merkuriev, Zapisky Nauk. seminarov LOMI, Leningrad, 77 (1978) 148.
25. S.P. Merkuriev, Lett. Math. Phys. 3 (1979) 141.
26. S.P. Merkuriev, Doklady AN SSR 241 (1978) 68.
27. S.P. Merkuriev, Teor. Mat. Fiz. 8 (1971) 235.
28. A.Y. Povzner, Doklady AN SSSR 104 (1955) 360.
29. T. Kato, Perturbation Theory for Linear Operators, Springer, Berlin (1976).
30. S.P. Merkuriev, C. Gignoux, A. Laverne, Ann. Phys. (N.Y.) 99 (1976) 30.
31. C. Chandler, The Coulomb Problem, Invited talk, Few-Body Conference, Eugene, Oregon (1980).
32. I.M. Gelfand, G.E. Shilov, Distributions, Moscow (1964).
33. I. Herbst, Comm. Math. Phys. 35 (1974) 181.
34. J.J. Benayoun, J. Chauvin, C. Gignoux and A. Laverne, Phys. Rev. Lett. 36 (1976) 1439.

35. W.M. Kloet, J.A. Tjon, Ann. Phys. (N.Y.) 79 (1973) 407.
36. H. Bateman, A. Erdélyi, Higher Transcendental Functions I, Mc Graw-Hill (1953).
37. L.D. Faddeev, P.P. Kulish, Teor. Mat. Fiz. 4 (1970) 153.
38. V.S. Buslaev, V.B. Matveev, Teor. Mat. Fiz. 2 (1970) 367.
39. S.P. Merkuriev, Proc. of the 1977 Europ. Symposium on Few-Particle Problems in Nuclear Phys., Potsdam, 94 (1977).
40. J. Arvieux, Nucl. Phys. A221 (1974) 253.

Acta Physica Austriaca, Suppl. XXIII, 111–155 (1981)

INVERSE SPECTRAL AND SCATTERING THEORY+

by

K. CHADAN
Lab. de Phys. Théor. et Hautes Energies
Univ. de Paris XI, 91 405 Orsay, France

I. INTRODUCTION

In these lectures, we shall try to give a résume of the techniques which have been devised for solving various spectral and scattering inverse problems. The invention of these techniques goes back to the 50's, and is mainly due to Gel'fand and Levitan, and Marchenko, with important contributions by Jost and Kohn, Faddeev, Newton and Sabatier, Regge, Loeffel, Martin, Cornille, Gasymov and Levitan,... All the references are given at the end.

We shall be concerned in these lectures with the non-relativistic scattering of a particle by a static potential, although most of the results have been generalized to Klein-Gordon or Dirac equations, and also to many-channel problems.

As is clear from its name, the inverse scattering problem is the converse of the direct scattering problem. The direct problem consists of calculating the phase shifts, the scattering amplitudes, and the binding energies (in short, the scattering data) from the Schrödinger equation (or any

+Lectures given at XX.Internationale Universitätswochen für Kernphysik, Schladming, Austria, February 17-26, 1981.

other equations of motion: Dirac, Klein-Gordon,...), assuming the potential to be known. In other words, we know the equations of motion, and we know the interaction. We then solve the equations and obtain the physical quantities of interest, i.e. scattering data. The inverse problem is just going back from the scattering data to the potential. A little thought is sufficient to convince you that in fact the major part of Physics consists in solving inverse problems: going from experimental facts and data to the fundamental laws governing natural phenomena. In this respect, the man who did solve an inverse problem for the first time was Isaac Newton who, from the Kepler's laws of motion of the planets was able to deduce his celebrated law of gravitation. His merits were incommensurably great because he did all of the theoretical work by himself: he invented the general framework (Newtonian Mechanics) and was able to find the interaction between heavenly bodies from astronomical data.

For many years, the inverse spectral and scattering theory in Quantum Mechanics, although very beautiful and elegant, was considered rather as a mathematical curiosity, because it was obvious from the beginning that there would never be enough experimental data to calculate the potential. And there were no obvious approximation schemes either, for calculating an approximate potential from insufficient scattering data. Because of this, very few people studied the inverse problems.

And then, all of a sudden, thanks to the great work of Gardner, Green, Kruskal, and Miura, in 1967, the inverse scattering theory reached the zenith of fame among mathematical physicists (and now mathematicians) because it was shown in the above work that one can solve some nonlinear partial differential equations (in this case the Korteweg-de Vries equation), which are important in various area of Physics, by solving a (linear) inverse scattering problem.

This led to the deepening of the concept of soliton and the discovery of its universality in many areas of Physics.

It is impossible to give, in three lectures, a complete description of all these, with all the proofs, and talk about the important work which has been done since ten years for extending the applications of the original work of Gardner et al. to new classes of nonlinear equations. We shall therefore content ourselves with inverse scattering theory.

The interested reader will then, hopefully, have no difficulty in pursuing his readings among the great number of papers devoted to the applications to various nonlinear equations.

Some attempts have been made very recently for generalizing the inverse scattering methods to field theory. We shall not discuss this more general case here, but we shall say a few words about confining potentials (Grosse and Martin, Quigg and Rosner).

We end this brief introduction with some heuristic remarks which are sometimes useful for getting an a priori idea about what one would need in order to have a well-defined inverse problem. Since most of our discussions will be limited to the one-dimensional case (either the radial equation on R^+, or the Schrödinger equation on R), let us begin with the radial equation for the ℓ-th wave,

$$\psi''_\ell(E,r) + [E - \frac{\ell(\ell+1)}{r^2}]\psi_\ell(E,r) = V(r)\psi_\ell(E,r) \quad , \qquad (1)$$

$$0 \leq r < \infty \quad ,$$

where $E = k^2$, k being the wave number, and $r \geq 0$. As is customary, we have put $\hbar = 2M = 1$, where M is the mass of the particle. The potential V is a real function, and we assume that it is such that the total Hamiltonian is self-adjoint, with a finite number of bound states (with negative energies), and that the usual scattering theory, with the well-known connection between the asymptotic form of the wave function and the cross section, is valid. A sufficient condition for all these to be true is

$$\int_0^\infty r|V(r)|dr < \infty \ . \tag{2}$$

Solving (1) for $k \geq 0$, with the physical boundary condition

$$\psi_\ell(E,0) = 0 \ , \tag{3}$$

we find that

$$\psi_\ell(E,r) \underset{r \to \infty}{=} e^{i\delta_\ell(k)} \sin[kr + \delta_\ell(k) - \tfrac{1}{2}\ell\pi] + o(1) \ . \tag{4}$$

Here, the phase shift $\delta_\ell(k)$, which is a real quantity, depends both on k and ℓ, $\ell = 0,1,\ldots$ Also, if the potential is attractive enough ($V < 0$), there may be bound states with energies $E_1^{(\ell)} < E_2^{(\ell)} < \ldots < E_{n_\ell}^{(\ell)} < 0$. Solving the direct problem is going from V(r) to the scattering data

$$\{\delta_\ell(k),\ \text{for all}\ k \geq 0,\ E_j^{(\ell)},\ j=1,2,\ldots,n_\ell \mid \ell = 0,1,2,\ldots\}$$

$$\equiv \{F(k,\cos\Theta);\ E_j(\leq 0),\ \text{for all}\ j\} \tag{5}$$

where F is the total scattering amplitude.

The inverse problem is going from these data to the potential. There are several kinds of inverse problems one may think of, each having its own interest. The first one is the inverse problem at fixed ℓ, which is to find the potential from one phase shift (fixed ℓ) given for all energies, plus the energies of the bound states of that angular momentum. Another one is to find the potential from all the phase shifts, given for one single energy.

In the fully three-dimensional case, when the potential is not spherically symmetric, one has to find the potential from the total amplitude. In the one-dimensional case, where one has

$$\psi''(E,x) + E\psi(E,x) = V(x)\psi(E,x) \quad , \qquad (6)$$
$$-\infty < x < \infty \quad ,$$

the S-matrix is a two by two matrix because one has reflection coefficients from the left and from the right, and transmission coefficients to the left and to the right. The potential itself is thought of as made of two parts, one for $x > 0$ and one for $x < 0$. This one-dimensional case is the one which has been found most useful in connection with solving nonlinear problems.

Anyway, it should be intuitively clear that for finding the potential, which is, modulo (2), an arbitrary real function of a real variable, we should need, as input, something of the same generality, i.e. another real function of one real variable (unless the potential belongs to a very restricted class, which we shall consider a little later). This is exactly the case of the phase shift for one ℓ and all the values of $k(>0)$. As we shall see, when the potential does not admit bound states in the considered angular momentum state, the connection between the $V(r)$ and $\delta(k)$ is indeed one to one.

When bound states are present, we must know also their energies and the normalizations of their wave functions. This is because, as it will become clear later, the fundamental ingredient in the solution of the inverse problem is the completeness of the set of solutions of the Schrödinger equation in the range $-\infty < E < \infty$. When there are no bound states, it is known that scattering states, which correspond to $E \geq 0$, are complete. In the presence of bound states, we must add their wave functions to the set of scattering states in order to achieve completeness. This is the profound reason why bound states, when they are present, enter into the construction of the potential.

Let us mention here, before continuing, that the quantities which are experimentally measured are the cross sections, not the scattering amplitudes. If we really want to calculate the potential from experimental data, we must first calculate the scattering amplitudes from the cross sections. This problem, which is highly nontrivial, has been thoroughly studied by A. Martin, R. Newton, H. Cornille, Sabatier,...

As a last example, we can consider the discrete case, i.e. the case where the spectrum is discrete. This may be the case either with a confining potential in an infinite domain, which we shall see later, or a nonnasty potential in a finite interval, which we shall consider now briefly.

It is well known that, for good potentials, the existence of continuum (or scattering) states is due to the unboundedness of the domain. For finite domains, with usual (Dirichlet, Neumann,...) boundary conditions, the spectrum of the Hamiltonian is discrete, with infinity as the only accumulation point. The simplest example is the Schrödinger equation

$$\phi'' + E\phi = V(x)\phi \quad , \tag{7}$$

where $a < x < b$, together with the boundary conditions

$$\phi(E,a) = \alpha, \ \phi'(E,a) = \beta; \ \phi(E,b) = \gamma, \ \phi'(E,b) = \delta \ , \tag{8}$$

α, β,... real. For good potentials, this problem has eigenvalues E_1, E_2,...E_n,... accumulating at $+\infty$. The inverse problem is to find the potential from the eigenvalues. It turns out, as we shall see, that one can find the potential uniquely provided one knows also the normalization constants

$$C_n^{-1} = \int_a^h \phi_n^2(x)\, dx \tag{9}$$

where ϕ_n is the n-th eigenfunction.

So far, in the scattering problem, we have not been assuming anything particular about the potential. Let us suppose we know a priori that the potential belongs to some particular class of functions. For instance, that it is an holomorphic function of r around the origin which does not reduce to a polynomial. In this case, the potential is equivalent to a countably infinite set of real numbers, which, for instance, we can take as the coefficients of its Taylor expansion. One may then ask whether the knowledge of another set of real numbers, such as the values of all the phase shifts at some fixed energy, would determine uniquely the potential. The answer is again yes if the set $\{\delta_\ell, \ell = 0,1,\ldots\}$ satisfies some compatibility condition.

In these lectures, we shall study only the variable energy cases in one dimension, which have been most useful in connection with nonlinear equations, and leave aside the case where the energy is fixed and one varies the angular momentum. This last problem, which is more useful for experimentalists doing phase shifts analysis, has been studied extensively by Sabatier, Loeffel, and many others. One usually obtains a whole class of long-range potentials with asymptotic tail $r^{-3/2}$ and oscillating, unless the phase shifts decrease asymptotically fast enough, in which case there exists, among the solutions of the inverse problem, one potential with asymptotic tail $r^{-2-\varepsilon}$.

II. THE GEL'FAND-LEVITAN METHOD FOR THE RADIAL SCHRÖDINGER EQUATION

A. The Regular Solution

This is the case of equation (1), and we assume the potential to satisfy (2). For simplicity, we treat the case of the S-wave ($\ell = 0$). For higher waves, the problem is, save for minor technical changes (replacing sine and cosine by Bessel and Neumann functions, etc.), exactly the same. The Gel'fand-Levitan technique for the inverse problem is based on the study of the so-called regular solution of the Schrödinger equation, i.e. the solution which satisfies the boundary condition

$$\phi(k,0) = 0 \ , \quad \phi'(k,0) = 1 \ , \tag{10}$$

where $E = k^2$, $k \geq 0$. Combining the differential equation with these boundary conditions, one is led to the study of the Volterra integral equation

$$\phi(k,r) = \frac{\sin kr}{k} + \int_0^r \frac{\sin k(r-r')}{k} V(r')\phi(k,r')dr' \ . \tag{11}$$

By differentiation, it is easily seen that the solution of this equation is a solution of the differential equation. We have therefore to show that the above integral equation has a unique solution which is twice differentiable. This is easily done (see the books of Newton, or De Alfaro and Regge, on scattering theory) by the usual method, which is to iterate the equation, starting from the free solution $\phi^{(0)}$,

$$\phi^{(0)} = \frac{\sin kr}{k} \ , \quad \left|\frac{\sin kX}{k}\right| \leq C \frac{X}{1+|k|X} e^{|\mathrm{Im} k|X} \ , \tag{12}$$

where C is an appropriate constant independent of k and X ($X \geq 0$), and study the convergence of the series. One finds easily that, under the assumption (2) on the potential, and whatever k may be in the finite complex plane, one has an absolutely and uniformly convergent series defining a unique solution ϕ, which satisfies the radial Schrödinger equation and the boundary conditions (10). Moreover, each term of the series being an entire function of k, one obtains the following results for the solution:

1. ϕ is an entire function of k for every fixed finite r.
2. Its asymptotic behavior for large k is given by

$$\phi(k,r) \underset{|k|\to\infty}{=} \frac{\sin kr}{k} + \frac{e^{|\mathrm{Im}k|r}}{k}\, o(1) \tag{13}$$

and

$$\lim_{|k|\to\infty} [\phi'(k,r) - \cos kr] = 0, \tag{14}$$

where the small term o(1) is uniform in r for all r > 0. This means that, under the assumption (2), one can neglect the potential when the energy becomes very large. In other words, for large values of k, ϕ is very close to the free solution $\phi^{(0)}$. It is also clear from (1) and (10), that ϕ is an even function of k. And since it is real for real values of k, we have, by the Schwarz principle,

$$\phi(-k^*,r) = \phi(k^*,r) = [\phi(k,r)]^* \tag{15}$$

where * means complex conjugate. The asymptotic behavior of ϕ for large r, and the properties of the phase shift δ as well as those of the Jost function will be established later, once we have studied the properties of the Jost solution.

B. The Jost Solution and the Jost Function

Instead of imposing boundary conditions at the origin, as we did for ϕ, we may choose the point at infinity. The solution f defined by the boundary conditions

$$\lim_{r\to\infty} e^{-ikr} f(k,r) = 1 , \tag{16}$$

$$\lim_{r\to\infty} e^{-ikr} f'(k,r) = ik , \tag{17}$$

is called the Jost solution. Again, one may study it by using the integral equation

$$f(k,r) = e^{ikr} - \int_r^\infty \frac{\sin k(r-r')}{k} V(r') f(k,r') dr' \quad . \tag{18}$$

Starting from

$$f^{(0)}(k,r) = e^{ikr} \tag{19}$$

and iterating the integral equation, we are led, in the half-plan $\text{Im}k \geq 0$, to an absolutely and uniformly convergent series defining a unique solution of the Schrödinger equation which satisfies indeed (16) and (17). Moreover, f, for every fixed finite value of r, is holomorphic in the open upper half-plane, continuous and bounded in the closed upper half-plane, and we have the bounds ($\text{Im}k \geq 0$)

$$|f(k,r)| \leq K\, e^{-\text{Im}k\, r} \tag{20}$$

where K is an appropriate constant. Now, using (20) in the integral equation, we get, successively, the asymptotic estimates

$$f(k,r) \underset{r\to\infty}{=} e^{ik r} + e^{-\text{Im}k\, r}\, o(1) \ , \tag{21}$$

$$f'(k,r) \underset{r\to\infty}{=} ik\, e^{ik r} + e^{-\text{Im}k\, r}\, o(1) \quad , \tag{22}$$

and for $r > 0$,

$$f(k,r) \underset{\substack{k \to \infty \\ \text{Im}k \geq 0}}{=} e^{ik r} + \frac{e^{-\text{Im}k\, r}}{|k|}\, o(1) \quad , \tag{23}$$

all uniform with respect to the "fixed" variable. If we wish to include $r = 0$, we must replace (23) by

$$f(k,r) \underset{k \to \infty}{=} e^{ik r} + e^{-\text{Im}k\, r}\, o(1) \quad , \tag{24}$$

where again o(1) is uniform with respect to r, $r \geq 0$.

We can also differentiate f with respect to k by using its perturbative series and the estimate previously given. It follows that f is continuously differentiable with respect to k in $\mathrm{Im}k \geq 0$, with the possible exception of $k = 0$. We also get, for all $r > 0$, $k \neq 0$,

$$|\dot{f} - ir\, e^{ik\,r}| \leq K \frac{e^{-\mathrm{Im}k\; r}}{|k|} , \quad \dot{f} = \frac{df}{dk} , \tag{25}$$

where K is independent of r.

It can also be easily checked on the perturbative series, that f has the following symmetry property:

$$f\,(-k^*,r) = [f(k,r)]^* \quad . \tag{26}$$

For real values of k ($\neq 0$), f(k,r) and f(-k,r) are two independent solutions. Indeed, it is easily seen from (21) and (22) that

$$W\{f(k,r),\ f(-k,r)\} = -2ik \neq 0 \tag{27}$$

where W is the Wronskian. In the same way as sine can be written as the sum of two exponentials, we can write, taking into account that ϕ is an even function of k,

$$\phi(k,r) = \frac{1}{2ik}\,(F(-k)f(k,r) - F(k)f(-k,r)) \quad . \tag{28}$$

This equation is valid in general for real values of k only.

The function F(k), which can be written as the Wronskian of f with ϕ,

$$F(k) = W\{f(k,r),\phi(k,r)\} \quad , \tag{29}$$

is called the Jost function and plays a very important

role in what follows. First of all, since both f and ϕ and their derivatives are holomorphic in k in Imk > 0, the same is true for F. It then follows, from (26), that

$$F(-k^*) = [F(k)]^*, \quad \text{Im}k \geq 0 \quad . \tag{30}$$

For real values of k, we then have

$$|F(-k)| = |F(k)| , \quad k \text{ real} \quad . \tag{31}$$

If we denote by $\delta(k)$ the phase of F(-k) for $k \geq 0$:

$$F(-k) = |F(\pm k)| e^{i\delta(k)} \quad , \tag{32}$$

we find, from (28) and (21), that

$$\phi(k,r) \underset{\substack{r \to \infty \\ k \text{ real}}}{=} \frac{|F(k)|}{k} \sin [kr + \delta(k)] + o(1) \quad . \tag{33}$$

Comparing this with (4) we see that the physical solution is simply ($\ell = 0$):

$$\psi \equiv \frac{k}{F(k)} \phi(k,r) \quad , \tag{34}$$

and that the phase shift is just minus the phase of F(k).

Another remarkable property of the Jost function is the intimate connection between its roots in the half-plane Imk > 0 and the bound state energies, i.e. energies at which ϕ belongs to $L^2(0,\infty)$. This can be proved in many ways. One way is to remember that the physical solution ψ satisfies the Fredholm equation

$$\psi(k,r) = \sin k r - \int_0^\infty \frac{\sin k r_<}{k} e^{ik r_>} V(r') \times \psi(k,r') dr' \quad . \tag{35}$$

It can then be shown that the solution of this equation is given by the usual Fredholm formula N/D, where D = F(k), i.e. F is just the Fredholm determinant. The connection between all the zeros of F(k) and the L^2-solutions of the homogeneous equation corresponding to (35) becomes then obvious. Since it is known that for "good" potentials all the binding energies (the discrete spectrum of the Hamiltonian) are real, it follows that if

$$F(k_o) = 0 \ , \qquad \mathrm{Im}\, k_o \geq 0 \ , \tag{36}$$

k_o is pure imaginary. It can also be shown that all the roots of F on the imaginary axis are simple. We leave aside here the case where $k_o = 0$, which, for the S-wave, corresponds to a resonance at zero energy.

On the other hand, F(k) cannot vanish on the real axis (except at k = 0), simply because (31) would then imply that ϕ given by (28) is identically zero, a fact which would contradict (10).

If we combine (29), taken for large values of r, with the integral equation (11) and the asymptotic estimates (21) and (22), we find

$$F(k) = 1 + \int_0^\infty e^{ik\,r}\, V(r)\phi(k,r)dr \ . \tag{37}$$

Again, it is easily seen on this integral representation, that F(k) is well defined for Imk $\geq$ 0 and holomorphic in Imk > 0. From the definition (29) and the asymptotic forms (13) and (23), one can deduce that

$$\lim_{\substack{k\to\infty \\ \mathrm{Im}k\geq 0}} F(k) = 1 \ . \tag{38}$$

This property has many consequences. One is that the zeros of F cannot accumulate at infinity without contradicting

(38). Since they cannot accumulate at $k = 0$ either (the potential is short range!), they are finite: the number of bound states is finite.

Another consequence of (38) is that we can define the phase shift which is, in general, defined modulo π in such a way that

$$\delta(\infty) = 0 \quad . \tag{39}$$

The Levinson theorem then reads, assuming again $F(0) \neq 0$,

$$\delta(0) - \delta(\infty) = n\pi \tag{40}$$

where n is the number of bound states. Notice also that, taking k real positive in (30), we obtain:

$$\delta(-k) = -\delta(k) \ , \qquad k \geq 0 \quad . \tag{41}$$

When k (real) is very large, one can show in fact that (2) entails

$$\int^{\infty} \frac{|\delta(k)|}{k} dk < \infty \ . \tag{42}$$

This will be useful later.

Let us also recall that by differentiating the Schrödinger equation with respect to k, we can show that if k_o corresponds to a bound state, we have for the normalization constant

$$C_o^{-1} = \int_o^{\infty} \phi^2(k_o,r)dr = \frac{-\dot{F}(k_o)}{2k_o f'(k_o,0)} \quad . \tag{43}$$

C. Eigenfunction Expansion

We have now all the ingredients to write down the eigenfunction expansion associated with the radial Schrödinger equation (1) on the interval $[0,\infty)$, together with the boundary conditions (10). Assuming again (2), one can prove the following

Theorem (Weyl-Titchmarsh):

Let $\phi_j(r)$, $j = 1,2,\ldots,n$, be the bound state wave functions. The functions $\phi(E,r)$, $E \geq 0$, and $\phi_j(r)$ form a complete orthogonal system, and the completeness relation for these can be written symbolically, using the Dirac delta function,

$$\frac{1}{\pi} \int_0^\infty \phi(E,r)\phi(E,t) \frac{E^{1/2}}{|F(E^{1/2})|^2} dE + \sum_{j=1}^{n} C_j \phi_j(r)\phi_j(t) = \delta(r-t), \tag{44}$$

where the normalization constants are given by formula (43) for each $k_j = i\gamma_j$, and we use the variable E instead of k.

In the free case when the potential vanishes, (44) reduces to the usual completeness relations for Fourier sine transforms

$$\frac{2}{\pi} \int_0^\infty \frac{\sin k r}{k} \frac{\sin k t}{k} k^2 dk = \delta(r-t) \quad . \tag{45}$$

We can rewrite the "completeness" relation (44) by using ψ given by (34). We get then

$$\frac{2}{\pi} \int_0^\infty \psi(k,r)\psi(k,t)dk + \sum_j \psi_j(r)\psi_j(t) = \delta(r-t) \quad , \tag{46}$$

ψ_j being now the solution of the homogeneous integral equation associated with (35), and

$$\int_0^\infty \psi_j^2(r)dr = 1 \quad . \tag{47}$$

Since we are dealing with continuous spectra, $\{\psi(k,r)\}$ are not elements of the Hilbert space, and therefore are not eigenfunctions in the usual sense. To attach meaning to them, we must consider them as kernels of integral transforms, exactly as (sin kr/k) is the kernel of the Fourier sine transform. We shall not enter into the details of such transformations since we shall not use them in this lecture.

The symbolic relations (44), (45) are of course to be understood in the sense of L^2- space. Given any function f(r) which is twice continuously differentiable and vanishes outside some finite interval not containing the origin, it can be shown that

$$f(r) = \frac{2}{\pi}\int_0^\infty |F(k)|^{-2} k^2 \phi(k,r) \left[\int_0^\infty f(t) \times \phi(k,t)dt\right]dk + \sum_j C_j \phi_j(r) \left[\int_0^\infty \phi_j(t)f(t)dt\right] \quad . \qquad (48)$$

Since these functions f(r) are dense in $L^2(0,\infty)$, it follows that (48) holds in the L^2 sense, and we obtain (44). For later use, we shall write it in the condensed form

$$\int_{-\infty}^{\infty} \phi(E,r)\phi(E,t)d\rho(E) = \delta(r-t) , \qquad (49)$$

where

$$d\rho(E) = \begin{cases} \dfrac{1}{\pi} \dfrac{E^{1/2}}{|F(E^{1/2})|^2} , & E \geq 0 \\[2ex] \sum_j C_j \, \delta(E-E_j) , & E < 0 \end{cases} \qquad (50)$$

is called the spectral density.

D. The Gel'fand-Levitan Integral Equation

As we saw before, the regular solution ϕ, for each value of r, is an entire function of the variable k with the asymptotic behaviour given by (13). On the basis of these, and using the celebrated Paley-Wiener theorem (or its variants) on the Fourier transform of entire functions, one can show that ϕ has the following integral representation :

$$\phi(k,r) = \frac{\sin k r}{k} + \int_0^r K(r,t) \frac{\sin k t}{k} dt \qquad (51)$$

where the kernel K is independent of k, the integral being meaningful under the condition (2). Inverting (51) we get

$$K(r,t) = -\frac{1}{\pi} \int_0^\infty (\phi(k,r) - \frac{\sin k r}{k}) \sin k t \; dk , \qquad (52)$$

the integral being meaningful for $0 < t < r$ by virtue of (13). At $t = r$, K is usually infinite, but it can be shown that it is L^1 in t, so that (51) makes sense.

We shall not give the proof of all these via the properties of ϕ and the Paley-Wiener theorem, but rather start from (51) as an Ansatz and show the existence of K directly. To this end, we first notice that

$$K(r,0) = 0 \quad . \qquad (53)$$

We introduce now (51) into both sides of (11), take the Fourier sine transform, assuming $t < r$, and change at will the order of integrations. The result is the integral equation

$$K(r,t) = \frac{1}{2} \int_{\frac{r-t}{2}}^{\frac{r+t}{2}} V(s) ds + \int_{\frac{r-t}{2}}^{\frac{r+t}{2}} ds \int_0^{\frac{r-t}{2}} \cdot$$

$$V(s+u) K(s+u, s-u) \, du \quad . \qquad (54)$$

We see immediately that both the inhomogeneous term and the kernel of this integral equation are meaningful if

$t < r$. We can therefore try to solve it by iteration, as is usually done, and see whether we can justify a posteriori the formal operations we have performed for its obtaining.

Let us assume first that the potential is integrable at the origin. It is then easily seen that the iteration leads to a convergent series as the unique solution of (54), and

$$|K(r,t)| < \frac{1}{2} \int_{\frac{r-t}{2}}^{\frac{r+t}{2}} |V(s)|ds \exp[\int_{0}^{\frac{r+t}{2}} u|V(u)|du] . \qquad (55)$$

This result shows that $K(r,t)$ is finite everywhere in $0 < t < r$, and permits us, with some lengthy but trivial extra work which we shall not reproduce here, to justify the formal steps. Putting $t = r$ in this equation we now deduce

$$K(r,r) = \frac{1}{2} \int_{0}^{r} V(t)dt \quad . \qquad (56)$$

This last result is, so far, the most important conclusion we obtain from (51).

Let us go back now to the general case where we assume only (2). From (55) it is easily seen that everything is again meaningful, except (56) because now $K(r,r)$ is in general infinite. However, (55) shows that $K(r,t)$ is L^1 in t up to $t = r$. This, in turn, shows that (51) is meaningful. Moreover, regularizing the potential at the origin, we get from (54)

$$K(R,R) - K(r,r) = \frac{1}{2} \int_{r}^{R} V(t)dt . \qquad (57)$$

In this formula no reference is made to the origin, and both sides are well-defined. We now make the regularization disappear by a limiting process. Since the r.h.s. is meaningful all the time, for $r > 0$, the same is true for the l.h.s., and (57) is therefore true for the unregularized potential. What happens at the end is that both $K(R,R)$ and $K(r,r)$ become infinite, but the difference remains finite. When the potential is a continuous function, we get

$$V(r) = 2\frac{d}{dr}K(r,r) \quad . \tag{58}$$

We can also consider (51) as a Volterra integral equation for $\sin kr/k$, with the inhomogeneous term ϕ and kernel K. Solving it, calling its reolvent kernel by $\tilde{K}$, we have

$$\phi^{(0)} \equiv \frac{\sin kr}{k} = \phi(k,r) + \int_0^r \tilde{K}(r,t)\phi(k,t)dt \quad . \tag{59}$$

Again, it can be shown that $\tilde{K}(r,t)$ exists and is finite for all $t < r$, and integrable in t at $t = r$.

We have now all the ingredients for deriving the celebrated Gel'fand-Levitan integral equation. To this end, we have to combine the integral representation (51) and (59) with the completeness relation (49)-(50). From (49) we have

$$\int_{-\infty}^{\infty} \phi(E,r)\phi(E,r')d\rho(E) = 0 \ , \qquad r \neq r' \quad . \tag{60}$$

We now multiply both sides of (59) by $\phi(E,r_1)$ and integrate with $d\rho(E)$. If $0 < r < r_1$, recalling (60), we obtain

$$\int_{-\infty}^{\infty} \frac{\sin\sqrt{E}\,r}{\sqrt{E}}\phi(E,r_1)d\rho(E) = 0 \quad . \tag{61}$$

Using now (51), we find with a slight change of notations, and for $t < r$,

$$\int_{-\infty}^{\infty} \frac{\sin\sqrt{E}\,r}{\sqrt{E}}\,\frac{\sin\sqrt{E}\,t}{\sqrt{E}}\,d\rho(E) + \int_0^r K(r,s)ds \int_{-\infty}^{\infty} \cdot$$

$$\frac{\sin\sqrt{E}\, s}{\sqrt{E}} \; \frac{\sin\sqrt{E}\, t}{\sqrt{E}} \, d\rho(E) = 0 \quad . \tag{62}$$

If we now use (definition of $d\sigma$)

$$d\sigma(E) = d\rho(E) - d\rho_o(E) = \begin{cases} d\rho(E) - d(\frac{2E^{3/2}}{3\pi}) , & E \geq 0 \\ d\rho(E) , & E < 0 \end{cases} \tag{63}$$

in the last equation and remember (45), we obtain the Gel'fand Levitan equation

$$K(r,t) + G(r,t) + \int_0^r K(r,s) G(s,t) ds = 0 , \tag{64}$$

where the kernel G is given by

$$G(r,t) = \int_{-\infty}^{\infty} \frac{\sin \sqrt{E}\, r}{\sqrt{E}} \; \frac{\sin \sqrt{E}\, t}{\sqrt{E}} \, d\sigma(E) \quad . \tag{65}$$

Remember that σ is the difference between the spectral measure ρ and the free measure ρ_o. The kernel G can also be written

$$G(r,t) = \frac{2}{\pi} \int_0^{\infty} \frac{\sin k\, r}{k} \, \frac{\sin k\, t}{k} \left[\frac{1}{|F(k)|^2} - 1\right] k^2 dk$$

$$+ \sum_j \frac{C_j}{4\gamma_j^2} \sinh \gamma_j\, r \sinh \gamma_j\, t \quad . \tag{66}$$

Equation (64), for each fixed r, is a Fredholm equation for K(r,t) in its second variable t. We have therefore to show that it has a unique solution under the restrictions on $d\rho(E)$ we found before. Once the solution is found, we have to see whether the potential we obtain has the desired properties. So there are two problems:

the unicity of K, and therefore V, and the properties of V. The unicity of the solution can be shown to be true in general. As for the properties of V, we shall find them later with the help of the Marchenko equation.

As is clear from (63) and (50), our input here is the Jost function, not the scattering data

$$\{\delta(k),\ \text{for all } k \geq 0;\ E_j,\ C_j,\ j = 1,2,\dots,n\}\ , \qquad (67)$$

where we have also included the normalization constants C_j. But these two quantities are related to each other by the integral representation

$$\underset{\mathrm{Im}k\geq 0}{F(k)} = \prod_j \left(1 - \frac{E_j}{E}\right) \exp\left[\frac{-2}{\pi} \int_0^\infty \frac{k'\delta(k')}{k'^2-k^2}\, dk'\right]\ , \qquad (68)$$

which is just the dispersion relation for log F(k) in terms of its imaginary part $-\delta(k)$. We can therefore conclude that, given the scattering data (67), provided that they satisfy (39) and (42), and of course the Levinson theorem (40), one can calculate in a unique way the potential. It turns out then that (42) implies that the potential satisfies (2). One can also show that the wave function calculated from (51) has the asymptotic form (33).

In all the above analysis, we have taken the reference case to be free case where the potential is zero. This is why (51), (63) and (65) are all given in terms of the input (scattering data) and of the free quantities $\phi^{(0)} = \sin kr/k$ and $d\rho_0$. Instead of having the reference potential zero, we can start from a given potential V_1 for which everything $(\phi_1, d\rho_1, \dots)$ is known, and compare ϕ, $d\rho$, ... with them. One is led then to the more general Gel'fand-Levitan equation:

$$\phi(k,r) = \phi_1(E,r) + \int_0^r K(r,t)\phi_1(E,t)\,dt \qquad (69)$$

and

$$K(r,t) + G(r,t) + \int_0^r K(r,s)G(s,t)ds = 0 , \qquad (70)$$

where now

$$G(r,s) = \int_{-\infty}^{\infty} (d\rho(E) - d\rho_1(E))\, \phi_1(E,r)\phi_1(E,s) \quad . \qquad (71)$$

One obtains then the potential from

$$V(r) = V_1(r) + 2\frac{d}{dr} K(r,r) \qquad (72)$$

which is the generalization of eq.(58).

III. THE MARCHENKO METHOD

This method uses the properties of the Jost solution instead of the regular solution. From the asymptotic formula (23) it is obvious that, for each fixed $r > 0$, the function

$$h(k,r) = f(k,r) - e^{ikr} , \qquad (73)$$

which is holomorphic in $\text{Im}k > 0$, belongs to $L^2(-\infty,\infty)$ for k real, and that

$$\int_{-\infty}^{\infty} |h(\sigma+i\tau)|^2 d\sigma \underset{\tau\geq 0}{=} O(e^{-2\tau r}) \quad . \qquad (74)$$

We can therefore apply directly to this function the celebrated Titchmarsh theorem on Fourier transforms of functions which are analytic in the upper half-plane, and get

$$A(r,t) \equiv h(t,r) = \frac{1}{2\pi} \int_{-\infty}^{\infty} [f(k,r) - e^{ikr}] \cdot$$

$$e^{-ikt} dk = 0 , \qquad t < r \quad . \qquad (75)$$

The inversion of this Fourier transform leads then to

$$f(k,r) = e^{ikr} + \int_r^\infty A(r,t)e^{ikt}dt \tag{76}$$

which is the analog of (51) for ϕ. This representation is valid everywhere in $\mathrm{Im}k \geq 0$, $A(r,t)$ being an L^2-function in t for all $r > 0$, $t > r$. Notice again that A is independent of k.

If we substitute now (76) into the integral equation (18) for f, and unfold the Fourier transform, we obtain the integral equation

$$A(r,t) = \frac{1}{2}\int_{\frac{r+t}{2}}^\infty V(s)ds - \int_{\frac{r+t}{2}}^\infty ds \int_0^{\frac{t-r}{2}} \cdot$$

$$V(s-u)A(s-u,s+u)du \ . \tag{77}$$

This equation, the analog of (54) for K, can now be solved by iteration. It admits a unique solution satisfying the bound

$$|A(r,t)| < \frac{1}{2}\int_{\frac{r+t}{2}}^\infty |V(s)|ds \exp\left[\int_r^\infty |u\ V(u)|du\right] \ . \tag{78}$$

All the formal steps in going from (76) to (77) are now easily justified. Making $t = r$, we obtain

$$A(r,r) = \frac{1}{2}\int_r^\infty V(t)dt \ . \tag{79}$$

This time, contrary to (56), we do not have to assume anything about the integrability of V at the origin. If the potential is a continuous function, we have

$$V(r) = -2\frac{d}{dr}A(r,r) \ . \tag{80}$$

To obtain the Marchenko integral equation, we start from the expansion theorem (49), assuming for simplicity that there are no bound states. Using (28) it can be written

$$\frac{1}{2\pi}\int_{-\infty}^{\infty} f(k,r)[f(-k,t) - S(k)f(k,t)]dk = \delta(r-t) , \qquad (81)$$

where

$$S(k) = e^{2i\delta(k)} = \frac{F(-k)}{F(k)} , \ k \text{ real} , \qquad (82)$$

is the S-matrix.

On the other hand, exactly in the same way we found the representation (59), we can derive also

$$e^{ikr} = f(k,r) + \int_r^{\infty} \bar{A}(r,t)f(k,t)dt . \qquad (83)$$

This can be done either by using the general theory of integral transforms with L^2- kernel $A(r,t)$ or, more simply, by considering (76) as a Volterra integral equation for e^{ikr} with the kernel A and the inhomogeneous term f. Because of (78) we can solve this equation and obtain (83) with $\bar{A}(r,t) \in L^2(r,\infty)$ in t for all $r > 0$.

Now combining (81) and (83) we get

$$\int_{-\infty}^{\infty} f(k,r)[e^{-ikt} - S(k)e^{ikt}]dk = 0 , \qquad t < r . \qquad (84)$$

If we substitute now in this equation the Jost solution f by its representation (76), we obtain

$$A(r,t) = A_o(r+t) + \int_r^{\infty} A(r,s)A_o(s+t)ds , \qquad t > r , \qquad (85)$$

where

$$A_o(t) = (2\pi)^{-1} \int_{-\infty}^{\infty} [S(k) - 1]\, e^{ikt}\, dk \quad . \tag{86}$$

When bound states are present, we obviously have

$$A_o(t) = (2\pi)^{-1} [S(k) - 1]e^{ikt}dk + \sum_j s_j\, e^{-\gamma_j t} \tag{87}$$

where the s_j's are given by $(E_j = -\gamma_j^2)$:

$$s_j = [\frac{-2i\gamma_j}{\dot{F}(i\gamma_j)f'(i\gamma_j,0)}]^{1/2} \quad . \tag{88}$$

Equation (85) with the kernel (86) is the Marchenko equation. It is somewhat simpler than the Gel'fand-Levitan equation in that here the kernel is directly related to $S(k) = \exp(2i\delta)$, whereas with the Gel'fand-Levitan equation we have first to calculate the Jost function and use it to calculate the spectral density $d\rho(E)$, which in turn is used in (63) and (65).

We now have two problems to study, namely, the unicity of the solution of the Marchenko equation, and the verification that the potential we obtain has the desired properties. We shall not study the first problem. It can be done along the same lines as for the Gel'fand-Levitan equation with some extra work. Let us mention here that one can show, quite generally, that the Marchenko method is equivalent to the Gel'fand-Levitan method, that is that the potential calculated by the formula (80) is identical to that calculated by (58), provided of course that the Jost function in (82) is the same as that in (50) and (66).

As for the properties of the potential, one can show, on the basis of (85) and (86), that in order to have (2), it is necessary and sufficient that the kernel $A_o(t)$ be differentiable and satisfy

$$\int_0^\infty t|A_0'(t)|dt < \infty \quad . \tag{89}$$

In fact, it can be shown that in many ways, $V(r)$ and $A_0'(t)$ have similar properties. Also, it can be shown that (89) is equivalent to assume that the phase shift is a continuous function of k and satisfies the integrability condition (42).

IV. ONE-DIMENSIONAL PROBLEM ON THE WHOLE REAL LINE

A. The Direct Problem

The scattering problem in one dimension on the entire real line:

$$\phi''(E,x) + E\phi(E,x) = V(x)\phi(E,x) \; , \tag{90}$$

$$-\infty < x < \infty \; ,$$

has some fundamental differences from the scattering with the radial equation merely because of the fact that now the origin (or any other finite point) does not play a role as boundary point. In the general case where the potential is not symmetric with respect to any point, it is intuitively clear that the points $+\infty$ and $-\infty$ play distinct roles because one can send a signal at $t = -\infty$ from one or the other and study how it evolves in the course of time. So, the problem is in fact similar to the two-coupled-channel case.

It turns out that to make a complete study of the problem, one needs the following conditions on the potential. First of all, one must assume, as usual, that V is locally L^1. At infinity, x^2V must be L^1. All these can be combined in the single condition

$$\int_{-\infty}^{\infty} [1 + x^2]|V(x)|dx < \infty \tag{91}$$

which we shall assume throughout.

We seek now scattering solutions ψ_1 and ψ_2 having the following asymptotic forms:

$$\psi_1(k,x) = \begin{cases} e^{ikx} + s_{12}(k)e^{-ikx} + o(1) & x \to -\infty \\ s_{11}(k)e^{ikx} & x \to +\infty \end{cases} , \quad (92)$$

$$\psi_2(k,x) = \begin{cases} s_{22}(k)e^{-ikx} & x \to -\infty \\ e^{-ikx} + s_{21}(k)e^{ikx} & x \to +\infty \end{cases} . \quad (93)$$

In the time-dependent description, the solution ψ_1 is the one which corresponds to a wave e^{ikx} incoming at $t = -\infty$ from the left (i.e., at $x = -\infty$ for $t = -\infty$), and propagating to the right. At $t = +\infty$, part of it is reflected to the left and gives rise to the term $s_{12}(k)e^{-ikx}$, where $s_{12}(k)$ is the reflection coefficient to the left, whereas another part is transmitted to the right and corresponds to $s_{11}(k)e^{ikx}$, $s_{11}(k)$ being the transmission coefficient to the right. The alternative notations $s_{11}(k) = t(k), s_{12}(k) = r(k)$ etc. are very frequently used in the literature (t for transmission coefficients, and r for reflection coefficients

Similarly for $\psi_2(k,x)$, where now $x = \pm\infty$ have been exchanged, $s_{22}(k)$ is the transmission coefficient to the left, and $s_{21}(k)$ the reflection coefficient to the right.

We call the matrix

$$S(k) = \begin{bmatrix} s_{11}(k) & s_{12}(k) \\ s_{21}(k) & s_{22}(k) \end{bmatrix} \quad (94)$$

the S-matrix of the equation (90). The question is: to what extent does the S-matrix determine the potential V? For this purpose, we must first study the properties of the S-matrix. It then turns out, in analogy with the Marchenko method, that analogous equations exist which would permit us to solve the inverse problem.

Before entering into a detailed study of the properties of the S-matrix, let us make the following heuristic remark. As a consequence of the general scattering theory, it follows from the reality of the potential that the S-matrix is unitary (conservation of probabilities, or of the number of particles) and must satisfy the symmetry condition

$$s_{11}(k) = s_{22}(k) \qquad (95)$$

which is due to time-reversal invariance. It follows that the S-matrix depends only on three independent real functions of k, $0 \leq k < \infty$. On the other hand, the potential $V(x)$, $-\infty < x < \infty$, may be considered as two real functions of x in the interval $[0,\infty)$. Since the potential determines completely the S-matrix it seems plausible that there should exist one relation among the three independent real elements of the S-matrix in terms of which it can be expressed. This additional property turns out to be the analyticity of the transmission coefficients $s_{11}(k) = s_{22}(k)$ in the upper half-plane $\mathrm{Im}k > 0$. We shall see that this entails that the whole S-matrix is essentially determined by the specification of one of the reflection coefficients $s_{12}(k)$ or $s_{21}(k)$.

As is well known, all the solutions of the Schrödinger equation can be obtained by linear combinations of some fundamental solutions satisfying precise boundary conditions. For these, we take

$$f_+(k,x): \quad \lim_{x\to\infty} e^{-ikx} f_+(k,x) = 1 \ ,$$
$$f_-(k,x): \quad \lim_{x\to-\infty} e^{ikx} f_-(k,x) = 1 \ . \tag{96}$$

These solutions are somewhat the analogue of the Jost solutions of the radial Schrödinger equation. They satisfy the integral equations

$$f_+(k,x) = e^{ikx} - \int_x^\infty \frac{\sin k(x-t)}{k} V(t) f_+(k,t)dt \ , \tag{97}$$

$$f_-(k,x) = e^{-ikx} + \int_{-\infty}^x \frac{\sin k(x-t)}{k} V(t) f_-(k,t)dt \ . \tag{98}$$

These equations being of the Volterra type, there is no difficulty, following the same procedure as for the Jost solutions, to prove that they admit unique solutions having properties similar to (21), (22), and (23)-(26): f_+ is holomorphic in Imk > 0, f_- in Imk < 0, and

$$f_\pm(k,x) = e^{\pm ikx} + O(\frac{1}{|k|}) e^{\mp \mathrm{Im}kx} \tag{99}$$

as $|k| \to \infty$ in the appropriate closed half-planes. We therefore have, as before,

$$\underset{\mathrm{Im}\,k\geq 0}{f_+(k,x)} = e^{ikx} + \int_x^\infty A_+(x,y) e^{iky} dy \ , \tag{100}$$

together with a similar representation for f_- in the lower half-plane Imk $\leq$ 0 with kernel A_-. Both A_+ and A_- are, for each fixed x, square-integrable in y by the Titchmarsh theorem.

Now writing

$$A_+(x,y) = \frac{1}{2} B_+ \left(x, \frac{y-x}{2}\right) \qquad (101)$$

the representation (100) becomes

$$f_+(k,x) = e^{ikx} \left[1 + \int_0^\infty B_+(x,y)\, e^{2iky}\, dy\right] \qquad (102)$$

where B_+ is again square integrable in y for each fixed x. Likewise

$$f_-(k,x) = e^{-ikx} \left[1 + \int_{-\infty}^{0} B_-(x,y)\, e^{-2iky}\, dy\right] \qquad (103)$$

where B_- is also square-integrable in y. These last representations will turn out to be more convenient here than the representation (100).

Using the above representations in the integral equations (97) and (98) and taking the inverse Fourier transforms, we obtain, in complete analogy with Section III, integral equations similar to (77) for B_+ and B_-, and bounds similar to (78). Again, it turns out that

$$B_+(x,0) = \int_x^\infty V(t)dt \ , \qquad (104)$$

$$B_-(x,0) = \int_{-\infty}^{x} V(t)dt \ . \qquad (105)$$

Thus

$$V(x) = -\frac{\partial}{\partial x} B_+(x,0) = \frac{\partial}{\partial x} B_-(x,0) \ . \qquad (106)$$

Here both equations are necessary for the study of the potential at the two ends of the line, B_+ for $x = +\infty$ and B_- for $x = -\infty$.

To go further, we need the properties of the S-matrix.

They are summarized in the following

<u>Theorem 1.</u> The coefficients $s_{ij}(k)$, $i,j = 1,2$, are continuous functions of k on the whole real axis and satisfy

$$s_{ij}(-k) = [s_{ij}(k)]^* \quad . \tag{107}$$

For large values of $|k|$,

$$s_{12}(k) = O(\tfrac{1}{k}), \quad s_{21}(k) = O(\tfrac{1}{k}) \quad , \tag{108}$$

$$s_{11}(k) = s_{22}(k) = 1 + O(\tfrac{1}{k}) \quad . \tag{109}$$

The function $s_{11}(k)$ is the limiting value on the real axis of a function analytic in the upper half-plane, with the exception of a finite number of simple poles at the points $k_j = i\chi_j = 1,2,\ldots,n$, on the imaginary axis. The residues at these poles are given by

$$\operatorname{Res} s_{11}(k)|_{k=i\chi_n} = i\gamma_n = [i \int_{-\infty}^{\infty} f_+(i\chi_n,x) f_-(i\chi_n,x)dx]^{-1} \quad . \tag{110}$$

The S-matrix with elements $s_{ij}(k)$ is a unitary matrix, and we have therefore

$$|s_{11}(k)|^2 + |s_{12}(k)|^2 = |s_{22}|^2 + |s_{21}|^2 = 1 \ , \tag{111}$$

$$s_{11} s_{21}^* + s_{12} s_{22}^* = 0 \quad . \tag{112}$$

The strict inequality

$$|s_{12}(k)| = |s_{21}(k)| < 1 \tag{113}$$

holds for all real $k \neq 0$. If $|s_{12}(0)| = 1$, then, necessarily,

$$s_{12}(0) = s_{21}(0) = -1 \quad . \tag{114}$$

More precisely, one gets then the following precise statement for $k \to 0$:

$$s_{11}(k) \simeq Lk \ , \qquad s_{12}(k) \simeq -1 + L_1 k \ ,$$

$$s_{21}(k) \simeq -1 + L_2 k \quad , \qquad (115)$$

where L ($\neq 0$), L_1, and L_2 are pure imaginary.

The properties of the S-matrix summarized in the above theorem make it possible to reconstruct all its elements from the knowledge of only one of the reflection coefficients s_{12} or s_{21}. Indeed, from (110), one has

$$|s_{11}(k)| = (1 - |s_{12}(k)|^2)^{1/2} \quad . \qquad (116)$$

Since this transmission coefficient is meromorphic in the upper half-plane, it is well known that it can be reconstructed with the help of Hilbert relations (dispersion relations) from its absolute value on the real axis. Thus, for $\mathrm{Im}k > 0$,

$$s_{11}(k) = \exp\left[\frac{1}{2i\pi}\int_{-\infty}^{\infty} \frac{\mathrm{Log}(1-|s_{12}(k')|^2)^{1/2}}{k'-k}\, dk'\right] \times \prod_{j=1}^{n} \left(\frac{k+i\chi_j}{k-i\chi_j}\right) , \qquad (117a)$$

and

$$s_{11}(k) = \lim_{\varepsilon \downarrow 0} s_{11}(k + i\varepsilon), \quad k \ \text{real} \quad . \qquad (117b)$$

This, when used in (112), leads to

$$s_{21}(k) = -\frac{s_{12}(-k)\, s_{11}(k)}{s_{11}(-k)} \quad . \qquad (118)$$

Remember that, for large values of $|k|$, we have (108)-(109).

B. The Inverse Problem

It was seen at the end of section III, with the help of the Marchenko method, that the properties of the potential at the origin are related to the properties of the Fourier transform of the S-matrix. Here we shall first develop the method for solving the inverse problem, quite in analogy with the Marchenko method. Then we shall study the properties of the S-matrix which are necessary and sufficient in order for the potential to satisfy the integrability conditions (91). That it is the Marchenko approach rather than that of Gel'fand and Levitan which should be used here for solving the inverse problem is merely the reflection of the fact that the origin $x = 0$ does not play any special role here, and that we are dealing exclusively with solutions which have prescribed asymptotic behaviors at $x = \pm\infty$, i.e. which are the analogue of the Jost solution in the radial case.

We have already established the integral representations (102) and (103) whose analogue for the Jost solution of the radial case was the starting point of the Marchenko method. We have therefore to find out the integral equations for the kernels B in terms of the scattering data (S-matrix, its poles (bound states), and their residues). We shall leave out the details of the derivations, which are similar to the radial case, and give only the results which are as follows: the kernels B_+ and B_- satisfy the integral equations

$$B_\pm(x,y) \pm \int_0^{\pm\infty} B_\pm(x,t)\Omega_\pm(x+y+t)dt$$

$$+ \Omega_\pm(x+y) = 0 , \qquad (119)$$

$y > 0$ for B_+, $y < 0$ for B_-,

with + signs for B_+, and - signs for B_-. The kernels are given by

$$\Omega_{\pm}(t) = \sum_j M_j^{(\pm)} e^{-\chi_j t} + A_{\pm}(t) , \tag{120}$$

where

$$M_j^{(\pm)} = [\int_{-\infty}^{\infty} f_{\pm}^2(i\chi_j,x)dx]^{-1} , \tag{121}$$

and

$$s_{21}(k) \equiv r_R(k) = \int_{-\infty}^{\infty} A_+(t)e^{2ikt}dt , \tag{122a}$$

$$s_{12}(k) \equiv r_L(k) = \int_{-\infty}^{\infty} A_-(t)e^{-2ikt}dt . \tag{122b}$$

The two equations (119) and (120) are the basic equations for solving the inverse problem, i.e. to find the potential V(x) from the knowledge of one of the reflection coefficients, say $s_{12}(k)$, for all real values of k, together with the poles $i\chi_j$ and the positive normalization constant $M_j^{(+)}$. Indeed, from $s_{12}(k)$ and the numbers χ_j, we can calculate $s_{11}(k)$ with the help of (117a). Knowing $s_{11}(k)$, we can calculate its residues $i\gamma_j$, (110), at the poles $i\chi_j$. We notice now that (118) gives us also $s_{21}(k)$. Since $s_{22} = s_{11}$, the S-matrix is completely determined. At the same time, we determine also $M_j^{(-)}$ from the identity

$$M_j^{(+)} M_j^{(-)} = \gamma_j^2 , \quad j = 1,2,\dots , \tag{123}$$

which is a consequence of (110) and (121), and the fact that at the poles $i\chi_j$, the two solutions f_+ and f_- are proportional. Knowing now s_{12}, s_{21}, $M_j^{(-)}$, we can calculate unambiguously the kernels $\Omega_{\pm}(t)$ with the help of (122a) and (122b). We can solve therefore the integral equations (119) to determine $B_{\pm}$, and then the potential V(x) from the formulae (106).

Concerning the properties of the potential, they can be deduced from the integral equations. The final result is summarized in the following

Theorem 2. The continuous 2x2 matrix S(k) is the S-matrix of the one-dimensional Schrödinger equation (90) whose potential satisfies (91) if and only if the following conditions are fulfilled:

- Unitarity: (111) and (112);
- Reality: (107);
- Symmetry: (95);
- Asymptotic behaviour as $|k| \to \infty$: (108) and (109);
- Analyticity: The function $s_{11}(k)$ is the boundary value of a function analytic in the half-plane $\mathrm{Im} k > 0$, where it has the asymptotic form $1+O(k^{-1})$, and a finite number of simple poles on the imaginary axis;
- The Fourier transforms $A_{\pm}(t)$ of $s_{21}(k)$ and $s_{12}(k)$ must satisfy the conditions

$$\int_a^{\infty}(1 + t^2)|A_+'(t)|dt < C(a) < \infty \tag{124}$$

and

$$\int_{-\infty}^{h} (1+t^2)|A_-'(t)|dt < C(h) < \infty \quad . \tag{125}$$

In the case when $s_{11}(0) = 0$, one should have (115).

We should mention in ending this section that it is clear, by comparing (122) with (86), that here the analogues of S-1 in the Marchenko formulation of section III are $s_{12}(k)$ and $s_{21}(k)$. All these quantities vanish in the limit of infinite k, and so the integrals make sense, at least for $t \neq 0$.

V. THE PURELY DISCRETE SPECTRUM

This case can be thought of as the limiting case of

the Gel'fand-Levitan equation previously studied when the continuous spectrum vanishes, and we are left with only the discrete spectrum (now infinite). This is the case, as we mentioned in the introduction, either with a good potential in a finite interval with precise boundary conditions at both ends, or with a confining potential in the infinite interval $[0,\infty)$, i.e. $V(\infty) = \infty$, and a precise boundary condition at the origin.

In the first case, the completeness relation leads to the spectral density

$$\frac{d\rho}{dE} = \sum_j C_j \; \delta(E-E_j) \tag{126}$$

in the presence of the potential, and to

$$\frac{d\rho_o}{dE} = \sum_j C_j^{(0)} \; \delta(E-E_j^{(0)}) \tag{127}$$

for the free case. One must therefore replace in (63) $d\rho$ and $d\rho_o$ by these new expressions, and use the new $d\sigma$ in the definition of the kernel G of (65) which reads now:

$$G(r,t) = \int_{-\infty}^{\infty} \phi^{(0)}(E,r)\phi^{(0)}(E,t)d\sigma(E) \quad . \tag{128}$$

Notice that both ϕ and $\phi^{(0)}$ should satisfy the same boundary conditions. The rest of the theory is quite similar, and the potential is again obtained from (64) and (58). Concerning the convergence problems, there is no difficulty, in the case of finite intervals, for proving that all the infinite series one is dealing with are convergent.

A related problem is when one gives two different spectra $\{E_j, E_j'\}$, corresponding to two different boundary conditions, instead of giving $\{E_j, C_j\}$. The two problems are closely related, and one can go from one set of data to the other easily. Indeed, one can show that the C_j are given by explicit formulae in terms of E_j and E'_j. It follows that $d\rho(E)$ can be calculated explicitely and uniquely from two spectra. This, in turn, means that two spectra determine uniquely the potential. For details, see the paper of Levitan and Gasymov, and below.

More interesting is the case of confining potentials. Here we have $V(\infty) = \infty$, and we are looking for solutions which belong to $L^2(0,\infty)$ and satisfy the boundary condition

$$\phi'(E,0) = \alpha\phi(E,0) \quad . \tag{129}$$

Here one must distinguish between the case where α is finite, and in particular zero, and the case where $\alpha = \infty$, which is a limiting case and corresponds to

$$\phi(E,0) = 0 \quad . \tag{130}$$

Consider first the case of finite α. It can be shown that the generic case is $\alpha = 0$, i.e.

$$\phi'(E,0) = 0 \quad . \tag{131}$$

Here we are dealing with Fourier cosine integrals. Since $V(\infty) = \infty$, there is, in general, for each value of the energy, a unique solution $\phi(E,x)$ which is L^2(i.e.vanishes) at $x = \infty$. The spectrum is obtained by imposing (129) or (131). We define again

$$d\rho(E) = \sum_j C'_j \; \delta(E-E'_j)dE \tag{132}$$

as the sum over all the eigenvalues, C'_j being defined by

$$C'^{-1}_j = \int_0^\infty \phi_j^2(x)dx \quad . \tag{133}$$

In the free case $(V = 0)$, we have only the pure continuous spectrum $[0,\infty)$, corresponding to Fourier cosine integral.

We define then

$$d\sigma(E) = \begin{cases} d\rho(E) , & E < 0 \\ d\rho(E) - \frac{1}{\pi} \sqrt{E}\, dE , & E \geq 0 \end{cases} , \tag{134}$$

and construct the kernel

$$F(x,y) = \lim_{A\to\infty} \int_{-\infty}^{A} d\sigma(E) \cos \sqrt{E}\, x \cos \sqrt{E}\, y \quad . \tag{135}$$

It can then be shown that the Gel'fand-Levitan equation

$$K(x,z) + F(x,z) + \int_0^x K(x,y)F(y,z)dy = 0 \tag{136}$$

has a unique solution from which one can calculate the potential by

$$V(x) = 2 \frac{d}{dx} K(x,x) \quad . \tag{137}$$

A sufficient condition under which the existence of the limit in (135) is ensured and the rest of the analysis is valid is that the potential should increase not faster than x^2 at infinity, and there are some indications that even x^4 may be an admissible behaviour. The conclusions are exactly the same for all finite values of α.

Consider now again (129) with two different (finite) values α_1 and α_2. We obtain now two spectra $\{E_j^{(1)}\}$ and $\{E_j^{(2)}\}$. It can be shown that one has, for all n,

$$C_n^{-1}(\alpha_1) = \int_0^\infty [\phi_n^{(1)}(x)]^2 dx = \frac{E_n^{(2)} - E_n^{(1)}}{\alpha_2 - \alpha_1} \prod_{k\neq n} \frac{E_k^{(2)} - E_n^{(1)}}{E_k^{(1)} - E_n^{(1)}} \quad . \tag{138}$$

This shows that two spectra corresponding to two boundary conditions determine completely the spectral density $d\rho_1(E)$, and therefore uniquely the potential.

The case where we impose (130) at the origin, and which is the most interesting for physical applications (Charmonium,...), is somewhat different and has been studied throughout by H. Grosse and A. Martin following the work of Quigg, Rosner and Thacker. In the case of Charmonium, and other particles of the family, the quantities one can obtain from experiment are the energies and the numbers

$$B_j = |\phi'(E_j,0)|^2 \quad , \tag{139}$$

assuming here that the wave function has been normalized:

$$\int_0^\infty \phi^2(E_j,x)dx = 1 \quad . \tag{140}$$

Remember that here we have $\phi_j(0) = 0$. However, it can again be shown that the numbers B_j, together with the energies E_j, determine uniquely the spectral density. Indeed, as we said before, for each value of E (real or complex), there is only one solution $\phi(E,x)$ which is L^2 at infinity and is therefore L^2 on the positive real axis. Consider now the function R(E) (very similar to the inverse of the R-matrix of Wigner) defined by

$$R(E) = \frac{\phi'(E,0)}{\phi(E,0)} \quad . \tag{141}$$

It can be easily shown, by combining the Schrödinger equation with its complex conjugate, that

$$\frac{\mathrm{Im}\ R(E)}{\mathrm{Im}\ E} = \frac{\int_0^\infty |\phi(E,x)|^2 dx}{|\phi(E,0)|^2} > 0 \quad . \tag{142}$$

This means that R(E) is a Herglotz function, as is also $- R^{-1}(E)$. It is also clear that R(E) is a (real) meromorphic function with only real zeros and poles because of the simple fact that the Hamiltonian, defined by $- \Delta + V$ together with either (130) or (131) is a self-adjoint operator in $L^2(0,\infty)$ and therefore that the only zeros of ϕ' or ϕ, i.e. the two spectra $\{E_j'\}$, $\{E_j\}$, are all real. The two sets are also interlaced, and both are bounded from below.

As is well-known, Herglotz functions cannot increase faster than $|E|$ and decrease slower than $|E|^{-1}$ as $|E| \to \infty$ in all directions, excluding the positive real axis. In fact, it can be shown in the present case that

$$R(E) \underset{E\to-\infty}{=} (-E)^{1/2} + O\left(\frac{1}{\sqrt{-E}}\right) \quad . \tag{143}$$

It follows then from the Phragmén-Lindelöf theorem that R(E) grows like $(-E)^{1/2}$ in all complex directions. This, in turn, implies that we have the (absolutely convergent) representations

$$-\frac{1}{R(E)} = \sum_j \frac{\phi^2(E_j',0)}{E_j'-E} \quad , \tag{144a}$$

$$\int_0^\infty \phi^2(E_j',x)dx = 1 \quad , \tag{144b}$$

and

$$R(E) = C + E\sum_j \frac{[\phi'(E_j,0)]^2}{E_j(E_j-E)} \quad , \tag{145a}$$

$$\int_0^\infty \phi^2(E_j,x)\,dx = 1 \quad . \tag{145b}$$

Before continuing, let us note that if we write (144) and (145) as Stieltjesintegrals, the measures (sums of δ-functions) are exactly the spectral measures for the inverse problems with (131) and (130) respectively. Indeed one can show, by calculating the Wronskian of two solutions with E and E + dE, that the residues are

$$\frac{d}{dE}\frac{1}{R(E)}\bigg|_{E=E_j} = \frac{\int_0^\infty \phi^2(E_j,x)\,dx}{[\phi'(E_j,0)]^2} \tag{146}$$

and

$$\frac{d}{dE} R(E)\bigg|_{E=E_j'} = -\frac{\int_0^\infty \phi^2(E_j',x)\,dx}{[\phi(E_j,0)]^2} \quad . \tag{147}$$

Consider now (145), corresponding to (130). We know the spectrum E_j and the constants B_j. In order to calculate the function R, there remains the unknown subtraction constant C in (145). However, it can be shown that C is not free and can be calculated from $\{E_j\}$ and $\{B_j\}$. It follows that we can calculate, through (145), the function R(E) from the above data. We know therefore the l.h.s. of (144), which means that we can calculate then the set E_j' and the residues, i.e. the spectral density for the inverse problem with (131). This in turn, when used in the Gel'fand-Levitan theory, determines the potential uniquely. Finally, the constant C is given by the limiting procedure

$$C = \lim_{N\to\infty} C_N, \quad C_N = \sum_{j=1}^{N} \frac{(\phi'(E_j,0))^2}{E_j} - \frac{2}{\pi}\sqrt{\bar{E}_N} \, ,$$

$$\bar{E}_N = \frac{1}{2}(E_N + E_{N+1}) \, , \tag{148}$$

which turns out, in practice, to be converging very fast.

Another procedure, whose proof has not yet been found, but seems to be correct because of its very fast convergence in all the examples which have been treated numerically, has been suggested by Quigg, Rosner, and Thacker, and is as follows. We consider the one-dimensional problem on the real line $(-\infty,\infty)$, assuming the potential to be even: $V(-x) = V(x)$.

Since $V(\pm\infty) = \infty$, we have odd and even energy levels given by $\{E_j\}$ and $\{E'_j\}$ of the half-line problem with (130) and (131), respectively. These two sets of energy levels are interlaced, the lowest one being always the first "odd" level E_1, corresponding to (130). We order the two sets together in increasing magnitude $\{E_1, E'_1, E_2, E'_2, \ldots\}$, and call the total set by $\{\tilde{E}_j\}$. We introduce now a reference energy E_{ON} between E_N and E_{N+1}, and write

$$\bar{E}_j = \tilde{E}_j - E_{ON} , \quad j = 1,2,\ldots \tag{149}$$

It turns out that a good choice would be

$$E_{ON} = \frac{1}{2} (E_N + E_{N+1}) \quad . \tag{150}$$

Among the above energy levels, there are exactly N which are negative, and we write them

$$\bar{E}_j = - \bar{\chi}_j^2 = \tilde{E}_j - E_{ON} , \quad j = 1,2,\ldots N \quad . \tag{151}$$

We consider now the scattering problem on the whole real line - a problem which was treated in detail in the previous section - with the following properties:

- it is reflectionless, i.e. $r(k) = 0$ for all k;
- it has N bound states with energies given by (151).

Note that these energies are all negative, as they should. Also, V being even, the two reflection coefficients are equal (= 0).

According to (120) and (123) we need, in principle, the set $\{\bar{E}_j, M_j^{(1)}\}$ to calculate $\bar{V}_N$. However, this potential being even, it can be shown that the normalization constants are given by

$$M_j^{(1)} = M_j^{(2)} = 2\bar{\chi}_j \prod_{n \neq j} \left| \frac{\bar{\chi}_n + \bar{\chi}_j}{\bar{\chi}_n - \bar{\chi}_j} \right| \quad . \tag{152}$$

We have therefore all the necessary ingredients for calculating the reflectionless potential $\bar{V}_N$ having N bound states with binding energies (151). It turns out then that the even confining potential V is very close to the potential augmented by E_{ON}:

$$\bar{V}_N(x) + \frac{1}{2}(E_N + E_{N+1}) \cong V(x) \tag{153}$$

for all values of x for which

$$V(x) < E_{ON} = \frac{1}{2}(E_N + E_{N+1}) \ , \tag{154}$$

i.e. up to the turning points. In practice, it is seen that in the allowed region the l.h.s. of (153) oscillates around the exact potential while staying very close. The expectation, supported by numerical calculations, is that

$$\bar{V}_N(x) \underset{N\to\infty}{\to} V(x) \tag{155}$$

in larger and larger intervals of the x-axis.

REFERENCES

- For general scattering theory, see:

R.G. Newton, Scattering Theory of Waves and Particles (Mc Graw Hill, 1966).

V. de Alfaro and T. Regge, Potential Scattering, John Wiley and Sons, 1965.

W.O. Amrein, J.M. Jauch, and K.B. Sinha, Scattering Theory in Quantum Mechanics, Benjamin, 1977.

M. Reed and B. Simon, Methods of Modern Mathematical Physics, vol. III, Academic Press, 1979.
H.M. Nussenzveig, Causality and Dispersion Relations, Academic Press, 1972.
B. Simon, Quantum Mechanics for Hamiltonians defined as quadratic forms, Princeton University Press, 1971.

- For the study of inverse problems, see:

Z.S. Agranovich and V.A. Marchenko, The Inverse Problem of Scattering Theory, English translation, Gordon and Breach, 1963.
K. Chadan and P.C. Sabatier, Inverse Problems in Quantum Scattering Theory, Springer Verlag, 1977. Russian translation, MIR, 1980. This book contains an almost complete list of references to original works of Gel'fand and Levitan, Marchenko, Jost and Kohn, Faddeev, Newton, Sabatier, Regge, Loeffel, Martin, Cronille,...
P. Deift and E. Trubowitz, Inverse Scattering on the Line, Comm. Pure. Appl. Math. 32 (1979) 121. This paper gives a method different from the original method of Faddeev.
R.G. Newton, Inverse scattering, I, One Dimension, J. Math. Phys. 21 (1980) 493; Three Dimensions, ibid. 21 (1980) 1698. See also his recent preprint.

- For the case of discrete spectrum, including confining potentials, see:

B.M. Levitan and M.G. Gasimov, Determination of a differential equation by two of its spectra, Russian Math. Survey 19 (1964) 1.
H. Grosse and A. Martin, Theory of the inverse problem for confining potentials, Nucl. Phys. B148 (1979) 413.
C. Quigg, J.L. Rosner and H.B. Thacker, Phys. Rev. D18 (1978) 274; ibid. 287.

- For applications to Q.F.T., see:

E.K. Sklyanin and L.D. Faddeev, Sov. Phys. Dokl. 23 (1978) 902.

J. Honerkamp, P. Weber and A. Wiesler, Nucl. Phys. B152 (1979) 266.

H. Grosse, Phys. Lett. B86 (1979) 287.

H.B. Thacker, Rev. Mod. Phys. 53 (1981) 256.

These papers, especially the last one, contain full references to the works of Zakharov and Shabat, Faddeev, Manakov, Sklyanin, Thacker, Kulish, etc..

- For applications to nonlinear partial differential equations, see:

C.S. Gardner, M.J. Greene, M.O. Kruskal and R.M. Miura, Phys. Rev. Lett. 19 (1967) 1095.

A.C. Scott, F.Y.F. Chu and D.M. McLaughlin, Proc. IEEE 61 (1973) 1443.

F. Calogero (Ed.), Nonlinear evolution equations solvable by the spectral transform (Pitman, London, 1978).

P.C. Sabatier (Ed.), Problèmes Inverses. Evolutions non linéaires (C.N.R.S., Paris, 1979).

Acta Physica Austriaca, Suppl. XXIII, 157–208 (1981)

TIME DELAY OF QUANTUM SCATTERING PROCESSES+

by

Ph. A. MARTIN
Laboratoire de Physique Théorique
Ecole Polytechnique Fédérale, Lausanne
Switzerland

CONTENTS

+Lectures given at the XX. Internationale Universitätswochen für Kernphysik, Schladming, Austria, February 17-26, 1981.

These lectures present various aspects of the theory of the time delay of scattering processes. Proofs are sketched in such a way that the reader may hopefully understand their basic mechanisms! Some points of rigor and additional mathematical developments have to be supplemented by the reader or can be found in the literature on the subject which is discussed at the end of these notes.

Since we will study mainly non-relativistic two-body scattering processes, we first summarize briefly the theory of simple scattering systems. In the following, we will consider both the time dependent formulation and the stationary formalism of scattering.

I. BRIEF SUMMARY OF TWO-BODY SCATTERING

a) Time Dependent Setting

Let $U_t = \exp(-iH_o t)$ and $V_t = \exp(-iHt)$ be the free and total evolution on an Hilbert space H of quantum states, with self adjoint generators H_o, H, being respectively the free and the total hamiltonian. H_o has an absolutely continuous spectrum $\Lambda \subset R$. The asymptotic condition

$$\lim_{t\to-\infty} \|(V_t^* U_t - \Omega_-)\phi\| = 0 \qquad (1a)$$

$$, \quad \phi \in H ,$$

$$\lim_{t\to\infty} \|(V_t^* U_t - \Omega_+)\phi\| = 0 \qquad (1b)$$

defines the wave operators $\Omega_\mp$. They are supposed to be complete, i.e. Range Ω_- = Range $\Omega_+ = H_{abs}$ where H_{abs} is the subspace of absolute continuity of H. We have the intertwining relation $H\Omega_\pm = \Omega_\pm H_o$, the scattering operator $S = \Omega_+^* \Omega_-$ is unitary and conserves energy, $[S,H_o] = 0$. The scattering state ψ_t corresponding to an incoming state ϕ is $\psi_t = V_t \Omega_- \phi$. From the asymptotic condition and the completeness, it follows that ψ_t behaves asymptotically in time as

$$\|\psi_t - \phi_t\| \to 0 \quad , \quad t \to -\infty \qquad (2a)$$

$$\|\psi_t - S\phi_t\| \to 0 \quad , \quad t \to \infty \quad , \qquad (2b)$$

where $\phi_t = U_t \phi$ is the freely moving incoming state.

b) Stationary Formalism (Potential Scattering)

We consider $H_o = -\frac{1}{2m}\Delta$ and $H = H_o + V(\underline{x})$ on $H = L^2(R^3)$, where the potential is assumed to be such that there exist eigenfunctions $\psi(\underline{k},\underline{x})$ given by the stationary solutions of the Schrödinger equation

$$(-\frac{1}{2m}\Delta + V(\underline{x}) - \lambda)\psi(\underline{k},\underline{x}) = 0 , \qquad \lambda = \frac{k^2}{2m}, \quad k = |\underline{k}| ,$$

with asymptotic spatial behaviour (outgoing wave)

$$\psi(\underline{k},\underline{x}) = e^{i\underline{k}\cdot\underline{x}} + f(\lambda,\hat{x},\hat{k}) \frac{e^{ikr}}{r} , \quad r \to \infty , \quad r = |\underline{x}| ;$$

$\hat{x},\hat{k}$ are the angles of $\underline{x}$ and $\underline{k}$, and $f(\lambda,\hat{x},\hat{k})$ is the scattering amplitude.

The time development of the scattering wave function is then explicitely given by

$$\psi_t(\underline{x}) = \frac{1}{(2\pi)^{3/2}} \int d^3k \ \psi(\underline{k},\underline{x}) e^{-i\frac{k^2}{2m}t} \ \phi(\underline{k}),$$

and $\psi(\underline{k},\underline{x})$ defines the kernel of the wave operator

$$\psi(\underline{k},\underline{x}) = (2\pi)^{3/2} (\underline{x}|\Omega_-|\underline{k}) \ . \tag{3}$$

If $V(\underline{x}) = V(r)$ is spherically symmetric, $\psi(\underline{k},\underline{x})$ has a development in radial waves

$$\psi(\underline{k},\underline{x}) = 4\pi \sum_{\ell m} i^\ell \frac{u_\ell(k,r)}{kr} (\hat{x}|\ell m)(\ell m|\hat{k}) \ , \tag{4}$$

where $(\hat{x}|\ell m)$ is the normalized spherical harmonic $Y_\ell^m(\hat{x})$. $u_\ell(k,r)$ is the regular solution of the radial equation

$$\left(-\frac{d^2}{dr^2} + \frac{\ell(\ell+1)}{r^2} + 2m\,V(r) - 2m\lambda\right)u_\ell(k,r) = 0 \tag{5}$$

with asymptotic behaviour as $r \to \infty$

$$u_\ell(k,r) = \frac{1}{2i}\left(S_\ell(\lambda)e^{i(kr-\ell\pi/2)} - e^{-i(kr-\ell\pi/2)}\right) + O(1) . \tag{6}$$

$S_\ell(\lambda) = e^{i\delta_\ell(\lambda)}$ is the scattering matrix element with angular momentum ℓ and $\delta_\ell(\lambda)$ the corresponding phase shift.

c) Heuristic Definition of Time Delay

We look at the scattering wave for large positive time and far away from scattering center. According to (2b), we have

$$\psi_t(\underline{x}) \simeq \frac{1}{(2\pi)^{3/2}} \int d^3k \ e^{i(\underline{k}\cdot\underline{x} - \frac{k^2}{2m}t)} (S\phi)(\underline{k}) \ , \quad t \to \infty .$$

Correspondingly, its spherical components $\psi_t^\ell(r)$ with angular momentum ℓ behave as

$$\psi_t^\ell(r) \simeq i^\ell\sqrt{\frac{2}{\pi}}\int k^2 dk\, \frac{u_\ell^o(k,r)}{kr}\, e^{-i\frac{k^2}{2m}t}\, S_\ell(\lambda)\phi^\ell(k)$$

$$\simeq \frac{1}{i\sqrt{2\pi}}\,\frac{1}{r}\int kdk\,(e^{i(kr-\frac{k^2}{2m}t+\delta_\ell(\lambda))} - (-1)^\ell e^{-i(kr+\frac{k^2}{2m}t)})\phi^\ell(k), \quad (7)$$

where we have introduced the asymptotic form (6) of the free radial function $u^o(k,r)$.

By the stationary phase argument, $\psi_t^\ell(r)$ is appreciably different from zero only for those values of (r,t) such that the arguments of the exponentials in (7) are stationary with respect to energy variations (assuming $\phi^\ell(k)$ real). Since both r,t are positive, the second term of (7) does not contribute. The first term gives a contribution if

$$\frac{d}{dk}\,(kr - \frac{k^2}{2m}t + \delta_\ell(\lambda)) = 0, \quad (8)$$

i.e.

$$r = V(t - \tau_\ell(\lambda)) \quad \text{with} \quad \tau_\ell(\lambda) = \frac{d}{d\lambda}\,\delta_\ell(\lambda)\,. \quad (9)$$

Since $v = \frac{k}{m}$ is the free radial velocity and $\delta_\ell(\lambda) = 0$ in case of free motion, it is natural to interprete $\tau_\ell(\lambda)$ in (9) as the delay experienced by the radial wave packet in presence of the interaction. (9) is called the Eisenbud-Wigner formula.

If the S-matrix has a pole located just below the real axis corresponding to a resonance of energy λ_o and width $\Gamma > 0$, we can approximate $S_\ell(\lambda)$ by $S_\ell(\lambda) \sim \frac{\alpha(\lambda)}{\lambda_o - i\Gamma - \lambda}$ for $\lambda \sim \lambda_o$ where $\alpha(\lambda)$ is a slowly varying function of energy. Therefore, $\tau_\ell(\lambda)$ takes the form

$$\tau_\ell(\lambda) = \mathrm{Im}(\frac{d}{d\lambda}\log S_\ell(\lambda)) \sim \frac{\Gamma}{(\lambda-\lambda_o)^2+\Gamma^2}\,, \quad \lambda \sim \lambda_o\,, \quad (10)$$

and $\tau_\ell(\lambda_o) = \frac{1}{\Gamma}$ gives the life time of the resonance.

The purpose of the next sections is to give a general and mathematically precise definition of time delay and to study its relation to the energy derivative of the phase shift. The basic concept entering in this definition is that of sojourn time.

II. SOJOURN TIME

Let P be a projection on some subspace PH of states in H. The probability of finding the scattering state ψ_t in PH at time t is $(\psi_t, P\psi_t) = \|P\psi_t\|^2$, and the total mean time spent by ψ_t in PH is $\int_{-\infty}^{\infty} dt \|P\psi_t\|^2$. This quantity (which can be finite or infinite depending on P and ψ_t) will be called the <u>sojourn time</u> (or transit, or residence time) of ψ_t in PH. A choice of P depends of the physical question at hand.

In two-body scattering, we take $P = P_\Sigma$ to be the projection on a region Σ in configuration space where the scattered particles are at finite distance of each other. Precisely, in potential scattering, we define

$$(P_\Sigma \phi)(\underline{x}) = \chi_\Sigma(\underline{x})\phi(\underline{x}) \quad , \quad \chi_\Sigma(\underline{x}) = \begin{cases} 1 \, , & \underline{x} \in \Sigma \\ 0 \, , & \underline{x} \in \Sigma \end{cases} \quad ,$$

for a bounded subset Σ of R^3.

We denote respectively by

$$T_\Sigma(\phi) = \int_{-\infty}^{\infty} \|P_\Sigma \psi_t\|^2 dt = \int_{-\infty}^{\infty} (\phi, \Omega_-^* V_t^* P_\Sigma V_t \Omega_- \phi) dt \tag{11}$$

and

$$T_\Sigma^o(\phi) = \int_{-\infty}^{\infty} \|P_\Sigma \phi_t\|^2 dt = \int_{-\infty}^{\infty} (\phi, U_t^* P_\Sigma U_t \phi) dt \tag{12}$$

the sojourn times in Σ of the scattering state and of the freely moving state $\phi_t = U_t\phi$ corresponding to the same incoming state ϕ.

It is also useful to consider the quadratic forms associated with $T_\Sigma(\phi)$ and $T^o_\Sigma(\phi)$:

$$T_\Sigma(\phi,\chi) = \int_{-\infty}^{\infty}(\phi,\Omega_-^* V_t^* P_\Sigma V_t \Omega_- \chi)dt = \int_{-\infty}^{\infty} (\phi, U_t^* \Omega_-^* P_\Sigma \Omega_- U_t \chi)dt \qquad (13)$$

where we have used the intertwining relation to obtain the last form in (13), and $T^o_\Sigma(\phi,\chi) = \int_{-\infty}^{\infty}(\phi,U_t^* P_\Sigma U_t\chi)dt$.

Finiteness of Sojourn Time

It is expected that sojourn times in finite regions are finite on the grounds that scattering states propagate away from any bounded region. In potential scattering this follows essentially from the usual $t^{-3/2}$ decay of the wave packet. Indeed, in the case of free motion, we have, using the explicit form of the free propagator,

$$(\underline{x}|U_t|\underline{y}) = \left(\frac{m}{2i\pi t}\right)^{3/2} e^{\frac{im|\underline{x}-\underline{y}|^2}{2t}} ,$$

$$||P_\Sigma\phi_t||^2 = \left(\frac{m}{2\pi|t|}\right)^3 \int_\Sigma d^3x |d^3y\, e^{\frac{im|\underline{x}-\underline{y}|^2}{2t}} \phi(\underline{y})|^2 \le |\Sigma| \left(\frac{m}{2\pi|t|}\right)^3 ||\phi||_1^2 ;$$

$||\phi||_1 = \int d^3x|\phi(\underline{x})|$ is the L^1 norm of $\phi(\underline{x})$.
This shows that $||P_\Sigma\phi_t||^2$ is integrable and that $T^o_\Sigma(\phi)$ is finite for the dense set of ϕ with finite L^1-norm. For scattering states, things are not so obvious and we give three different methods to show the finiteness of the time integral (11).

a) Time Dependent Method

The time dependent method is based on the asymptotic behaviour (2ab) of the scattering state. We notice that

$$\|P_\Sigma\psi_t\| = \|P_\Sigma(\psi_t-\phi_t) + P_\Sigma\phi_t\| \leq \|\psi_t-\phi_t\| + \|P_\Sigma\phi_t\| \quad .$$

Therefore, $\|P_\Sigma\psi_t\|^2$ will be integrable at $t \to -\infty$ if, in addition to $\|P_\Sigma\phi_t\|$, $\|\psi_t-\phi_t\|$ is integrable at $t \to -\infty$. Similarly, since $\|P_\Sigma\psi_t\| \leq \|\psi_t-S\phi_t\| + \|P_\Sigma S\phi_t\|$, $\|P_\Sigma\psi_t\|^2$ is integrable at $t \to \infty$ if $\|\psi_t-S\phi_t\|$ and $\|P_\Sigma S\phi_t\|$ have the same property. Thus we get[+]

Proposition 1: Assume that $\|P_\Sigma\phi_t\| \in L^1(-\infty,\infty)$ and

(i) $\|\psi_t - \phi_t\| \in L^1(-\infty,0)$ (14a)

(ii) $\|P_\Sigma S\phi_t\|$ and $\|\psi_t - S\phi_t\| \in L^1(0,\infty)$; (14b)

then $T_\Sigma(\phi)$ is finite.

It is useful to know that (14ab) follow when the asymptotic condition (1a,b) holds with an integrable rate of convergence and the S-operator has some smoothness in its energy dependence. In order to formulate the needed smoothness property (see (16) below), it is convenient to assume that H_o has a simple spectrum, so that S becomes a function of H_o only. Then $H = L^2(\Lambda,d\lambda)$ and H_o, S act as multiplication operators by λ and $S(\lambda)$:

$$(H_o\phi)(\lambda) = \lambda\phi(\lambda) \quad , \qquad (S\phi)(\lambda) = S(\lambda)\phi(\lambda) \quad .$$

(In fact, we have always this situation with a spherically symmetric potential when we work on a subspace $H_{\ell m}$ with fixed angular momentum.)

[+]In prop.1,2 square integrability would be sufficient to slow finitness of $T_\Sigma(\phi)$. However, integrability is used to show the existence of time delay in prop. 4,5.

<u>Proposition 2:</u> Let ϕ be of compact support on Λ and $\rho(\lambda)$ a C_0^∞ function on Λ such that $\rho\phi = \phi$.

Assume that

$$\|(V_t^* U_t - \Omega_-)\phi\| \in L^1(-\infty,0), \tag{15a}$$

$$\|(V_t^* U_t - \Omega_+)\phi\| \in L^1(0,\infty), \tag{15b}$$

and

$$\int d\alpha\ (1 + |\alpha|)|\tilde{S}_\rho(\alpha)| < \infty \tag{16}$$

where $\tilde{S}_\rho(\alpha)$ is the Fourier transform of $S_\rho(\lambda) = S(\lambda)\rho(\lambda)$. Then (14a) and (14b) hold. (Notice that (16) is true if $S(\lambda)$ is twice continuously differentiable.)

<u>Proof:</u> (14a) is identical with (15a). To show (14b), we use the fact that one can write $S_\rho = \int d\alpha\ \tilde{S}_\rho(\alpha)\ U_\alpha$ in the functional calculus of H_0. Thus we have

$$\int_0^\infty \|(V_t\Omega_- - U_tS)\phi\|dt = \int_0^\infty \|(V_t\Omega_+ - U_t)\int d\alpha\ \tilde{S}_\rho(\alpha)U_\alpha\phi\|dt$$

$$\leq \int d\alpha|\tilde{S}_\rho(\alpha)|\ |\int_0^\infty \|(V_{t+\alpha}\Omega_+ - U_{t+\alpha})\phi\|dt$$

$$\leq \int d\alpha|\tilde{S}_\rho(\alpha)|\ (\int_0^\infty \|(V_t\Omega_+ - U_t)\phi\|dt + \int_0^\alpha \|(V_t\Omega_+ - U_t)\phi\|dt)$$

$$\leq \int_0^\infty \|(V_t\Omega_+ - U_t)\phi\|dt\ \int d\alpha|\tilde{S}_\rho(\alpha)| + 2\|\phi\|\int d\alpha|\alpha||\tilde{S}_\rho(\alpha)| < \infty\ .$$

One shows in the same way that $\|P_\Sigma S\phi\| \in L^1(-\infty,\infty)$ when $\|P_\Sigma\phi_t\| \in L^1(-\infty,\infty)$.

In potential scattering with spherical symmetry, the hypothesis of prop. 2 hold for a large class of potentials.

Proposition 3: Let $H_o = -\frac{1}{2m}\Delta$, $H = H_o + V$, V spherically symmetric with $V(\underline{x}) \in L^2_{loc}(R^3)$ and $V(r) = O(\frac{1}{r^{5/2+\varepsilon}})$, $\varepsilon > 0$, $r \to \infty$. Then (15a,b) and (16) hold true for the dense set of ϕ belonging to $C^\infty(R^3)$ and such that $\hat{\phi}(0) = $
$= \frac{1}{(2\pi)^{3/2}} \int \phi(y) d^3y = 0$.

Proof: We show (15a). The standard Jauch-Cook estimate gives

$$||(V_t^* U_t - \Omega_-)\phi|| = ||\int_{-\infty}^{t} \frac{d}{ds} V_s^* U_s \phi ds|| \leq \int_{-\infty}^{t} ds ||VU_s\phi|| \, . \tag{17}$$

Using again the explicit form of the free propagator and the fact that $\hat{\phi}(0) = 0$ we get

$$||VU_t\phi||^2 = \left(\frac{m}{2\pi|t|}\right)^3 \int d^3x |V(\underline{x})|^2 \left|\int d^3y \left(e^{\frac{im|\underline{x}-\underline{y}|^2}{2t}} - 1\right)\phi(\underline{y})\right|^2$$

$$= \left(\frac{m}{2\pi|t|}\right)^3 \left(\frac{m}{2|t|}\right)^{2\mu} \int d^3x |V(\underline{x})|^2 \left|\int d^3y \left(\frac{e^{\frac{im|\underline{x}-\underline{y}|^2}{2t}}}{\left(\frac{m|\underline{x}-\underline{y}|^2}{2t}\right)^{\mu}}\right) |\underline{x}-\underline{y}|^{2\mu}\phi(\underline{y})\right|^2$$

$$\leq \frac{C_1}{|t|^{3+2\mu}} \int d^3x |V(\underline{x})|^2 \int d^3y (|\underline{x}|^{2\mu} + |\underline{y}|^{2\mu}) |\phi(\underline{y})|^2$$

$$\leq \frac{C_2}{|t|^{3+2\mu}} ||(1 + |\underline{x}|^{2\mu})V||_2^2 \tag{18}$$

where we have used $|\frac{e^{ix}-1}{x^\mu}| \leq 2$ and $|\underline{x}-\underline{y}|^{2\mu} \leq 2(|\underline{x}|^{2\mu} + |\underline{y}|^{2\mu})$ for $0 \leq \mu \leq 1$. Choosing $\mu = \frac{1}{2} + \frac{\varepsilon}{4}$, $\varepsilon>0$, $||(1 + |\underline{x}|^{1+\varepsilon/2})V||_2$ is finite with our hypothesis on V; (17) combined with (18) gives (15a). The proof of (15b) is identical provided that $S(\lambda)$ has the property (16). (16) results in turn of a partial wave analysis.

If we choose for Σ a sphere of radius R and the

potential is central, all quantities involved in the definition of the sojourn time are reduced by the rotation group. Therefore, we can apply the above analysis to the restriction of the scattering system to each subspace $H_{\ell m}$, thus showing the finiteness of sojourn time for the class of potentials defined in prop. 3.

b) The Abstract Method

The abstract method relies on the fact that certain operators belong to the trace class. The basic tool is the following theorem.

Theorem 1: Let U_t be a unitary group with self adjoint generator H_o having absolutely continuous spectrum Λ, and Γ a trace class operator. Then

(i) $\int_{-\infty}^{\infty} dt(f,U_t^* \Gamma U_t g) < \infty$ for f, g belonging to a dense set D in H.

(ii) There exists an essentially unique family of trace class operators Γ_λ acting on the components H_λ of the direct integral $H = \int^{\oplus} H_\lambda \, d\lambda$ which diagonalizes H_o, such that

$$\int_\Lambda (\phi_\lambda, \Gamma_\lambda \psi_\lambda) d\lambda = \int_{-\infty}^{\infty} (\phi, U_t^* \Gamma U_t \psi) dt , \tag{19}$$

$$\frac{1}{2\pi} \int_\Lambda d\lambda \; Tr_\lambda \; \Gamma_\lambda = Tr \; \Gamma ,$$

$$\frac{1}{2\pi} \int_\Lambda \|\Gamma_\lambda\|_{\lambda,1} \, d\lambda \leq \|\Gamma\|_1 \quad . \tag{20}$$

Here Tr, Tr_λ, $\|\cdot\|_1$, $\|\cdot\|_{\lambda,1}$ are respectively the trace and tracenorm in H and in H_λ.

Proof: We verify the theorem when $\Gamma\phi = (f,\phi)g$ is of rank one. By the definition of the direct integral, ϕ is represented by a family $\{\phi_\lambda \in H_\lambda, \lambda \in \Lambda\}$ and $(\phi,\psi) =$

$\int_\Lambda (\phi_\lambda,\psi_\lambda)_\lambda \, d\lambda$ where $(\phi_\lambda,\psi_\lambda)_\lambda$ is the scalar product in H_λ; thus we have

$$(U_t f,g) = \int_\Lambda d\lambda \; e^{i\lambda t}(f_\lambda,g_\lambda)_\lambda \quad .$$

Let D be the set of ϕ such that $\|\phi_\lambda\|$ is a bounded function of λ. Then, for $\phi \in D$, $|(\phi_\lambda,g_\lambda)| \leq \|\phi_\lambda\| \, \|g_\lambda\| \leq C\|g_\lambda\|$ is a square integrable function of λ (since $\int_\Lambda \|g_\lambda\|^2 d\lambda = \|g\|^2 < \infty$).

Therefore

$$\int_{-\infty}^{\infty} (\phi,U_t^* \Gamma U_t \psi)dt = \int_{-\infty}^{\infty} (U_t\phi,g)(f,U_t\psi)dt$$

$$= 2\pi \int_\Lambda (\phi_\lambda,g_\lambda)_\lambda (f_\lambda,\psi_\lambda)_\lambda \; d\lambda$$

$$= \int_\Lambda (\phi_\lambda,\Gamma_\lambda\psi_\lambda)_\lambda \; d\lambda < \infty \; .$$

The second equality follows from Parseval's relation for the scalar product of L^2-function, thus showing (i). The third equality results of the definition $\Gamma_\lambda\psi_\lambda = 2\pi(f_\lambda,\psi_\lambda)g_\lambda$ on H_λ, establishing (19). Γ_λ being itself of finite rank on H_λ is trace-class, and (20) follows from the identity

$$\mathrm{Tr}\,\Gamma = (f,g) = \int_\Lambda (f_\lambda,g_\lambda)_\lambda \; d\lambda = \frac{1}{2\pi}\int_\Lambda \mathrm{Tr}_\lambda \; \Gamma_\lambda \; d\lambda \; .$$

The proof is completed by representing general trace class operators as limit of finite rank operators.

We apply now part (i) of the theorem to the sojourn time $T_\Sigma(\phi)$. Notice on (13) that $T_\Sigma(\phi)$ would be finite if $\Omega_-^* P_\Sigma \Omega_-$ were trace class. This is not the case, but we have the following lemma.

<u>Lemma 1:</u> Let $H_o = -\frac{\Delta}{2m}$, $H = H_o + V$ with V relatively bounded with respect to H_o. Then

(i) $R_o^*(z)\Omega_-^* P_\Sigma \Omega_- R_o(z)$ is trace class with $R_o(z)=(H_o-z)^{-1}$, $\mathrm{Im}\; z \neq 0$.

(ii) $E_o(\Delta)\Omega_-^* P_\Sigma \Omega_- E_o(\Delta)$ is trace class, where $E_o(\Delta)$ is a spectral projection of H_o on a bounded subset Δ of Λ.

Proof: In momentum representation $P_\Sigma R_o(z)$ has the kernel

$$(\underline{k}_1 | P_\Sigma R_o(z) | \underline{k}_2) = P_\Sigma(\underline{k}_1 - \underline{k}_2)\left(\frac{k_2^2}{2m} - z\right)^{-1} \quad \text{with}$$

$$P_\Sigma(\underline{k}) = \frac{1}{(2\pi)^{3/2}} \int_\Sigma e^{i\underline{k}\cdot\underline{x}} \, d^3x \quad .$$

Clearly $\int |P_\Sigma(\underline{k})|^2 d^3k = |\Sigma|$ by Parseval identity. This implies that the Hilbert-Schmidt norm $\int d^3k_1 \int d^3k_2 |(\underline{k}_1|P_\Sigma R_o(z)|\underline{k}_2)|^2$ of $P_\Sigma R_o(z)$ is finite. By the resolvent identity for $R(z) = (H-z)^{-1}$, $P_\Sigma R(z) = P_\Sigma R_o(z) - P_\Sigma R_o(z) V R(z)$ is also Hilbert-Schmidt since $VR(z)$ is bounded. Therefore, $R^*(z) P_\Sigma R(z) = (P_\Sigma R(z))^* P_\Sigma R(z)$ being the product of two Hilbert-Schmidt is trace class. We obtain (i) from the intertwining relation and the fact that Ω_- is bounded.

We can write $E_o(\Delta) = E_o(\Delta)(H_o - z)R_o(z)$ in (ii) and obtain the result from (i) and the fact that $E_o(\Delta)(H_o-z)$ is bounded for compact Δ.

The lemma shows immediately that $T_\Sigma(\phi)$ is finite for any ϕ with compact support on Λ, since then $E_o(\Delta)\phi = \phi$ for some bounded Δ, and the integrand in (13) is of the form $(\phi, U_t^* \Gamma U_t \chi)$ with $\Gamma = E_o(\Delta)\Omega_-^* P_\Sigma \Omega_- E_o(\Delta)$ being trace class.

One should emphasize that this proof is very general because it assumes nothing more on V that H_o-boundedness and existence of wave operators. It implies in particular that sojourn time of Coulomb scattering states are finite for a dense set of incoming states. Moreover, this method applies as well to non-local and non-central interactions.

c) H-Smoothness

The concept of H-smoothness is expressed by the following definition. Let $E(\Delta)$ be a spectral projection of a selfadjoint operator H on $\mathcal{H}$. Then a bounded operator B on $\mathcal{H}$ is H-smooth on Δ if

$$\int_{-\infty}^{\infty} \|Be^{-iHt}\psi\|^2 dt \le C\|\psi\|^2 \text{ for all } \psi \in E(\Delta)\mathcal{H} .$$

Thus finitness of $T_\Sigma(\phi)$ follows if one can show that P_Σ is H-smooth on the absolutely continuous subspace of $\mathcal{H}$.

A relevant theorem for us is

Theorem 2: Assume that the potential $V(\underline{x})$ is bounded and $V(\underline{x}) = O(\frac{1}{|\underline{x}|^{1+\varepsilon}})$. Let $F(\underline{x})$ be a bounded function which is $O(\frac{1}{|\underline{x}|^{1+\varepsilon}})$, $|\underline{x}| \to \infty$; then $|F(\underline{x})|^{1/2}$ is H-smooth on finite intervals Δ in the absolutely continuous spectrum of H.

The theorem shows in particular that P_Σ is H-smooth on $\mathcal{H}_{abs}$ for potentials decaying faster than Coulomb. The proof of H-smoothness requires to work out detailed commutator estimates. When it can be applied, this technique gives explicit upperbounds on the sojourn time in terms of the potential.

d) Energy Shell Sojourn Time

The definition (13) of $T_\Sigma(\phi,\chi)$ and the group property of U_t show that $T_\Sigma(U_\tau\phi,U_\tau\chi) = T_\Sigma(\phi,\chi)$ for all τ, i.e. the quadratic form $T_\Sigma(\phi,\chi)$ commutes with the free evolution. This implies that the sojourn time has energy shell components $T_\Sigma(\lambda)$, $\lambda \in \Lambda$, on $\mathcal{H}_\lambda$ such that $T_\Sigma(\phi,\chi) = \int_\Lambda d\lambda\,(\phi_\lambda, T_\Sigma(\lambda)\chi_\lambda)_\lambda$. In fact, the existence of $\{T_\Sigma(\lambda)\}$ is insured by part (ii) of theorem 1 (choosing again ϕ of compact support on Λ and $\Gamma = E_o(\Delta)\Omega_-^* P_\Sigma \Omega_- E_o(\Delta)$ being trace class operators as in part (ii) of lemma 1). We deduce from (20) that

$$\frac{1}{2\pi}\int_{\Delta}\mathrm{Tr}\, T_{\Sigma}(\lambda)d\lambda = \mathrm{Tr}\, E_o(\Delta)\Omega_-^{*}P_{\Sigma}\Omega_-E_o(\Delta)$$

$$= \mathrm{Tr}\, (P_{\Sigma}\Omega_-E_o(\Delta))^{*}(P_{\Sigma}\Omega_-E_o(\Delta))$$

$$= \mathrm{Tr}\, P_{\Sigma}\Omega_-E_o(\Delta)\Omega_-^{*}P_{\Sigma}$$

$$= \mathrm{Tr}\, P_{\Sigma}E(\Delta)P_{\Sigma} \quad . \tag{21}$$

We have used the cyclicity of the trace and the intertwining relation. Since $E(\Delta)$ is now a spectral projection of H, (21) gives a simple interpretation of $T_{\Sigma}(\lambda)$: $\frac{1}{2}\mathrm{Tr}_{\lambda}\, T_{\Sigma}(\lambda)$ is the density of states of H at the value λ of the total energy and lying in the subspace $P_{\Sigma}H$.

It is instructive to perform explicitely the energy shell reduction in the case of potential scattering, with $H_o = -\frac{\Delta}{2m}$. Then the direct integral decomposition of $L^2(R^3)$ with respect to H_o reads $\int_0^{\infty} H_{\lambda}^{\oplus} d_{\lambda}$, with $H_{\lambda} = L^2(S)$ for each $\lambda \geq 0$: $L^2(S)$ is the set of square integrable functions on the surface of the unit sphere S. For any $\phi \in L^2(R^3)$ we set $\phi_{\lambda}(\hat{k}) = (2m^3\lambda)^{1/4}\phi(\underline{k}) \in L^2(S)$, $\hat{k} \in S$ for fixed $\lambda = \frac{|\underline{k}|^2}{2m}$, and verify that $(\phi,\psi) = \int_0^{\infty}(\phi_{\lambda},\psi_{\lambda})_{\lambda}\, d\lambda$ with $(\phi_{\lambda},\psi_{\lambda})_{\lambda} = \int_S d^2k\, \phi_{\lambda}(\hat{k})\psi_{\lambda}(\hat{k})$. If Γ is an operator on $L^2(R)^3$ with continuous kernel $(\underline{k}_1|\Gamma|\underline{k}_2)$, one has

$$\int_{-\infty}^{\infty}(\phi,U_t^{*}\Gamma U_t\psi)dt = \int_0^{\infty}(\phi_{\lambda},\Gamma_{\lambda}\psi_{\lambda})d\lambda.$$

Here Γ_{λ} is the operator on $L^2(S)$ with kernel

$$(\hat{k}_1|\Gamma_{\lambda}|\hat{k}_2) = 2\pi m k_1(\underline{k}_1|\Gamma|\underline{k}_2),\ k_1 = |\underline{k}_1| = |\underline{k}_2| = \sqrt{2m\lambda}\ . \tag{22}$$

With (22), (3) and the definition (13) of the sojourn time, we can easily express the kernel of its energy shell components $T_{\Sigma}(\lambda)$ in term of the eigenfunction $\psi(\underline{k},\underline{x})$:

$$(\hat{k}_1|T_{\Sigma}(\lambda)|\hat{k}_2) = 2\pi m k_1(\underline{k}_1|\Omega_-^{*}P_{\Sigma}\Omega_-|\underline{k}_2)$$

$$= \frac{k_1 m}{(2\pi)^2} \int_{\Sigma} d^3x \ \psi^*(\underline{k}_1,\underline{x})\psi(\underline{k}_2,\underline{x}) \ . \tag{23}$$

For a spherically symmetric potential, we take for Σ a sphere of radius R and find from (23) and (4) the energy shell components $T_R^{\ell}(\lambda) = (\ell m|T_R(\lambda)|\ell m)$ in terms of the radial waves

$$T_R^{\ell}(\lambda) = \frac{4m}{k} \int_0^R |u_{\ell}(k,r)|^2 dr \ , \qquad k = \sqrt{2m\lambda} \ . \tag{24}$$

This formula will be useful to study the asymptotic behaviour of the sojourn time for large R (see Section III c).

It has also the following direct interpretation in the stationary point of view on scattering. Let $\psi_{\ell}(\underline{x}) = \frac{u_{\ell}(k,r)}{r}(\hat{x}|\ell m)$ be the stationary scattering wave at fixed angular momentum ℓ. Then

$$\int_{|\underline{x}|\leq R} |\psi(\underline{x})|^2 d^3x = \int_0^R |u_{\ell}(k,r)|^2 dr$$

is proportional to the integrated density of particles in the sphere Σ_R. Moreover, the corresponding incoming radial component of the current is for large R

$$j_R^{in}(\underline{x}) = \frac{1}{2im}\left(\psi_{\ell}^{in^*}(\underline{x})\frac{d}{dr}\psi_{\ell}^{in}(\underline{x}) - \psi_{\ell}^{in}(\underline{x})\frac{d}{dr}\psi_{\ell}^{in^*}(\underline{x})\right) \sim$$

$$\sim -\frac{k}{4m}\frac{1}{R^2}|(\hat{x}|\ell m)|^2 \ , \quad |\underline{x}| = R \ ,$$

where $\psi_{\ell}^{in}(\underline{x}) \sim -\frac{1}{2i} e^{-i(kr-\ell\pi/2)}(\hat{x}|\ell m)$ is the incoming part of $\psi_{\ell}(\underline{x})$ (see (6)). This gives an incoming flux in Σ_R which is $R^2 \int_S d^2\hat{x} j_R^{in}(\underline{x}) = \frac{k}{4m}$. Thus $T_R^{\ell}(\lambda)$ for large R can be viewed as <u>the integrated density of scattered particles in Σ_R by unit of incoming flux.</u>

III. THE TIME DELAY

The difference between the sojourn times of the scattered particles and the freely moving particles in Σ

$$\tau_{\Sigma}(\phi) = T_{\Sigma}(\phi) - T_{\Sigma}^{0}(\phi) \tag{25}$$

represents the delay in Σ caused by the interaction. $\tau_{\Sigma}(\phi)$ is therefore called <u>the time delay for the region Σ and the incoming state ϕ.</u> As Σ approaches the whole space R^3 both $T_{\Sigma}(\phi)$ and $T_{\Sigma}^{0}(\phi)$ diverge. However, it is expected that their difference $\tau_{\Sigma}(\phi)$ has a limit as $\Sigma \to R^3$ when the interaction is sufficiently short-ranged. Thus

$$\tau(\phi) = \lim_{\Sigma\to R^3} \tau_{\Sigma}(\phi) \tag{26}$$

will be called <u>the time delay of the scattering process with incoming state ϕ</u> whenever the limit exists. We present now various methods to show the existence of the limit (26).

a) Time Dependent Method

One uses the asymptotic condition to show that $\tau(\phi)$ (when it exists) <u>depends only on the S-operator.</u>

<u>Proposition 4:</u> Assume that $\|\psi_t-\phi_t\| \in L^1(-\infty,0)$, $\|P_{\Sigma}S\phi_t\|$ and $\|\psi_t-S\phi_t\| \in L^1(0,\infty)$; then

$$\tau(\phi) = \lim_{\Sigma\to R^3} \int_0^{\infty} (\phi_t,(S^*P_{\Sigma}S-P_{\Sigma})\phi_t)dt$$

whenever the limit exists.

<u>Proof:</u>

$$\tau_{\Sigma}(\phi) = \int_{-\infty}^{0}((\psi_t,P_{\Sigma}\psi_t)-(\phi_t,P_{\Sigma}\phi_t))dt+\int_0^{\infty}((\psi_t,P_{\Sigma}\psi_t)-(S\phi_t,P_{\Sigma}S\phi_t))dt$$

$$+ \int_0^{\infty}((S\phi_t,P_{\Sigma}S\phi_t) - (\phi_t,P_{\Sigma}\phi_t))dt \quad . \tag{27}$$

The integrands of the two first terms of (27) vanish as $P_\Sigma \to I$. Moreover, they are majorized by integrable functions uniformly in Σ for $t \leq 0$ and $t \geq 0$ respectively since

$$|(\psi_t,P_\Sigma\psi_t)-(\phi_t,P_\Sigma\phi_t)| \leq |((\psi_t-\phi_t),P_\Sigma\phi_t)|+|(\psi_t,P_\Sigma(\psi_t-\phi_t))| \leq 2\|\psi_t-\phi_t\|$$

and similarly

$$|(\psi_t,P_\Sigma\psi_t) - (S\phi_t,P_\Sigma S\phi_t)| \leq 2\|\psi_t - S\phi_t\| \quad .$$

The result follows from dominated convergence.

Now we relate $\tau(\phi)$ to the energy derivative of the S-operator.

<u>Proposition 5:</u> Let ϕ be of compact support on Λ and assume all the hypothesis of prop. 2. Then $\lim_{\Sigma\to R^3} \tau_\Sigma(\phi) = \tau(\phi)$ exists and

$$\tau(\phi) = -i(\phi, S^* \frac{dS}{dH_o} \phi) \quad .$$

<u>Proof:</u> Notice that by prop. 2, the result of prop. 4 is true. Choosing $\rho(\lambda)$ as in prop. 2 and writing again $S_\rho = \int \tilde{S}_\rho(\alpha) U_\alpha \, d\alpha$ we have

$$\tau(\phi) = \lim_{\Sigma\to R^3} \int_o^\infty (S\phi_t,[P_\Sigma,S]\phi_t)dt$$

$$= \lim_{\Sigma\to \mathbf{R}^3} \int_o^\infty dt \int d\alpha \, \tilde{S}_\rho(\alpha)((S\phi_t,P_\Sigma\phi_{t+\alpha}) - (S\phi_{t-\alpha},P_\Sigma\phi_t)) \qquad (28)$$

$$= - \lim_{\Sigma\to R^3} \int d\alpha \, \tilde{S}_\rho(\alpha) \int_{-\alpha}^{o} (S\phi_t,P_\Sigma\phi_{t+\alpha})dt \quad . \qquad (29)$$

The exchange of the α and t integrals is possible because the integrand in (28) is majorized by the integrable function $|\tilde{S}_\rho(\alpha)|(\|\phi\|\|P_\Sigma\phi_t\| + \|\phi\|\|P_\Sigma S\phi_t\|)$ in both variables. The last line results of the change of variable $t - \alpha \to t$ in the second integral. Taking now the limit $P_\Sigma \to I$ in (29) we get by dominated convergence (the integrand of (29) is majorized by $|\alpha\tilde{S}_\rho(\alpha)|\,\|\phi\|^2$)

$$\tau(\phi) = -(S\phi, \int d\alpha\ \alpha\ S_\rho(\alpha)\ U_\alpha\phi)$$

$$= -i(S\phi, \frac{dS_\rho}{dH_o}\phi) = -i(\phi, S^* \frac{dS}{dH_o}\phi) \quad . \tag{30}$$

If the potential is central and $\{\Sigma\}$ is a sequence of spheres centered at the origin, we can apply prop. 4 and 5 to the restriction of the scattering system to each subspace $H_{\ell m}$ with fixed angular momentum. Therefore, the conditions on the potential under which this derivation can be performed are exactly the same as those of prop. 3, i.e. potentials decaying as $O(\frac{1}{r^{5/2+\varepsilon}})$. The energy shell components of τ are then obviously $\tau_\ell(\lambda) = -iS_\ell^*(\lambda)\frac{d}{d\lambda} S_\ell(\lambda) = \frac{d}{d\lambda}\delta_\ell(\lambda)$, which is the Eisenbud formula (9).

b) Abstract Method

The abstract method is based on the condition that the difference of the total and the free resolvent $R_z - R_{oz} = (H-z)^{-1} - (H_o-z)^{-1}$ belongs to the trace class of operators.

It enables to treat non-spherically symmetric potentials and also non-local interactions. One uses in an essential way the theory of the spectral displacement function due to Krein and Birman. We summarize it in the two following theorems. The first one gives an integral representation of $\mathrm{Tr}(R_z - R_{oz})$.

Theorem 3: Let (H, H_o) be a pair of selfadjoint operators on H with $R_z - R_{oz}$ trace class, $\mathrm{Im}\ z \neq 0$. Then there exists a function $\xi(\mu)$ on $\mathbb{R}$ such that

(i) $\xi(\mu)(1+\mu^2)^{-1} \in L^1(-\infty,\infty)$

(ii) $\mathrm{Tr}\,(R_z - R_{oz}) = \int_{-\infty}^{\infty} \frac{\xi(\mu)}{(\mu-z)^2}\, d\mu$.

$\xi(\mu)$ is called the spectral displacement function for the pair (H,H_o).

The second theorem connects $\xi(\mu)$ with the phase of the S-operator.

Theorem 4: Under the hypothesis of theorem 3, the components S_λ of S in the direct integral $H = \int_\Lambda^\oplus H_\lambda d\lambda$ which diagonalizes H_o are of the form $S_\lambda = I_\lambda + G_\lambda$ with G_λ trace class on H_λ. Furthermore, one has $\mathrm{Tr}_\lambda \Delta(\lambda) = 2\pi\xi(\lambda)$, $\lambda \in \Lambda$ with $\Delta(\lambda) =$ $= -i \log S_\lambda = -i \log (I_\lambda + G_\lambda)$.

It should be noted that since G_λ is trace class, $\Delta(\lambda)$ is a well defined trace class operator on H_λ and $S_\lambda = e^{i\Delta(\lambda)}$. For this reason $\Delta(\lambda)$ is called the phase shift operator. We know from the results of section (IId) that we have energy shell time delay operators $\tau_\Sigma(\lambda) = T_\Sigma(\lambda) - T^o_\Sigma(\lambda)$ which are of trace class for bounded regions Σ. Then the main result obtained with the help of theorems 2 and 3 is

Proposition 6: Assume that $R_z - R_{oz}$ belongs to the trace class and that the subspace of bound states of H is finite dimensional. Then

$$\lim_{\Sigma\to R^3} \mathrm{Tr}_\lambda\, \tau_\Sigma(\lambda) = \frac{d}{d\lambda} \mathrm{Tr}_\lambda\, \Delta(\lambda) \quad .$$

The limit and the derivative have to be understood in the sense of distributions.

Sketch of proof: In order to single out the contribution of bound states, we write $R_z = R_z^b + R_z^{ac} = (H^b - z)^{-1} +$ $+ (H^{ac} - z)^{-1}$ where R_z^b and R_z^{ac} are the restriction of R_z to the subspaces H_b, H_{ac} of bound states and of absolute continuity of H. From the part (i) of lemma 1 and (20) in part (ii) of theorem 1, we find for $\mathrm{Im}\, z \neq 0$

$$\frac{1}{2\pi} \int_{\Lambda} \frac{\mathrm{Tr}_\lambda \tau_\Sigma(\lambda)}{(\lambda-z)^2} d\lambda = \mathrm{Tr}(R_{oz}\Omega_-^* P_\Sigma \Omega_- R_{oz}) - \mathrm{Tr}(R_{oz}P_\Sigma R_{oz}) =$$

$$= \mathrm{Tr}(\Omega_- R_z P_\Sigma R_z \Omega_- - R_{oz}P_\Sigma R_{oz}) = \mathrm{Tr}(R_z^{ac}P_\Sigma R_z^{ac} - R_{oz}P_\Sigma R_{oz}). \quad (31)$$

To obtain (31) have used the intertwining relation and the fact that Ω_- maps H isometrically onto H_{ac}.

The assumption of finite dimensionality of H_b implies that $R_z^{ac} - R_{oz}$ is still trace class, and we will apply theorems 2 and 3 to the pair (H^{ac},H_o) (because of the asymptotic completness (H^{ac},H_o) defines the same scattering operator as (H,H_o)). Since $R_z^{ac} - R_{oz}$ is trace class and $P_\Sigma \to I$ strongly, the limit of the right hand side of (31) exists (indeed $A_n \to A$ strongly and B trace class implies $A_nB \to AB$ in the trace norm) and we have

$$\lim_{\Sigma\to R^3} \frac{1}{2\pi} \int \frac{\mathrm{Tr}_\lambda\ \tau_\Sigma(\lambda)}{(\lambda-z)^2} d\lambda = \mathrm{Tr}((R_z^{ac})^2-(R_{oz})^2) = \frac{d}{dz}\mathrm{Tr}(R_z^{ac}-R_{oz}) \quad (32)$$

(exchange of trace and derivative is allowed by the trace class property).

Now it follows from theorem 2 that

$$\frac{d}{dz}\mathrm{Tr}(R_z^{ac}-R_{oz}) = \frac{d}{dz}\int_{-\infty}^{\infty} d\mu \frac{\xi(\mu)}{(\mu-z)^2} = -\int_{-\infty}^{\infty} d\mu\, \xi(\mu) \frac{d}{d\mu}\frac{1}{(\mu-z)^2} \quad . \quad (33)$$

Combining (32) and (33) leads to

$$\lim_{\Sigma\to R^3} \int \frac{\mathrm{Tr}_\lambda\ \tau_\Sigma(\lambda)}{(\lambda-z)^2} d\lambda = -2\pi \int_{-\infty}^{\infty} d\lambda\ \xi(\lambda) \frac{d}{d\lambda}\frac{1}{(\lambda-z)^2} \quad . \quad (34)$$

A technical step consists in showing that (34) holds for any C^∞ function $g(\lambda)$ in place of $\frac{1}{(\lambda-z)^2}$, i.e.

$$\lim_{\Sigma\to R^3} \int \mathrm{Tr}_\lambda \tau_\Sigma(\lambda) g(\lambda) d\lambda = -2\pi \int_{-\infty}^{\infty} \xi(\lambda) \frac{d}{d\lambda} g(\lambda) d\lambda .$$

Choose now $g(\lambda)$ with support in Λ. Since $2\pi\xi(\lambda) = \mathrm{Tr}_\lambda \Delta(\lambda)$ for $\lambda \in \Lambda$ by theorem 3, this gives the result of the proposition.

(30) generalizes the Eisenbud-Wigner in several respects: one can treat non-spherically symmetric potentials and even non-local interactions; in all cases, the total time-delay is related to the derivative of the average phase shift $\mathrm{Tr}_\lambda\ \Delta(\lambda)$. Moreover, a smooth energy dependence of S_λ is not assumed (this why (30) holds only in the sense of distributions). In potential scattering $R_z - R_{oz}$ belongs to the trace class if $V(\underline{x}) \in L^1 \cap L^2$, corresponding to a fall off $V(\underline{x}) = O(\frac{1}{|x|^{3+\varepsilon}})$ at infinity.

c) Stationary Method

It is interesting to investigate the asymptotic behaviour of the sojourn time $T_\Sigma(\phi)$ for large spatial regions Σ. This can easily be done starting from the expression (24) for the energy shell components $T_R^\ell(\lambda)$ in term of radial waves (the present considerations apply to potential scattering with spherical symmetry and $\Sigma = \Sigma_R$ = sphere of radius R).

Differentiating the radial equation (5) with respect to energy gives

$$- \frac{d^2}{dr^2} \frac{d}{d\lambda} u(k,r) = (2m(\lambda - V(r)) - \frac{\ell(\ell+1)}{r^2}) \frac{du(k,r)}{d\lambda} + 2mu(k,r) . \quad (35)$$

Combining (5) and (35), one finds the identity

$$|u(k,r)|^2 = -\frac{1}{2m}\frac{d}{dr}(u^*(k,r)\frac{d^2u(k,r}{drd\lambda} - \frac{du(k,r)}{d\lambda}\frac{du^*(k,r)}{dr}) .$$

Since u(k,0) = 0, (u(k,r) is the regular solution of the radial equation) we have

$$T_R^\ell(\lambda) = \frac{4m}{k}\int_0^R |u(k,r)|^2 dr = \frac{2}{k}(\frac{du}{d\lambda}\frac{du^*}{dr} - u^* \frac{d^2u}{drd\lambda})|_{r=R} . \quad (36)$$

For large R (36) can be calculated explicitely using the asymptotic form (6) of u(k,r) with the result

$$T_R^\ell(\lambda) = \frac{2R}{v} - iS_\ell^*(\lambda)\frac{d}{d\lambda}S_\ell(\lambda) - \frac{1}{2\lambda}\sin(2kR-\ell\pi+\delta_\ell(\lambda)) + o(1) ; \quad (37)$$

$v = \frac{k}{m}$ is the radial velocity.

For the free sojourn time we have obviously the same expression with

$$T_R^{o\ell}(\lambda) = \frac{2R}{v} - \frac{1}{2\lambda}\sin(2kR - \ell\pi) + o(1) . \quad (38)$$

The divergent term $\frac{2R}{v}$ in (37) and (38) is the same: it represents simply <u>the time needed by a classical particle to cross Σ_R along its diameter.</u> Therefore, we find again that $\tau_R^\ell(\lambda) = T_R^\ell(\lambda) - T_R^{o\ell}(\lambda)$ converges weakly as $R \to \infty$ to $\tau^\ell(\lambda) = \frac{d}{d\lambda}\delta_\ell(\lambda)$, i.e.

$$\lim_{R\to\infty} \int \tau_R^\ell(\lambda)g(\lambda)d\lambda = \int \tau^\ell(\lambda)g(\lambda)d\lambda$$

for smooth functions $g(\lambda)$, since the oscillating terms in (37) and (38) do not contribute to the weak limit because of the Riemann Lebesgue lemma.

Notice that according to the discussion of section (IId), $\tau_R^\ell(\lambda)$ can be viewed as the <u>excess or defect of the particle density in the interaction region by unit of incoming flux.</u>

d) Geometrical Aspects of Time Delay

We have based our study of time delay on the concept of sojourn time of the wave function in some large spatial region. However, there is an alternative definition of time delay, in close analogy to the classical one, which can be found by purely geometrical considerations. To motivate this definition, we recall the classical construction of time delay.

Let $\{\underline{x}(t),\underline{p}(t)\}$ be a classical scattering trajectory and Σ_R a sphere of radius R containing the support of the interaction (assuming that the potential is of finite range for simplicity):

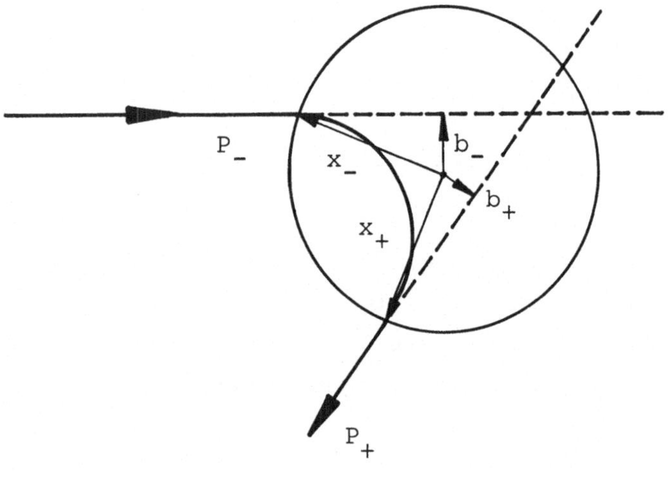

Fig. 1.

Denote by $-t_-(t_+)$ the times at which the particle enters (leaves) the sphere Σ_R, $\underline{x}_- = \underline{x}(-t_-)$ $(\underline{x}_+ = \underline{x}(t_+))$ the corresponding positions and $\underline{p}_- = \underline{p}(-t_-)$ $(\underline{p}_+ = \underline{p}(t_+))$ the incoming (outgoing) momentum. (We take the particle to be in Σ_R at $t = 0$ so that t_-, $t_+ \geq 0$.) If we define $\tilde{\underline{x}}(t) = \underline{x}(t) - \frac{\underline{p}(t)}{m} t$, we have

$$\tilde{\underline{x}}_- = \underline{x}_- + t_- \frac{\underline{p}_-}{m} , \quad \tilde{\underline{x}}_+ = \underline{x}_+ - t_+ \frac{\underline{p}_+}{m} \quad . \tag{39}$$

Multiplying these equations scalarly by $\underline{p}_-$ ($\underline{p}_+$) and solving for t_- and t_+ gives the sojourn time $T_R = t_- + t_+$ in terms of the parameters of the trajectory,

$$T_R = m\left(\frac{\tilde{\underline{x}}_- \cdot \underline{p}_-}{|\underline{p}_-|^2} - \frac{\tilde{\underline{x}}_+ \cdot \underline{p}_+}{|\underline{p}_+|^2}\right) + m\left(-\frac{\underline{x}_- \cdot \underline{p}_-}{|\underline{p}_-|^2} + \frac{\underline{x}_+ \cdot \underline{p}_+}{|\underline{p}_+|^2}\right) \quad . \tag{40}$$

Let $\underline{b}_-$ ($\underline{b}_+$) be the impact parameters corresponding to the incoming (outgoing) free parts of the trajectory; then one sees on Fig. 1, that

$$\underline{x}_- \cdot \underline{p}_- = -|\underline{p}_-|\sqrt{R^2 - b_-^2} \ , \quad \underline{x}_+ \cdot \underline{p}_+ = |\underline{p}_+|\sqrt{R^2 - b_+^2}$$

so that the second term of (40) behaves as $\frac{2Rm}{|\underline{p}|}$ as $R \to \infty$ ($|\underline{p}_-| = |\underline{p}_+| = |\underline{p}|$ by energy conservation). Since a free particle would spend the time $T_R^o = \frac{2m}{|\underline{p}|} \sqrt{R^2 - b_-^2} \sim \frac{2mR}{|\underline{p}|}$ in Σ_R, we get for the time delay

$$\tau = \lim_{R\to\infty} (T_R - T_R^o) = \frac{m}{|\underline{p}|^2} \left(\lim_{t\to-\infty} \tilde{\underline{x}}(t) \cdot \underline{p}(t) - \lim_{t\to+\infty} \tilde{\underline{x}}(t) \cdot \underline{p}(t)\right) \tag{41}$$

whenever the limits exist.

We now form the quantum mechanical analogue of the expression (41). If $\underline{x}$ and $\underline{p}$ are the position and momentum operators of the particle at $t = 0$, $H_o = \frac{|\underline{p}|^2}{2m}$, $H = H_o + V$, we will have

$$\underline{x}(t) = V_t^* \, \underline{x} \, V_t \ ,$$

$$\tilde{\underline{x}}(t) = V_t^*(\underline{x} - \frac{\underline{p}}{m}t)V_t = V_t^* U_t \underline{x} U_t^* V_t ,$$

$$\underline{p}(t) = V_t^* \underline{p} V_t = V_t^* U_t \underline{p} U_t^* V_t \quad . \tag{42}$$

Therefore, the appropriate quantum mechanical observable corresponding to $\tilde{\underline{x}}(t)\cdot\underline{p}(t)$ is

$$\tilde{D}(t) = V_t^* \, U_t \, D \, U_t^* \, V_t$$

where we take for D the hermitian combination $D = \frac{1}{2}(\underline{x}\cdot\underline{p}+\underline{p}\cdot\underline{x})$.

Finally (41) leads to the following quantum mechanical definition of τ for an incoming state ϕ:

$$(\phi, 2H_o\tau\phi) = \lim_{t\to-\infty} (\Omega_-\phi, \tilde{D}(t)\Omega_-\phi) - \lim_{t\to\infty} (\Omega_-\phi, \tilde{D}(t)\Omega_-\phi) \quad . \tag{43}$$

Since the operators $\underline{x}(t)$ and $\underline{p}(t)$ (42) evolve in the Heisenberg picture, we have taken their average in (43) in the scattering state $\psi = \Omega_-\phi$ at time $t = 0$. D is the generator of the dilation group on $L^2(R^3)$:

$$e^{i\alpha D}\underline{p}e^{-i\alpha D} = e^{-\alpha}\underline{p}, \quad e^{i\alpha D}\underline{x}e^{-i\alpha D} = e^{\alpha}\underline{x} \quad . \tag{44}$$

Therefore, (43) shows that τ is closely related to the asymptotic form of $\tilde{D}(t)$ as $t \to \pm\infty$.

It remains to establish that (43) is identical with the Eisenbud-Wigner formula. This is easily done by using the properties of the dilation group. First we get from the asymptotic condition (1a,b)

$$(\phi, 2H_o\tau\phi) = (\phi, (D-S^*DS)\phi) = (\phi, S^*[S,D]\phi) \quad . \tag{45}$$

Assuming for simplicity that H_o has simple spectrum so that $S = S(H_o)$ (i.e. working in a subspace of fixed angular momentum for spherically symmetric potential), and $S(H_o)$

differentiable, we can compute

$$[S(H_o),D] = i \frac{d}{d\alpha} (e^{i\alpha D} S(H_o) e^{-i\alpha D})|_{\alpha = 0}$$

$$= i \frac{d}{d\alpha} S(e^{-2\alpha} H_o) = -2i\, H_o \frac{dS(H_o)}{dH_o} \quad .$$

When this is introduced in (45), one recovers the usual formula for τ.

IV. QUALITATIVE PROPERTIES OF TIME DELAY

a) Causal Bounds

Classical causality says that <u>the particle cannot leave the interaction region before entering it:</u> for a finite range interaction with support in the sphere Σ_a, we must have $T_a \geq 0$, T_a given by (40). Thus causality implies (see preceeding section)

$$\tau = T_a - T_a^o \geq - T_a^o = - \frac{2\sqrt{a^2-b^2}}{v} \geq - \frac{2a}{v} \quad . \tag{46}$$

The corresponding quantum mechanical causality principle is simply the fact that the sojourn time $T_R(\phi)$ (11) is non-negative. From this principle, it is also possible to work out lower bounds for the quantum mechanical time delay. The situation is particularly simple in the case of a s-wave ($\ell = 0$ angular momentum) when the potential has finite range. Indeed, since $V(r) = 0$, $r \geq a$, the s-radial function $u(k,r)$ will coincide with its asymptotic form for all $r \geq a$ (with $r = a$ included by the continuity of $u(k,r)$ in a),

$$u(k,r) = \frac{1}{2i} (S_o(\lambda) e^{ikr} - e^{-ikr}) \; ; \; r \geq a \quad . \tag{47}$$

Therefore, the evaluation (36), (37) of $T_a(\lambda)$ is not only asymptotic, but exact, giving now

$$T_a(\lambda) = \frac{2a}{v} + \tau(\lambda) - \frac{1}{2\lambda}\sin(2ka + \delta_o(\lambda)) \geq 0$$

and therefore

$$\tau(\lambda) \geq -\frac{2a}{v} + \frac{1}{2\lambda}\sin\,(2ka + \delta_o(\lambda)) \quad . \qquad (48)$$

There is an additional term in (48) compared to the classical bound (46). This term is small at high energy when the particle hardly sees the scatterer.

For higher angular momentum similar lower bounds can be derived. They are less simple because of the more complicated form of the radial wave in the external region $r \geq a$ (although known in terms of spherical Bessel functions). However, when the potential has not finite range, no simple lower bounds have been found.

b) Connexion with the Virial

In order to discuss the sign and the magnitude of the time delay caused by a given potential (i.e. to know if the interaction is responsible for advancement or retardation), it is useful to express τ in terms of the potential itself. This can be done by relating τ to the virial $V(\underline{x})+\frac{1}{2}\underline{x}\cdot\underline{\nabla}V(\underline{x})$, starting from the definition of τ given in section (IIId). With the notations of section (IIId), one has the following operator identity:

$$\tilde{D}(-t) - \tilde{D}(t) = D(-t) - D(t) + \left(\frac{|\underline{p}(-t)|^2}{m} + \frac{|\underline{p}(t)|^2}{m}\right)t \qquad (49)$$

and

$$D(-t) - D(t) = -\int_{-t}^{t} \frac{d}{ds}\, D(s)ds$$

$$= - i \int_{-t}^{t} V_t^{*}[H,D]\, V_t \, dt$$

$$= -4Ht + \int_{-t}^{t} V_s^{*}(2V + \underline{x}\cdot\underline{\nabla}V)V_s ds \quad . \tag{50}$$

The last line results of

$$[\frac{|\underline{p}|^2}{2m} + V(\underline{x}),D] = - i \left(\frac{|\underline{p}|^2}{m} - \underline{x}\cdot\underline{\nabla}V(\underline{x})\right) \quad .$$

When (49) and (50) are inserted in (43), one gets

$$(\phi,2H_o\tau\phi) = \lim_{t\to\infty} 2 \int_{-t}^{t} (\psi_s,(2V + \underline{x}\cdot\underline{\nabla}V)\psi_s)ds$$

$$+ \lim_{t\to\infty} 2t((\psi_{-t},H_o\psi_{-t})+(\psi_t,H_o\psi_t)-2(\phi,H_o\phi)) \tag{51}$$

where $\psi_t = V_t\Omega_-\phi$ is the scattering state at time t. By the asymptotic condition (2ab), the limit of the second term vanishes and we are left with the following representation of the time delay (up to an energy factor):

$$(\phi,H_o\tau\phi) = \int_{-\infty}^{\infty} (\psi_t,(2V + \underline{x}\cdot\underline{\nabla}V)\psi_t)dt \quad . \tag{52}$$

To establish the condition of validity of this formal derivation, one has to show the existence of both limits in (51). Writing $H_o = H-V$, the second term of (51) is identical with $\lim_{t\to\infty} 2t((\psi_{-t}V\psi_{-t}) + (\psi_t,V\psi_t))$ so that both limits will exist if $(\psi_t,V\psi_t)$, $(\psi_t,\underline{x}\cdot\underline{\nabla}V\psi_t)$ are $O(\frac{1}{t^{1+\varepsilon}})$, $t \to \infty$. Since $|(\psi_t,V\psi_t)| \leq \| |V|^{1/2}V_t\Omega_-\phi\|^2$ this amounts essentially to show that $|V|^{1/2}$ and $|\underline{x}\cdot\underline{\nabla}V|^{1/2}$ are H-smooth on H_{abs}. But in view of Theorem 2 (section IIc), the latter property will hold if $V(\underline{x})$ and $\underline{x}\cdot\underline{\nabla}V(\underline{x})$ are $O(\frac{1}{|x|^{1+\varepsilon}})$, $|\underline{x}| \to \infty$.

Notice that from (52) one can write down the corresponding representation for the energy shell components of τ, as in (23), (24). For a spherically symmetric potential one finds

$$\tau^{\ell}(\lambda) = \frac{4m}{k\lambda}\int_{0}^{\infty} dr\,|u_{\ell}(k,r)|^{2}\left(V(r) + \frac{r}{2}\frac{dV(r)}{dr}\right) \quad . \tag{53}$$

(53) gives some qualitative information on the behaviour of $\tau^{\ell}(\lambda)$: $\tau^{\ell}(\lambda)$ is likely to be small if $\frac{dV(r)}{dr} < 0$, i.e. if V(r) is monotonously decreasing so that there are no potential barriers to trap the particle near the origin. In particular $\tau^{\ell}(\lambda)$ is negative all energies if $V(r) + \frac{r}{2}\frac{dV(r)}{dr} < 0$, a condition which is fulfilled by a strongly repulsive potential without barriers like $\frac{1}{r^{\gamma}}$, $\gamma>2$, and vanishes when $\gamma = 2$.

In fact $\tau^{\ell}(\lambda)$ is governed by the combined behaviour of $V(r) + \frac{r}{2}\frac{dV(r)}{dr}$ and of the radial wave in (53). Large value of $\tau^{\ell}(\lambda)$ can be produced by a pole in the analytic continuation of $S^{\ell}(\lambda)$ as indicated in section (Ic). However, when Levinson's theorem holds (64) shows that $\tau^{\ell}(\lambda)$ cannot be positive for all energies.

c) Time Delay and Coulomb Scattering

One knows that the time delay for infinite space region (26) is finite for potentials decaying faster than the Coulomb potential. In fact, the time delay is finite whenever the time integral (51) is finite. As said in the preceeding section, the integral (51) exists if $V(\underline{x}) = O\left(\frac{1}{|\underline{x}|^{1+\varepsilon}}\right)$ and $V(\underline{x})$ has no oscillations at infinity i.e. $\underline{x}\cdot\underline{\nabla}V(\underline{x}) = O\left(\frac{1}{|\underline{x}|^{1+\varepsilon}}\right)$.

However, the time delay of Coulomb scattering is infinite as in the classical case. A simple way to see it is to evaluate the asymptotic behaviour of Coulomb sojourn time $T_R^{\ell}(\lambda)$ as $R \to \infty$. If one sets $V(r) = \frac{Ze^2}{r}$ in the radial equation (5), it is well known that the asymptotic behaviour (6) of $u_\ell(k,r)$ has to be modified to

$$u_\ell(k,r) \stackrel{\sim}{=} \frac{1}{2i}\left(S_\ell(\lambda)\, e^{i\left(kr - \frac{Zme^2}{k}\log 2kr - \frac{\ell\pi}{2}\right)} - \right.$$

$$\left. - e^{-i\left(kr - \frac{Zme^2}{k}\log 2kr - \frac{\ell\pi}{2}\right)}\right) . \tag{54}$$

When (54) is used in (36), one finds instead of (37) that the dominant contribution to $T_R^\ell(\lambda)$ is

$$T_R^\ell(\lambda) \simeq \frac{1}{v}\left(2R + \frac{Ze^2}{\lambda}\log 2kR\right) , \quad R \to \infty ,$$

so that $\tau_R^\ell(\lambda) \simeq \frac{Ze^2}{\lambda v}\log 2kR$ has the typical logarithmic divergence.

But it is worth noting that the time delay of a short range perturbation of Coulomb scattering relative to the Coulomb sojourn time is finite. Consider $H = H_c + V$ where V is a short range perturbation of the Coulomb hamiltonian $H_c = H_o + \frac{Ze^2}{r}$. Let $T_\Sigma^c(\phi)$ be the pure Coulomb sojourn time in Σ and $T_\Sigma(\phi)$ be the sojourn time corresponding to the full hamiltonian H. The relative time delay in Σ is now defined by

$$\tau_\Sigma^c(\phi) = T_\Sigma(\phi) - T_\Sigma^c(\phi) \quad .$$

In order to have the existence of $\lim_{\Sigma\to R^3} \tau_\Sigma^c(\phi)$ is it sufficient that $R_z - R_z^c = (H-z)^{-1} - (H_c-z)^{-1}$ belongs to the trace class, according to the general theory of section (IIIb). But this is the case as soon as V is H_o-bounded and $(H_o + V - z)^{-1} - (H_o - z)^{-1}$ is of trace class. Indeed the resolvent identity implies

$$R_z - R_z^c = -R_z^c V R_z = R_z^c(H_o - z)\left[(H_o+V-z)^{-1} - (H_o-z)^{-1}\right](H_o+V-z)R_z \; .$$

By the H_o-boundedness of V and $\frac{Ze^2}{r}$, $(H_o+V-z)R_z$ is bounded and $R_z^c(H_o-z)$ has a bounded extension, so that $R_z-R_z^c$ belongs to the trace class.

V. TIME DELAY IN THE OPTICAL MODEL AND IN MULTICHANNEL SCATTERING

The concept of time delay can be generalized to inelastic scattering processes where the colliding particle may undergo changes of internal states or produce new fragments. We shall always restrict our attention to the case where the incoming channel consists only of two fragments. A new feature comes from the fact that during the collision, in addition to advance or retardation, the particles can branch to other open channels. Both effects have to be taken into account in the definition of time delay. We study first the simpler model in which the effects of the new open channels are phenomenologically taken into account by a dissipative part of the interaction.

a) The Optical Model

The interaction is of the form $V = V_1 + iV_2$ where V_1 and V_2 are selfadjoints and V_2 is a negative operator $((\phi,V_2\phi)\leq 0$, ϕ in the domain of $V_2)$. The full hamiltonian $H = H_o + V$ generates a semi group of contractions $V_t = e^{-iHt}$ for positive time, i.e. $\|V_t\| \leq 1$, $t \geq 0$. The time dependent formalism can be developed quite similarly to that of elastic scattering.

The wave operators are now defined by the strong limits

$$\lim_{t\to\mp\infty} e^{i(H_o+V_1)t\mp V_2 t} e^{-iH_o t}\phi = \Omega_{\mp}\phi \quad .$$

They are no more isometries, but contractions: $\|\Omega_{\mp}\| \leq 1$. Ω_- intertwines V_t and U_t: $V_t\Omega_- = \Omega_- U_t$ for all t. (It can be shown that for $t < 0$, V_t is well defined on the range of Ω_-.) The S-operator $S = \Omega_+^*\Omega_-$ is also contractive, $\|S\| \leq 1$, and conserves energy, $[S,H_o] = 0$. The scattering state $\psi_t = V_t\Omega_-\phi = \Omega_- U_t\phi$ behaves as the freely moving state $U_t\phi$, $t \to -\infty$, and as $SU_t\phi$, $t \to \infty$, i.e. ψ_t verifies (2a) and (2b). The new point is that $\|\psi_t\|^2 \leq 1$ for $\|\phi\| = 1$, that is, probability is not conserved in the course of the time because of absorption.

The formulae of the stationary formalism (section (Ib)) are still valid with the difference that the potential $V(x) = V_1(\underline{x}) + iV_2(\underline{x})$ is complex with $V_2(\underline{x}) \leq 0$ and that the phase shift $\delta_\ell(\lambda)$ has a positive imaginary part, i.e. $|S_\ell(\lambda)| \leq 1$.

With this setting, the sojourn time of the scattering state in Σ can be defined exactly as in (11,13), and its finitness is established with the methods of section (IIb). In particular, it has energy shell components which are again given by (24) in case of spherical symmetry.

In the definition of time delay, one must take into account that there is an apparent retardation due to the fact that by absorption, the particle has a probability of presence less than one. Therefore, $T_R(\lambda)$ should not be compared with the full free sojourn time $T_R^o(\lambda) \simeq \frac{2R}{v}$, but with $\frac{2R}{v}$ reduced by the fraction of time $\gamma(\lambda)$ during which the particle has effectively been present. Since the probability of having the particle present at time t is $\|\psi_t\|^2 \leq 1$, the fraction of time during which it participates to the scattering process is in view of (2a), (2b)

$$\lim_{T\to\infty} \frac{1}{2T} \int_{-T}^{T} \|\psi_t\|^2 dt = \frac{1}{2}(\|\phi\|^2 + \|S\phi\|^2) = (\phi,\gamma\phi) \leq 1 \quad \text{with}$$

$$\gamma = \frac{1}{2}(I + S^*S) \quad .$$

The energy shell components $\gamma(\lambda) \leq I_\lambda$ give the average time of presence of the particle per unit time when the incident kinetic energy is λ. Thus, we define the energy shell components of the time delay for finite regions by

$$\tau_R(\lambda) = T_R(\lambda) - \gamma(\lambda)\,\frac{2R}{v} \quad . \tag{55}$$

Since $T_R^o(\lambda) \simeq \frac{2R}{v}$, $R \to \infty$, we notice that (55) can as well be written as

$$\tau_R(\lambda) = T_R(\lambda) - \frac{1}{2}(T_R^o(\lambda) + S_\lambda^* T_R^o(\lambda) S_\lambda)$$

and this is also equivalent with the following definition of τ:

$$\tau_R(\phi) = \int_{-\infty}^{\infty} \|P_{\Sigma_R}\psi_t\|^2 dt - \frac{1}{2}\int_{-\infty}^{\infty} (\|P_{\Sigma_R} U_t\phi\|^2 + \|P_{\Sigma_R} U_t S\phi\|^2) dt \quad . \tag{56}$$

The stationary method of section (IIc) can be generalized to the present situation to show the existence of the infinite space time delay (the methods of Section (IIab) are not suited because they both use at some point that the S-operator is unitary).

It follows from the definition (55) that the time delay for a sphere Σ_R, energy λ and angular momentum ℓ, is given by

$$\tau_R^\ell(\lambda) = \frac{4m}{k}\int_0^R |u_\ell(k,r)|^2 dr - \frac{1}{2}(1 + |S_\ell(\lambda)|^2)\,\frac{2mR}{k} \quad . \tag{57}$$

The following identity is valid when V is complex,

$$|u(k,r)|^2 = -\frac{1}{2m}\frac{d}{dr}\left(u^* \frac{d^2u}{dr\,d\lambda} - \frac{du}{d\lambda}\frac{du^*}{dr}\right) + (V-V^*)u^* \frac{du}{d\lambda} \quad ,$$

and can be used in (57) in conjunction with (6) to find the asymptotic behaviour of $\tau_R^\ell(\lambda)$ as $R \to \infty$.

The final result is

$$\tau^{\ell}(\lambda) = \lim_{R\to\infty} \tau_R^{\ell}(\lambda) = \mathrm{Im}[S_\ell^*(\lambda)\frac{dS_\ell(\lambda)}{d\lambda} - \frac{8m}{k}\int_0^\infty V_2(r)u_\ell^*(k,r)\frac{du_\ell(k,r)}{d\lambda}dr]$$

$$=(\mathrm{Im}\frac{d}{d\lambda}\log S_\ell(\lambda))|S_\ell(\lambda)|^2 + \frac{4im}{k}\int_0^\infty dr V_2(r)|u_\ell(k,r)|^2 \frac{d}{d\lambda}\log(\frac{u_\ell(k,r)}{u_\ell^*(k,r)}) \tag{58}$$

where the limit has to be understood in the weak sense.

Both terms in (58) are easily interpreted in the neighbourhood of resonance, where $S_\ell(\lambda) \simeq \frac{\alpha(\lambda)}{\lambda_o - i\Gamma - \lambda}$, $u_\ell(k,r) = \frac{\beta(\lambda,r)}{\lambda_o - i\Gamma - \lambda}$ with $\alpha(\lambda)$ and $\beta(\lambda,r)$ slowly varying for $\lambda \sim \lambda_o$. The first term of (58) behaves as $\frac{\Gamma}{(\lambda-\lambda_o)^2+\Gamma^2}|S_\ell(\lambda)|^2$, which can be referred as to giving the partial line width of the resonance. Using the formula

$$1-|S_\ell(\lambda)|^2 = -\frac{8m}{k}\int_0^\infty V_2(r)|u_\ell(k,r)|^2 dr$$

the second term is approximately $\frac{\Gamma}{(\lambda-\lambda_o)^2+\Gamma^2}(1 - |S_\ell(\lambda)|^2)$, so that both terms sum up to give $\tau(\lambda = \lambda_o) \simeq \frac{1}{\Gamma}$, the total life time of the resonance.

b) Time Delay in Multichannel Scattering

We discuss briefly the notion of time delay in multichannel processes. The level of discussion in this paragraph is entirely formal. We consider a N-body hamiltonian $H = H_o + \sum_{i<j} V_{ij}$, H_o being the N-body kinetic energy. We always assume that the coordinates of the center of mass have been removed in H, so that H acts on $\mathcal{H} = L^2(R^{3N-3})$. Channel subspaces $\mathcal{H}_\alpha = P^\alpha \mathcal{H}$ are labelled by partitions α of the N particles into clusters with a specification of the eigenfunction of the clusters. With each channel, we have the

free evolution $U_t^\alpha = e^{-iH^\alpha t}$ of the corresponding fragments and the wave operators $\Omega_\pm^\alpha$ on H_α, $\Omega_\pm^\alpha = \underset{t\to\pm\infty}{\text{s-lim}}\, V_t\, U_t^\alpha\, P^\alpha$. They intertwine H and H^α: $H\Omega_\pm^\alpha = \Omega_\pm^\alpha H^\alpha$, and their ranges $F_\pm^\alpha = \Omega_\pm^\alpha \Omega_\pm^{\alpha *}$ are pair-wise orthogonal: $F_\pm^\alpha\, F_\pm^\beta = \delta_{\alpha\beta}\, F_\pm^\alpha$. Asymptotic completeness reads $\sum_\alpha F_+^\alpha = \sum_\alpha F_-^\alpha$. The channel components of the S-matrix are the operators $S_{\beta\alpha} = \Omega_+^{\beta *}\, \Omega_-^\alpha$ from H_α to H_β. The $S_{\beta\alpha}$ conserve energy, $H^\beta\, S_{\beta\alpha} = S_{\beta\alpha}\, H^\alpha$, and the asymptotic completeness implies the unitarity relation

$$\sum_\gamma S_{\gamma\beta}^*\quad S_{\gamma\alpha} = \delta_{\alpha\beta}\quad P^\alpha\ .$$

The full scattering state $\psi_t = V_t \Omega_-^\alpha \phi_\alpha$, $\phi_\alpha \in H_\alpha$, corresponding to an incoming state in channel α, behaves asymptotically in time as

$$\|\psi_t - U_t^\alpha\, \phi_\alpha\| \to 0, \qquad t \to -\infty, \tag{59a}$$

$$\|\psi_t - \sum_\beta\, U_t^\beta\quad S_{\beta\alpha}\, \phi_\alpha\| \to 0, \qquad t \to \infty, \tag{59b}$$

which is the multichannel generalization of (2a), (2b).

With these definitions, we are in position to define the time delay. Consider a projection P_Σ on $L^2(R^{3N-3})$ on a region Σ where all N particles are at finite distance from each other (for instance Σ is a 3N-3 dimensional sphere of radius R). As usual, the sojourn time of the full scattering state in Σ is $T_\Sigma(\phi_\alpha) = \int_{-\infty}^{\infty} \|P_\Sigma \psi_t\|^2 dt$. We notice that the fraction of time spent by the scattering state in a given channel β is by (59a,b)

$$\lim_{T\to\infty} \int_{-T}^{T} \|P^\beta \psi_t\|^2 dt = (\phi_\alpha, \gamma_{\beta\alpha}\phi_\alpha)$$

with

$$\gamma_{\beta\alpha} = \frac{1}{2}\,(\delta_{\alpha\beta}\quad P^\alpha + S_{\beta\alpha}^*\quad S_{\beta\alpha})\ .$$

Thus, a natural generalization of the definition (56) to

multichannel processes is

$$\tau_\Sigma(\phi_\alpha) = \int_{-\infty}^{\infty}\|P_\Sigma\psi_t\|^2 dt$$

$$- \frac{1}{2}[\int_{-\infty}^{\infty}dt\|P_\Sigma U_t^\alpha\phi_\alpha\|^2 + \sum_\beta \int_{-\infty}^{\infty}dt\|P_\Sigma U_t^\beta S_{\beta\alpha}\phi_\alpha\|^2] \quad . \tag{60}$$

$\tau(\phi_\alpha) = \lim_{\Sigma\to R^{3N-3}} \tau_\Sigma(\phi_\alpha)$ can be worked out. One remarks first that, as in proposition 4, $\tau(\phi_\alpha)$ will <u>depend only on the $S_{\beta\alpha}$ components.</u> Indeed, one can write

$$\tau_\Sigma(\phi_\alpha) = \frac{1}{2}\int_0^\infty dt(\sum_\beta\|P_\Sigma U_t^\beta S_{\beta\alpha}\phi_\alpha\|^2 - \|P_\Sigma U_t^\alpha\phi_\alpha\|^2)$$

$$+ \frac{1}{2}\int_{-\infty}^{0}dt(\|P_\Sigma U_t^\alpha\phi_\alpha\|^2 - \sum_\beta \|P_\Sigma U_t^\beta S_{\beta\alpha}\phi_\alpha\|^2)$$

$$+ \int_0^\infty dt(\|P_\Sigma\psi_t\|^2 - \sum_\beta \|P_\Sigma U_t^\beta S_{\beta\alpha}\phi_\alpha\|^2)$$

$$+ \int_{-\infty}^{0}dt(\|P_\Sigma\psi_t\|^2 - \|P_\Sigma U_t^\alpha\phi_\alpha\|^2) \quad . \tag{61}$$

The integrands of the two last integrals converge to zero as $P_\Sigma \to I$ because of the unitarity relation. Moreover one can argue that they are uniformly majorized by integrable functions in time if the rate of the asymptotic relation (59a,b) is integrable. Thus, the two last terms vanish as $P_\Sigma \to I$. The two first terms of (61) involve only the free channel evolutions U_t^α and the $S_{\beta\alpha}$. They can be analyzed explicitely assuming sufficient smoothness of the kernels of the involved operators. One arrives at the following expression for the energy shell time delay :

$$\tau(\lambda) = i\sum_\beta S_{\beta\alpha}^*(\lambda)\frac{d}{d\lambda}S_{\beta\alpha}(\lambda) \quad . \tag{62}$$

$S_{\beta\alpha}(\lambda)$ are the "energy shell" parts of $S_{\beta\alpha}$ (since $S_{\beta\alpha}$ conserves energy). (62) is the multichannel generalization of the Eisenbud-Wigner formula.

It is interesting to see that the geometrical definition of time delay of Section (IIId) can be generalized to the multichannel situation. Let D^α be the generator of the dilation group on the channel subspace $P^\alpha H$, and $\tilde{D}^\alpha(t) = V_t^* U_t^\alpha D^\alpha U_t^{\alpha *} V_t$. Then the generalization of (43) is

$$(\phi_\alpha, 2H^\alpha \tau \phi_\alpha) = \lim_{t\to-\infty} (\Omega_-^\alpha \phi_\alpha, \tilde{D}^\alpha(t)\, \Omega_-^\alpha \phi_\alpha)$$

$$- \lim_{t\to\infty} \sum_\beta (\Omega_-^\alpha \phi_\alpha, \tilde{D}^\beta(t) \Omega_-^\alpha \phi_\alpha)$$

where ϕ_α is the incoming state in the channel P^α . Using the asymptotic condition and the unitarity relation, this gives as in (45)

$$(\phi_\alpha, 2H^\alpha \tau \phi_\alpha) = (\phi_\alpha, (D^\alpha - \sum_\beta S_{\beta\alpha}^* \; D^\beta \; S_{\beta\alpha})\phi_\alpha)$$

$$= \sum_\beta (\phi_\alpha, S_{\beta\alpha}^* \; (S_{\beta\alpha} \; D^\alpha - D^\beta \; S_{\beta\alpha})\phi_\alpha) \quad .$$

Since D^α generates a dilation of the kinetic energy in channel α we get as in the single channel case

$$S_{\beta\alpha}(\lambda) D^\alpha - D^\beta S_{\beta\alpha}(\lambda) = -2i\,\lambda \frac{d}{d\lambda} S_{\beta\alpha}(\lambda)$$

and this leads again to (62).

VI. LEVINSON'S THEOREM

a) Heuristics

The Levinson's theorem in its usual formulation

asserts that for a spherically symmetric potential V(r) producing a phase shift $\delta_\ell(\lambda)$ and n_ℓ bound states of angular momentum ℓ, one has

$$\delta_\ell(0) - \delta_\ell(\infty) = 2\pi n_\ell \quad . \tag{63}$$

Clearly (63) is equivalent with a sum rule for the time delay

$$\frac{1}{2\pi}\int_0^\infty \tau^\ell(\lambda)d\lambda + n_\ell = 0 \quad . \tag{64}$$

(63) or (64) is proven under the conditions $\int_0^\infty r^\alpha V(r)dr < \infty$, $\alpha = 1,2$, with the help of the analytic properties of the Jost functions. We do not repeat the standard proof here, but we will establish an analoguous sum rule for the total time delay $Tr_\lambda\ \tau(\lambda)$, a quantity which is defined as well for non central or non local interactions. Notice that in the spherically symmetric case, one has $Tr_\lambda\ \tau(\lambda) = \sum_{\ell=0}^{\infty}(2\ell+1)\tau^\ell(\lambda)$, $n = \sum_{\ell=0}^{\infty}(2\ell+1)n_\ell$ so that one could think that the relation (64) would hold with $Tr_\lambda\tau(\lambda)$ and the total number of bound states n in place of $\tau_\ell(\lambda)$ and n_ℓ. In fact, this is not the case.

In order to understand the origin of the sum rule satisfied by $Tr_\lambda\tau(\lambda)$ let us recall (see Section (IId) that

$$\lim_{\Sigma\to R^3}\frac{1}{2\pi}Tr_\lambda\tau_\Sigma(\lambda) = \lim_{\Sigma\to R^3}\frac{1}{2\pi}Tr_\lambda(T_\Sigma(\lambda) - T_\Sigma^0(\lambda)) = \frac{1}{2\pi}Tr_\lambda\tau(\lambda)$$

is the change in the density of states produced by the interaction. Moreover, if E(a,b) ($(E_0(a,b))$ denotes a spherical projection of H (H_0) for the interval (a,b), it follows formally from (21) that

$$\frac{1}{2\pi}\int_0^\lambda Tr_\mu\tau(\mu)d\mu = \lim_{\Sigma\to R^3}Tr[P_\Sigma(E(0,\lambda)-E_0(0,\lambda))P_\Sigma] = Tr(E(0,\lambda)-E_0(0,\lambda)) . \tag{65}$$

If H_o has absolutely continuous spectrum in $[0,\infty)$ and the interaction produces a finite number n of negative energy bound states, i.e. $E_o(0,\lambda) = E_o(-\infty,\lambda)$, $E(0,\lambda) = E(-\infty,\lambda) - E(-\infty,0)$ with $\mathrm{Tr}\, E(-\infty,0) = n$, one deduces from (65) that

$$\frac{1}{2\pi}\int_0^\lambda \mathrm{Tr}_\mu\, \tau(\mu)d\mu + n = \mathrm{Tr}(E(-\infty,\lambda) - E_o(-\infty,\lambda)) \quad . \tag{66}$$

We would find a sum rule identical to (64) if $\mathrm{Tr}(E(-\infty,\lambda) - E_o(-\infty,\lambda))$ would vanish as $\lambda \to \infty$. We know that both $E(-\infty,\lambda)$ and $E_o(-\infty,\lambda)$ tend strongly to I as $\lambda \to \infty$, thus the difference of these spectral projections converges strongly to zero, but not in the trace norm! Therefore, we do not expect the left hand side of (66) to vanish as $\lambda \to \infty$. This means that in order to obtain a sum rule, we have to take properly into account the high energy behaviour of $\mathrm{Tr}_\lambda \tau(\lambda)$.

Usually, the high energy behaviour of scattering quantities is given by the Born approximation. Let us compute the Born approximation of $\mathrm{Tr}_\lambda \tau(\lambda) = \frac{d}{d\lambda}\mathrm{Tr}_\lambda \Delta(\lambda)$. One has $\Delta(\lambda) = -i\log(I_\lambda + G_\lambda)$, with the operator G_λ given by

$$(\hat{k}_1|G_\lambda|\hat{k}_2) = -i\,\frac{k_1 m}{(2\pi)^2}\int d^3x\, e^{-i\underline{k}_2\underline{x}}\, V(\underline{x})\,\psi(\underline{k}_1,\underline{x}),\ |\underline{k}_1| = |\underline{k}_2| = \sqrt{2m\lambda}$$

where $\psi(\underline{k},\underline{x})$ is the stationary scattering wave function. Therefore, one finds in first order in $V(\underline{x})$

$$(\hat{k}_1|\Delta(\lambda)|\hat{k}_2) \simeq -i(\hat{k}_1|G_\lambda|\hat{k}_2) = -\frac{k_1 m}{(2\pi)^2}\int d^3x\, e^{-i\sqrt{2m\lambda}(\hat{k}_1-\hat{k}_2)\cdot\underline{x}}\, V(\underline{x})$$

implying $\mathrm{Tr}_\lambda \Delta_\lambda \simeq \int d^2k(\hat{k}|\Delta(\lambda)|\hat{k}) = -\frac{k_1 m}{\pi}\int V(\underline{x})d^3x$, and finally

$$\mathrm{Tr}_\lambda \tau(\lambda) \simeq -\frac{1}{\sqrt{\lambda}}\frac{1}{\pi}\left(\frac{m^3}{2}\right)^{1/2}\int V(\underline{x})d^3x \quad . \tag{67}$$

(67) shows clearly that the Born contribution to $\mathrm{Tr}_\lambda \tau(\lambda)$ is not integrable in energy and therefore, the integral in the left hand side of (66) does not converge as $\lambda \to \infty$. These considerations indicate also why sum on angular momentum and energy integration cannot be exchanged in (64). We discuss now the proper formulation of Levinson's theorem.

b) Sum Rules for the Total Time Delay

According to the preceeding discussion, one may expect to obtain a convergent sum rule if we substract to $\mathrm{Tr}_\lambda \tau(\lambda)$ its high energy behaviour given by (67). This is indeed the case and the Levinson's theorem takes the form

$$\frac{1}{2\pi} \int_0^\infty \left(\mathrm{Tr}_\lambda \tau(\lambda) + \frac{\nu}{\sqrt{\lambda}}\right) d\lambda + n = 0 \tag{68}$$

where $\nu = \frac{1}{\pi} \left(\frac{m^3}{2}\right)^{1/2} \int V(\underline{x})\,dx$ and n is the total number of bound states (counting each as often as its degeneracy).

A method to obtain (68) consists in writing a dispersion relation for the quantity Tr Q(z) with the operator Q(z) defined by

$$Q(z) = R_z - R_{oz} + R_{oz} V R_{oz} \quad . \tag{69}$$

R_z and R_{oz} are the resolvents associated with H and H_o respectively, and $-R_{oz}VR_{oz}$ is precisely the Born term in the expansion of $R_z - R_{oz}$ in powers of V.

The assumptions on the potential are

(i) $V(\underline{x})$ belongs to $L^1 \cap L^2$ (so that $R_z - R_{oz}$ is in the trace class, and Tr Q(z) and $\mathrm{Tr}_\lambda\ \tau(\lambda)$ are well defined).

(ii) The boundstates of H have negative energy and are finite in number. Two more specific assumptions will be made in the course of the derivation of (68).

It follows from (i) and (ii) that $Q(z)$ is analytic in the z plane, except on the half line $\{\mathrm{Im}\ z = 0, \mathrm{Re}\ z \geq 0\}$ and at the isolated real points $\lambda_j < 0$ corresponding to the bound states of H. Consider the contour C defined in the figure:

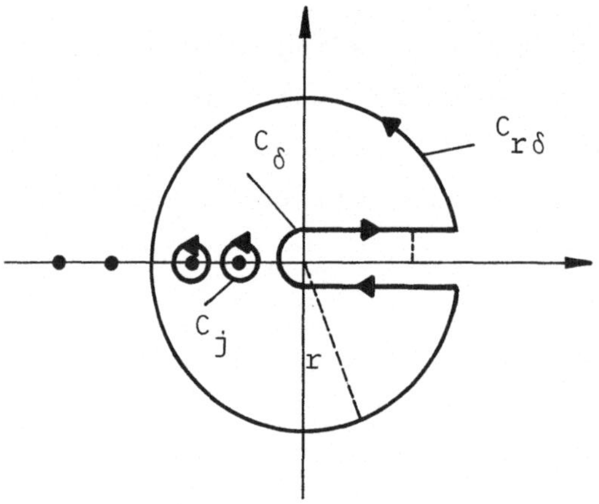

Fig. 2.

By the Cauchy theorem one has

$$\int_C \mathrm{Tr}\ Q(z)dz = \sum_j \int_{C_j} \mathrm{Tr}\ Q(z)dz \quad . \tag{70}$$

We evaluate the contributions of all pieces of the contours in (70).

<u>1. Contribution of the small circles C_j</u>

At $z = \lambda_j$, R_{oz}, $R_{oz}VR_{oz}$ are analytic, but $R_z \sim \frac{P_j}{\lambda_j - z}$, $z \to \lambda_j$, has a simple pole, with P_j being the projection on the eigensubspace of λ_j. Thus

$$\int_{C_j} \mathrm{Tr}\ Q(z)dz = \int_{C_j} \frac{\mathrm{Tr}\ P_j}{\lambda_j - z} = -2i\pi\nu_j$$

where ν_j is the multiplicity of the eigenvalue λ_j.

2. Contribution of the large circle $C_{r\delta}$

This contribution vanishes as $r \to \infty$ uniformly with respect to δ. This result follows from trace norm estimates which are summarized in the following lemma.

Lemma. α) There exist constants c and $d < 1$ such that $\| R_{oz}(VR_{oz})^n \|_1 \leq cd^n$ for $|z| \geq r_o$ and $|\mathrm{Im}\, z| \geq \delta$.

β) There exist constants c and $d < 1$ such that for $n \geq 2$ $|\mathrm{Tr}(VR_{oz})^n| \leq cd^n$, $|\mathrm{Re}\, z| \geq r_o$ and $\lim_{|\mathrm{Re} z| \to \infty} \mathrm{Tr}(VR_{oz})^n = 0$. The bound and the limit hold uniformly with respect to Im z.

The lemma is proven by estimation of trace norms and Hilbert-Schmidt norms of the involved operators with the help of the explicit form of the kernel (74) of the free resolvent.

We can use part α) of the lemma to write, for r large enough,

$$Q(z) = \sum_{n=2}^{\infty} (-1)^n \mathrm{Tr}(R_{oz}(VR_{oz})^n), \quad z \in C_{r\delta}$$

as a trace-norm convergent series and we have

$$\int_{C_{r\delta}} \mathrm{Tr} Q(z) = \sum_{n=2}^{\infty} (-1)^n \int_{C_{r\delta}} \mathrm{Tr}(R_{oz}(VR_{oz})^n) .$$

Moreover, it is not difficult to establish

$$\mathrm{Tr}(R_{oz}(VR_{oz})^n) = \frac{1}{n} \frac{d}{dz} \mathrm{Tr}(VR_{oz})^n, \quad \mathrm{Im}\, z \neq 0, \text{ and therefore}$$

$$\int_{C_{r\delta}} \mathrm{Tr}(R_{oz}(VR_{oz})^n) = \frac{1}{n}(\mathrm{Tr}(VR_{o,r-i\delta})^n - \mathrm{Tr}(VR_{o,r+i\delta})^n) .$$

Then part β) of the lemma enable us to conclude that

$$\lim_{r\to\infty} \int_{C_{r\delta}} dz\ \mathrm{Tr}(R_{oz}(VR_{oz})^n) = 0 \text{ for } n \geq 2 \text{ and}$$

$$\lim_{r\to\infty} \int_{C_{r\delta}} dz\ \mathrm{Tr}\ Q(z) = 0 \quad \text{uniformly with respect to } \delta.$$

3. Contribution of the half circle C_δ

We assume that $\mathrm{Tr}\ Q(z) = o(\frac{1}{|z|})$, $z \to 0$, (there are no zero energy bound states) so that $\int_{C_\delta} \mathrm{Tr}\ Q(z)dz$ vanishes as $\delta \to 0$.

4. Contribution of the positive real axis

The integral along the two segments above and below the real positive axis is

$$2i \int_0^r \mathrm{Im}\ \mathrm{Tr}Q(\lambda+i\delta)d\lambda = 2i \int_0^r d\lambda\ \mathrm{Im}(\mathrm{Tr}(R_{\lambda+i\delta}-R_{o,\lambda+i\delta}) +$$

$$+ \mathrm{Tr}(R_{o,\lambda+i\delta}\ VR_{o,\lambda+i\delta})) . \tag{71}$$

The following key lemma relates the boundary value of $\mathrm{Tr}(R_z-R_{oz})$ on the positive real axis to the time-delay:

<u>Lemma:</u> Let $z = \lambda + i\delta$ and λ_j the eigenvalues of H (repeated according to their multiplicities). Then

$$\mathrm{Im}\ \mathrm{Tr}(R_z-R_{oz}) = \frac{1}{2\pi} \int_0^\infty \frac{\delta \mathrm{Tr}_\mu \tau(\mu)}{(\mu-\lambda)^2+\delta^2} d\mu - \sum_j \frac{\delta}{(\lambda_j-\lambda)^2+\delta^2} . \tag{72}$$

<u>Proof:</u> Proceeding as in (31) (in the proof of prop. 6), we write

$$\frac{1}{2\pi}\int_0^\infty \frac{Tr_\mu \tau_\Sigma(\mu)}{(\mu-z^*)(\mu-z)}\,d\mu = Tr(R_{oz^*}(\Omega_- P_\Sigma \Omega_- - P_\Sigma)R_{oz})$$

$$= Tr(R_{z^*}P_\Sigma R_z - R_{oz^*}P\,R_{oz}) - Tr((I-\Omega_-\Omega_-^*)R_{z^*}P_\Sigma R_z) \quad .$$

Therefore, we get taking the limit $\Sigma \to R^3$

$$\frac{1}{2\pi}\int_0^\infty \frac{Tr_\mu \tau(\mu)}{(\mu-z^*)(\mu-z)}d\mu = Tr(R_{z^*}R_z - R_{oz^*}R_{oz}) - Tr((1-\Omega_-\Omega_-^*)R_{z^*}R_z) \quad . \tag{73}$$

The result of the lemma follows when we use in (73) the resolvent identity $(z^*-z)R_{z^*}R_z = R_{z^*}-R_z$ and the fact that $(1-\Omega_-\Omega_-^*)$ is the projection on the subspace of bound states.

Moreover, one can calculate explicitely the quantity $Tr\ R_{oz}\ VR_{oz}$ with the help of the kernel representation of R_{oz}:

$$(\underline{x}|R_{oz}|\underline{y}) = \frac{m}{2\pi}\frac{e^{i\sqrt{2mz}|\underline{x}-\underline{y}|}}{|\underline{x}-\underline{y}|} \quad (Im\sqrt{z} > 0), \tag{74}$$

$$Tr\ R_{oz}VR_{oz} = \left(\frac{m}{2\pi}\right)^2 \int d^3x \int d^3y\ \frac{e^{2i\sqrt{2mz}|\underline{x}-\underline{y}|}}{|\underline{x}-\underline{y}|^2}\ V(\underline{y})$$

$$= \left(\frac{m}{2\pi}\right)^2 \int d^3x\ \frac{e^{2i\sqrt{2mz}|x|}}{|x|^2} \int V(\underline{y})\,d^3y$$

$$= \frac{i}{\pi}\left(\frac{m}{2}\right)^{3/2} \frac{1}{\sqrt{z}}\ \ d^3y\ V(\underline{y}) = \frac{i}{2\sqrt{z}}\,\nu \quad . \tag{75}$$

Inserting (72) and (75) in (71) gives

$$2i\int_0^r d\lambda\,\mathrm{Im}\,\mathrm{Tr}Q(\lambda+i\delta)=2i\int_0^r d\lambda\left[\frac{1}{2\pi}\int_0^\infty d\mu\frac{\delta\,\mathrm{Tr}_\mu\tau(\mu)}{(\mu-\lambda)^2+\delta^2}+\frac{1}{2}\mathrm{Im}\left(\frac{i}{\sqrt{\lambda+i\delta}}\nu\right)\right]$$

$$-\,2i\ \delta\sum_j\int_0^r\frac{d\lambda}{(\lambda_j-\lambda)^2+\delta^2}\quad. \tag{76}$$

When we take first the limit $\delta \to 0$ and then let $r \to \infty$ in (76), we obtain (formally $\lim_{\delta\to 0}\frac{1}{\pi}\frac{\delta}{(\lambda-\mu)^2-\delta^2}=\delta(\lambda-\mu)$)

$$\lim_{r\to\infty}\lim_{\delta\to 0}\int_0^r d\lambda\ \mathrm{Im}\ \mathrm{Tr}\ Q(\lambda+i\delta)\ =\ i\int_0^\infty\left(\mathrm{Tr}_\lambda\tau(\lambda)\ +\ \frac{\nu}{\sqrt{\lambda}}\right)d\lambda\quad. \tag{77}$$

The existence of above limit can be rigorously established under the additional assumption that

$$\int_0^\infty\left|\mathrm{Tr}_\lambda\tau(\lambda)\ +\ \frac{\nu}{\sqrt{\lambda}}\right|d\lambda\ <\ \infty\quad. \tag{78}$$

Collecting the non-vanishing contributions to the contour integral in the limit $\delta \to 0$ and $r \to \infty$, (i.e. those of the bound states and (77)),gives the sum rule (68).

Half Bound States

Since $\mathrm{Tr}_\lambda\tau(\lambda) = \frac{d}{d\lambda}\mathrm{Tr}_\lambda\Delta(\lambda)$, the Levinson relation (68) can be written in a form similar to (63):

$$\mathrm{Tr}_0\Delta(0)\ -\ \lim_{\lambda\to\infty}\ (\mathrm{Tr}_\lambda\Delta(\lambda)\ +\ 2\sqrt{\lambda}\nu)\ =\ 2\pi n\quad. \tag{79}$$

It is known in the spherically symmetric case that there are exceptional cases in which the relation (63) for the S-wave ($\ell = 0$) has to be modified to

$$\delta_0(0)\ -\ \delta_0(\infty)\ =\ 2\pi(n_0\ +\ 1/2)\ . \tag{80}$$

This exceptional case occurs when the Jost function of the s-wave vanishes at threshold. Because of the additional term $\frac{1}{2}$ in (80), one speaks of the occurence of a "half bound state". In fact, half bound states can also appear in (68) or (79), replacing n by $n + \frac{1}{2}$. The general form of the Levinson's theorem is then

$$\mathrm{Tr}_0 \Delta(0) - \lim_{\lambda\to\infty}(\mathrm{Tr}_\lambda \Delta(\lambda) + 2\sqrt{\lambda}\nu) = 2\pi(n + \frac{1}{2} q) \qquad (81)$$

where q = 1 if there is a half bound state at threshold and q = 0 otherwise.

In our preceeding analysis half bound states have been excluded by the assumption that the contribution of the small half circle C_δ around z = 0 vanishes. This may be the case or not, but there is no explicit condition on the potential insuring the existence or the non-existence of half bound states. Notice that for non-spherically symmetric potentials half bound states do not appear to be in relation with specific values of the angular momentum. The Levinson's theorem in its generalized form (81) can be proven under the hypothesis that $V(\underline{x})$ is bounded and $V(\underline{x}) = O(\frac{1}{|x|^{3+\varepsilon}})$ (implying that $V \in L^1 \cap L^2$). This alternative proof follows the same lines as for central potentials, using Fredholm theory and analytic properties of the Fredholm determinant of the S-matrix.

REFERENCES

Section Ia,b,c : Your favorite book on scattering theory. The interpretation of the energy derivative of the phase shift as time delay appears first in Eisenbud, Princeton dissertation (1948), and Wigner [1].

Section II: The introduction of sojourn time in relation with time delay appears in [2,3]. The concept has received a precise formulation in Jauch-Marchand [4] and Jauch-Sinha-Misra [5].

a) The time dependent method follows Martin [6] and Amrein-Jauch-Sinha [7] chap.7.2. Proofs of prop. 2 and 3 can be found in [8].
b) The abstract method: see [5]. Theorem 1 is a special case of a more general theorem using concepts of abstract harmonic analysis [9].
c) H-Smoothness: the concept has been introduced by Kato [10]. See Reed-Simon [11] Chap. XIII 7. Theorem 2; upper bounds on sojourn time and detailed commutator estimates are established by Lavine [12,13].
d) Energy shell sojourn time occurs in [4,5]. The definition (24) of sojourn time in terms of eigenfunctions was introduced by Smith in his treatment of time delay [3]. Spectral properties of sojourn time are emphasized in [14].

Section III: The definition of time delay (25) occurs in [2,4,5].

a) The time dependent method follows [6,7,8].
b) The abstract method has been developped first in [5]. The proofs of theorems 3 and 4 involve beautiful mathematical developments and are due to Krein [15,16] and Birman-Krein [17]. (The reader can verify them without difficulty when $H-H_o = V$ is a perturbation of rank one.) A detailed account of Krein's theorem is found in [18].
c) The stationary method: This derivation is due to Smith [3].
d) The content of this subsection is taken from Narnhofer [19].

Other geometrical aspects of time delay are incorporated in the "angular time delay". "Angular time delay" $\omega_\lambda(\hat{k}_1,\hat{k}_2)$ for initial direction $\hat{k}_1$ and final direction $\hat{k}_2$ could be defined as the derivative of phase of the corresponding S-matrix element:

$$\omega_\lambda(\hat{k}_1,\hat{k}_2)= \frac{d}{d\lambda}\arg(\hat{k}_1|S_\lambda|\hat{k}_2)=\mathrm{Re}\left(-i\,\frac{1}{(\hat{k}_1|S_\lambda|\hat{k}_2)}\,\frac{d}{d\lambda}(\hat{k}_1|S_\lambda|\hat{k}_2)\right).$$

One recovers the usual time delay by summing $\omega_\lambda(\hat{k}_1,\hat{k}_2)$ on all final directions $\hat{k}_2$ with the corresponding scattering probabilities $|(\hat{k}_1|S_\lambda|\hat{k}_2)|^2$:

$$\int d^2\hat{k}_2 |(\hat{k}_1|S_\lambda|\hat{k}_2)|^2 \omega_\lambda(\hat{k}_1,\hat{k}_2) = \mathrm{Re}(-i\int d^2\hat{k}_2 (\hat{k}_1|S_\lambda^*|\hat{k}_2)\frac{d}{d\lambda}(\hat{k}_2|S_\lambda|\hat{k}_1)) =$$

$$= (\hat{k}_1|\tau(\lambda)|\hat{k}_1) \quad .$$

Aspects of angular time delay are discussed in [20,21,22,23].

Section IV:

a) The existence of the causal bound (48) has been shown by Wigner [1]. See also [3,24,25,26]. Causal bounds for higher angular momentum are obtained in [27] in terms of time delay for hard sphere and other known functions.

b) The connexion with the virial is rigorously established in [12] (see also [28]). We present here the derivation found in [19].

c) It is known that Coulomb wave operators can be defined with the help of a modified free evolution

$$\tilde{U}_t = e^{-i(H_o t - f(H_o)\log t)} ,$$

[29],[7] chap. 13, [11] chap. XI. 9.
One can then define a "time-delay" $\tilde{\tau}(\phi)$ by replacing the usual free evolution comparison term by the modified free motion. This can be done for instance in the derivation of section (IIId), by using $\tilde{U}_t$ instead of U_t everywhere. With this new definition, $\tilde{\tau}(\lambda)$ is the derivative of the Coulomb phase shift. However, this procedure does not give a clear interpretation of $\tilde{\tau}(\phi)$ as time-delay.

Section V: The content of a) is taken from [30]. For a formulation of scattering with dissipative interactions, see [31].

b) The present treatment of time delay in multichannel scattering is given by Bollé-Osborn in [32]. (See also [33] for the case of the three body problem.) The geometrical method is found in [19]. Formula (62) appears already in [3].

Section VI:

a) Levinson's theorem appears in [34]. Proofs of Levinson's theorem for partial waves can be found in [35] p. 356, [11] Chap. XI.8 and [36].

b) We follow Osborn-Bollé [37]. A detailed study of Levinson's theorem and of the question of half bound states by the method of Fredholm theory is presented in [38]. Extension of Levinson's theorem to non-local interaction by abstract stationary methods has been obtained by Dreyfus [39,40]. See also [41,42]. Sum rules for higher order energy moments of the time delay can also be obtained [43,44,45]. These sum rules have application in the calculation of the virial expansion in quantum statistical mechanics.

1. E.P. Wigner, Phys. Rev. 98 (1955) 145.
2. M.L. Goldberger, K.M. Watson, Collision Theory, Wiley, New York (1964) 485.
3. F.T. Smith, Phys. Rev. 118 (1960) 349.
4. J.M. Jauch, J.P. Marchand, Helv. Phys. Acta 40 (1967) 217.
5. J.M. Jauch, K.B. Sinha, B. Misra, Helv. Phys. Acta 45 (1972) 398.
6. Ph.A. Martin, Comm. Math. Phys. 47 (1976) 221.
7. W. Amrein, J.M. Jauch, K. Sinha, Scattering Theory in Quantum Mechanics, Benjamin, Reading Mass. (1977).
8. K. Gustafson, K. Sinha, On the Eisenbud-Wigner formula for time-delay, Lett. in Math. Phys. 4 (1980) 381.
9. Ph.A. Martin, B. Misra, J. of Math. Phys. 14 (1973) 997.
10. T. Kato, Math. Ann. 162 (1966) 258.

11. M. Reed, B. Simon, Methods of Mathematical Physics III, Academic Press, New York (1978).
12. R. Lavine, in Scattering Theory in Mathematical Physics, Ed. Lavita and Marchand, Reidel (1974).
13. R. Lavine, Constructive Estimates in Quantum Scattering, Univ. of Rochester preprint (1976).
14. L. W. MacMillan, T.A. Osborn, Annals of Phys. 126 (1980) 1.
T.Y. Tsang, T.A. Osborn, Nucl. Phys. A247 (1975) 43.
D. Bollé, T.A. Osborn, Phys. Rev. D11 (1975) 3417.
15. M.G. Krein, Mat. Sb. 33 (1953) 75 (Russian).
16. M.G. Krein, Soviet Math. Dokl. 3 (1962) 707.
17. M.S. Birman, M.G. Krein, Soviet. Math. Dokl. 3 (1962) 740.
18. K. Sinha, On the theorem of M.G. Krein, Univ. of Geneva preprint (1975).
19. H. Narnhofer, Phys. Rev. D22 (1980) 2387.
20. W. Brenig, R. Haag, Fortschr. Phys. 7 (1959) 183.
21. M. Froissart, M.L. Goldberger, K.M. Watson, Phys. Rev. 131 (1963) 2820.
22. H.M. Nussenzveig, Phys. Rev. D6 (1972) 1534.
23. D. Bollé, T.A. Osborn, Phys. Rev. D13 (1976) 299.
24. I. Saavedra, Nuclear Physics 29 (1962) 137.
25. H.M. Nussenzveig, Phys. Rev. 177 (1969) 1848.
26. H.M. Nussenzveig, Causality and Dispersion Relations, Academic Press, New York (1972).
27. J. Nowakowski, T.A. Osborn, Nuclear Physics A249 (1975) 301.
28. J.O. Hirschfelder, Phys. Rev. A19 (1979) 2463.
29. J. Dollard, J. Math. Phys. 5 (1964) 729.
30. Ph.A. Martin, Nuovo Cim. 30B (1975) 217.
31. E.B. Davies, Comm. Math. Phys. 71 (1980) 277.
32. D. Bollé, T.A. Osborn, J. Math. Phys. 20 (1979) 1121.
33. D. Bollé, T.A. Osborn, J. Math. Phys. 16 (1975) 1533.
34. N. Levinson, Kgl. Danske Videnskab. Salskab. Mat. Fys. Medd. 25 (1949) 9.

35. R. Newton, Scattering Theory of Waves and Particles, McGraw Hill, New York (1966).
36. J.M. Jauch, Helv. Phys. Acta 30 (1957) 143.
37. T.A. Osborn, D. Bollé, J. Math. Phys. 18 (1977) 432.
38. R. Newton, J. Math. Phys. 18 (1977) 1348.
39. T. Dreyfus, J. of Math. Analysis and Applic. 63 (1978) 666.
40. T. Dreyfus, Helv. Phys. Acta 51 (1978) 131, J.Phys. A9 (1976) L187.
41. V.S. Buslaev, Phys. Dokl. 7 (1962) 295.
42. A. Martin, Nuovo Cimento 7 (1958) 607.
43. D. Bollé, Ann. Phys. 121 (1979) 131.
44. D. Bollé, H. Smeesters, Phys. Lett. 62A (1977) 290.
45. T.A. Osborn, R.G. Froese, S.F. Howes, Sum Rule Dynamics, Univ. of Manitoba preprint.

Acta Physica Austriaca, Suppl. XXIII, 209–233 (1981)

FINITENESS OF TOTAL CROSS-SECTIONS[+]

by

A. MARTIN
CERN - Geneva, Switzerland

ABSTRACT

We derive an optimal condition for the finiteness of total cross-sections in potential scattering at any given energy, and by copying Froissart's trick for elementary particles, explicit bounds on amplitudes and cross-sections in the spherically symmetric case. We also study the coupling constant dependence of the cross-sections for potentials of a given sign by using analyticity properties with respect to this coupling constant. This paper contains several new unpublished results.

1. INTRODUCTION

In physics, we are familiar with situations in which the scattering amplitude and/or the total cross-section may be infinite. This is, for instance, the case with Rutherford scattering where the total cross-section <u>and</u>

[+]Lectures given at the XX. Internationale Universitätswochen für Kernphysik, Schladming, Austria, February 17-26, 1981.

the forward amplitude are infinite, both in classical and quantum mechanics. However, for interactions which decrease much faster than the Coulomb interaction, a striking difference will appear between the classical and the quantum cases. In classical mechanics, if your interaction is not strictly vanishing beyond a certain distance, the total cross-section will be divergent, even for instance if V(r), the potential, decreases exponentially. On the contrary, in quantum mechanics, we all know of simple cases where the total cross-section is finite, for instance scattering by a Yukawa potential $\exp(-\mu r)/r$. Naturally, one would like to know more than that and to see more clearly when the scattering amplitude is finite, including the forward direction, and when the total cross-section is finite. The interest of theoreticians in this problem is in waves. One of the first waves was in the sixties in the group of Professor Jauch in Geneva with some outsiders like myself. A renewal of interest has started recently with the paper of Amrein and Pearson [1] which contains references to previous works. This paper stimulated my interest in the problem. Here at this school we have another of the main contributors, Volker Enss, who in papers with B. Simon [2] discusses both the finiteness and the coupling constant dependence of the total cross-sections by using the "geometrical methods" described here. Unfortunately, I shall be unable to present here a review talk and most of the time I shall restrict myself to my own approach. In particular, I shall not describe the seemingly very powerful method of Combes and Guez [3].

Maybe I should crudely describe the answers we should expect in three dimensions. To get a good partial wave scattering amplitude, we know that it is enough to have a potential decreasing somewhat faster than $1/r$. To get a finite total cross-section, we need a potential decreasing somewhat faster than $1/r^2$, and to get a finite forward scattering amplitude we need a decrease faster than $1/r^3$.

Let me now describe the plan of my lectures. First I want to describe to you the strategy which was used by Froissart [4] to get a bound on elementary particle scattering amplitudes outside the framework of interactions by potentials. By adapting this method to the case of potentials, we derive bounds on amplitudes and total cross-sections. Then we describe a refined version of this approach in which both the bound and the finiteness are controlled by a unique quantity in the case of spherical symmetry. We also discuss the case of non-spherical symmetry, first for total cross-sections averaged over the direction of the incident beam. We show that the same quantity controls the finiteness of the total cross-section and explain why we believe our condition to be optimal. We also prove strict finiteness without averaging over the incident direction, at the price of an extra condition. Generally speaking, we try to avoid the averaging process which seems necessary in the other methods of approach.

The last part of the lectures is devoted to a new, as yet unpublished, method of studying the coupling constant dependence of the total cross-section, which consists of using the analyticity and positivity properties of the forward scattering amplitude with respect to the coupling constant for a potential with one given sign. In some cases, this method gives the best possible accessible results. An amusing by-product is what we call a "Pomeranchuk theorem": we show that if we take a compact positive potential, which is known to give a total cross-section approaching the hard core cross-section for $g \to +\infty$, the corresponding hole obtained by taking g negative produces a sequence of total cross-sections for a sequence of g's $\to -\infty$ which also approach the hard core value.

2. THE METHOD OF FROISSART IN ELEMENTARY PARTICLE SCATTERING AND A POTENTIAL SCATTERING ANALOGUE

For a moment we shall leave the framework of potential scattering and look at the case of elementary particle scattering in which one extensively uses unitarity and analyticity to get a bound on the scattering amplitude and the total cross-section.

The scattering amplitude for two spin 0 particles can be written, with the relativistic normalization:

$$F(s,\cos\theta) = \frac{\sqrt{s}}{2k}\sum_{0}^{\infty}(2\ell+1)f_\ell(s)P_\ell(\cos\theta) \,, \qquad (1)$$

s = square of the centre of mass energy

k = centre-of-mass momentum

θ = c.m. scattering angle

$P_\ell(x)$ = Legendre polynomial.

Unitarity is expressed by a condition on partial waves:

$$\mathrm{Im}\, f_\ell(s) \geq |f_\ell(s)|^2 \,, \qquad (2)$$

from which we get

$$0 \leq \mathrm{Im}\, f_\ell(s) \leq 1 \qquad (3)$$

at high energies, and

$$|f_\ell(s)| \leq 1 \quad . \qquad (4)$$

$\mathrm{Im}\, f_\ell(s)$ appears in the expression of the absorptive part

$$A(s,\cos\theta) = \frac{\sqrt{s}}{2k}\sum_{0}^{\infty}(2\ell+1)\,\mathrm{Im}\, f_\ell(s)P_\ell(\cos\theta) \,, \qquad (5)$$

and via the optical theorem in the total cross-section

$$\sigma_{Total} = \frac{2\pi}{k\sqrt{s}} A(s,1) = \frac{4\pi}{k^2} \sum_{o}^{\infty} (2\ell+1) \operatorname{Im} f_\ell(s) \quad . \tag{6}$$

Condition (3) is not sufficient to get a bound on (6) because we have an infinite sum. We need a control on large angular momentum partial wave amplitudes. Analyticity is the second ingredient we need. We demand that the absorptive part be continued beyond $\cos\theta = 1$ with a convergent Legendre polynomial expansion. Specifically we need

$$A(s, \cos\theta = 1 + \frac{4m^2}{k^2}) < s^2 \tag{7}$$

where m is a non-zero mass typical of the theory.

Inequality (7) has been obtained by Froissart by postulating the Mandelstam representation [4]. I have later shown that it is only a consequence of microcausality, positivity and the existence of a minimum mass in the theory [5]. To get a bound on $\operatorname{Im} f_\ell$ from (7), one uses the fact that each separate term of the series, being positive, is bound by s^2 and a lower bound on Legendre polynomials:

$$P_\ell(1 + \frac{4m^2}{2k^2}) > c_\varepsilon \exp((1-\varepsilon) \frac{2m\ell}{k}), \quad \varepsilon > 0 \ . \tag{8}$$

So we get

$$\operatorname{Im} f_\ell(s) < 1 \ , \tag{9}$$

$$\operatorname{Im} f_\ell(s) < c_1 \, s^2 \exp(-\frac{c_2 \ell}{k}) \ , \tag{10}$$

and we have to take the best of the two. It is easy to see that the critical value of ℓ for which we switch from (9) to (10) is

$$L = c\,k \log s \quad . \tag{11}$$

Beyond L, Im f_ℓ decreases exponentially and so does f_ℓ. So we get

$$\sigma_{Total} < \frac{4\pi}{k^2}\left(\sum_0^L (2\ell+1) + \text{"Tail"}\right) \lesssim \text{const}(\log s)^2 \qquad (12)$$

and

$$|F(s,\cos\theta)| < \text{const. } s(\log s)^2 ,$$

$$-1 < \cos\theta < +1 \quad . \qquad (13)$$

Now the parenthesis on elementary particle scattering is closed and we want to find an analogue in potential scattering to get bounds on amplitudes and total cross-sections. What I present here is an updated version of material contained in lectures I gave at the University of Washington in Spring 1964 [6]. We have, of course, to restrict ourselves to the case of spherically symmetric potentials. We change the normalization of the amplitude:

$$F(k^2,\cos\theta) = \frac{1}{k}\sum_0^\infty (2\ell+1) f_\ell(s) P_\ell(\cos\theta) , \qquad (14)$$

$$\text{Im } f_\ell(s) = |f_\ell(s)|^2$$

or (15)

$$f_\ell(s) = e^{i\delta_\ell} \sin\delta_\ell , \; \delta_\ell \text{ real} ,$$

$$\sigma_{Total} = \frac{4\pi}{k} A(k^2,\cos\theta = 1) = \frac{4\pi}{k^2}\sum_0^\infty (2\ell+1)\,\text{Im } f_\ell(s) \quad . \qquad (16)$$

The content of unitarity, expressed by (15), is not different from what we had in the relativistic case. What there remains to do is to find a condition which ensures that the partial wave amplitudes decrease for large ℓ. Here we are interested in getting conditions as weak as possible on the long range decrease of the interaction. In fact, the most interesting application of these considerations is probably atomic and molecular physics. Therefore, we shall not impose the exponential decrease of the partial wave amplitude, which would, in fact, be equivalent to an exponential decrease of the potential.

The reduced, radial Schrödinger equation

$$\left(-\frac{d^2}{dr^2} + \frac{\ell(\ell+1)}{r^2} - k^2 + V(r)\right) u_\ell(r) = 0 \tag{17}$$

can be transformed into the integral form

$$u_\ell = j_\ell(kr) + \frac{1}{k}\int_0^\infty K_\ell(kr,kr')V(r')u_\ell(r')dr' \tag{18}$$

where

$$j_\ell(x) = \sqrt{\frac{\pi x}{2}}\, J_{\ell+1/2}(x) , \tag{19}$$

$$K_\ell(x,x') = \sqrt{\frac{\pi x}{2}}\sqrt{\frac{\pi x'}{2}}\, J_{\ell+1/2}(x<)H^{(1)}_{\ell+1/2}(x>) \tag{20}$$

with $x< = x$ if $x < x'$, $x< = x'$ if $x > x'$. J and H are standard Bessel functions. The scattering amplitude is then given by

$$e^{i\delta_\ell}\sin\delta_\ell = -\frac{1}{k}\int_0^\infty j_\ell(kr)V(r)u_\ell(r)dr \quad . \tag{21}$$

To study the integral equation (18), we use a separable bound on the kernel (20) which is an updated version of the bound obtained in Seattle [7],

$$|K_\ell(x,x')| < 0.8\ \frac{\sqrt{xx'}}{(\ell+1/2)^{2/3}} \quad . \tag{22}$$

This allows us to get an upper bound on the right-hand side of (18) in which the integral is factored out, and to obtain, after multiplication by $\sqrt{r}V(r)$ and integration,

$$\int |u_\ell(r)||V(r)|\sqrt{r}\,dr$$

$$< \int |j_\ell(kr)|\sqrt{r}|V(r)|dr$$

$$+ \frac{0.8}{(\ell+1/2)^{2/3}} \int r|V(r)|dr \cdot \int |u_\ell(r)||V|\sqrt{r}\,dr , \tag{23}$$

from which it is easy to get a bound on the scattering amplitude if the condition

$$0.8 \cdot (\ell + 1/2)^{-2/3} \int r|V(r)|dr < 1 \tag{24}$$

is satisfied. This bound is

$$|e^{i\delta_\ell} \sin \delta_\ell| < \frac{|e^{i\delta_\ell} \sin \delta_\ell|_{\text{Born},|V|}}{1 - \frac{0.8}{(\ell+1/2)^{2/3}} \int_0^\infty r|V(r)|dr} \tag{25}$$

with

$$|e^{i\delta_\ell} \sin \delta_\ell|_{\text{Born},|V|} = \frac{1}{k}\int |j_\ell(kr)|^2|V(r)|dr \quad . \tag{26}$$

We see now that what replaces the critical L of the relativistic case given by (11) is

$$L = 3 \; [\int r|V(r)|dr]^{3/2} \quad . \tag{27}$$

Then we get bounds on the total cross-section and the scattering amplitude, using $|f_\ell|$ and Im $f_\ell < 1$ for $\ell < L$ and (25) for $\ell \geq L$:

$$|F(k^2,\cos\theta)| < \frac{L^2}{k} + \frac{2}{k} \sum_L^\infty (2\ell+1)|\sin \delta_\ell|_{\text{Born},|V|} , \tag{28}$$

$$-1 \leq \cos\theta \leq +1 ,$$

$$\sigma_{\text{Total}} < \frac{4\pi}{k^2} L^2 + \frac{8\pi}{k^2} \sum_{L}^{\infty} (2\ell+1) |\sin \delta_\ell|^2_{\text{Born},|V|} \quad . \tag{29}$$

In (28) and (29) the series (which are not at all negligible, as opposed to the relativistic case) can be majorized by replacing the sum from L to ∞ by a sum from O to ∞. It is then possible to get closed expressions:

$$\frac{1}{k} \sum_{0}^{\infty} (2\ell+1) |\sin \delta_\ell|_{\text{Born},|V|} = \int \frac{d^3x}{4\pi} |V(x)| ,$$

and

$$\sum_{0}^{\infty} (2\ell+1) |\sin \delta_\ell|^2 =$$

$$\int \frac{d \cos \theta}{2} \left[\sum_{0}^{\infty} (2\ell+1) P_\ell (\cos \theta) \frac{1}{k} \int j_\ell^2 (kr) |V| dr\right]^2$$

$$= \int \frac{(\sin k|x-x'|)^2}{|x-x'|^2} |V(x)||V(x')| \frac{d^3x}{4\pi} \frac{d^3x'}{4\pi} = \frac{k^2}{4\pi} \sigma_B \quad .$$

In conclusion, we get the bounds

$$|F(k^2, \cos \theta)| < \frac{g[\int r|V(r)|dr]^3}{k} + 2\int r^2 |V(r)| dr , \tag{30}$$

$$-1 \leq \cos \theta \leq 1 ,$$

and

$$\sigma_{\text{Total}} < \frac{4\pi}{k^2} \cdot g \, [\int r |V(r)| dr]^3 + 4 \, I \tag{31}$$

where

$$I = \int \frac{|V(x)||V(x')|}{|x-x'|^2} \frac{d^3x}{4\pi} \frac{d^3x'}{4\pi} \quad . \tag{32}$$

In what follows, I will play an important role. If I is

finite the potential belongs to the Rollnik class.

Let us remark that if $V = gv$ and if the various integrals exist, the bounds we have obtained imply

$$|F(k^2, \cos\theta)| < \text{const}\cdot g^3 \ ,$$

$$\sigma_{\text{Total}} < \text{const}\cdot g^3 \quad . \tag{33}$$

It is easy to see that as far as the total cross-section is concerned, we need the convergence of I and $\int r|V(r)|dr$, which is ensured if $V(r)$ decreases slightly faster than r^{-2} at infinity. How much faster? We will postpone this question until the next section where we shall get a bound on σ_{total} in terms of I alone.

3. AN "OPTIMAL" CONDITION FOR THE FINITENESS OF σ_T

Here we remain in the spherically symmetric case and consider only the problem of total cross-sections, not of amplitudes. We want to replace the two conditions of the previous sections by only one, which is the finiteness of I. The trick which was used in Ref.[8] is to find a bound on

$$C_\ell(V) = \frac{1}{k^2}\int r^2 dr\, r'^2 dr' |V(r)||K_\ell(kr,kr')|^2 |V(r')| \tag{34}$$

in terms of I and ℓ.

Suppose you can do this. Take Eq.(18), multiply by $|u_\ell||V|$ and integrate over r. Then use the Schwarz inequality in the right-hand side,

$$\int|u_\ell(r)|^2|V(r)|\,dr < [\int|u_\ell|^2|V|dr \cdot \int j_\ell^2|V|dr]^{1/2}$$

$$+ \int|u_\ell(r)|^2|V(r)|dr \cdot [C_\ell(V)]^{1/2} \quad . \tag{35}$$

Then, if for large enough $\ell, C_\ell(V)$ is less than one, you can keep the integral $\int u^2|V|dr$ under control and get bounds on the partial wave amplitude. In Ref.[8] we notice that

$$K_\ell(|x|,|x'|) = -\frac{i|x||x'|}{2}\int_{-1}^{+1}\frac{e^{i|x-x'|}}{|x-x'|}P_\ell(\cos\theta)\,d\cos\theta,$$

where θ is the angle between x and x'. Hence, by Schwarz inequality

$$|K_\ell(|x|,|x'|)|^2 < |x|^2|x'|^2\int\frac{d\cos\theta}{2|x-x'|^2}\cdot\frac{1}{2\ell+1} \tag{36}$$

which leads to

$$C_\ell(V) < \frac{I}{2\ell+1} \quad . \tag{37}$$

However, while writing these notes the author realized that it is possible to appreciably improve (36). From the new estimates of Ref.[1] on $K_\ell(x,x')$ we have

$$K_\ell(|x|,|x'|) < 0.8\,\frac{\operatorname{Im} f(|x|,|x'|)}{(\ell+1/2)^{2/3}} \quad . \tag{38}$$

We now have

$$\operatorname{Im} f(|x|,|x'|) < \left[\frac{|x||x'|}{2}\log\frac{|x|+|x'|}{||x|-|x'||}\right]^{1/2} \quad . \tag{39}$$

The bracket on the right-hand side of (39) is precisely proportional to the right-hand side of (36) and we get

$$|K_\ell(|x|,|x'|)|^2 < \frac{0.64}{(\ell+1/2)^{4/3}}\,|x|^2|x'|^2\int\frac{d\cos\theta}{2|x-x'|^2}$$

and hence

$$C_\ell(V) < \frac{I\cdot 0.64}{(\ell+1/2)^{4/3}} \quad . \tag{40}$$

So if $\ell > L_o$ given by

$$L_o + 1/2 = (0.64 \cdot I)^{3/4} \quad , \tag{41}$$

we get a bound on the partial wave amplitude,

$$|e^{i\delta_\ell} \sin \delta_\ell| < \frac{|e^{i\delta_\ell} \sin \delta_\ell|_{\text{Born}, |V|}}{1 - 0.8 \dfrac{\sqrt{I}}{(\ell+1/2)^{2/3}}} \quad , \tag{42}$$

and from there we get a bound on the total cross-section ,

$$\sigma_T < \frac{4\pi}{k^2} \left(L^2 + \frac{I}{\left(1 - 0.8 \dfrac{\sqrt{I}}{(L+1/2)^{2/3}}\right)^2}\right) \quad , \tag{43}$$

for any L integer $> L_o$. Taking for instance $L = I^{3/4}$ we get

$$\sigma_T < \frac{4\pi}{k^2} [I^{3/2} + 25 \cdot I] \quad . \tag{44}$$

Like the bound (31) this corresponds to $\sigma_T <$ const. $\cdot\ g^3$ if $V = gv$.

We claim that the finiteness of I is an optimal condition. Naturally, we speak of the long distance part of the integrand. It is clear that if the potential contains hard obstacles in a compact region, this will not make the cross-section infinite. Suppose, however, that this is not the case. Suppose also that the potential is not oscillating. Then the Born approximation

$$\frac{4\pi}{k^2} \int \frac{\sin^2 k|x-x'|}{|x-x'|^2} |V(x')| |V(x)| d^3x\, d^3x'$$

is not very different from I/2. If I diverges, the Born

approximation diverges too and this may lead to an infinite total cross-section. We shall prove this in a specific case. Assume

$$\begin{aligned} &\text{i)} \quad I = +\infty \\ &\text{ii)} \quad \int |V(r)|\, r^{1/2} dr < \infty \\ &\text{iii)} \quad V(r) \geq 0 \;. \end{aligned} \tag{45}$$

A potential like $(1+r^{1.99})^{-1}$ will satisfy these conditions. Under condition (i) one can still manage to control the large ℓ behaviour of partial wave amplitudes. Indeed, from the inequalities [7]

$$|K_\ell(x,x')| < .8\cdot(\ell+1/2)^{-2/3}\sqrt{xx'} \;,$$

$$|K_\ell(x,x')| < 1.2\cdot(\ell+1/2)^{1/3}$$

one deduces

$$|K_\ell(x,x')| < (\ell+1/2)^{-1/6}(xx')^{1/4} \tag{46}$$

and, by the same machinery which leads to Eq.(23), one gets (for simplicity, we take here k = 1)

$$\int |u_\ell(r)||V(r)|\, r^{1/4} dr \;\, [1-(\ell+1/2)^{-1/6}\int |V(r)|\, r^{1/2} dr]$$

$$< \;\; [\int j_\ell^2(r)\,|V(r)|\,dr\cdot\int |V(r)|\, r^{1/2} dr]^{1/2} \quad , \tag{47}$$

from which it is not difficult to get, for

$$\ell > (\int |V(r)|\, r^{1/2} dr)^6 \;,$$

$$|\sin\delta_\ell| > |\sin\delta_\ell|_{\text{Born}}\;\cdot$$

$$\cdot\;\frac{1-2(\ell+1/2)^{-1/6}\int r^{1/2}V(r)\,dr}{1-(\ell+1/2)^{-1/6}\;\int r^{1/2}V(r)\,dr} \tag{48}$$

and hence, taking $L = 4096\cdot\int V(r)\, r^{1/2} dr$,

$$\frac{k^2}{4\pi}\,\sigma_{\text{Total}} > -\frac{2}{3}\sum_{0}^{L}(2\ell+1)\sin^2\delta_{\ell,\text{Born}}$$

$$+\frac{2}{3}\sum_{0}^{\infty}(2\ell+1)\,\sin^2\delta_{\ell,\text{Born}} \quad . \tag{49}$$

Each $\sin\delta_{\ell,\text{Born}}$ is finite under the condition (ii) in Eq.(45), but the last series, which, except for a factor $\sin^2(|\vec{x}-\vec{x}'|)$ in the integrand, coincides with I, is divergent. The only way to save the convergence would be to have on oscillating V with carefully chosen oscillations. We believe that the technical assumption that $r^{1/2}V$ is integrable near the origin is inessential.

4. EXTENSION TO THE NON-SPHERICALLY SYMMETRIC CASE

First of all we must remember that if the potential has no symmetry, the total cross-section at a given energy will depend on the direction of the incident beam. It is clear for instance that if you take a flat disc, the cross-section will be maximum if the disc is perpendicular to the beam and very small (classically zero) if the disc is parallel to the beam. Under these circumstances, most authors introduce the notion of averaged total cross-section:

$$\bar{\sigma}_T(k) = \int \frac{d\Omega_k}{4\pi}\,\sigma_T(\vec{k}) \quad . \tag{50}$$

We shall also use this notion in the first part of this section, but later on we shall try to get conditions for the existence of σ_T for precise energy and direction of the incident beam.

Here we cannot eliminate the lower partial waves in which resonances can occur and therefore we shall modestly look for a situation in which the Born series converges.

We have

$$\psi_{\vec{k}}(x) = \phi_{\vec{k}}(x) + \int \frac{d^3x'}{4\pi} \frac{e^{ik|x-x'|}}{|x-x'|} V(x')\psi_{\vec{k}}(x') \tag{51}$$

where $\phi_{\vec{k}}$ represents a plane wave, or

$$\psi_{\vec{k}}(x) = \phi_{\vec{k}}(x) + \int \frac{d^3x'}{4\pi} G_k(x,x')V(x')\phi_{\vec{k}}(x') \tag{52}$$

where $G_k(x,x')$ is the Green's function satisfying the integral equation

$$G_k(x,x') = \frac{e^{ik|x-x'|}}{|x-x'|} + \int \frac{d^3x''}{4\pi} \frac{e^{ik|x-x''|}}{|x-x''|} V(x'')G_k(x'',x') . \tag{53}$$

The total cross-section is given by

$$\sigma_T(\vec{k}) = -\frac{1}{4\pi k}\mathrm{Im} \int e^{-i\vec{k}\cdot\vec{x}} V(x)G_k(x,x')V(x')e^{i\vec{k}\vec{x}'} d^3x\, d^3x' , \tag{54}$$

and the average total cross-section by

$$\bar{\sigma}_T(k) = -\frac{1}{4\pi k} \mathrm{Im} \int d^3x\, d^3x' \frac{\sin k|x-x'|}{k|x-x'|} V(x)G_k(x,x')V(x') . \tag{55}$$

Let us prove that if I, defined by Eq.(32), is less than unity, the Born series for G converges. Take the square of the modulus of both sides of (53), multiply by $|V(\vec{x})||V(\vec{x}')|$ and integrate. If we call

$$K = \int |V(x)||G_k(x,x')|^2|V(x')|d^3x\, d^3x' \tag{56}$$

and repeatedly use Schwarz inequalities, we get

$$K < I + 2\, I\, K^{1/2} + K\, I^{1/2}$$

which implies, if I < 1,

$$K^{1/2} < \frac{2I^{1/2}}{1 - I^{1/2}} \quad . \tag{57}$$

On the other hand, application of the Schwarz inequality to (55) gives

$$\bar{\sigma}_T(k) < \frac{4\pi}{k} \sqrt{IK}$$

$$< \frac{8\pi}{k} \frac{I}{1 - I^{1/2}} \quad . \tag{58}$$

The averaged total cross-section is therefore bounded if I is less than unity.

We believe that in fact the condition I < 1 is not really essential. If I is finite, we know that $G_k(x,x')$ exists, except perhaps in a set of measure zero. Enss, during this series of lectures, pointed out to me that since what decides the finiteness of the potential is the outer part of the potential, one can always remove from the potential what is inside a sufficiently large sphere of radius R, so that I < 1.

We would now like to discuss the finiteness of total cross-sections for a precise incident direction [9]. If we expand the total cross-section in a series we see that the first term is

$$-\frac{1}{k} \operatorname{Im} \int e^{-i\vec{k}\cdot\vec{x}} V(x) \frac{e^{ik|x-x'|}}{4\pi|x-x'|} V(x') \, e^{i\vec{k}\cdot\vec{x}'} \, d^3x \, d^3x'$$

$$= \frac{1}{k} \int \cos(\vec{k}\cdot(\vec{x}-\vec{x}'))V(x) \frac{\sin(k|x-x'|)}{4\pi|x-x'|} V(x') d^3x \, d^3x'$$

which is bounded (except for a factor 1/k) by

$$J = \int |V(x)| \frac{1}{|x-x'|} |V(x')| d^3x \, d^3x' \quad . \tag{59}$$

This indicates that we should supplement the condition I finite by J finite if we want to guarantee that $\sigma_T(\vec{k})$ is finite without averaging. We shall now prove this statement in the case I < 1 where the perturbation series for G_k converges. Let us look at the nth term of the perturbation expansion of the cross-section. It is bounded by

$$\sigma_n = \int |V(x_1)| \frac{1}{4\pi|x_1-x_2|} |V(x_2)| \ldots |V(x_n)| d^3x_1\, d^3x_2 \ldots d^3x_n \ . \tag{60}$$

To majorize σ_n in terms of I and J we multiply the integrand by

$$\frac{|x_1-x_2|^{1/2}+|x_2-x_3|^{1/2}+\ldots+|x_{n-1}-x_n|^{1/2}}{|x_1 - x_n|^{1/2}} \geq 1 \ .$$

In this way, we get a sum of terms

$$\begin{aligned} \sigma_n &< \mathrm{Tr}\ H\ H\ \underbrace{K\ K\ \ldots\ K}_{n-2} \\ &+ \mathrm{Tr}\ H\ K\ H\ K\ \ldots\ K \\ &+ \text{etc.}\ldots \end{aligned} \tag{61}$$

with

$$H = |V(x)|^{1/2} \frac{1}{4\pi|x-x'|^{1/2}} |V(x')|^{1/2}, \tag{62}$$

$$K = |V(x)|^{1/2} \frac{1}{4\pi|x-x'|} |V(x')|^{1/2} \ , \tag{63}$$

and then

$$\begin{aligned} \sigma_n &< (n-2)\ \mathrm{tr}(H^2)\cdot(\mathrm{tr}\ K^2)^{\frac{n-2}{2}} \\ &= (n-2)\,(4\pi)^2\ J\cdot I^{\frac{n-2}{2}} \end{aligned} \tag{64}$$

so that σ_n is finite if J and I are finite and, in addition, the series $\sum\sigma_n$ converges for $I < 1$.

It is very difficult to know if the additional condition J finite is too strong or not. If V is dominated by a spherically symmetric potential W,

$$|V(x)| < W(|x|) ,$$

J will be finite if $W(|x|)$ decreases faster than $|\vec{x}|^{-2,5}$. This means that the condition J finite is weaker than the requirement that $|V(x)|$ should be integrable. If I is less than unity and J finite, and if $\int V(x)d^3x$ diverges, we have a situation where the total cross-section is finite for any direction, but the forward amplitude is infinite. Indeed, it is trivial to see that the conditions $I < 1$ and $J < \infty$ ensure not only the finiteness of the total cross-section, but also that of the scattering amplitude amputated from the Born term. If the Born term is infinite the forward amplitude is therefore infinite also.

5. THE COMPLEX g APPROACH TO TOTAL CROSS-SECTIONS

We now want to propose a new approach to study the growth of the potential with the coupling constant, based on analyticity properties of the forward scattering amplitude with respect to the coupling constant. We take the potential to be

$$V = gv$$

and, for reasons which will become clear later, we take

$$v \leq 0 .$$

Then, if $\text{Im}\, g > 0$, $\text{Im}\, V < 0$. We shall, in addition, assume that

$$I = \frac{1}{(4\pi)^2} \int \frac{v(x)v(x')}{|x-x'|^2} d^3x \, d^3x'$$

is finite, and also in the first part we shall assume that the forward scattering amplitude exists:

$$\int |v(x)| d^3x < \infty \quad . \tag{65}$$

Then, the potential for Im g > 0 is an absorptive potential, familiar to nuclear physicists who want to describe, for instance, neutron scattering by nuclei, with the special restriction that the real and imaginary parts are proportional. Everybody believes that such potentials are all right: the S matrix exists, etc. I believe that it is not terribly difficult to prove rigorously that $F(g, \cos\theta = 1)$, the forward amplitude, is analytic in the upper half plane in g. In the case of a spherically symmetric potential, the strategy I would use is to decompose into partial waves whose asymptotic form is

$$e^{-i(kr-\frac{\ell\pi}{2})} + \eta_\ell \, e^{i(kr-\frac{\ell\pi}{2})}$$

and prove, using the Poincaré theorem, that η_ℓ is analytic in g. The absorptive character of V guarantees that $|\eta_\ell| \leq 1$, and therefore that the S matrix has no poles. It remains to resum the partial wave amplitudes. This may be tedious but causes no real problem. This being said, the total cross-section is now made of two terms,

$$\sigma_T = \sigma_{elastic} + \sigma_{Abs} \quad , \tag{66}$$

$$\sigma_T = \frac{4\pi}{k} \, \text{Im} \, F(g, \cos\theta = 1) = -\frac{4\pi}{k} \, \text{Im} \int \phi^*_{\vec{k}} V \psi_{\vec{k}} \, d^3x \quad , \tag{67}$$

where ϕ_k is a plane wave and ψ_k the exact wave function. $\sigma_{elastic}$ is clearly positive and

$$\sigma_{Abs} = -\frac{4\pi}{k}\int|\psi_k|^2 \text{ Im } V d^3x \tag{68}$$

is positive if Im V is ≤ 0. This guarantees that the total cross-section is positive. The conclusion is that F(g), the forward scattering amplitude, is analytic in Im g ≥ 0 with Im F(g) ≥ 0. Such a function is called a Herglotz function. For a discussion of Herglotz functions I refer to the book of Shohat and Tamarkin, "The Problem of Moments" [10].

If F is a Herglotz function, it admits the representation

$$F(g) = A+Bg + \frac{1}{\pi}\int_{-\infty}^{+\infty} \frac{(C+gg')\text{Im } F(g')}{(C+g'^2)(g'-g)} dg' , \tag{69}$$

with A, B, C real,

$$B > 0 , \quad C > 0 , \text{ arbitrary} ,$$

and the properties

i) $\frac{F(g)}{g} \to B \quad$ for $g \to \infty , \quad 0 < \text{Arg } g < \pi ,$

ii) $\int_{-\infty}^{+\infty} \frac{\text{Im } F(g')dg'}{D^2+g'^2} < \infty .$

In our case, we can take C = 0 because Im F(g) behaves like g^2 at $g \to 0$, and the representation reduces to

$$F(g) = A + Bg + \frac{g}{\pi}\int_{-\infty}^{+\infty} \frac{\text{Im } F(g')dg'}{g'(g'-g)} . \tag{70}$$

Property i) is precisely what we want. It shows that, in an averaged sense, $\sigma_T(g)$ increases slower than g, or more exactly $g(\log g)^{-1}$.

Conclusion:

If

$$\int |v| d^3x < \infty \quad \text{and}$$

$$I < \infty \; ,$$

$$\int_{-\infty}^{+\infty} \frac{\sigma_T(g')dg'}{D^2+g'^2} < \infty \; . \tag{71}$$

The case where the forward amplitude is infinite

If $F(g_1 \cos\theta = 1)$ is infinite, it remains that if I is finite, what we can write as "$F-F_{Born}$", averaged over incident directions, still exists. To make things precise, let us define $V_\varepsilon = V\exp(-\varepsilon r)$ with the intention of letting ε go to zero. Then

$$\lim_{\varepsilon\to 0} (F_\varepsilon - F_{\varepsilon,Born}) = \lim_{\varepsilon\to 0} (F_\varepsilon - g\, f_{\varepsilon,Born})$$

exists. We define

$$\phi_\varepsilon = \frac{F_\varepsilon - g\, f_{\varepsilon,Born}}{g^2} \; . \tag{72}$$

ϕ_ε exists at $g^2 = 0$. ϕ_ε behaves like $1/g$ for g going to ∞ in complex directions and therefore satisfies the integral representation

$$\phi_\varepsilon = \frac{1}{\pi} \int_{-\infty}^{+\infty} \frac{\operatorname{Im} F_\varepsilon(g')dg'}{g'^2(g'-g)} \; . \tag{73}$$

On the other hand, $\phi_\varepsilon \to \phi$ for $\varepsilon \to 0$, at least inside the (finite) radius of convergence of the power series of the amplitude in g around $g = 0$; (the radius is > 0 because $I < \infty$). Take the imaginary part of (73) for some complex value of g inside the radius of convergence. We have

$$\text{Im}\ \phi_\varepsilon(g) = \frac{\text{Im}\ g}{\pi} \int_{-\infty}^{+\infty} \frac{\text{Im}\ F_\varepsilon(g')dg'}{g'^2|g'-g|^2} \quad . \tag{74}$$

$\text{Im}\ \phi_\varepsilon(g) \to \text{Im}\ \phi(g)$ for $\varepsilon \to 0$. On the other hand, notice that the integrand in (74) is positive. The integrand has a limit as $\varepsilon \to 0$. We are in a case of dominated convergence and

$$\text{Im}\ \phi(g) \geq \frac{\text{Im}\ g}{\pi} \int_{-\infty}^{+\infty} \frac{\text{Im}\ F(g')dg'}{g'^2|g'-g|^2} \quad .$$

The conclusion is

$$\int_{-\infty}^{+\infty} \frac{\sigma_{\text{Total}}(g')dg'}{g'^2(g'^2+C^2)} < \infty \quad . \tag{75}$$

This corresponds to an average increase slower than $g^3(\log g)^{-1}$. This result is not as good as that of Enss and Simon for potentials decreasing like $r^{-2-\varepsilon}$, which is that σ_{total} cannot increase faster than g^2. However, the condition $I < \infty$ contains potentials behaving like $r^{-2}(\log r)^{-1/2-\varepsilon}$. It is not yet excluded that for such potentials one could have such an increase for particular energies.

Finally, we want to illustrate the use of this complex g technique by an "analogue" of the "Pomeranchuk theorem". Let me remind you that in elementary particle scattering, the particle-particle amplitude and particle-antiparticle amplitude are connected by crossing symmetry and correspond to boundary values of the same analytic function for energies approaching $\pm\infty$. Then, the "Pomeranchuk theorem" says that

$$\lim_{\text{Energy} \to \infty} \left[\sigma_{\text{Total}}\binom{\text{particle}}{\text{-particle}} - \sigma_{\text{Total}}\binom{\text{particle}}{\text{-antiparticle}}\right] = 0 \ .$$

If the limit does not exist, zero belongs to the set of limiting values of the cross-section.

Our analogue is the following. Take the case of a potential gv of <u>compact support,</u> v being purely negative. Then, as noticed for instance by Enss and Simon, for $g \to -\infty$ the total cross-section tends to a limit which is the cross-section of a hard core. Incidentally, this limit is not necessarily reached monotonously, even though the phase shifts (Kato monotony) are monotonous functions of g. Anyway, the limit exists. The limit of the full amplitude exists as well. So we have

$$F(g) \to c \quad \text{for} \quad g \to -\infty \ .$$

Intuitively, we have higher and higher obstacles as g increases, and in the end the detailed shape of the potential does not count, but only its support:

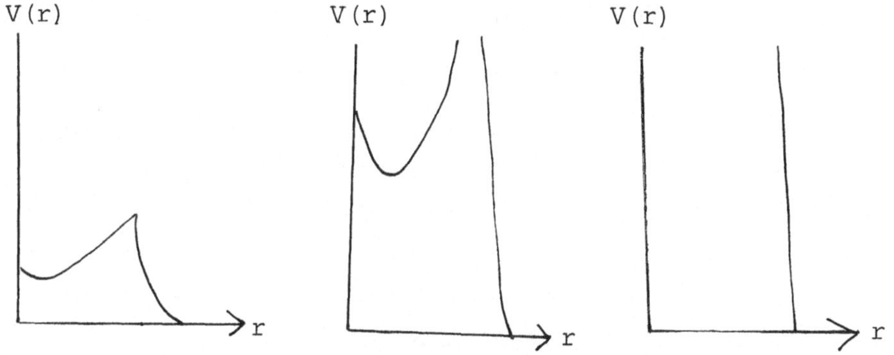

The question you can ask yourself is what happens for $g \to \infty$, i.e., when you have a deeper and deeper hole.

Let us first notice that if F(g) is Herglotz and if $F(g) \to c$ for $g \to -\infty$, $F(g) \to c$ for $g \to \infty$ in <u>any complex direction.</u> The proof is as follows:

$$|\exp i\, F(g)| = \exp(-\, \mathrm{Im}\, F(g)) \leq 1 \ .$$

So exp(iF(g)) is bounded in Im F(g) > 0 and has a limit for $g \to -\infty$. A small modification of Montel's theorem says that exp(iF(g)) has exactly the same limit in any complex direction. Naturally, if F(g) has a limit for $g \to +\infty$, this limit is necessarily again the same and we would have

$$\lim_{g \to +\infty} (\sigma(g) - \sigma(-g)) = 0 .$$

In fact this is not so. Lim $\sigma(g)$ for $g \to +\infty$ does not exist. However, we have a statement completely analogous to the Pomeranchuk theorem [11]:

$$\int_0^\infty \frac{\sigma_T(g) - \sigma_T(-g)}{g} dg$$

converges. Therefore, 0 belongs to the set of limiting values of $\sigma_T(g) - \sigma_T(-g)$. Harald Grosse has checked that this is the case for a spherical square well. As $g \to +\infty$, one finds that $\sigma_T(g)$ stays very close to the hard core value except for some occasional spikes of the order of $g^{1/2}$.

It has been pointed out to me by Dr. Enss that, for a cubic well, one gets oscillations of the cross-section from zero to twice the hard core value as $g \to +\infty$.

REFERENCES

1. W.O. Amrein and D.C. Pearson, J. Phys. A12 (1979) 1469; W.O. Amrein, D.C. Pearson and K.B. Sinha, Scattering Theory in Quantum Mechanics, Reading, Benjamin (1977). This paper contains references to previous work.
2. V. Enss and B. Simon, Phys. Rev. Letters 44 (1980) 319 and 764;
   V. Enss and B. Simon, to appear in Comm. Math. Phys.;
   V. Enss and B. Simon, preprint to appear in Classical,

Semi-Classical and Quantum Mechanical Problems in Mathematics, Chemistry and Physics, editors K.Gustafson and W.P. Reinhardt, Plenum (1980/81).

3. J.M. Combes and M. Guez, private communication, to be published.
4. M. Froissart, Phys. Rev. 123 (1961) 1053.
5. A. Martin, Nuovo Cimento 42 (1966) 930.
6. A. Martin, Lectures at the University of Washington, Seattle (1964), unpublished.
   See also A. Martin, Nuovo Cimento 23 (1962) 641;
   A. Martin, Nuovo Cimento 31 (1964) 1229.
7. K. Chadan and A. Martin, appendix of Comm. Math. Phys. 70 (1979) 1.
8. A. Martin, Comm. Math. Phys. 69 (1979) 89;
   Notice that in the present lectures we give an improved version in which the total cross-section is bounded by $I^{3/2}$ for large I, instead of I^2.
9. A. Martin, Comm. Math. Phys. 73 (1980) 79.
10. J.A. Shohat and J.D. Tamarkin, "The Problem of Moments", American Mathematical Society, New York (1943).
11. See for instance, A. Martin, Nuovo Cimento 39 (1965) 704.

Acta Physica Austriaca, Suppl. XXIII, 235–328 (1981)

ANALYTICITY PROPERTIES OF THE S-MATRIX: HISTORICAL SURVEY AND RECENT RESULTS IN S-MATRIX THEORY AND AXIOMATIC FIELD THEORY[+]

by

D. IAGOLNITZER

DPh-T, CEN Saclay, 91191 Gif-sur-Yvette, France

ABSTRACT

An introduction to recent works, in S-matrix theory and axiomatic field theory, on the analysis and derivation of momentum-space analyticity properties of the multi-particle S-matrix is presented. It includes an historical survey, which outlines the successes but also the basic difficulties encountered in the sixties in both theories, and the evolution of the subject in the seventies.

CONTENTS

[+]Lectures given at XX. Internationale Universitätswochen für Kernphysik, Schladming, Austria, February 17-26, 1981.

INTRODUCTION

The aim of these lectures is to provide an introduction to recent works on the analysis and derivation of the momentum-space analytic structure of multiparticle collision amplitudes between sets of massive particles with short-range interactions. We shall not attempt here to give technical details (for a more precise presentation, see [1,2]), but we wish instead to describe the general evolution of the subject, starting from the traditional approaches and results of the sixties, outlining their successes, but also their limitations and difficulties, and then introducing the more recent investigations carried out in the seventies in close connection with related mathematical developments.

Recent results from the viewpoint of mathematical physics concern mainly, either in S-matrix theory or in axiomatic field theory, the study of the physical-region analytic structure, to which we shall therefore mostly devote our attention. The physical region of a given process is the set of all real initial and final energy-momenta variables subject to the mass-shell conditions and to energy-momentum conservation (see Eq.(3) at the end of this section). For a two-body equal-mass process (two initial and two final particles with mass $\mu > 0$), it reduces, in terms of the usual squared center-of-mass energy variable $s = (p_1+p_2)^2$, to the region s real, $s \geq 4\mu^2$ (with some further conditions on the momentum transfer t not specified here): see Fig. 1 in Sect. 2. The study of the physical-region structure may therefore seem very limited, since one knows that analyticity on the complex mass-shell far away from the physical region is needed, already in the two-body case, for crossing, dispersion relations,... It has, however, proved to have its own physical and mathematical interest in the general multiparticle case and can be considered as a basic starting point for further developments. This has already appeared in a number of heuristic studies in S-matrix theory, e.g. the derivation in the sixties of crossing, hermitean analyticity and related results (see Sect.4), and in the seventies of multiparticle dispersion relations [3] and of multi-Regge theory [4]. It is also the viewpoint of

S-matrix theorists [5] that the understanding of the physical-region analytic structure should indeed be one of the basic ingredients , together with further topological assumptions, for investigating modern theories of particle physics (including quarks, gluons,...), and interesting developments have been carried out in this direction. This is, however, another subject and I will not argue about, or discuss it here.

In Sect. 1, the situation for the two-body S matrix is briefly reviewed.

In Sect. 2, we then come to the multiparticle case and describe general physical-region and analyticity properties, to be made more precise and completed in Sect. 3: (i) analyticity outside $+\alpha$-Landau surfaces and plus $i\varepsilon$ rules at $+\alpha$-Landau points, (ii) à la Cutkosky local discontinuity formulae around the $+\alpha$-Landau singularities and related "à la Feynman" representations, which give detailed information on the S-matrix structure. We also introduce corresponding decompositions of the S-matrix in terms of "generalized Feynman integrals" associated with graphs with one internal line and triangle graphs in the simple case, which is that treated in recent works both in S-matrix theory and axiomatic field theory (see Sect. 5,6), of a $3 \rightarrow 3$ equal-mass process below the 4-particle threshold. Properties (i) and (ii) were already suggested to a large extent at the beginning of the sixties on the basis of perturbation theory (i.e. the study of analyticity properties of Feynman integrals), and further heuristic arguments. Some errors made at that time have, however, to be corrected.

It soon appeared that these analyticity properties should be independent of perturbation theory, and the aim of the various works we shall describe was to get more precise and general statements as well as a satisfactory physical understanding of these properties, to check their internal consistency (in particular with unitarity), and to derive them, as far as possible, from basic axioms. Progress in axiomatic field theory was stopped in the sixties, in the multiparticle case, by both conceptual and technical difficulties that required a deeper analysis of the general structure of the theory and new mathematical methods. As a consequence, results on analyticity properties (on the mass-shell) have been obtained only recently in that framework. We shall come back to it later. The first important developments in the multiparticle case are those of the approach usually called Analytic S-matrix theory, which is independent of field theory and is based on unitarity and on Chew's basic principle of "maximal analyticity" [6]. Its heuristic successes in the sixties, as well as its difficulties, in particular in the derivation [7-9] which is a basic part in this program of the physical-region discontinuity formulae, will be outlined in Sect. 4. This derivation is based on the analysis of unitarity equations, starting from general preliminary assumptions on scattering functions associated in the physical region with the idea of "maximal analyticity": analyticity outside Landau surfaces and plus $i\varepsilon$ rules. These assumptions are completed in [9] by assuming from the outset analyticity outside the $+\alpha$-parts of the Landau surfaces, as follows e.g. from macrocausality: see below, whereas this is in [7,8] part of the results one aims to prove. The difficulties encountered are crucial as we shall explain on a simple example. They were in the sixties either ignored in [7,8] (and as a consequence, the proofs given are crucially not correct as they stand, in spite of the actual value from many viewpoints of these works), or were treated by ad hoc, more or less precise assumptions in [9]. A more satisfactory understanding of the problems and partial solutions have been obtained in the seventies, as described below. This requires, however, the preliminary introduction of macrocausality and of essential support theory, or of hyperfunction theory.

In the same time as analytic S-matrix theory was developed, it soon became as a matter of fact clear [10-19] (as already foreseen in some earlier investigations [20] for the two-body S-matrix) that physical-region analyticity properties were intimately linked with the macroscopic space-time description of processes in terms of multiple scattering (see in particular [10]), and they were shown in the second part of the sixties to be essentially equivalent [18,19] to general properties of macrocausality and macrocausal factorization. These properties, in the general form given in [18,19] that allows this equivalence, are refined versions and generalizations of the cluster properties considered [12,13], or proved in field theory [14], in the middle of the sixties, and are the general outcome of the developments of the subject in various directions [15-17]. Macrocausality is a general causality property in terms of particles, in macroscopic space-time. (It does not directly follow from microcausality which refers to underlying fields at the microscopic level: see discussion later.) It states exponential fall-off properties under space-time dilation, or more generally factorization properties, of transition probabilities between appropriate sequences of initial and final displaced wave functions (corresponding to particles that are well localized asymptotically along classical trajectories), for non-causal configurations of these particles, or of subgroups of particles, i.e. when they cannot be linked causally via real stable intermediate particles in accordance with classical ideas. It is shown [18] to be (essentially) equivalent to analyticity of scattering functions outside the $+\alpha$-Landau surfaces and to the plus $i\varepsilon$ rules. Macrocausal factorization is a stronger factorization property applying to the case of causal configurations, and is in turn equivalent (in simple cases) to the discontinuity formulae and related results, such as, for the $3 \rightarrow 3$ S-matrix below the 4-particle threshold, the decomposition in terms of generalized Feynman integrals of Sec. 2.3.

These results are described in Sect. 3.1 (macrocausality) and 3.2 (macrocausal factorization) where the general and natural mathematical statements [21] of these properties in terms of the notion of essential support introduced below are also given. These statements cover situations where the usual plus $i\varepsilon$ rule or discontinuity formulae cannot apply (e.g. some of the points that lie on several $+\alpha$-Landau surfaces). They can be expressed equivalently in terms of the notion of singular spectrum of hyperfunction theory, as proposed in [22,23], in view of the equivalence (see below and Appendix) of the two mathematical notions, and are also called micro-analyticity property and microlocal discontinuity formulae, following a suggestive language (first used in hyperfunction theory) which refers to the fact that, according to these statements, certain directions (in the space of space-time displacements of the particles) are "non-singular" (= outside the essential support, or singular spectrum) directions of relevant distributions at $+\alpha$-Landau points. Note, however, that this is a pure question of mathematical language which adds nothing by itself and does not refer to the notion of microcausality or microlocality of field theory.

Before giving indications on the mathematical part (which will play a more crucial role in further results: see below), let us summarize the situation as it thus appears at the end of the sixties. As we have seen (Sect. 3), the physical-region analytic structure that can be reasonably expected is equivalent to, and can thus be derived from macrocausality and macrocausal factorization if the latter are considered as basic properties of the theory. Macrocausal factorization has a satisfactory physical interpretation, but is a strong property compared e.g. to the general property of macrocausality, and is not likely to be independent of macrocausality and unitarity. From an axiomatic viewpoint, it is thus desirable to derive it, if possible, together with the related analytic structure, from these more basic principles. This program is a closely related, more precise version of that already outlined above (= derivation of the physical-region discontinuity formulae). In the alternative philosophy of "maximal analyticity" (= accept the strongest analyticity properties that appear to be the "simplest ones" compatible with unitarity), the aim (which is to some extent complementary to the previous one) is rather to check the internal consistency, from various viewpoints, of the physical-region analytic structure previously discussed with unitarity. Finally, if one goes back to axiomatic field theory, then both macrocausality and macrocausal factorization, or the related analyticity properties, have to be derived from the basic axioms of the theory. The various results obtained in the seventies in S-matrix theory (from the two viewpoints mentioned above) and in axiomatic field theory will be outlined below and in more detail in Sec. 4,5 and in Sect. 6 respectively. All make use of the results of closely related developments of essential support theory, or of hyperfunction theory.

Essential support theory was developed in the first part of the seventies by a collaboration of the present author with J. Bros, starting from the earlier results of [18]: see [24] (and references therein). An outline of the basic notions and results is given here in Sect. 1 of the Appendix. The essential support of a distribution f is, at each real point, a cone with apex at the origin in the space of dual variables, composed of "singular directions" along which a generalized local Fourier transform of f does not fall off exponentially in a well defined sense. It is shown to characterize by duality in a well defined way the real points where f is analytic, or is the boundary value of an analytic function, and more generally possible decompositions of f into sums of boundary values of analytic functions (obtained from directions that may depend on the real point considered). It leads to simple and powerful generalizations of edge-of-the-wedge theorems, and to general theorems on the analytic structure (= essential support) of products of distributions, integrals, restrictions,... that extend previous results involving distributions that are boundary values of analytic functions. These theorems will be usefully applied to the unitarity-type, on-mass-shell integrals that will occur either in S-matrix theory or in axiomatic field theory. They do not cover, however, certain "u = O" situations that will play a crucial role in S-matrix theory. The more recent u = O results of [25,26] are based on a new regularity property R on the way rates of exponential fall off tend to zero where directions of the essential support (= causal directions in the application) are approached. They are outlined in Sect. 2 of the Appendix.

In Sect. 3 of the Appendix, we introduce hyperfunction theory [27], which was developed independently by very different methods, but led in the first part of the seventies to notions and results which turned out to be very closely related to those of essential support theory described in Sect. 1[+]. In particular, the notions of essential support and singular spectrum characterize analyticity properties in the same way, and coincide for distributions. Hyperfunction theory does not play a crucial role in these lectures since most of the results described have been obtained within the more simple framework of essential support theory. Powerful results [28] on Feynman integrals have, however, been obtained in connection with the further notion of <u>holonomicity</u> [27], or regular <u>holonomicity</u> [29], introduced in hyperfunction theory and briefly presented also in the Appendix. This same notion

[+]Part of these results have been, as a matter of fact, inspired by the analogous ones of [27].

and related results have also led to some recent results, to be mentioned in Sect. 5,7 on the u = 0 problem and on the nature of the S-matrix singularities.

We now return to S-matrix theory. The developments of essential support theory have first led in the first part of the seventies to improve in various ways the previous analysis of [9], and to remove in particular one of its technical assumptions (= "patching assumption") in view of the (u $\neq$ 0) structure theorem on unitarity-type integrals of [30]. The results obtained are outlined at the end of Sect. 4. They are still crucially based, however, besides unitarity and macrocausality on an ad hoc assumption of "mixed-α cancellation" or more generally of "separation of singularities in unitarity equations". The property of "separation of singularities" is believed to hold and has been conversely checked in a number of cases from the discontinuity formulae and related results, in a heuristic [31] or more precise [32] way. This is satisfactory to some extent, in particular from the viewpoint of "maximal analyticity" ideas (= checking internal consistency) mentioned earlier. From a deductive, axiomatic viewpoint, "separation of singularities" cannot, however, be considered as a basic axiom and should itself be derived rather than used as an ad hoc crucial assumption. The recent works on this subject are described in Sect. 5.

The problem can be decomposed into two parts. The first one is the u = 0 problem which à priori prevents one from obtaining any result at all at any point on the analytic structure of some of the unitarity-type integrals

(such as )

encountered even in the simplest cases. It has been solved in [25] on the basis of a regularity property corresponding physically to a refined form of macrocausality (see in this connection the previous mathematical discussion). Another approach to this problem [33,34], which is more in the spirit of checking internal consistency, makes use of stronger holonomicity assumptions. A discussion of both approaches is given in Sect. 5.1. Even though the u = 0 problem is solved, other apparently crucial problems remain if "separation of singularities" is not assumed. They have been solved [35,36], as will be explained in Sect.5.2, for the 3 → 3 S-matrix below the 4-particle threshold, with the help of a weak "no sprout" analyticity assumption that goes slightly beyond macrocausality (and is probably linked with refined macrocausality).

In Sect. 6 we return to axiomatic field theory. We first describe the results derived in [37] in the framework of the "linear program", on the basis of microcausality and the spectral condition. (Our presentation will follow the more direct derivation given in Part IV of [1].) The results obtained at that stage in the multiparticle case are, however, only very weak forms of macrocausality or analyticity properties: scattering functions are not known to be analytic outside the +α-Landau surfaces, or even to be boundary values of well defined analytic functions, except in a small part of the physical region (whereas this is expected, and proved from macrocausality, almost everywhere:see Sect. 2). It soon appeared in the seventies that the situation should be improved (as was already the case in the sixties for the derivation of second-sheet analyticity for the two-body S-matrix) by the further exploitation of the off-shell unitarity or extended unitarity equations arising in field theory from the axiom of asymptotic completeness of the relevant fields, together with regularity assumptions needed to avoid à la Martin pathologies. The work [38] we shall describe in Sect. 6 is the outcome, using methods of essential support theory, of a program proposed long ago by V. Glaser. It is based on a direct exploitation of the above equations and has led to the desired results (= decompositions in terms of generalized Feynman integrals), up to some technical limitations, again for the 3 → 3 S-matrix below the 4-particle threshold. The method used shows clearly how, starting from the support properties associated in space-time with microcausality, one obtains at the end the exponential fall-off properties, in terms of intermediate particles, associated with macrocausality and macrocausal factorization. The latter are obtained in a somewhat stronger form that applies to the Green functions themselves, i.e. the off shell versions of the S-matrix in field theory, and reduce to those previously described for the S-matrix on the mass-shell.

An alternative important approach in axiomatic field theory is that developed by J. Bros and coworkers. It is based on the introduction of Bethe-Salpeter-type equations and irreducible kernels. A number of basic preliminary results in that approach were obtained in the seventies and have also led recently to results on the 3 → 3 S-matrix below the 4-particle threshold that are closely related to those described in Sect. 6. The work of Bros makes use of somewhat different mathematical methods and provides "semi-global" results. It is described in detail in his lectures [39] and is thus omitted here.

The approach of axiomatic field theory is potentially more powerful than that of S-matrix theory since (at least in principle) one might ultimately derive in a precise way global analyticity properties on the complex mass-shell, beyond the physical region, generalizing those of the two-body case. However, only limited results have been obtained so far in this direction (see the lectures by Bros). On the other hand, all preceding considerations on field theory take place in the framework of the standard Wightman-type axioms of local fields or observables and of the corresponding Haag-Ruelle asymptotic theory. This framework might have to be considerably modified, in view of recent theoretical considerations, whereas the analyticity properties of the S-matrix should remain essentially unchanged.

In the final section (Sect.7) we review recent investigations on the nature of the Landau singularities, in connection in particular with a conjecture [40] by M. Sato on holonomicity. We first outline results [23] that hold for graphs with only single or double lines and confirm (in space-time dimension 4) this conjecture. The investigations of [4] on the 3-particle thresholds (and more generally basic m-particle normal thresholds), which give precise information on the nature of the singularity in a simplified theory, suggest, however, that holonomicity cannot be expected in general. As conjectured on the other hand in [34], the S-matrix should be at best an infinite sum of à la Feynman regular holonomic terms near singularities associated with graphs including sets of an odd number > 1 of lines between two vertices (e.g. the three-particle threshold in a two-body process).

We conclude this introduction with some notations. We consider for simplicity a theory with only spinless particles (the difficulties due to spin being unessential in the physical region). As already mentioned all masses are strictly positive.

Being given a $m \rightarrow m'$ process, the corresponding momentum-space S-matrix kernel $S_{m,m'}(p_1,\ldots,p_m;p_{m+1},\ldots,p_{m+m'})$, where the indices $1,\ldots,m$ and $m+1,\ldots,m+m'$ label the initial and final particles respectively, is defined in the space $M_{m,m'}$ of all real $(p_1,\ldots,p_{m+m'})$ subject to the mass-shell conditions

$$p_k^2 = \mu_k^2, \quad (p_k)_o > 0 , \qquad k = 1,\ldots,m+m' , \tag{1}$$

where $p_k^2 = (p_k)_o^2 - \vec{p}_k^{\,2}$, $(p_k)_o$ and $\vec{p}_k$ being the energy and momentum parts of the ν-vector p_k, and $\mu_k > 0$ is the mass of particle k. As the kernel of a bounded operator, $S_{m,m'}$ is in particular a well defined tempered distribution. The same is also true for the connected S-matrix kernels S^c $(p_1,\ldots,p_{m+m'})$ which are defined from the non-connected ones by standard recursive relations (see e.g. [1], Ch.I).

Finally, by energy-momentum conservation, these kernels can be written (if the p_k are not all colinear) in the form:

$$S^{(c)}_{m,m'}(p_1,\ldots,p_{m+m'}) = s^{(c)}_{m,m'}(p_1,\ldots,p_{m+m'})\,\delta^{\nu}\Big(\sum_{i=1}^{m} p_i - \sum_{j=m+1}^{m+m'} p_j\Big) \tag{2}$$

where $s^{(c)}_{m,m'}$ is now a well defined distribution in the physical region $M_{m,m'}$ of the process:

$$M_{m,m'} = \{(p_1,\ldots,p_{m+m'});\ p_k \text{ real},\ p_k^2 = \mu_k^2,\ (p_k)_o > 0\ ,\ \forall k,\ \sum_{i=1}^{m} p_i = \sum_{j=m+1}^{m+m'} p_j\}\ . \tag{3}$$

The connected kernel $s^c_{m,m'}$ of a given process will be denoted $f_{m,m'}$ and is called the scattering function of the process. The indices m,m' will be left implicit most of the time.

1. THE TWO-BODY S-MATRIX

Important progress was made, as well known, in the first part of the sixties in the derivation of general momentum-space analyticity properties of the two-body S-matrix, on the complex mass-shell, such as crossing and dispersion relations. The latter have been proved rigorously in axiomatic field theory in [42]. Concerning the physical region, the above results entail in particular (see Fig.1) that the two-body scattering function f is the boundary value of an analytic function $\underline{f}$ from the directions Im s > 0.

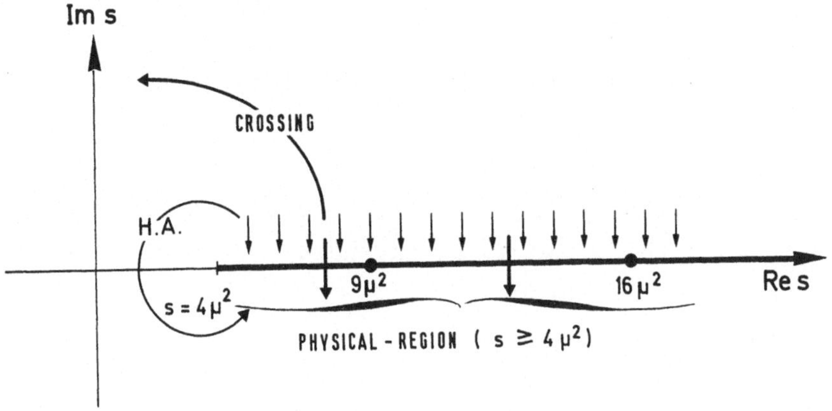

Fig. 1

The study of the physical-region analytic structure was also completed by the derivation [43] of "second-sheet analyticity" across the region $4\mu^2 < s < 9\mu^2$ (i.e. the fact that $\underline{f}$ can be analytically continued through this part of the real axis, from the region Im s > 0), with possible poles in the second sheet associated with unstable particles. As noticed and emphasized by A. Martin [44], a regularity assumption (e.g. the continuity of f) is needed to avoid otherwise possible pathologies corresponding physically to the accumulation, near points of the real axis, of an infinite number of unstable particles with arbitrarily small widths (i.e. of poles arbitrarily close to the real axis, in the region Im s < 0). The same analysis also shows the two-sheeted square-root type nature of the singularity of $\underline{f}$ (considered in view of Lorentz invariance as a function of $s = (p_1+p_2)^2$ and e.g. of $t = (p_1-p_3)^2$) at the two-particle threshold $s = 4\mu^2$, in space-time dimension 4. It is based on the physical unitarity equation valid below the 3-particle threshold and on the property of hermitean analyticity, which transforms it into an integral relation between f and the determination $f^{(1)}$ of $\underline{f}$ obtained after one turn around $s = 4\mu^2$ along the path shown in Fig. 1. This is outlined in Sect. 7.2 as an introduction to the recent investigations on three-particle thresholds.

A related analysis with similar results was also given in the sixties by J. Bros [45] on the basis of the Bethe-Salpeter type equation, and can be found (on the basis of the unitarity equation) in Sect. 1 of [36]. The results are in these two works extended to the off-shell four-point Green function.

Before coming to the multiparticle case, we notice that the analysis of the physical region structure is, at that stage, far from complete. It does not include the existence of analytic continuations across the real regions $9\mu^2 < s < 16\mu^2$, $16\mu^2 < s < 25\mu^2, \ldots$, which is expected heuristically (see Sect. 2.4) and is also equivalent to macrocausality for two-body processes (see Sect. 3). It provides à fortiori no information on the nature of singularities at the thresholds $s = (n\mu)^2$, $n \geq 3$: this will be discussed in Sect. 7.2.

2. PHYSICAL-REGION ANALYTICITY PROPERTIES OF THE MULTIPARTICLE S-MATRIX: HEURISTIC DESCRIPTION

The following physical-region analyticity properties are suggested by perturbation theory and heuristic arguments. As described in Sect. 3, they can be derived from macrocausality and macrocausal factorization. They refer to the scattering function f (see definition at the end of the Introduction) of any given process.

2.1 Analyticity outside $+\alpha$-Landau surfaces and plus $i\varepsilon$ rules

A first property of f is analyticity outside the $+\alpha$-Landau surfaces of connected graphs.

We recall that, being given a topological multiple scattering graph G, e.g.

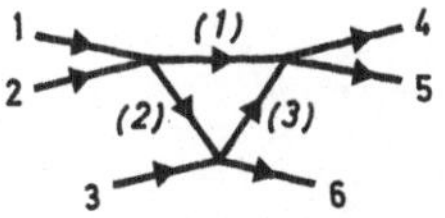

in a 3 → 3 process, a physical-region point $P = (P_1, \ldots, P_{m+m'})$ belongs by definition to L(G), resp. to its $+\alpha$-part $L^+(G)$, if there exist on-mass-shell energy-momenta K_ℓ and real scalars α_ℓ, resp. positive α_ℓ, for each internal line ℓ of G, with $\sum_\ell \alpha_\ell^2 \neq 0$, such that energy-momentum conservation be satisfied at each vertex, and such that the loop equations

$$\sum_{\ell \in L} \varepsilon(\ell)\, \alpha_\ell\, K_\ell = 0 \qquad (4)$$

be satisfied for each closed loop L of G($\varepsilon(\ell) = \pm 1$ depending on the respective orientations of the line ℓ in G and in L). For the above triangle graph, (4) reduces to $\alpha_1 K_1 - \alpha_2 K_2 - \alpha_3 K_3 = 0$.

The $+\alpha$-Landau surfaces are in the two-body case (m=m'=2) the "thresholds" $s = (n\mu)^2$, n = 2,3,... associated with the graphs:

(We consider here a theory with only one mass μ.)

Their equations are more complicated in general and involve combinations of initial and final variables (e.g. p_1+p_2, p_4+p_5, p_3-p_6 for the above triangle graph). The main $+\alpha$-Landau surfaces are still, however, almost everywhere smooth (= analytic) submanifolds of the physical region of codimension one (i.e. their dimension is one unit smaller than that of the physical region), and they divide it into various sectors [16]. This is schematically represented in Fig. 2, which is the analogue of the real axis $s \geq 4\mu^2$ of the two-body case with its surfaces $s = (n\mu)^2$, $n = 2,3,\ldots$

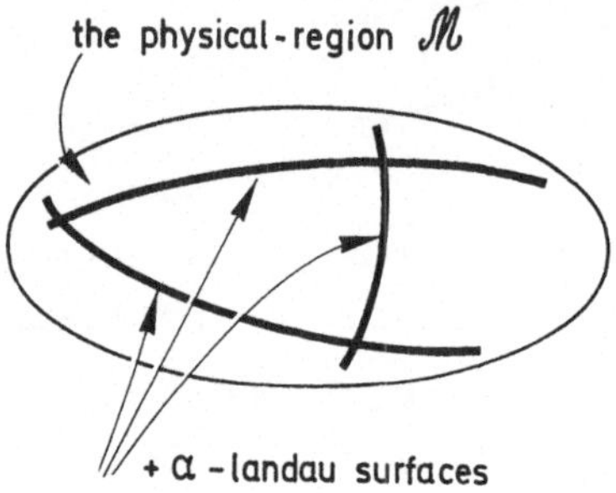

Fig. 2

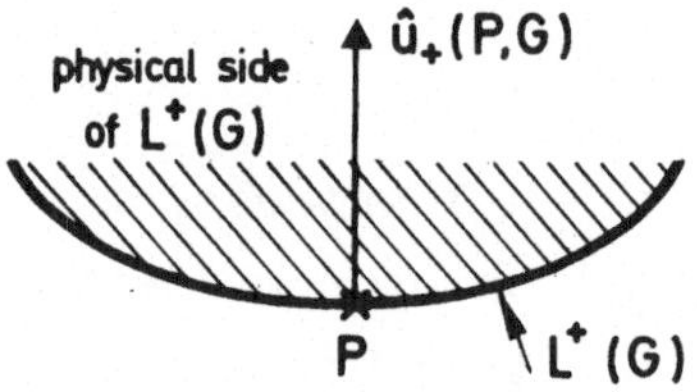

Fig. 3

In the neighborhood of any given $P = (P_1,\ldots,P_{m+m'})$, each one of these surfaces divides the physical region into two sides; one of these is well characterized by certain convexity properties of the surface [17] (in a larger ambient space), and also by the "causal direction" $\hat{u}_+(P;G)$ introduced in Sect. 3: see Fig.3. It is called the physical side of $L^+(G)$. By definition points "above the threshold of $L^+(G)$" are those of the physical side. In the two-body case, the physical side of the surface $s = (n\mu)^2$ is the region $s \geq (n\mu)^2$.

The situation shown in Fig. 2 is of course schematic. One also encounters situations which look locally like those of Fig. 4,

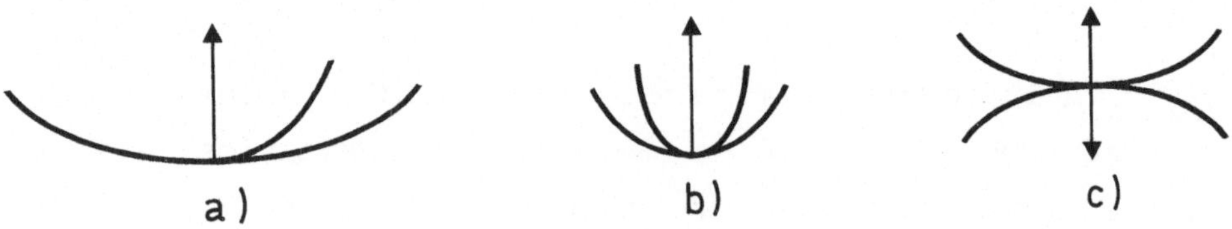

Fig. 4

and there are also points P where the surfaces $L^+(G)$ are no longer smooth submanifolds, and surfaces of codimension higher than one. An example of the latter is the surface $L^+(G)$ of a tree graph such as

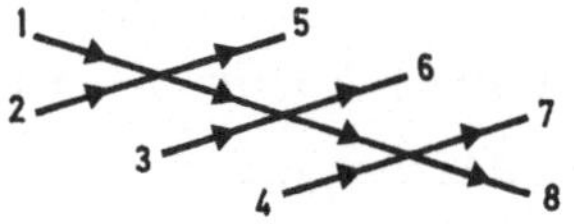

in a $4 \to 4$ process which lies at the intersection of $L^+(G_1)$ and $L^+(G_2)$, with

G_1 = 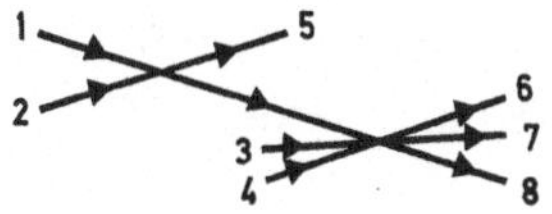

, G_2 = 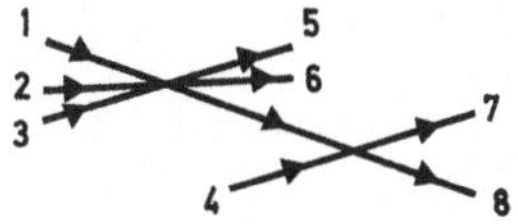

.

Cases where $L^+(G)$ is no longer a smooth submanifold are similarly those of points P that lie on the boundary of $L^+(G)$, where G is not a tree graph, and on several surfaces $L^+(G')$, $L^+(G")$,... where G', G" are various "contractions" of G (Example a) of Fig. 4 is obtained when only one contraction of G is involved).

Finally, surfaces of graphs G such as

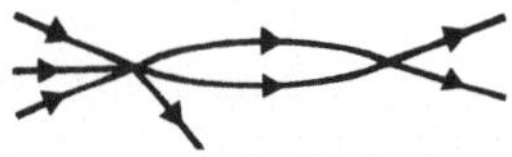

which correspond to so-called M_o points (= some initial, or some final, p_k are colinear) are always of codimension larger than one in the physical region, in terms of the momentum variables (at least if the dimension ν of space-time is larger than 2): indeed a condition such as $\vec{p}_1 = \vec{p}_2$ is a set of ν-1 conditions $(p_1)_\mu = (p_2)_\mu$, $\mu = 1,\dots,\nu-1$. In the two-body case, M_o points are the points of the set $s = 4\mu^2$. This set corresponds moreover to all p_k colinear, a case for which the physical region M is itself a singular submanifold of $R^{\nu(m+m')} \equiv R^{4\nu}$. As well known, the situation is improved there by considering f as defined in the space of new variables such as the Lorentz invariants s,t.

We shall not discuss in detail the corresponding considerations at M_o points in the multiparticle case.

Plus iε rules

The usual "plus iε rule" is the assertion that f can be analytically continued from one side of a +α-Landau surface to the other side via "infinitesimal plus iε distortions' on the complex mass-shell (= complexified manifold $\underline{M}$ of the physical region M). Equivalently, f is locally, in the neighborhood of a point P of a surface $L^+(G)$, the boundary value (possibly in the sense of distributions) of an analytic function $\underline{f}$ (defined in a domain of $\underline{M}$) from "plus iε" imaginary directions.

The plus iε directions are the general version of the directions Ims > 0 of the two-body case. They depend, however, in general on the surface and on the point P considered. If we consider a surface $L^+(G)$ which is locally near P a real analytic codimension-one surface and a system of real analytic coordinates $q = (q_1,\ldots,q_\rho)$, $\rho = 1,\ldots, 3(m+m')-4$, chosen such that $q_1 = 0$ and $q_1 \geq 0$ represent locally $L^+(G)$ and its physical side, then the plus iε directions are the directions of the half-space $\mathrm{Im}q_1 > 0$, i.e. one has near q = 0 (which represents P):

$$f(q) = \lim_{\substack{\varepsilon=(\varepsilon_1,\ldots,\varepsilon_\rho)\to 0 \\ \varepsilon_1 > 0}} \underline{f}(q+i\varepsilon) \quad . \tag{5}$$

The plus iε directions can also be defined (Sect.3) as those dual to the causal direction $\hat{u}_+(P;G)$ (= the direction $u_1 > 0$, $u_2 = \ldots = u_\rho = 0$ in the above local coordinate system).

By analytic continuation via plus iε distortions

around the various $+\alpha$-Landau surfaces encountered, one is led to the conclusion (to be somewhat corrected: see below) that the scattering function f is, as in the two-body case, the boundary value in its physical region of a well defined analytic function $\underline{f}$ (defined in a domain of the complex mass-shell): the boundary value is still analytic outside the $+\alpha$-Landau surfaces and is obtained at each $+\alpha$-Landau point from the corresponding plus $i\varepsilon$ directions.

This conclusion is to be corrected because of the occurrence of points that lie on several $+\alpha$-Landau surfaces and of M_o points. We exclude the latter from the present discussion. In the case of points P where several "related" surfaces of a graph G and of contracted graphs are involved, the set of plus $i\varepsilon$ directions may be smaller than before (see Sect.3), but this makes no basic problem. However, f can no longer be expected to be locally the boundary value of an analytic function if P lies on several non-related surfaces (or sets of surfaces) with conflicting (e.g. opposite) plus $i\varepsilon$ directions. This occurs for instance in a 3 → 3 process, in the simple case of all points $P = (P_1,\ldots,P_6)$ such that $P_1 = P_6$, $P_2 = P_4$, $P_3 = P_5$. These points all belong to the surfaces $L^+(G')$, $L^+(G'')$ of the two graphs:

$G' =$ 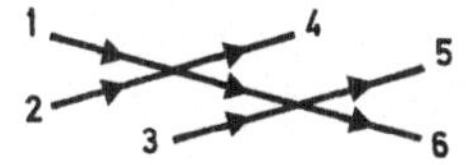

$G'' =$ 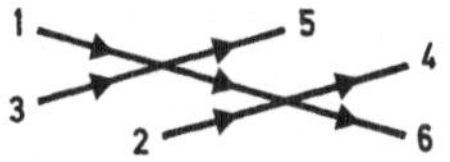

But $L^+(G')$ and $L^+(G'')$ have (as easily seen in terms of space-time diagrams: see Sect. 3) opposite "causal directions" and correspondingly opposite plus $i\varepsilon$ directions at P: this is the situation shown in Fig. 4c). In such cases, the correct way of stating the "plus $i\varepsilon$ rule" at P is to require that f be locally a <u>sum</u> of different boundary values of analytic functions from respective plus $i\varepsilon$ directions. In space-time dimension $\nu > 2$, all points where the usual plus $i\varepsilon$ rule does not hold lie in submanifolds of codimension > 1 in the physical region. Hence, the previous statement on the existence of a well defined analytic function $\underline{f}$ from which f is the boundary value in the physical region is still obtained "almost everywhere" (i.e. at all other points).

In dimension $\nu = 2$, special phenomena occur. In particular, the set of points P mentioned above ($P_1 = P_6$, $P_2 = P_4$, $P_3 = P_5$) has codimension one. See further discussion in [46] and the remark that concludes Sect. 3.

2.2 Discontinuity formulae and à la Feynman local representation

Let us consider a surface $L^+(G)$ which in the neighborhood of a point P is a smooth codimension-one surface, and let P lie on no other $+\alpha$-Landau surface. The plus $i\varepsilon$ rule entails that f can be analytically continued from one side of $L^+(G)$ to the other side via a "plus $i\varepsilon$" distortion. A first content of the discontinuity formula at P, if it holds, is that f can also be analytically continued around $L^+(G)$, from the non physical side of $L^+(G)$, via an opposite "minus $i\varepsilon$" distortion: see Fig. 5.

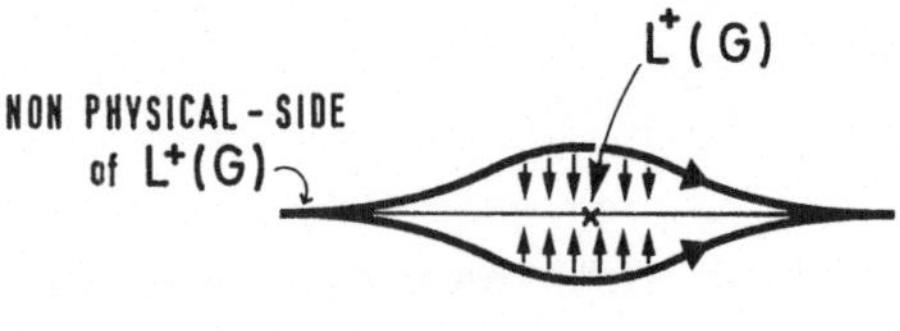

Fig. 5

We shall denote by $f^{(L)}$ the (minus $i\varepsilon$) boundary value of this minus $i\varepsilon$ analytic continuation of f. Then the usual discontinuity formula (when it holds) asserts that locally

$$\delta^{\nu}(\sum p_i - \sum p_j) \times (f - f^{(L)}) = D$$

$$= \delta^{\nu}(\sum p_i - \sum p_j) \times d \qquad (6)$$

where D is a certain on-mass-shell integral associated with G, each vertex being replaced by a connected S-matrix kernel: this is the analogue of Cutkosky formula for Feynman integrals, in which constants (times energy-momentum conservation δ^{ν}-functions) are associated to each vertex. We give below two examples in a 3 → 3 process:

a) G = 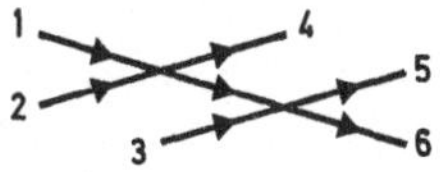

The surface $L^+(G)$ is simply defined by $k^2 = \mu^2$, $k_o > 0$, where $k = p_1 + p_2 - p_4$ $(= p_5 + p_6 - p_3)$. The plus iε directions are the directions $\mathrm{Im}\, k^2 > 0$. The physical side is $k^2 > \mu^2$. Then:

$D(p_1 \ldots p_6)$ = 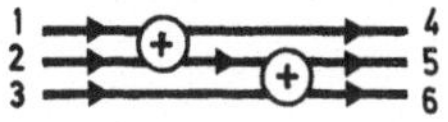

≡

$$\equiv \int S^c_{2,2}(p_1,p_2;k,p_4)\, S^c_{2,2}(k,p_3;p_5,p_6)\, \delta(k^2-\mu^2)\, \theta(k_o)\, d^4k \qquad (7)$$

which gives simply locally, in view of the δ-functions contained in each kernel $S^c_{2,2}$:

$$d(p_1,\ldots,p_6) = f_{2,2}(p_1,p_2;p_4,k)\, f_{2,2}(k,p_3;p_5,p_6)\, \delta(k^2-\mu^2) \quad (8)$$

where $k \equiv p_1 + p_2 - p_4$ in the r.h.s. of (8).

In this case, the discontinuity is concentrated on the surface $L^+(G)$ $(f = f^{(L)}$ at $k^2 > \mu^2)$ and the singularity is correspondingly easily shown to be necessarily a pole with a factorized residue:

$$f(p_1,\ldots,p_6) = \frac{1}{2i\pi}\,\frac{a(p_1,\ldots,p_6)}{k^2-\mu^2+i\varepsilon} \tag{9}$$

with $k = p_1+p_2-p_4$, $a(p_1,\ldots,p_6)\Big|_{k^2=\mu^2} =$

$$= f_{2,2}(p_1,p_2;p_4,k)\,f_{2,2}(k,p_3;p_5,p_6)\ .$$

b) Let G = 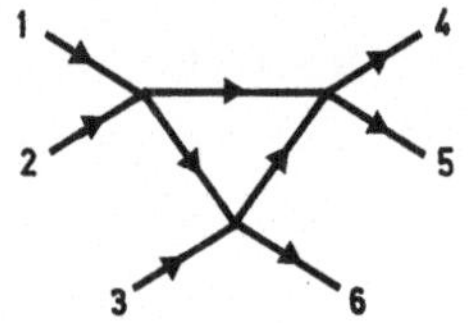

.

Then:

$D(p_1\ \ldots\ ,p_6)$ = 

$\equiv$

$$\equiv \int S^c_{2,2}(p_1,p_2;k_1,k_2)\,S^c_{2,2}(k_2,p_3;k_3,p_6)\,S^c_{2,2}(k_1,k_3;p_4,p_5)$$

$$\times \prod_{\ell=1,2,3} \delta(k^2-\mu^2)\,\theta((k_\ell)_o)\,d^\nu k_\ell\ . \tag{10}$$

D is a well defined-distribution (see Sect. 7.1). From the δ^ν-functions contained in the kernels $S^c_{2,2}$, an overall δ-function $\delta^\nu(\sum_{i=1}^{3} p_i - \sum_{j=4}^{6} p_j)$ can again be extracted. The distribution d is no longer concentrated on $L^+(G)$. It is on the other hand (and this is a general feature) automatically concentrated on the <u>physical side</u> of $L^+(G)$, as it must if it is to be the discontinuity of f: this is because the region defined (locally) from the Landau equations <u>without the loop equations</u> (4) is precisely in general the physical side of $L^+(G)$; d is therefore zero on the non-physical side in view of the mass-shell δ-functions $\delta(k_\ell^2-\mu_\ell^2)$ and of the conservation δ^ν functions contained in each $S^c_{2,2}$.

Discontinuity formulae of the type just described

can be expected only for graphs G with only single lines, i.e. at most one line between any pair of vertices. Otherwise, they have to be modified either by replacing some of the scattering functions associated with each vertex by appropriate analytic continuations (the simplest case is the discontinuity around $s = 4\mu^2$ in the two-body case: see Sect. 7), or by including on each set α of multiple lines a box S_α^{-1} that represents, in a theory with only one mass μ, the inverse in H_α of the restriction of S to the subspace H_α of states with $|\alpha|$ or more particles. Note that 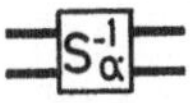

coincides with the non-connected kernel 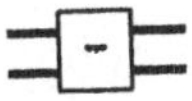 of S^{-1} for $|\alpha| = 2$. (Hence the discontinuity  around $s = 4\mu^2$ in the two-body case is also equal to

 .)

This is no longer true for $|\alpha| > 2$. The factor S_α^{-1} arises from the general algebraic analysis of [9]. Its origin is easily seen on the example (given in Ch. III of [21]) of the discontinuity around the 3-particle threshold $s = 9\mu^2$ in a two-body process (not to be confused with the total discontinuity, equal to

around $s = 4\mu^2$). In fact the unitarity equation reads in the region $s < 16\mu^2$:

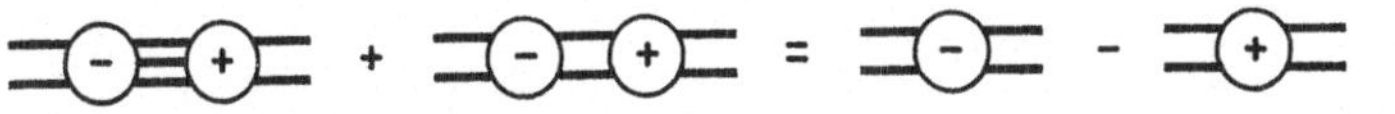 , (11)

where the minus bubbles stand for connected kernels of $-S^{-1} = -S^\dagger$. The last term in the r.h.s. of (11) vanishes for $s < 9\mu^2$. If we then consider a minus $i\varepsilon$ analytic continuation of Eq.(11) around $s = 9\mu^2$, starting from the region $(4\mu^2) < s < 9\mu^2$, we obtain (since $f^{(-)}$ is known by unitarity to be a minus $i\varepsilon$ boundary value):

 (12)

The comparison of (11) and (12) shows that the discontinuity 

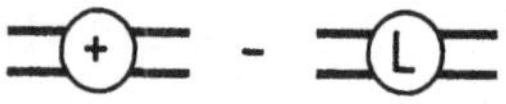

must be equal, after on-mass-shell convolution on the right

with 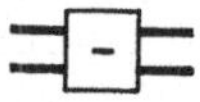 , to  .

The solution D = 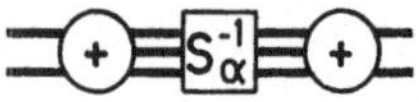

satisfies this condition since by easy calculations:

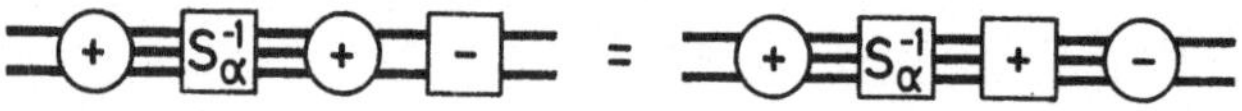 , (13)

and since

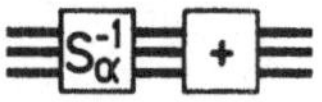 is the identity $I_{3,3}$

in view of the definition of S_α^{-1}.

The discontinuity formula gives in principle information on the nature of the singularity, at least if the structure of the "elementary" bubbles occurring in D in the integration domain is already known: see Sect. 7 where the case of the triangle graph of example b) is in particular treated (Sect. 7.1).

A la Feynman local representation

In case such as examples a), b) for a 3 → 3 process below the 4-particle threshold (and away from M_0 points), the scattering functions associated with each vertex remain analytic in integration domains. One can then easily extract from d (see [36]) a "plus iε" part d_+ which is well defined locally modulo an analytic function through the decomposition theorems of essential support or hyperfunction theory, and is locally the plus $i\varepsilon$ boundary value of an analytic function. One has correspondingly

$$f \simeq d_+ \tag{14}$$

in the neighborhood of P, where $\simeq$ means modulo analytic backgrounds. In example a), d_+ is merely the r.h.s. of (9). In example b), it can be shown that it coincides on the mass-shell, modulo analytic backgrounds, with the "G-convolutions" or "generalized Feynman integrals" that are well defined in axiomatic field theory (where the "bubbles" are not restricted to the mass-shell) with Feynman-like propagators rather than mass-shell δ-functions associated with each internal line: see [39,38]. Similar "generalized Feynman integrals" can also be defined locally, modulo analytic backgrounds, in terms of pure on-mass-shell scattering functions, by introducing off-shell extrapolations of these functions and appropriate cut-off factors off-the-mass-shell: see e.g. [34]. (The same analysis can also be made in the framework of essential support theory.)

In cases of graphs with multiple lines, the situation is more complicated, and a local representation of f may involve an infinite number of Feynman-like contributions, as appears e.g. in the simplified model of the 3-particle threshold in a 3 → 3 process treated in [41]: see Sect.7.2.

2.3 The 3 → 3 S-matrix below the four-particle threshold

Let us consider the 3 → 3 S-matrix, in a theory with only one type of particle of mass $\mu > 0$, and restrict our attention to the region

$$R = \{p=(p_1,\dots,p_6), p \in M,\ (3\mu)^2 < s < (4\mu)^2,\ p \notin M_0\}. \tag{15}$$

In that region, one encounters exactly 18+α-Landau surfaces $L^+(G_\beta)$ which are all real analytic codimension-one surfaces: 9 are associated with the graphs

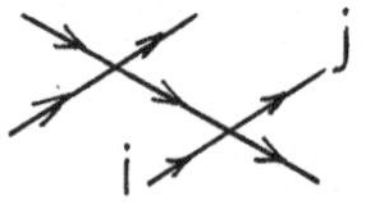

,

i = 1,2,3, j = 4,5,6, and 9 with triangle graphs

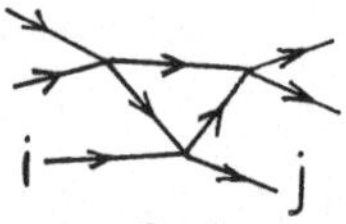

.

The local structure of f at points P that lie on only one of these surfaces has been studied in Sect. 2.3. One encounters, however, also situations such as those shown in Fig. 4b), 4c) (e.g. in the latter case all points P such that $P_1 = P_6$, $P_2 = P_4$, $P_3 = P_5$ already mentioned in Sect. 2.2). The 18 surfaces $L^+(G_\beta)$ are not "related" at any point in the region considered. We shall in this section admit on heuristic bases that f is in the whole region R a sum of corresponding terms $d_{\beta,+}$:

$$f \simeq \sum_\beta d_{\beta,+} \tag{16}$$

where ≃ means again modulo analytic backgrounds in R. The $d_{\beta,+}$ have been already introduced locally, but they can be defined in the same way in the whole region R from the respective discontinuities d_β along $L^+(G_\beta)$: they are uniquely defined modulo analytic backgrounds in R, each $d_{\beta,+}$ being analytic in R outside $L^+(G_\beta)$, and being along $L^+(G_\beta)$ a corresponding "plus iε" boundary value: see [36]. They again coincide (on the mass-shell, modulo analytic backgrounds) with the G-convolutions or generalized Feynman integrals of field theory. We write correspondingly (16) in the form:

,(17)

where the plus signs above each internal line refer to

Feynman-like propagators.

Formula (16) can be derived from, and is in fact equivalent [2] in R to macrocausality and macrocausal factorization (Sect. 3). It has been derived on the other hand in S-matrix theory [36] from refined macrocausality and unitarity (plus the no-sprout assumption): see Sect.5, and in field theory [38,39] up to some technical limitations: see Sect. 6, related results being given in the lectures by Bros.

Note: internal lines ——— refer in general in S-matrix theory to mass-shell δ-functions, whereas they usually refer in field theory (e.g. in the lectures by Bros and in [38]) to Feynman-like propagators: formulae such as (17) are then written without + signs above the internal lines.

3. MACROCAUSALITY, MACROCAUSAL FACTORIZATION AND PHYSICAL-REGION ANALYTIC STRUCTURE

3.1 Macrocausality, $+\alpha$-Landau singularities and plus $i\varepsilon$ rules

Being given a physical process $m \to m'$, let us consider initial and final wave functions ϕ_k and space-time translation vectors a_k which transform each ϕ_k into $\phi_k^{(ak)}$:

$$\phi_k^{(a_k)}(p_k) = \phi_k(p_k)e^{-ip_k \cdot a_k} \quad .$$

Here, $p_k \cdot a_k = (p_k)_o(a_k)_o - \vec{p}_k \cdot \vec{a}_k$ where $(a_k)_o$ and $\vec{a}_k$ are the time and space components of a.

If the vectors a_k are chosen to be equal within various subgroups K of initial and final particles, a general cluster property in space-time asserts the factorization of the transition probability W into the product of the partial transition probabilities W_K:

$$W - \prod_K W_K \to 0 \tag{18}$$

in the limit when all $|a_K - a_{K'}| \to \infty$, $K \neq K'$, $|a| = a_o^2 + \vec{a}^2$.

The factorization property (18) is equivalent [47,1] to a corresponding property of transition amplitudes (without phases), the latter being in turn equivalent [12] (see also Ch.I of [1]) to the decrease at infinity of the <u>connected</u> amplitude $S^c(\{\phi_k^{(a_k)}\})$ whenever one or more $|a_k - a_{k'}| \to \infty$, and to the corresponding cluster property in momentum space: the connected S-matrix kernel $S^c_{m,m'}(p_1,\ldots,p_{m+m'})$ contains no partial energy-momentum conservation δ-function between subgroups of particles. No further information on regularity or analyticity properties of scattering functions is obtained, this being due to the absence of information on the rate of fall-off in (18): as discussed in detail in Ch.I of [1], a strong rate of fall-off cannot be expected in general because particles are not necessarily localized in space-time in a sufficiently sharp way, even in an asymptotic sense, and because the various subgroups K can still be linked by causal effects (such as real stable intermediate particles and à priori possibly others), in the limit when the $|a_K - a_{K'}| \to \infty$. This may occur even though the $a_K - a_{K'}$ are space-like (e.g. $(a_K)_o = 0, \forall K$: even in the best case particles cannot be localized in approximately bounded space-time regions, but will "occupy" full velocity cones or space-time trajectories).

Macrocausality will be a sharper cluster property in space-time according to which the rate of fall-off in (18) is exponential (in a well defined sense) under appropriate conditions. To obtain sharp localization properties in space-time of the particles, modulo exponential fall-off, it is convenient to put $a_k = \tau u_k$ for each k, where τ will be the parameter that will tend to infinity, and to consider gaussian-type displaced wave functions, with widths shrinking with τ, of the form [15,18]:

$$\phi_{k,\tau}^{(P_k,u_k)}(p_k) = \chi_k(p_k)\, e^{-\gamma\tau(\vec{p}_k-\vec{P}_k)^2}\, e^{-i\tau p_k\cdot u_k} , \qquad (19)$$

where P_k is a given on-mass-shell energy-momentum, χ_k is locally analytic at P_k, and $\gamma > 0$. When $\tau \to +\infty$, it can then be shown that the particle with wave function (19) is asymptotically localized, in a space-time coordinate system scaled to τ (i.e. each x is replaced by x/τ), along the classical trajectory (P_k,u_k) parallel to P_k and passing through u_k (see Fig.6) modulo well specified exponential fall-off properties.

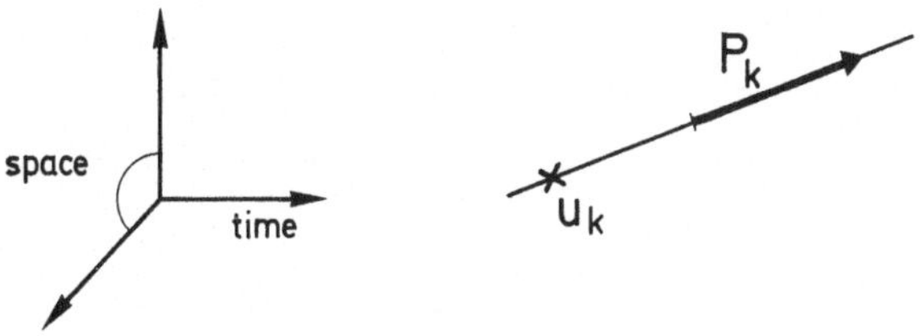

Fig. 6

This result [18] is the analogue of a previous one [48,14] according to which a particle with wave function $\chi_k(p_k)e^{-i\tau\, p_k\cdot u_k}$, where χ_k is C^∞ (infinitely differentiable) and has a (small) support around P_k, is asymptotically localized, in the above scaled coordinate system, in the velocity cone $V_k^{(u_k)}$ composed of all trajectories (p_k,u_k), $p_k \in$ support of χ_k, modulo <u>rapid</u> fall-off properties (i.e. faster than any inverse power) outside $V_k^{(u_k)}$. The rapidity of this fall-off comes (by Fourier transform) from the C^∞ character of χ_k. However, this result is not easily extended to get exponential fall-off because any function χ_k with compact support has at least C^∞ singularities on the boundary of its support even though it is analytic inside the latter: these singularities will produce by Fourier transform rapid, not exponential fall-off. The presence of the gaussian factor

$$e^{-\gamma\tau(\vec{p}_k - \vec{P}_k)^2}$$

in (19) allows one to get exponential fall-off with τ (with a rate at least proportional to γ for small γ) outside the trajectory (P_k, u_k) whenever χ_k is locally analytic at P_k. This type of property is also at the basis of the further results described below on the links between macrocausal and analyticity properties and of the developments of essential support theory.

Macrocausality is, in its most simple form, the assertion (away from M_o points P) that the transition probability between the wave functions (19) should fall off exponentially in the $\tau \to \infty$ limit (with again a rate of fall-off at least proportional to γ for small γ) if $u = \{u_k\}$ is not "causal" at $P = \{P_k\}$, i.e. if the initial and final particles cannot be linked causally via a network of intermediate real stable particles, = if there exists no classical multiple scattering diagram $D_+(P,u)$ with initial and final trajectories (P_k, u_k). This assertion corresponds physically to the previous asymptotic localization properties of the particles and to the idea of short-range of interactions (= all effects of transfer of energy-momentum that cannot be associated with stable real particles fall off exponentially with distance, hence with τ). It excludes implicitly à la Martin pathologies (= infinite number of unstable particles with arbitrarily small widths, which would produce arbitrarily small rates of exponential fall-off).

An example of a diagram D_+ is shown in Fig. 7 with K_1, K_2, K_3 on mass-shell, energy-momentum conservation at each vertex, and $c-a = \alpha_1 K_1$, $c-b = \alpha_3 K_3$, $b-a = \alpha_2 K_2$, $\alpha_1, \alpha_2, \alpha_3 > 0$. A somewhat stronger statement of macrocausality is the assertion that the transition probability

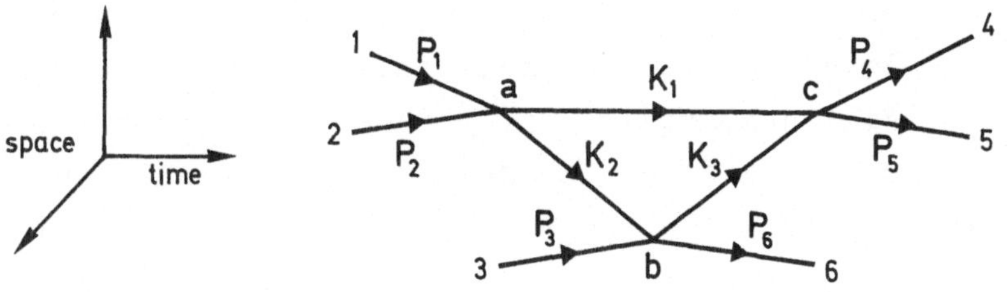

Fig. 7

should factorize into the product of partial W_K (modulo exponential fall-off properties) if there is a $D_+(P,u)$, but at most composed of disconnected parts linking together the initial and final trajectories of various subgroups K. (Note that the u_k need not here coincide within each subgroup K.) It can be shown [19] that this latter property is in turn equivalent (apart from exceptional cases) to the assertion that the connected amplitude between the wave functions (19) falls off exponentially (with again a rate at least proportional to γ for small γ) if there exists no connected $D_+(P,u)$.

Analyticity properties of scattering functions [18,21]

If P is not a $+\alpha$-Landau point of a connected graph, then by definition of the $+\alpha$-Landau surfaces (see Sect.2), one checks easily, as first explicitly noticed in [11], that there exists no non-trivial $D_+(P,u)$ (D_+ is trivial if it includes no internal line, all initial and final trajectories meeting at a common point in space-time). It is then proved that macrocausality at P (in its last form) is equivalent to the local analyticity of the scattering function f at P.

If P belongs to only one surface $L^+(G)$ of a connected

graph G which is locally a real analytic codimension-one surface, then one checks [16] that there exists one and only one non-trivial D_+ whose topological structure is G, modulo global space-time translations and dilations by > 0 coefficients. Equivalently there is only one non-trivial causal u at P modulo addition of trivial ones and multiplication of all u_k by a common $\rho > 0$; u is trivial at P if it is of the form $u_k = \lambda_k P_k + a$, where λ_k is an arbitrary real scalar and a is independent of k. (Addition of $\lambda_k P_k$ does not change the trajectory k, and addition of a amounts to a global translation of all trajectories together.) A set $u = \{u_k\}$ thus defined modulo addition of trivial sets at P, which is equivalent to a configuration of trajectories (P_k, u_k) modulo global space-time translations, defines a point in the <u>cotangent space</u> $T_P^* M$ at P to the physical region manifold M (= the dual space of the tangent space $T_P M$), if the scalar product $\langle p,u \rangle$ of points $p = \{p_k\}$ and $u = \{u_k\}$ in $R^{\nu(m+m')}$ is defined by:

$$\langle p,u \rangle = \sum_k \varepsilon_k \, p_k \cdot u_k \,, \tag{20}$$

$\varepsilon_k = -1$ if k is initial, $\varepsilon_k = +1$ if k is final. Indeed the gradients at P of the functions $p_k^2 - \mu_k^2$ and $(\Sigma p_i - \Sigma p_j)_\mu, \mu = 0,1,2,\ldots,\nu-1$ (whose vanishing defines M in $R^{(m+m')\nu}$) are the vectors $\lambda_k P_k$ and a. The <u>causal direction</u> $\hat{u}_+$ (P;G) is the direction defined in $T_P^* M$ by the causal u at P. It is then proved that macrocausality at P is equivalent to the plus $i\varepsilon$ rule at P described in Sect. 2.

If P is a non-M_o point that lies on several $+\alpha$-Landau surfaces, the set of causal directions at P is the union of those associated with each set of related surfaces involved at P. For each set of related surfaces, the causal directions are those of a closed convex salient cone (which does not reduce in general to a single direction) and the plus $i\varepsilon$ directions are those of the dual cone. The plus $i\varepsilon$ rules of Sect. 2 are then again equivalent to macrocausality.

At M_o points P, the statement of macrocausality requires a somewhat more detailed analysis. It is unchanged (apart from exceptional situations) if vertices "at infinity" in some direction are admitted in the definition of the diagrams D_+: at such a vertex, the trajectories of the incoming and outgoing trajectories must be all parallel, but need not coincide. Following ideas of [23], they are still required to satisfy a law of "angular momentum conservation": an example is shown in Fig. 8 for a vertex involving two incoming and two outgoing equal-mass particles 1,2 and 3,4:

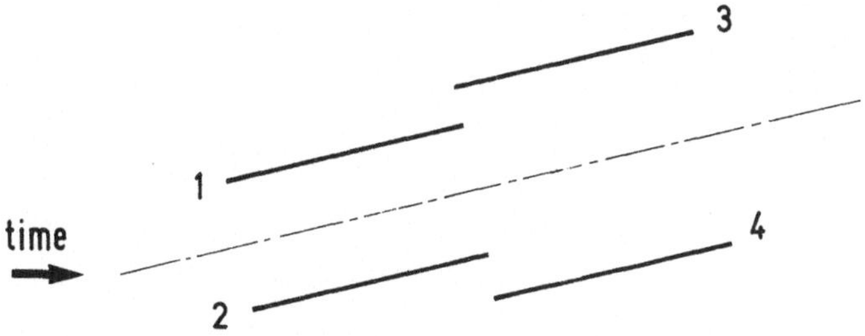

Fig. 8

In the neighborhood of M_o points, macrocausality entails by itself no simple analyticity property of f, the situation being improved if one relaxes, as in field theory (see Remark 2 at the end of Sect. 6), the mass-shell constraints.

We conclude this subsection with the following statement of macrocausality (in its last form on connected amplitudes) which covers all cases treated above and directly follows [21,1], in terms of the notion of essential support, from the very definition of the latter and some technical considerations. An analogous "micro-analyticity postulate" in terms of the notion of singular spectrum of hyperfunction theory (see Appendix) was proposed independently in [22] and by M. Sato. As already mentioned in the Introduction, the essential support and singular spectrum coincide.

Macrocausality

The set of singular directions of S^c, resp. of the scattering function f, at any physical region point P is contained in the set of causal directions corresponding to configurations of external trajectories of connected diagrams D_+ with external energy-momenta P_k, resp. to relative configurations of external trajectories (= defined modulo global space-time translations).

3.2 Macrocausal factorization, discontinuity formulae and related results

Macrocausal factorization [19,22] applies to the causal case, i.e. when there exists one (or more) D_+ (P,u). If there is only one D_+ (P,u) or several D'_+ (P,u) that can be obtained from a unique D_+ by replacing (when possible) certain interactions at certain vertices by no interaction or by several subinteractions (see [2]), and if there is no set of more than one line between any pair of vertices, it says (apart from exceptional cases: see [2]) that the transition amplitude between the wave functions (19) is equal, modulo exponential fall-off in the $\tau \to \infty$ limit (in a neighboring cone of u and in the same sense as before), to the integral, over all possible real intermediate states, of a product of scattering amplitudes associated with each vertex of D_+. If D_+ is e.g. the diagram of Fig. 7, then this integral is the action of $D^{(n.c)}$ between the wave functions under consideration, where

$D^{(n.c)}(p_1 \ldots, p_6) =$ 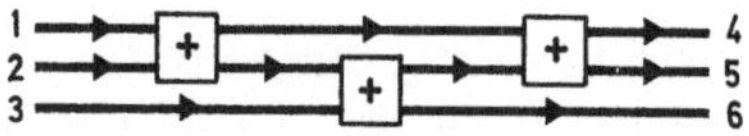 (21)

is defined in the same way as D in (10), except that the plus boxes represent non-connected S-matrices.

If D_+ includes sets α of multiple lines, a box S_α^{-1}

is to be inserted on each set (see definition in Sect. 2). Finally, if there is more than one D_+ (apart from trivial changes, such as insertion of new vertices on sets of multiple lines), a sum of corresponding contributions is to be considered.

In usual cases, such as those of Sect. 2.3, this factorization property yields a similar property for the connected amplitude, in which the kernels associated with each vertex are replaced by connected S-matrices and in which there is now one term for each possible connected diagram. The cases, e.g. those of Sect. 2.3, where this last result applies, include non-trivial ones in which $S_{3,3}$, or the kernels $S_{2,2}$, do not coincide with $S^c_{3,3}$, or with the kernels $S^c_{2,2}$, in the whole integration domain. The result for the connected amplitude may involve several terms D_β, even if there is only one term $D^{(n.c.)}$.

The previous statements, when they hold, are again readily expressed in terms of essential support properties (and hence again equivalently of singular spectrum properties), called also "microlocal discontinuity formulae". If $\hat{u}_+$ denotes the direction defined in $T^*_P M$ by u, then

$$\hat{u}_+ \notin ES_P(s - d^{(n.c.)}) \quad (22)$$

resp.

$$\hat{u}_+ \notin ES_P(f - d) \quad (23)$$

where $d^{(n.c.)}$ in (22), resp. d in (23), is possibly to be replaced by a sum of corresponding terms, as described previously.

A la Cutkosky discontinuity formulae

We finally explain how the usual discontinuity formula of Sect. 2 follows in simple cases, e.g. those of Sect. 2.3, from the above "microlocal" statements. In such cases, $ES_P(d)$ is known (e.g. through the application of the structure theorem: see Sect. 4) to be composed at most of $\hat{u}_+(P;G)$ and of the opposite direction $\hat{u}_-(P;G)$. This result combined with macrocausality ($\hat{u}_+$ is the only singular direction at P of f) and macrocausal factorization ($\hat{u}_+ \notin ES_P(f-d)$) shows that $\hat{u}_-(P;G)$ is the only singular direction of f-d. Hence f-d is locally the <u>minus</u> $i\varepsilon$ boundary value $f^{(L)}$ of an analytic function $\underline{f}^{(L)}$. Since d = 0 on the non-physical side of $L^+(G)$, $f = f^{(L)}$ there. Hence $\underline{f}^{(L)}$ appears explicitly to be a minus $i\varepsilon$ analytic continuation of f and d appears explicitly to be the discontinuity $f - f^{(L)}$ of f. Q.E.D.

The equivalence in the region R of Sect. 2.3 of macrocausality and macrocausal factorization with the decomposition (16) is proved also by simple arguments of essential support theory: see [2] (or similar ones of hyperfunction theory).

<u>Remark:</u> In certain two-dimensional space-time models, macrocausality and macrocausal factorization yield a factorization property of the multiparticle momentum-space S matrix itself into a product of two-body S-matrices (see [46]). This is due to very special features occurring in dimension 2, which make possible (and natural) this factorization property. In dimension > 2, similar results are not consistent with the general structure of this S-matrix that has been described: macrocausal factorization is a factorization property, not of the S-matrix itself, but of transition amplitudes between displaced wave functions, and the factorization that takes place (in causal cases) crucially depends on the way particles are displaced from each other.

4. THE ANALYTIC S-MATRIX THEORY OF THE SIXTIES AND RELATED RESULTS OF THE FIRST PART OF THE SEVENTIES

The analytic S-matrix theory of the sixties is based on Chew's principle of maximal analyticity. Although the works carried out at that time mix the two problems, they can be divided into two categories. The first one is concerned with the derivation of the physical-region analytic structure of the multiparticle S-matrix. The second one, which we shall briefly discuss first, makes use of the knowledge of the physical region structure to derive properties such as crossing, the spin-statistics theorem, hermitean analyticity and "extended" unitarity (which state relations analogous to unitarity but involving analytic continuations of scattering functions), TCP,..., see [7,49,50], a general presentation being given in Ch. IV of [1]. These derivations are based on the assumption that scattering functions can be analytically continued away from the physical region in the most simple way one may imagine, and on the physical-region discontinuity formulae and related results, including in particular the pole structure (9) and corresponding expressions of scattering functions as products of poles, with residues factorizing again into a product of scattering functions associated with each vertex in the case of tree graphs with more than one internal line.

To discuss e.g. crossing it is convenient to use, as in field theory, notations in which energy-momentum conservation for each given process reads

$$\sum_{k=1}^{m+m'} p_k = 0, \text{ the energies } (p_k)_0 \text{ being} > 0$$

if k is initial and < 0 if k is final. The physical regions of various "crossed" processes are then various disconnected real parts of the complex mass-shell manifold $\underline{M}^{(n)} =$
$= \{(p_1,\ldots,p_n), \text{all } p_k \text{ complex}, p_k^2 = \mu_k^2 \ \forall k = 1,\ldots,n,$

$\sum_k p_k = 0\}$, corresponding to various values of m,m' satisfying m+m'=n. Each scattering function $f_{m,m'}$ is known (physical-region structure) to be the boundary value in its physical region of an analytic function $\underline{f}_{m,m'}$ (see Sect. 2), crossing is the assertion that there is a unique analytic function $\underline{f}^{(n)}$, defined in a domain of the complex mass-shell, with which all functions $\underline{f}_{m,m'}$ of crossed processes coincide. This is established in the S-matrix approach by "embedding" the scattering functions under consideration in higher order processes in a way such that they appear as various factors in the residues of various physical-region pole-singularities of the latter. The "simplicity" assumptions on analytic continuation upon which these proofs rely are, however, different to control and they should rather be considered as interesting heuristic arguments. There has been no basic progress on this problem since the sixties and we shall thus not discuss it in more detail here.

The remainder of this section is now devoted to the first part, i.e. the derivation of the physical-region structure. In the traditional approaches [7-9] this is done in the following steps:

a) It is first argued that "maximal analyticity", when applied to the physical region, should provide analyticity outside Landau surfaces of connected graphs. This is because equations derived from unitarity ($SS^{-1} = SS^{\dagger} = 1$) and from the decomposition of the S-matrix into its connected parts involve relations (see e.g. Eq. (25) below) between connected kernels of S and $S^{-1} = S^{\dagger}$ and other "bubble diagram functions" (= on-mass-shell convolution integrals whose bubbles are connected kernels of S or S^{-1}). Singularities of the latter may à priori arise from the integration procedure. By considering the set of all these equations, it is argued that singularities of the S-matrix should be at most those generated in that way, i.e. the only singularities should be those of phase-space integrals (= integrals associated with multiple scattering graphs, with constants times conservation δ-functions at each vertex and mass-shell δ-functions, rather than Feynman propagators, for each internal line).

Remark: Although the singularities of phase-space integrals are contained in the Landau surfaces (defined in Sect. 2.1), this statement as it stands is not of much interest because some Landau "surfaces" cover the entire physical region. This is e.g. the case for the graph

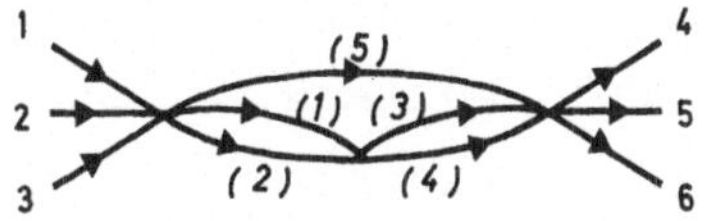

in a 3 → 3 (equal-mass) process: being given any physical-region point $p = (p_1, \ldots, p_6)$, there always exist on-mass-shell energy-momenta $k_1, \ldots, k_5$ for the internal lines $1, \ldots, 5$ such that $k_1 = k_2 = k_3 = k_4$, $k_1 + k_2 + k_5 = \sum_{i=1}^{3} p_i$. It is then sufficient to choose $\alpha_1 = \alpha_2 = -\alpha_3 = -\alpha_4$ $(\neq 0)$, $\alpha_5 = 0$. To extract actual information from the idea explained above of maximal analyticity, a more refined analysis of the singularities of phase-space integrals is needed. This is non-trivial because one already encounters in this study the "u = 0" problem. The general study of phase-space integrals carried out recently in [33] shows that their singularities are restricted to codimension-one modified Landau surfaces. For simplicity, we still speak below of "Landau surfaces".

The idea of maximal analyticity is interpreted as providing moreover the existence of an analytic continuation of scattering functions from one side of a singularity surface to the other side. Some non-complete arguments of internal consistency lead to postulate that, at least along the $+\alpha$-Landau surfaces, the rule of analytic continuation is always the plus iε rule.

An alternative way of getting rid of the previous difficulty concerning Landau surfaces and to establish in the same time general plus iε rules is to make recourse to macrocausality (see Sect. 3.1) which provides moreover analyticity outside the +α-parts of the Landau surfaces. This is done e.g. in [9]. (In the spirit of "maximal analyticity", macrocausality is used there only to complement previous arguments.)

b) The important part of the program is then to derive the detailed structure at +α-Landau points in the form of the discontinuity formulae and related results (and to show analyticity along the "mixed-α" parts of the Landau surfaces if the latter has not been assumed).

To that purpose one considers unitarity equations in the neighborhood of a given +α-Landau surface $L^+(G)$, and one tries to extract information on the scattering function f considered from the analysis of the various terms involved: see illustration below. The methods of the Cambridge group [7,8] and of [9] are somewhat different, but the general idea is the same. The Cambridge group starts from unitarity equations as they hold on the physical side of $L^+(G)$ and compares them with minus iε analytic continuations of the equation that holds on the non-physical side. The method of [9] consists in first using repeated applications of unitarity to transform the original unitarity equations into equations of the form:

 $= D + \Sigma$ other bubble diagram functions , (24)

where 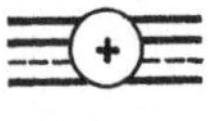

is as before the connected S-matrix kernel of the process and D is the bubble diagram function associated with the graph G considered (e.g. the r.h.s. of (10) in the case of the triangle graph of Sect. 2.2). The term d will then appear explicitly as the discontinuity $f-f^{(L)}$ of f if one can prove that the last sum in the r.h.s. of (24) is in the neighborhood of P the boundary value of an analytic function from minus $i\varepsilon$ directions (after factorization of overall conservation δ-functions): this follows easily from the fact that d = 0 on the non-physical side of $L^+(G)$ and that f is itself a plus $i\varepsilon$ boundary value.

In order to illustrate these methods and the crucial difficulties encountered, we consider the simple case of the graph

G = 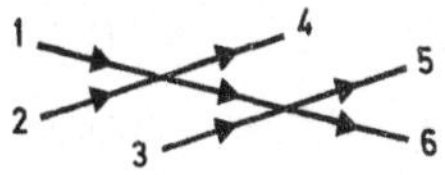

in a 3 → 3 process, for a theory with only one type of particle. We moreover restrict our attention to the region $s < (4\mu)^2$ $(s = (p_1+p_2+p_3)^2)$. The unitarity equation in that region, directly derived from $SS^{-1} = 1$ and decompositions of $S_{3,3}$, $(S^{-1})_{3,3}$ into connected parts, is:

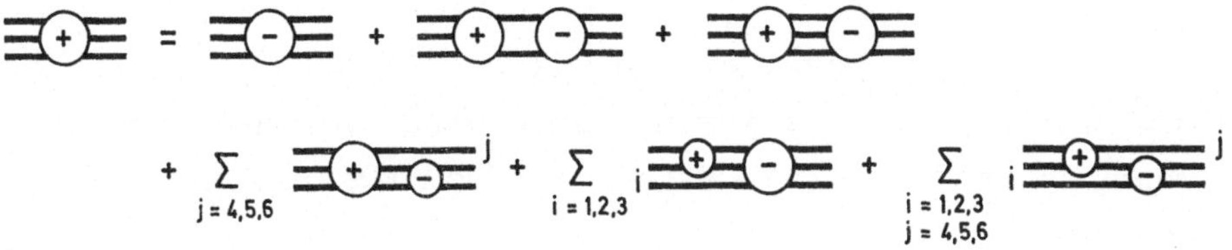

(25)

where the plus and minus bubbles stand for connected kernels of S and $-S^{-1}$.

All terms involved contain an overall energy-momentum conservation δ-function. The analyticity properties of interest below refer to the distributions obtained in the physical region after factorizing out these δ-functions. By an abuse of language, we shall still write them diagrammatically in the same way.

Olive's proof [7] (unfortunately not correct: see below)

In the r.h.s. of (25), one encounters one term, namely

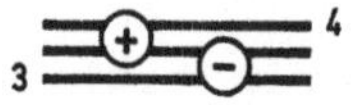 which is explicitly singular along $L^+(G)$

(and in fact contains δ-function singularity $\delta(k^2-\mu^2)$ on that surface, $k = p_1+p_2-p_4$).

The term  and hence also the term 

are along $L^+(G)$ plus $i\varepsilon$ boundary values. Let H be the sum of all other terms in the r.h.s. of (25). It is claimed in [7] that these terms are either analytic or minus $i\varepsilon$ boundary values

(like )

along $L^+(G)$. Let us then rewrite (25) in the form:

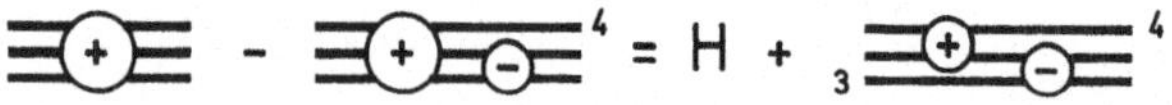 (26)

On the non physical side ($k^2 < \mu^2$) of $L^+(G)$, the last term in the r.h.s. is absent. By a minus $i\varepsilon$ analytic continuation around $L^+(G)$ (starting from the region $k^2 < \mu^2$) one thus obtains:

 , (27)

H being unchanged since it was asserted (see above) to be a minus $i\varepsilon$ boundary value.

Substracting (27) from (26), "multiplying" on the right both sides of the equation obtained by 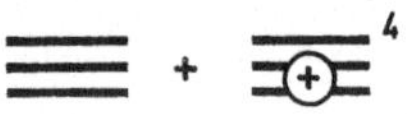

(in the sense of on-mass-shell convolutions,

 being the kernel of the identity) ,

and using two-particle unitarity below the 3-particle threshold,

 ,

one gets in a simple way:

 (28)

which is the desired discontinuity formula.Q.E.D.

A slightly different proof, based on the same assertion on H, can also be given: multiplication on the right by

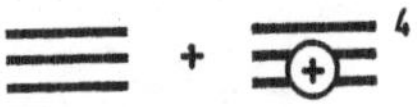

is made directly on Eq.(26), thus giving

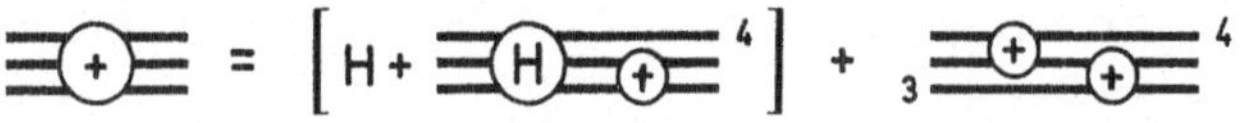 . (29)

If H is along $L^+(G)$ a minus $i\varepsilon$ boundary value, the same result follows for

 . Hence 

appears explicitly as the discontinuity, as explained below Eq.(24). This proof, given in [21,51], has the advantage that the existence of the minus $i\varepsilon$ continuation of

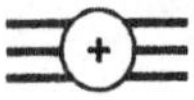

around $L^+(G)$ (from the region $k^2 < \mu^2$) needs not be assumed, but is itself established as a byproduct.

It has, however, been recognized recently, as explained in detail in [35], that Olive's original assertion according to which the terms occurring in H are analytic or minus iε boundary values along $L^+(G)$ is crucially non-correct: several of the terms involved, such as

,

can definitely <u>not</u> be expected to be analytic or minus iε boundary values at various points of $L^+(G)$. And even if Olive's assertion was proved at some point P of $L^+(G)$, this would not be sufficient to prove the discontinuity formula near P: in order to carry out the last step, Eq.(27) is needed not only at P but also at all points $p = (P_1,\ldots P_4, P_5', P_6')$ with P_5', P_6' on mass-shell, $P_5'+P_6'=P_5+P_6$, since these points occur in integration domains after multiplication on the right by

 .

But whatever P is, a term such as

cannot be expected to be analytic or a minus iε boundary value at those points p such that $P_5' = P_3$, $P_6' = P_1+P_2-P_4$. A related comment applies to the second proof: some terms occurring in

 , such as 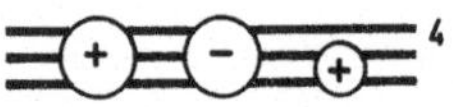 ,

cannot be expected to be analytic or minus iε boundary values at any point of $L^+(G)$. Although several of the individual terms in H do not satisfy Olive's assertion, one may try to make the proof correct by showing that the sum H itself of these various terms, or

H + 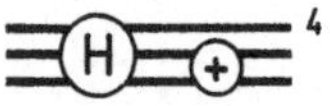

is a minus iε boundary value. This cannot be expected for H itself, but it can be expected, and will precisely be <u>proved</u> in Sect. 5.2, for a modified sum (in which the bubble

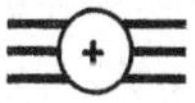

in the two terms of the l.h.s. of (26) is replaced by the contribution $F_{3,4}$ to 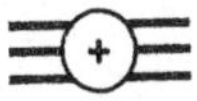

associated with the surface $L^+(G)$, and also for

(at points P lying on no other +α-Landau surface).

To understand the problems, the preliminary study of the analyticity properties of bubble diagram functions is needed. The analyticity properties assumed for each bubble have been described in paragraph a). The scattering functions $f^{(-)}$ associated with minus bubbles satisfy, by a direct use of unitarity ($S^{-1} = S^{\dagger}$), minus iε rules at +α-Landau points (and are analytic outside Landau, or +α-Landau) surfaces. In the methods used in the sixties [49,7-9], one tries to establish analyticity domains of the bubble diagram functions (with respect to the external variables, on the complex mass-shell, on which they depend) by eliminating all δ-functions in the integrand and by considering distortions of the (real) integration domain in complex space, in a way such that the individual functions associated with each bubble b all remain in their analyticity domain. A first difficulty appears because one always encounters in integration domains (even in the simplest cases) values of the internal on-mass-shell energy-momenta corresponding to points where some of the individual f_b or $f_b^{(-)}$ are no longer boundary values of analytic functions. This difficulty, which was ignored in the works of the Cambridge group, is treated in [9] by an ad hoc technical assumption stated there in a more or less precise way (patching assumption). It was overcome in the first part of the seventies by the methods of essential support theory (described in the Appendix-Sect. 1), or of hyperfunction theory, in which this assumption is no longer needed. They provide a $u \neq 0$ structure theorem (first proved in [30], and in a completely similar way in the framework of hyperfunction theory in [23]), which is a more precise and somewhat more general formulation of previous statements of [49,8,9] and in particular of the idea that singularities are associated with space-time Landau diagrams. This theorem is described below. It still gives no information at "$u = 0$" points: this second difficulty is a general aspect of the fact that the method previously outlined, based on distortions of contours in complex space, gives à priori no result if there exists no distorted contour such that all individual functions $\underline{f}_b$ or $\underline{f}_b^{(-)}$ remain analytic. As recognized recently, $u = 0$ points, far from being exceptional, cover in somes cases, e.g. for the term

the entire physical region (see Sect. 5.1). The $u \neq 0$ structure theorem will, however, be sufficient for the purposes of the present section.

<u>$u \neq 0$ structure theorem [30]</u>

Let F_B be the bubble diagram function associated with a bubble diagram B, with $F_B = f_B \times \delta^{\nu}(\Sigma p_i - \Sigma p_j)$. We shall denote here, on the other hand, by $F_b = f_b \times \delta((\Sigma p_i - \Sigma p_j)_b$ the corresponding distributions associated with the various plus or minus bubbles b. A diagram ε_B will be a collection of subdiagrams ε_b (associated with each bubble b) that "fit together". Each ε_b is a configuration of space-time trajectories, associated with the incoming and outgoing lines of b, corresponding to a point in the essential support of F_b at the point P_b representing the set of energy-momenta of these trajectories. The diagrams ε_b "fit together" if the trajectories associated in ε_{b_1}, ε_{b_2} with a common internal line of B joining b_1 and b_2 coincide: these trajectories are then identified as a common internal trajectory of ε_B. Finally P is a u = 0 point of B if there exists a "u = 0" diagram ε_B whose all internal trajectories pass through a common point, while some internal trajectories do not pass through this point.

The association of points in the essential support of F_B or F_b (resp. of f_B or f_b), with configurations (resp. relative configurations) of external trajectories of B or of b, is made in the way already explained in Sect. 3.1.

The general $u \neq 0$ structure theorem says this: if $P = \{P_k\}$ is not a u = 0 point of B, then F_B is a well defined distribution and the only possible singular directions of F_B, resp. of f_B, at P are those associated with the configurations, resp. relative configurations, of external trajectories of diagrams ε_B with external energy-momenta P_k.

In the cases under consideration, the ε_b are by virtue of macrocausality configurations of external trajectories of diagrams D_b which are classical diagrams if b is a plus bubble, or opposite diagrams is b is a minus bubble. Hence the diagrams ε_B are diagrams D_B with internal lines which are either the internal lines of subdiagrams D_b or the original internal lines associated with the internal lines of B. The α_ℓ's associated with the internal lines ℓ as in Sect. 3.1 are positive, resp. negative, if ℓ is the internal line of a subdiagram D_b and if b is a plus bubble, resp. a minus bubble. They are arbitrary (and can in particular be equal to zero) for the original internal lines. (They can be infinite if vertices at infinity are involved.)

Assumption of mixed-α cancellation or of separation of singularities in unitarity equations

The difficulties that have been mentioned in Olive's proof are already present; independently of the u = 0 problem as a matter of fact, the configuration of external trajectories of several mixed-α diagrams D_B (= diagrams with both positive and negative α_ℓ) coincide with the configurations of external trajectories of the causal diagrams D_+ at various points P of $L^+(G)$. Hence $\hat{u}_+(P;G)$ is expected to be a singular direction of the corresponding f_B at these points, whereas the only singular direction should be the opposite direction $\hat{u}_-(P;G)$ if these f_B were minus iε boundary values. This fact also prevents one from using in a simple way general results of essential support theory (or hyperfunction theory) which are general versions of edge-of-the-wedge theorems (see Appendix). A more refined analysis will then be needed: see Sect. 5.2.

All difficulties, including also the u = 0 problem, have been treated in [9] through the idea, and a corresponding ad hoc assumption of "mixed-α cancellation", that singularities associated with Landau diagrams with basically different topological structures should be "independent". (The crucial role played by this assumption was recognized only more recently.) Coming back to (24), it appears that all possible singularities of the terms in the last sum are associated with mixed-α negative-α diagrams, whereas the terms

 and D involve no mixed-α diagram.

This is seen in a particularly simple way on the example of Eq.(29). "Mixed-α cancellation" thus entails that the mixed-α diagrams can be ignored: the singularities they may produce must cancel among themselves since the singularities of other terms in (24) are associated with diagrams

with different structures. The sum $\sum$ is then easily shown to be a minus iε boundary value at P and the discontinuity formula follows, as has been explained below (24).

The analysis of [9] and the somewhat more satisfactory and general analysis of the seventies that has completed it apply to general graphs with only single lines. In the case of graphs with multiple lines, they also provide the desired results, involving the boxes S_{α}^{-1}: the derivation makes use, however, in general of infinite series and is of a formal nature.

"Separation of singularities in unitarity equations" according to the structure of the Landau diagrams involved has been formulated in a general way in [32,23]. It has no à priori basis (and is not a consequence of the general "edge-of-the-wedge" results of essential support or hyperfunction theory). As mentioned in the Introduction, it is, however, believed to hold and has been conversely derived in a number of cases [31,32] from the discontinuity, or microlocal discontinuity formulae. It will be proved directly in Sect. 5.2 for the 3 → 3 S-matrix below the 4-particle threshold in relevant cases.

Remark: "Separation of singularities" will correspond in general to the fact that the nature of the singularities should be different if they correspond to diagrams with different structures. (For instance two plus iε boundary values of analytic functions that would correspond to a pole-singularity and to a branch point cannot cancel each other.) Note, however, that this independence can also be expected in cases where the nature of the singularities will turn out to be the same. E.g. in space-time dimension 2, the singularities arising for f from the four connected +α-diagrams where the notation indicates that a vertex may be present or not (= diagrams whose topological structure is a triangle or a graph with one internal line), all coincide

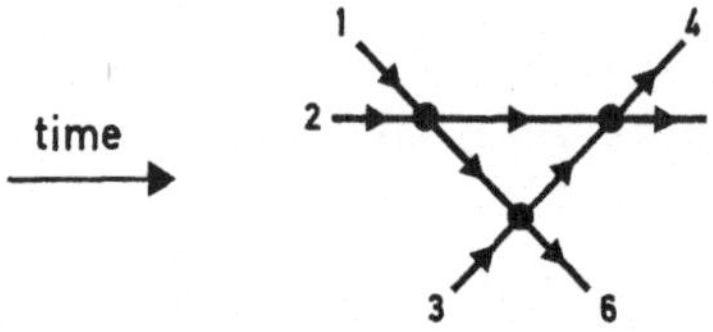

in momentum-space with the surface $p_1=p_6$, $p_2=p_5$, $p_3=p_4$, the singular (= causal) direction is the same at any point of that surface, and the nature of the singularity (= a pole) is also the same (see Sect. 7.1). They are, however, still expected to have independent cancellations with other terms in the unitarity equation.

5. S-MATRIX THEORY: RECENT RESULTS

5.1 The u = 0 problem

The first difficulty one encounters if "separation of singularities" is not assumed is the u = 0 problem already mentioned. Being given e.g. the term 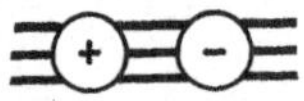

and any physical region point $P = (P_1,...,P_6)$, there always exist u = 0 diagrams D_B obtained by putting together subdiagrams D_{b_1}, D_{b_2} associated with each bubble: the subdiagram D_{b_1} may be for instance that shown in Fig. 9a, or if P is above the four-particle threshold, that shown in Fig. 9b. The diagram D_{b_2} has in each case the same internal trajectories (1), (2), (3), and has outgoing trajectories 4, 5, 6 passing through the origin. In the second case, D_{b_2} has the same two vertices: (2), (3) meet, give three internal intermediate particles that travel together backward in time to the origin, where they meet (1).

The works carried out either in essential support theory or in hyperfunction theory show clearly that no

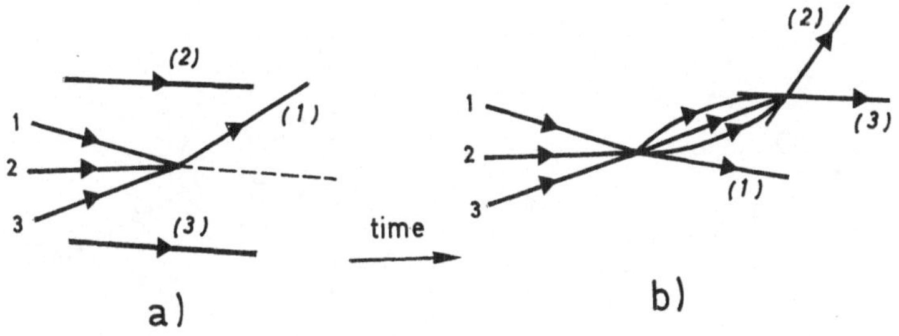

Fig. 9

result on analyticity properties can be obtained at u = 0 points P if one starts only from the essential support, or singular spectrum, property associated with macrocausality (and unitarity) for the individual bubbles. The problem is treated in [25] on the basis of a regularity property R (see Appendix) of the connected S-matrix which corresponds physically to a refined form of macrocausality, as briefly explained below on an example. The u = 0 theorem proved in [25] is similar to the u ≠ 0 structure theorem of Sect. 4, except that certain limiting procedures must be considered in general.

Property R gives information on the way rates of exponential fall-off tend to zero when causal directions are approached. Let us in fact consider a set of directions that are all non-causal at P and at all points p in a neighborhood N of P of width α, but approach a causal direction. A simple example of such a set is obtained by considering a causal diagram, e.g. that of Fig. 10a), and by "opening" one of its vertices (Fig. 10b)) in a way such that trajectories 1, 2, 4 do not meet.

If the second vertex is taken to infinity, the directions $\hat{u}$ defined by the configurations of external trajectories of the diagrams of Fig. 10b) are all non-causal, but tend to the causal direction $\hat{u}_+$ defined by the diagram of Fig. 10a).

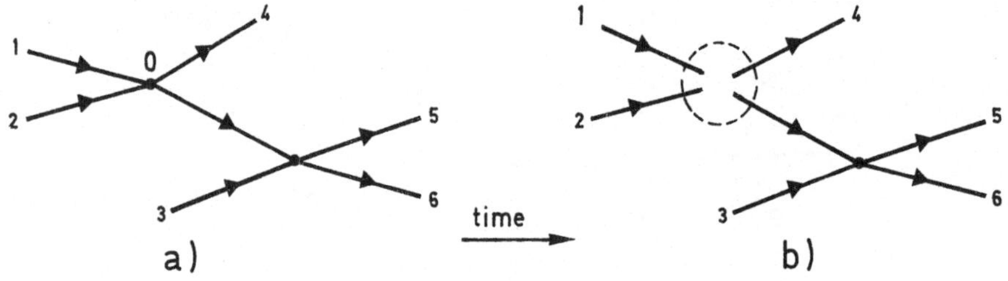

Fig. 10

Macrocausality and general results of essential support theory ensure exponential fall-off factors $e^{-\alpha\gamma\tau}$ for all directions of such a set, with the same uniform $\alpha > 0$ and for all $\gamma > 0$ sufficiently small. The maximal value of γ depends, however, on the position of the second vertex. The main content of property R is that it is at least proportional to the distance to the essential support, which in the above example remains constant and corresponds in fact to the opening of the first vertex. This means physically that one always keeps the exponential fall-off properties arising from the uniform opening of that vertex.

To give some idea of the content of property R in terms of analyticity properties, let us consider for simplicity a point P of a smooth codimension-one surface $L^+(G)$. Macrocausality is then equivalent at P to the plus $i\varepsilon$ rule. By definition, the plus $i\varepsilon$ directions are, in the local coordinate system already considered in Sect. 2.1, the directions of the half-space $\mathrm{Im}\, q_1 > 0$. The plus $i\varepsilon$ rule asserts that the analytic function $\underline{f}$ from which f is the boundary value is analytic for complex $q = \mathrm{Re}q + i\,\mathrm{Im}q$ such that Req belong to a neighborhood of the origin and Imq belongs to a domain of the form shown in Fig.11a):

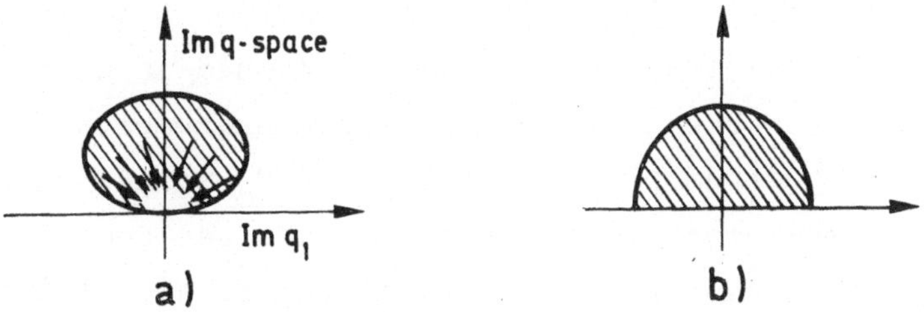

Fig. 11

By definition, f satisfies the no sprout property at P if this domain is of the form shown in Fig. 11b), the important fact being the shape of the domain along the $\mathrm{Im}q_1$-axis, near q = 0. The no sprout property at P corresponds to the exclusion of otherwise possible pathologies. It is a very weak aspect of the idea of "local maximal analyticity" at P. The latter says that f can be analytically continued locally around the singularity, the surface $q_1 = 0$ being the only singularity in complex space in the (possibly multisheeted) domain generated by analytic continuation. "Local maximal analyticity" is itself a very weak aspect of the results one ultimately aims to prove (see Sect. 7).

The regularity property R is clearly closely related to (and can probably be derived from) the no sprout property. In fact, it is known (see Appendix) that property R at P on the scattering function f is indeed (essentially) equivalent to the no sprout property. However, the problem is not exactly the same here, since the regularity property R associated with refined macrocausality (and used in the proof of the u = 0 structure theorem of [25]) refers to the connected S matrix, containing the energy-momentum conservation δ-function and expressed in terms of momentum variables. More work would be needed to get precise statements on its general content in terms of analyticity properties and to check its general validity.

The approach to the u = 0 problem of [33,34]

is different. In [33] a general result on phase-space integrals is obtained, as already mentioned in Sect. 4. It has been used there first to justify statements on the singular spectrum of the S-matrix at M_o points (namely the angular-momentum conservation law: see Sect. 3.1). A conjecture on the singular spectrum of actual bubble diagram functions is then proposed on the basis of these results. This u = 0 conjecture involves, as the u = 0 theorem of [25], certain limiting procedures which are, however, different.

Recent mathematical results [52] on products of regular holonomic functions at u = 0 points have more recently led to corresponding results [34] on bubble diagram functions based on holonomicity assumptions on scattering functions. Namely, it is assumed that the latter should be, near any physical-region point, regular holonomic functions or more generally (finite or infinite) sums of regular holonomic contributions, each of which satisfying further properties close to those of one of the Feynman integrals that are singular at P. (The latter are known [28] to be in particular regular holonomic.) Infinite sums are clearly required in some situations (see Sect. 7.2), in which case convergence is also assumed. The u = 0 result that follows from this assumption involves limiting procedures similar to those of [33]. The exact link between the results of [25] and [34] is not yet known. The assumptions made in [34] appear to be much stronger and more detailed, and the result may therefore be possibly more refined in some situations.

The u = 0 results of [34] are of interest mainly in the philosophy of "maximal analyticity" in which one wishes to check internal consistency. In fact, the assumption used is close to the final results one aims to <u>derive</u> in an axiomatic, deductive approach, i.e. the discontinuity (or microlocal discontinuity) formulae. We have already explained in Sect. 2.2, 2.3 that the latter are in fact equivalently stated in simple cases in terms of à la Feynman integrals, and the results presented in Sect. 7.2 show that this is again true in a simplified model at three-particle thresholds, where an infinite convergent sum of à la Feynman terms is obtained. More work is needed here also to obtain precise and general results on the validity of this assumption.

5.2 Macrocausality, unitarity and the 3 → 3 S-matrix below the 4-particle threshold

In the case of a 3 → 3 process below the 4-particle threshold, the results of [25], or also of [34], indicate that the u = 0 problem occurring for

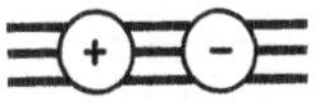

and other terms can be ignored: the essential support of these terms is still given by the u ≠ 0 structure theorem. As explained in Sect. 4, crucial difficulties still remain if "separation of singularities" is not assumed. They have been solved in [35] at points P of surfaces $L^+(G)$ of graphs with one internal line that lie on no other surface, with the help of a "no sprout" assumption on f (= f satisfies the no sprout property described in Sect. 5.1). These results have been extended in [36] with the help of a no sprout assumption that is a slight extension of the previous one (see below) and by a somewhat different method. The following result is then obtained:

<u>Theorem:</u> "Refined macrocausality, unitarity and the no sprout assumption imply the decomposition (16) of the 3 → 3 S-matrix in the region R as a sum of à la Feynman integrals associated with graphs with one internal line and triangle graphs".

Refined macrocausality is used, as explained above, to solve the u = 0 problem and will play no further role below. It is probably closely related with the no sprout assumption, as already mentioned in Sect. 5.1. We now introduce the latter in a more precise way in the form needed in the proof of the theorem. Macrocausality and the general results of essential support, or hyperfunction, theory entail (non-unique) decompositions of f below the four-particle threshold of the form

$$f = \sum_{\beta} f_{\beta} \tag{30}$$

where each f_β is, in R, analytic outside $L^+(G_\beta)$ and is along $L^+(G_\beta)$ a plus $i\varepsilon$ boundary value. (The $L^+(G_\beta)$ are the 18+α-Landau surfaces encountered in R: see Sect.2.3.)

<u>No sprout assumption:</u> "There exists a decomposition of the form (30) such that each f_β satisfies the no sprout property at all points P of $L^+(G_\beta)$ in R".

The distributions f_β provided by the no sprout assumption are now uniquely defined in R modulo analytic backgrounds, whereas more serious ambiguities were possible in (30). This fact follows (see [36]) from the following lemma [35] which plays an important role below and is a consequence of Bremerman's continuity theorem.

<u>Lemma:</u> "Let h be a distribution defined in a real domain containing a (smooth, codimension-one) surface $L^+(G)$, analytic outside $L^+(G)$ and satisfying the plus $i\varepsilon$ rule along $L^+(G)$. If h is analytic at one point of $L^+(G)$ and if it satisfies the no sprout property at all points of $L^+(G)$, then it must be analytic at all points of $L^+(G)$".

In other words, h cannot be analytic at some points of $L^+(G)$ and singular at other points, if it satisfies the no sprout property.

We now return to the proof of the theorem. Let us consider Eq. (25) and let us group together contributions to the various terms whose singularities can be expected to cancel among themselves, if one believes to "separation of singularities":

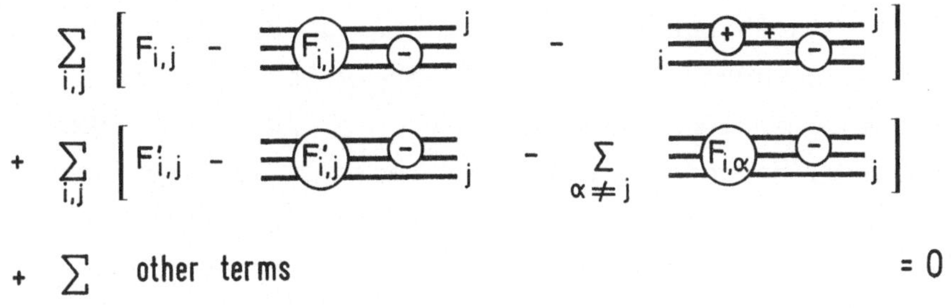

$$
\sum_{i,j} \left[F_{i,j} - \;\fbox{diagram}\; - \;\fbox{diagram}\; \right]
$$

$$
+ \sum_{i,j} \left[F'_{i,j} - \;\fbox{diagram}\; - \sum_{\alpha \neq j} \fbox{diagram} \right]
$$

$$
+ \sum \text{ other terms} = 0 \tag{31}
$$

In (31), $F_{i,j}$ and $F'_{i,j}$ denote the contributions to

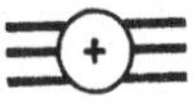

associated with the surfaces $L^+(G_{i,j})$ of the graphs

$$
G_{i,j} = \;\fbox{diagram}\; \quad \text{and} \quad G'_{i,j} = \;\fbox{diagram}
$$

respectively, and

is a "$+ i\varepsilon$" contr. to ,

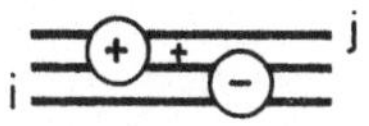

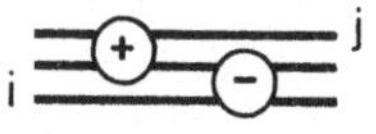

well defined in R modulo an analytic background as in Sect. 2.2.

Let us first consider a given pair (i,j) and a corresponding bracket $H_{i,j} = h_{i,j} \, \delta^{\nu}(p_1 + p_2 + p_3 - p_4 - p_5 - p_6)$ in the first sum of the l.h.s. of (31). By direct inspection, the three terms in $h_{i,j}$ are analytic in R outside $L^+(G_{i,j})$ and are plus $i\varepsilon$ boundary values along $L^+(G_{i,j})$; in other words, their essential support at any P of $L^+(G_{i,j})$

in $\hat{u}_+(P;G_{i,j})$. Their singularities arise from diagrams D_B of the form

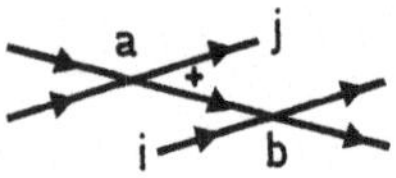

,

or from the same diagrams with two more internal trajectories passing through b (which have a "related" topological structure). A detailed analysis shows (in space-time dimension > 2) that $\hat{u}_+(P;G_{i,j})$ is absent from the essential support of all other terms in (31), at least at some points P of $L^+(G_{i,j})$. This is not true at points that lie in a certain open subset or in other lower-dimensional submanifolds of $L^+(G_{i,j})$, where $\hat{u}_+(P;G_{i,j})$ is indeed the relative configuration of external trajectories of some mixed-α diagrams. The equality (31) ensures, however, that $\hat{u}_+(P;G_{i,j})$ is absent from $ES_P(h_{ij})$ at the previous points. Since it was the only possible singular direction of the terms that contribute to $h_{i,j}$, and hence of $h_{i,j}$, one concludes that $h_{i,j}$ is in fact <u>analytic</u> at these points. On the other hand, one checks by direct inspection that the three terms in $h_{i,j}$, and hence $h_{i,j}$ itself, satisfy the no sprout property. The lemma quoted above then entails that $h_{i,j}$ is analytic in the whole region R.

Multiplication, on the right, of $H_{i,j}$ by

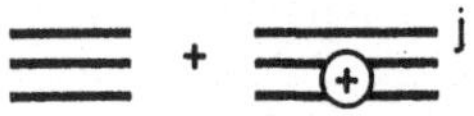

gives

$H'_{i,j} = F_{i,j} -$ 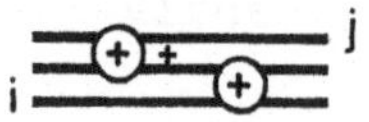

.

One checks on the other hand that analyticity of $h_{i,j}$ in R entails analyticity of $h'_{i,j}$. Hence:

$F_{i,j} \simeq$ 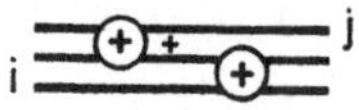

(32)

where ≃ is defined as in (16),(17).

Once (32) is established (for each pair (i,j)), a similar analysis with only minor changes allows one to determine $F'_{i,j}$, again modulo analytic backgrounds in R: see [36]. (The no sprout property for

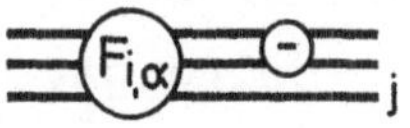

follows e.g. from the results of Sect. 7.1, since the discontinuity of that term is

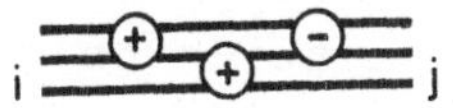

The change of the two-body plus bubble on the right by a minus bubble does not modify the results since both $f_{2,2}$ and $f^{(-)}_{2,2}$ are analytic in integration domains.) The decomposition (16) is thus established in R. Q.E.D.

<u>Remark:</u> A proof similar to above cannot be obtained directly from Eq. (29) because, as already mentioned in Sect. 4, $\hat{u}_+$ $(P;G_{i,j})$ is a singular direction of terms such as

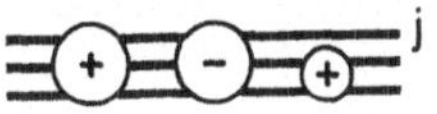

at <u>all</u> points P of $L^+(G_{i,j})$.

6. AXIOMATIC FIELD THEORY: MICRO-CAUSALITY, OFF-SHELL UNITARITY AND THE 3 → 3 S-MATRIX BELOW THE 4-PARTICLE THRESHOLD

The basic quantities of axiomatic field theory for the purposes of this section are the connected "chronological functions" τ_c which are, for each n, well defined distributions in the space R^{4n} of n space-time vector variables $x_1,\ldots,x_n$ and are in fact (possibly regularized) connected, "amputated" vacuum expectation values of the chronological product $T(x_1,\ldots,x_n)$ of n field operators $A(x_1),\ldots,A(x_n)$. Their Fourier transforms $\tilde{\tau}_c$ are correspondingly defined in the space R^{4n} of energy-momentum variables $p_1,\ldots,p_n$. In view of the translation invariance of τ_c under translation of all x_k by a common space-time vector a, $\tilde{\tau}_c$ contains an energy-momentum conservation δ^ν-function $\delta^\nu(\sum_{k=1}^{n} p_k)$. It can be shown (see below), as first established in [14], that, being given any process $m \to m'$, with $m + m' = n$, the distribution $\tilde{\tau}_c$ can be restricted (as a distribution) to the mass-shell $M_{m,m'} = \{p = (p_1,\ldots,p_n);\ p_k^2 = \mu^2\ \forall k, (p_k)_o < 0,\ k = 1,\ldots,m,\ (p_k)_o > 0,\ k = m + 1,\ldots,n\}$, and the following relation holds:

$$S^c_{m,m'}(p_1,\ldots,p_m;p_{m+1},\ldots,p_n) = \tilde{\tau}_c(-p_1,\ldots,-p_m;p_{m+1},\ldots,p_n)\Big|_{M_{m,m'}} \tag{33}$$

We need also to consider functions $(\tau_c)_I$, where I is a subset of $(1,\ldots,n)$, which are defined in a way similar to τ_c, except that $T(x_1,\ldots,x_n)$ is replaced by the product of $T(x(J))$ and $T(x(I))$, where J is the complement of I in $(1,\ldots,n)$ and $x(I)$, resp. $x(J)$, denote the sets of points x_k, $k \in I$, resp. $k \in J$.

Microcausality asserts that the commutator $[A(x), A(y)]$ of two field operators vanishes if $(x-y)^2 < 0$, where $x^2 = x_o^2 - \vec{x}^2$, i.e. if x-y is space-like, and entails (see [14,37]) that:

$$\tau_c(x_1,\ldots,x_n) = (\tau_c)_I(x_1,\ldots,x_n) \text{ if } (x_1,\ldots,x_n) \notin \Sigma_I \tag{34}$$

where $\Sigma_I = \{(x_1,\ldots,x_n), x_j - x_i \gtrsim 0$ for any pair of points x_i in $x(I)$ and x_j in $x(J)\}$. The condition $x_j - x_i \gtrsim 0$ means either that $x_j - x_i$ is space-like ($(x_j-x_i)^2 < 0$), or belongs to the cone $\bar{V}^+$ (= set of points x in space-time such that $x^2 \geq 0$, $x_o \geq 0$).

The spectral condition is the assertion that the spectrum of the energy-momentum operator of the theory is contained (for a theory with only one type of particles of mass $\mu > 0$) in the union of the origin (corresponding to the "vacuum"), of the positive-energy mass-shell hyperboloid $V^+(\mu)$ ($p^2 = \mu^2$, $p_o > 0$), corresponding to one-particle states, and of the continuum $\bar{V}^+(2\mu)$ ($p^2 \geq (2\mu)^2$, $p_o > 0$). It entails [14,37] that $(\tilde{\tau}_c)_I$ has its support in the region

$$\sum_{j \in J} p_j \;(= - \sum_{i \in I} p_i) \in \bar{V}^+(\mu) \cup \bar{V}(2\mu), \text{ if } 1 < |J| < n-1$$

$$\in \bar{V}^+(2\mu) \;, \qquad \text{if } |J| = 1 \text{ or } n-1. \tag{35}$$

Being given a point P in $M_{m,m'}$, let S_P be the set of all I such that $(\tilde{\tau}_c)_I$ vanishes in the neighborhood of P. It is then easily seen that

$$ES_P(\tilde{\tau}_c) \subset \bigcap_{I \in S_P} \Sigma_I \quad . \tag{36}$$

This follows from the following elementary lemma (see e.g. [2]): If the Fourier transform $\tilde{f}$ (here $\tau_c - (\tau_c)_I$) of a distribution f (here $\tilde{\tau}_c - (\tilde{\tau}_c)_I$) vanishes outside a cone C with apex at the origin, then the generalized Fourier transform of f at any point P falls off exponentially (in the appropriate sense) outside C, i.e. $ES_P(f) \subset C$. (A different proof is given in [37].)

By standard results of essential support theory (see Appendix, Sect. 1), (36) entails that the restriction of $\tilde{\tau}_c$ to $M_{m,m'}$ does exist (away from M_o points), and it gives information on the essential supports of $S^c_{m,m'}$, hence of the scattering function $f_{m,m'}$ at any point P of $M_{m,m'}$: they are contained respectively in the sets associated with all possible configurations of trajectories (P_k, x_k) corresponding to all points $(x_1, \ldots, x_n)$ in $ES_P(\tilde{\tau}_c)$, resp. relative configurations of such trajectories.

Four-point function (two-body processes)

For a two-body process (m = m' = 2, n = 4), the result (36) reduces, at any physical region point P, to:

$$ES_P(\tilde{T}_c) \subset \{x = (x_1,\dots,x_4), x_1=x_2,\ x_3=x_4,\ x_3-x_1 \in \bar{V}^+\} \quad (37)$$

where $\bar{V}^+$ is the set of points x in space-time such that $x^2 \geq 0$, $x_o \geq 0$. By Lorentz invariance, this result can moreover be improved, namely the condition $x_3-x_1 \in \bar{V}^+$ can be replaced by $x_3-x_1 = \lambda(P_1+P_2)$, $\lambda \geq 0$. The possible points in $ES_P(\tilde{T}_c)$ and $ES_P(S^c_{2,2})$ are represented in Fig.12. (Equivalently $\tilde{T}_c$ is the boundary value of an analytic function from the directions Im s > 0.)

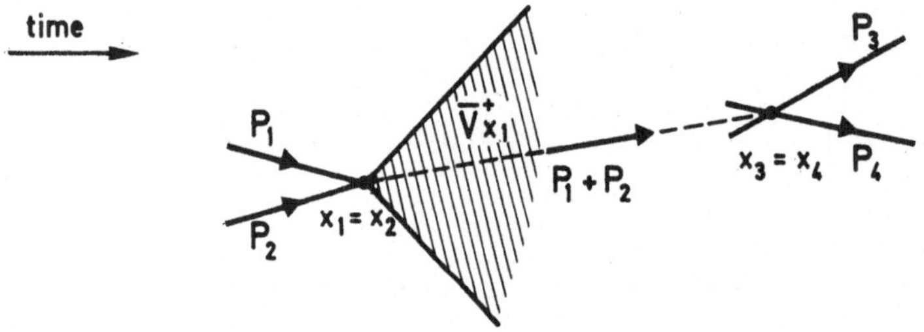

Fig. 12

Let us consider a point P in the region $(2\mu)^2 < s < (3\mu)^2$. There, a better result is expected according to the macrocausality ideas, namely $x_1=x_2=x_3=x_4$ (= equivalently $\tilde{T}_c$ is analytic at P), since energy-momentum cannot be transferred from $x_1=x_2$ to $x_3=x_4$ by stable particles if $x_3 \neq x_1$. This result can be obtained (see Sect. 1 and 7.2) by a regularity assumption which eliminates a priori possible à la Martin pathologies. We admit below correspondingly that:

$$ES_P(\tilde{T}_c) \subset \{(x_1,x_2,x_3,x_4), x_1=x_2=x_3=x_4\} \quad \forall P \text{ s.t. } (2\mu)^2 < s < (3\mu)^2 . \quad (38)$$

Equivalently, $\tilde{\tau}_c$ after factorization of its e.m.c. δ^4-function, and $f_{2,2}$ are analytic at P.

<u>Five-point function</u> (2 → 3 or 3 → 2 processes)

We next consider a 2 → 3 process in the region $s < (4\mu)^2$ $(s = (p_1+p_2)^2 = (p_3+p_4+p_5)^2)$. In this case, (36) entails that

$$ES_P(\tilde{\tau}_c) \subset C_3 \cup C_4 \cup C_5 \quad , \tag{39}$$

where

$$C_k = \{x = (x_1,\dots,x_5);\ x_1=x_2, x_i=x_j,\ x_i-x_k \in \bar{V}^+,\ x_k-x_1 \in \bar{V}_+\} \tag{40}$$

k = 3,4,5, (i,j,k) = (3,4,5). The cone C_3 is represented in Fig. 13a):

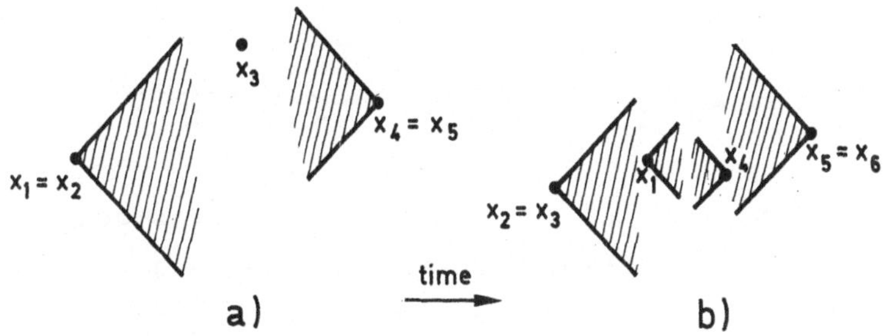

Fig. 13

This result, which cannot be improved in a simple way by Lorentz invariance, is very far from the essential support expected from the macrocausality ideas, which corresponds in the region $s < 4\mu^2$, and away from M_0 points, to $x_1=x_2=x_3=x_4=x_5$. We explain below how it can be improved by using the further axiom of <u>asymptotic completeness,</u> together with the previous result (38) on the four-point function.

Let $(\tilde{\tau}_c)_{i,j} = \tilde{\tau}_c - (\tilde{\tau}_c)_I$, where $I = (1,2,k)$, $(i,j,k) =$
$= (3,4,5)$, and let

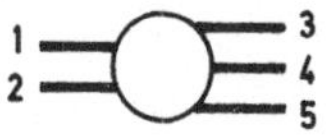

and

represent diagrammatically $\tilde{\tau}_c$ and $(\tilde{\tau}_c)_{i,j}$ respectively. Microcausality and the spectral condition yield the following information which is the analogue of (39):

$ES_P($  $) \subset C_i \cup C_j \cup C_k^-$, (41)

where $C_k^- = \{x=(x_1,\ldots,x_5),\ x_1=x_2,\ x_i=x_j,\ x_k-x_1 \in \bar{V}^+,$
$x_k-x_i \in \bar{V}^+\}$.

Asymptotic completeness yields on the other hand the following relations in the region $s < (4\mu)^2$, obtained by introducing in the appropriate place a complete set of intermediate states and by standard analysis of axiomatic field theory (see e.g. [53]):

 , (42)

where the last term in the right-hand side is defined in the same way as bubble diagram functions as a convolution integral over internal on-mass-shell four-momenta (with the measure $\delta(k_\ell^2-\mu^2)\,\theta((k_\ell)_o)\,d^4k_\ell$ for each internal line). The only difference is that the external four-momenta are not restricted to the mass-shell. In view of (38), (41), the standard $u \neq 0$ results of essential support theory (or hyperfunction theory) on products and integrals (see Appendix) show that this integral is well defined, and give information on its essential support, for $P_i \neq P_j$. Theorems 2,3,4 of the Appendix yield in fact a $u \neq 0$ structure theorem which is very close to that described in Sect. 4. The diagram ε_b are now configurations $\{x_k, x_\ell\}_b$ of points in space-time corresponding to points in the essential support of the distribution F_b (defined without mass-shell constraints) at points $\{P_k, P_\ell\}_b$, where the P_k are the on-mass-shell external energy-momenta and the P_ℓ are on-mass-shell internal energy-momenta in the integration domain. A diagram ε_B is correspondingly a configuration of points in space-time obtained as a collection of ε_b that fit together in the sense that

$$(x_\ell)_2 - (x_\ell)_1 = \lambda_\ell P_\ell$$

where λ_ℓ is an arbitrary real scalar, for any internal line ℓ of B joining bubbles 1 and 2. A u = 0 point P of B is such that there exists a diagram ε_B whose all <u>external</u> points x_k are at the origin, whereas some internal points are not there. The u ≠ 0 structure theorem at u ≠ 0 points P then says that F_B (resp. f_B) is well defined and that its essential support is at most the set of points $(x_1,\ldots,x_n)$ corresponding to configurations (resp. relative configurations) of external points of diagram ε_B.

A systematic use of the relations (42) for the various values of k = 3,4,5, yields moreover, by successive applications (see [38]) of the general "edge-of-the-wedge" results of essential support theory, the

<u>Theorem</u>: "ES_P () $\subset C = \{x; x_1=x_2, x_3=x_4=x_5,$

$$x_3-x_1 \in \bar{V}^+\} \qquad (43)$$

at any point P of $M_{2,3}$ such that $s < (4\mu)^2$, $(P_i+P_j-P_k)^2 \neq$ $\neq \mu^2$ where i,j,k is any permutation of (3,4,5)."

The exclusion of the codimension-one submanifolds $(P_i+P_j-P_k)^2 = \mu^2$ (which contain the M_o points) in the results obtained so far in [38] is due to reasons that we shall not discuss here in detail. Lorentz invariance implies now, as for the four-point function, an improvement of (43) in which the condition $x_3 - x_1 \in \bar{V}^+$ is replaced by $x_3 - x_1 = \lambda(P_1 + P_2)$, $\lambda \geq 0$. The situation is then the same as that shown in Fig. 12 for the $2 \to 2$ processes, except that $x_3 = x_4$ is replaced by $x_3 = x_4 = x_5$. To reobtain macrocausality (see above), a regularity assumption for $2 \to 3$ processes would again be needed to avoid the a priori possible à la Martin pathologies.

Six-point function ($3 \to 3$ processes)

We shall now consider, as in Sect. 5, a $3 \to 3$ process in the region $s < (4\mu)^2$. Generalized Feynman integrals, such as

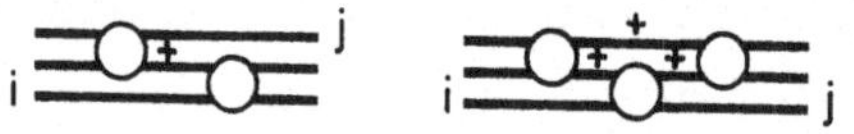

are defined in a natural way in axiomatic field theory (see [38],[39]) where the bubbles

are not defined a priori on-mass-shell, by associating Feynman-type propagators to each internal line. It can be checked that they coincide, after restriction to the mass-shell, and modulo analytic backgrounds, with the $(D^C_\beta)_+$ introduced in Sect. 2.

The analogue of (39), which follows again from microcausality and the spectral condition, is here:

$ES_P($ 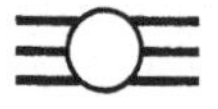 $- \sum_{\substack{i=1,2,3 \\ j=4,5,6}}$ 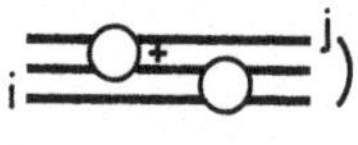 $) \subset \bigcup_{\substack{s=1,2,3 \\ t=4,5,6}} C^t_s$, (44)

where e.g. $C^4_1 = \{x, x_2 = x_3, x_5 = x_6, x_1 - x_2 \in \bar{V}^+, x_5 - x_4 \in \bar{V}^+, x_4 - x_1 \in \bar{V}^+\}$ (Fig. 13b)).

The asymptotic completeness equations are the three equations (42) (with three lines instead of two on the left of the $3 \to 3$ bubbles) and three analogous equations involving terms

 , $(i,j,k) = (1,2,3)$, $k = 1,2,3$.

As previously, microcausality and the spectral condition yield information also on

 , or  .

The full use of all previous informations, together with the standard $u \neq 0$ results, now yields [38]:

<u>Theorem.</u> "ES_P ( $- \Sigma_i$ 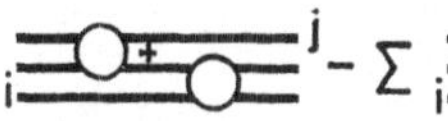 $- \Sigma_i$ 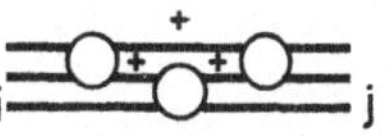)

$$\subset \{x; x_1=x_2=x_3,\ x_4=x_5=x_6,\ x_4-x_1 \in \bar{V}^+\} \qquad (45)$$

at any point P of $M_{3,3}$ such that $s < (4\mu)^2$, $(P_i+P_j-P_k)^2$, where (i,j,k) is any permutation of $(1,2,3)$ or of $(4,5,6)$."

The condition $x_4-x_1 \in \bar{V}^+$ can, as previously, be replaced by $x_4-x_1 = \lambda(P_1+P_2+P_3)$, $\lambda \geq 0$ by Lorentz invariance. A supplementary regularity assumption (for $3 \to 3$ processes) would be needed to avoid à la Martin pathologies, i.e. to replace this latter condition by $x_1 = \ldots = x_6$, and thus to reobtain the decomposition (17), equivalent to macrocausality and macrocausal factorization, on the mass-shell. As in the case of the five-point function, the submanifolds $(P_i+P_j-P_k)^2 = \mu^2$ are excluded so far in the results of [38].

<u>Remarks:</u> 1) Although this has not yet been established so far, it is believed that the terms

 , or  ,

after restriction to the mass-shell and factorization of their e.m.c. δ-function, can be obtained from the scattering function $f_{3,3}$ by analytic continuation, on the complex mass-shell, around two-particle thresholds. The relations (42), or the analogous relations for the six-point function then appear as (off-shell) "extended unitarity" equations between scattering functions and their analytic continuations.

In S-matrix theory, these analytic continuations, and the mass-shell versions of Eq. (42) or of its analogues for $3 \to 3$ processes are not known at the outset, although one would like to reobtain them at a later stage. Hence only physical, on-mass-shell unitarity has been used in Sect. 5.

2) The results that have been described suggest that macrocausality and macrocausal factorization properties analogous to these described in Sect. 3 should hold in general for the Green functions. For instance, macrocausality would be stated as follows: the only points $(x_1,\ldots,x_n)$ in $ES_P(\tilde{\tau}_c)$ at any physical-region point P are those for which there exists a classical diagram D_+ (without "vertices at infinity") with external trajectories (P_k,x_k), each point x_k being moreover precisely placed at the interaction vertex of D_+ in space-time that involves the external particle k. By restriction of the Green functions to the mass-shell, the macrocausality property of the S-matrix is reobtained. The essential support property of the Green functions is however stronger in view of the last condition mentioned above on the location of the points x_k. In particular, it is still at M_o points a causality condition, whereas this is essentially lost at these points after restriction to the mass-shell. For instance, in a two-body equal mass process, the essential support of $\tilde{\tau}_c$ when $P_1=P_2(=P_3=P_4)$, $P \in M$, is still of the form $x_1=x_2$, $x_3=x_4$, $x_3-x_1=\lambda(P_1+P_2)$, $\lambda \geq 0$, and entails analyticity from the directions Im s > 0, whereas one only obtains, by restriction to the mass-shell,

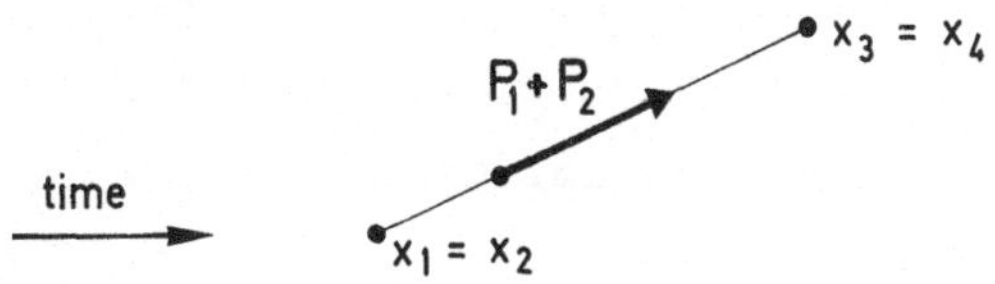

Fig. 14

the essential support shown in Fig. 8. The latter coincides with that of $(S^{-1})^{c}_{2,2}$ at the points P considered, and yields no simple analyticity property of $f_{2,2}$.

7. NATURE OF LANDAU SINGULARITIES

In Sect. 7.1, we present results on the nature of $+\alpha$-Landau singularities of graphs G with single lines. These results, derived from the discontinuity formulae, apply to situations in which individual scattering functions occurring in the term D that represents the discontinuity remain analytic in integration domains. They have been obtained in [23] on the basis of general results of the theory of holonomic functions. We shall present here a direct proof, independent of that theory. More refined results have also been obtained in [23] in situations in which individual scattering functions do not necessarily remain analytic but in which their analytic structure is known (= that determined previously) in the integration domains. They apply to $+\alpha$-Landau points that lie on several related surfaces of graphs with single lines. Similar results apply also under some conditions and in space-time dimension 4, to graphs with possibly double lines (= sets of two lines between some vertices). They are based on the well known squared-root-type nature of two-particle thresholds. Results of [23] include for instance the behavior of the S-matrix in the neighborhood of a point where the Landau surface of a triangle graph in a 3 → 3 process meets that of a "self-energy" graph obtained by contracting one of the internal lines: see the end of Sect. 7.1.

In Sect. 7.2, we first come back to two-particle thresholds in two-body processes. The proof of the square-root-type nature of the singularity is well known and is technically more simple than that of the previous results. However, it is conceptually more difficult in some sense because the structure of the individual bubbles in the discontinuity formula is not known in advance: one has in fact to solve an integral Fredholm-type equation between f and its analytic continuation around $s = 4\mu^2$. We shall see, following [41], that the singularity is of a different nature if the dimension ν of space-time is odd, and we shall describe the more general results of [41] on m-particle thresholds in $m \to m$ processes in a simplified theory in which interactions involving less than m initial or m final particles are neglected. The nature of the singularity depends on the parity of $(m-1)(\nu-1)$: if $(m-1)(\nu-1)$ is odd, it is again of a square-root type. Otherwise, e.g. at $\nu=4$, $m=3$, the scattering function is shown to be locally, under some general conditions, an infinite convergent sum

$$\sum_n a_n(p)[z^{((m-1)\nu-m-1)/2} \ln z]^n \ , \ z = s-(m\mu)^2, \ s = (\sum_{i=1}^{m} p_i)^2,$$

of à la Feynman terms with locally analytic coefficients. A decomposition of the solution allows one to express f (after resummation) in terms of non-holonomic functions of the form 1/ln.

7.1 Graphs with single or double lines

We first consider singularities associated with graphs with single lines, i.e. with one line at most between any pair of vertices. The most simple case is that of example a) of Sect. 2.2: the discontinuity formula (7) entails, as already mentioned, that the singularity is a pole with a factorized residue. In the case of example b) (in space-time dimension ν), let us consider e.g. a physical-region point $P = (P_1,\ldots,P_6)$ of $L^+(G)$ in the region $(3\mu)^2 < s = (p_1+p_2+p_3)^2 < (4\mu)^2$, $p \notin M_o$: it can then be checked that the scattering functions $f_{2,2}$ associated with each bubble remain analytic in integration domains when p varies in a neighborhood of P (the incoming energy variables s_b all remain in the region $(2\mu)^2 < s_b < (3\mu)^2$). In the neighborhood of P, formula (10) then yields:

$$d(p) = \int a(p,k) \prod_{j=1}^{2\nu+3} \delta(f_j(p,k))dk \tag{46}$$

where

$$k = k_1, k_2, k_3, \quad dk = \prod_{\ell=1,2,3} d^\nu k_\ell ,$$

the f_j are the arguments of the three mass-shell δ-functions and of 2ν conservation δ-functions associated with two of the bubbles (the third one giving rise to the overall δ^ν-function factored out in D), and a remains analytic in the integration domain.

Being given a real analytic local coordinate system $q = (q_1,\ldots,q_1)$ of M near P, with $q_1 = 0$ and $q_1 > 0$ representing locally $L^+(G)$ and its physical side, one can choose (see [54]) $2\nu+2$ of the $2\nu+3$ functions f_j (relabeled below $f_1,\ldots,f_{2\nu+2}$) and $\nu-2$ other analytic functions $x_1(q,k),\ldots,x_{\nu-2}(q,k)$ such that:

(i) $$f_{2\nu+3}(q,k) = q_1 - \sum_{i=1}^{\nu-2} [x_i(q,k)]^2 , \tag{47}$$

(ii) the change of variables $q, k \to q, f_1,\ldots,f_{2\nu+2}, x_1,\ldots,x_{\nu-2}$ produces a regular jacobian $J(q,f_1,\ldots,f_{2\nu+2}, x_1,\ldots,x_{\nu-2})$. Hence, after elimination of the first $2\nu+2$ δ-functions:

$$d(q) = \int a'(q,x)\,\delta(q_1 - \sum x_i^2)\,dx \tag{48}$$

where $x = (x_1,\ldots,x_{\nu-2})$, and $a' = a \times J\big|_{f_1=\ldots=f_{2\nu+2}=0}$ is analytic in the integration domain.

The presence of the δ-function $\delta(q_1-\sum x_i^2)$ in the r.h.s. of (48) is not surprising and is in fact fully consistent with the fact that d is identically zero (see Sect. 2) on the non-physical side ($q_1 < 0$) of $L^+(G)$. The fact that the r.h.s. of (48) is a well defined distribution (locally) is already known from various more general considerations (see e.g. [25]). It follows here from the fact that $\delta(q_1 - \sum x_i^2)$ is itself a well defined distribution of q_1,x. By definition:

$$\begin{aligned} \langle d,\phi\rangle &= \int \delta(q_1-\sum x_i^2)\; a'(q,x)\,\phi(q)\;dq\;dx \\ &\equiv \int \phi(\sum x_i^2,q_2,\ldots,q_\rho)\; a'(\sum x_i^2,q_2,\ldots,q_\rho)\,dq_2\ldots dq_\rho\;dx \end{aligned} \tag{49}$$

for any C^∞ test function with support in a sufficiently small neighborhood of $q = 0$.

In the case $\nu = 2$, i.e. $\nu-2 = 0$, there are no variables x, and d is explicitly of the form:

$$d(q) = a'(q)\,\delta(q_1) \quad . \tag{50}$$

When $\nu \geq 3$, one checks as explained below that

$$d(q) = \theta(q_1)\,q_1^\beta\; a''(q) \;, \quad \beta = \frac{\nu-4}{2} \quad , \tag{51}$$

where $a''(q) = \int a'(q;|x|,\Omega)\,d\mu(\Omega)\big|_{|x|=\sqrt{q_1}}$ is locally analytic at $q = 0$. (Here x has been replaced by $|x|$ and angle variables Ω. Although $|x|$ is taken equal to $\sqrt{q_1}$, one recovers a locally analytic function of q after integration over Ω.)

Note that $\beta \geq -\frac{1}{2}$ if $\nu \geq 3$, in which case the r.h.s. of (51) is a well defined distribution.

Proof of (51)

The result is first easily obtained in the region $q_1 > 0$, where the r.h.s. of (48) is in fact a well defined function, by considering the new variables $t = \sum x_i^2$ and Ω: this change of variables introduces the factor $t^\beta \equiv q_1^\beta$. At $q_1 < 0$, one has obviously $d = 0$; one finally checks easily that $\langle d,\phi\rangle$ as defined in (49) is indeed equal to $\int \theta(q_1) q_1^\beta\, a''(q)\phi(q)\, dq$ even when the support of ϕ contains $q = 0$; Q.E.D..

The edge-of-the-wedge theorem shows that two distributions that admit the same discontinuity differ at most by a locally analytic function. Hence one obtains the following results for the scattering function f near $q = 0$:

$$\beta = -1: \qquad f(q) = a_1(q) \frac{1}{q_1+i\varepsilon} + a_2(q) \quad , \tag{52}$$

$$\beta = 0,1,2,\ldots: \qquad f(q) = a_1(q) q_1^\beta \ln q_1 + a_2(q) \quad , \tag{53}$$

$$\beta = -\frac{1}{2},\frac{1}{2},\frac{3}{2},\ldots: \ f(q) = a_1(q) q_1^\beta + a_2(q) \quad , \tag{54}$$

where $\beta = \frac{\nu-4}{2}$ and a_1, a_2 are locally analytic functions. In fact, one checks easily that the functions $a_1(q)q_1^\beta$, or $a_1(q)q_1^\beta \ln q_1$ admit the discontinuity $a''(q)\theta(q)q_1^\beta$ (or $a'(q)\delta(q_1)$ if $\beta = -1$) for appropriate choices of a_1 in terms of a'' or a'. The difference of the cases (53), (54) is simply that q_1^β is analytic in the first case, and hence does not change after turning around $q_1 = 0$, whereas it changes its sign if β is half-integer.

The same analysis applies equally to more general graphs G with only single lines, at generic points P of real analytic codimension-one surfaces $L^+(G)$ such that all scattering functions associated with each bubble remain analytic in integration domains. The results (52), (53), (54) are obtained in the same way with

$$\beta = \frac{\nu\ell - m - 1}{2} \tag{55}$$

where ℓ is the number of independent closed loops and m is the number of internal lines ($\beta \geq -1$ in the cases under consideration). Results of the same type are also obtained in [23] for graphs G with sets of doubles lines, in space-time dimension $\nu = 4$ and under conditions such that the only singularities of individual scattering functions are the corresponding two-particle thresholds. It is shown in fact that the boxes $S_\alpha^{-1} \equiv S^{-1}$ on each set of double lines can be removed, the scattering functions at each vertex being then replaced by functions that remain analytic in the integration domains.

The more refined results of [23] include e.g. the following local form of f in the neighborhood of a point P where the Landau surface of the triangle

G=
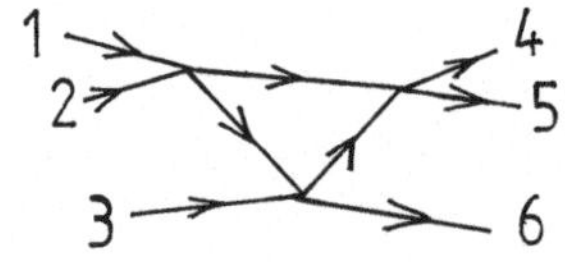

meets that of the "self-energy" graph

G_1=
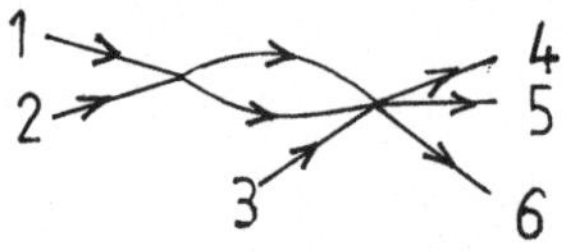

In a coordinate system where $L^+(G_1)$ is given by $q_1 = 0$ and $L^+(G)$ by $q_1 = q_2^2$, one has:

$$f = [a_1(q)\sqrt{q_1+i0} + a_2(q)]\log(\sqrt{q_1+i0} + q_2)$$

$$+ a_3(q)\sqrt{q_1+i0} + a_4(q) \qquad (56)$$

where $a_1,\ldots,a_4$ are analytic at $q = 0$.

7.2 Graphs with multiple lines: recent investigations

We first come back to the 2-particle threshold in a two-body process. The unitarity equation reads below the 3-particle threshold:

(57)

We assume here (as proved in field theory) that the two-body scattering function f, expressed in terms of the variables s, t, is still near $s = 4\mu^2$ the plus $i\varepsilon$ boundary value of an analytic function $\underline{f}$ from the directions Im $s > 0$ and that $\underline{f}$ can be analytically continued around $z = 0$, $z = s - 4\mu^2$. Hermitian analyticity asserts that the boundary value $f^{(1)}$ of the analytic continuation of $\underline{f}$ around $z = 0$ coincides with $f^{(-)}$ and thus transforms (57) into an integral relation between f and $f^{(1)}$. Fredholm theory and regularity assumptions then ensure that $\underline{f}$ satisfies "local maximal analyticity" near $z = 0$, i.e. admits a multivalued analytic (or meromorphic) continuation around $z = 0$: more precisely, $\underline{f}$ is analytic (or meromorphic) in a certain covering of $V - \{z = 0\}$ where V is some complex neighborhood of the origin.

The integral relation between f and $f^{(1)}$ gives for each partial wave f_ℓ:

$$f_\ell - f_\ell^{(1)} = C_\nu\, s^{-\frac{1}{2}}\,(s-4\mu^2)^{\frac{\nu-3}{2}}\, f_\ell\, f_\ell^{(1)} \text{ at } z > 0 \qquad (58)$$

where $f_\ell^{(1)}$ is the determination of $\underline{f}_\ell$ obtained at $z > 0$ after one turn.

If ν is even, the factor $(s-4\mu^2)^{(\nu-3)/2}$ changes its sign after one turn and one easily recovers the two-sheeted, square -root type nature of the singularity at $z = 0$: by analytic continuation around $z = 0$, Eq.(58) is reproduced except that f_ℓ is replaced by $f_\ell^{(2)}$. If ν is odd, this factor is analytic and one finds a behavior of the form [41]

$$f_\ell(s) = \frac{1}{a(s) \times \frac{1}{2i\pi} \ln(4\mu^2-s) + b_\ell(s)} \tag{59}$$

where b_ℓ is a uniform function and $a = C_\nu \; s^{-1/2}(s-4\mu^2)^{(\nu-3)/2}$. To see this, one defines $g_\ell = \frac{1}{a(s) f_\ell}$. Eq.(58) then yields

$$g_\ell^{(1)} - g_\ell = 1 \quad .$$

Hence g_ℓ is equal to $\frac{1}{2i\pi} \ln\,(4\mu^2-s)$ modulo a uniform function. If $\nu \geq 5$, and if f is assumed to remain bounded when $z = 0$ is approached, $1/b_\ell$ is locally analytic at $z = 0$.

In the simplified theory of m-particle thresholds in $m \to m$ processes, the unitarity equation reads:

(60)

We again assume that, near the threshold $z = s-(m\mu)^2 = 0$, f is the plus $i\varepsilon$ boundary value of an analytic function $\underline{f}$ which can be continued around $z = 0$, and that $f^{(1)}$ co-incidings with $f^{(-)}$. Eq.(60) is thus transformed again into an integral relation between f and $f^{(1)}$. Under regularity conditions, $\underline{f}$ satisfies again "local maximal analyticity".

If $(m-1)(\nu-1)$ is odd, one shows again the two-sheeted, square-root type nature of the singularity at $z = 0$. In fact let us rewrite Eq. (60) at $z > 0$, with obvious notations, in the form one obtains

$$f - f^{(1)} = f * f^{(1)} \tag{61}$$

and let us consider an analytic continuation of this equation. After one turn around $z = 0$

$$f^{(1)} - f^{(2)} = -f^{(1)} * f^{(2)} \tag{61a}$$

where the minus sign arises from the change of sign of the factor z^{β}, $\beta = \frac{(m-1)\nu-m-1}{2}$ that arises from the mass-shell integration after elimination of δ-functions. Eq. (61a) is identical to (61) except that f is replaced by $f^{(2)}$. Arguments of Fredholm theory then entail that $f = f^{(2)}$. Q.E.D.

If $(m-1)(\nu-1)$ is even, the factor z^{β} does not change its sign. For simplicity, let f remain bounded when one approaches $z = 0$ in any finite number of sheets (see [41] for the more general case) and let us consider the case $\beta > 0$ ($\nu \geq 3$ if $m \geq 3$). Then one proves [41]:

Theorem. "The general solution of (61) is of the form

$$f = \sum_n u^{*(n+1)}(p) \left[\frac{i}{2\pi} \ln ((m\mu)^2 - s)\right]^n \tag{62}$$

where u is locally analytic of all variables at $z = s - (m\mu)^2 = 0$. Conversely, any function f of the form (62) with u locally analytic satisfies (61)".

In (62), $u^{*(n+1)} = \underbrace{u * u * u \ldots * u}_{n+1}$.

Proof: Let us introduce the kernel u through the equation

$$f = u + \frac{i}{2\pi} \ln((m\mu)^2 - s) \; f * u \quad . \tag{63}$$

It is the analogue, when $(m-1)(\nu-1)$ is even, of Zimmerman's K matrix: In fact it is proved in [41] that (63) defines u in terms of f through Fredholm theory and that local analyticity of u at $z = 0$ is equivalent to Eq. (61). The series in the r.h.s. of (62), which is the Neumann series of (63) (at the value $\lambda = 1$ of the Fredholm paramter), is convergent for $\beta > 0$ in view of the factors $(z^\beta)^n$ that arise in $u^{*(n+1)}(p)$ from n mass-shell convolutions. Hence f is equal to the sum of this series. The converse part is also checked easily. Q.E.D.

Finally, although the sum (62) cannot be made explicitly, hermitian analyticity allows one (see [41]) to introduce at $z < 0$ orthogonal projectors $E_i(z,\Omega,\Omega')$, where Ω,Ω' are angle variables for the initial and final sets of energy-momenta ($E_i * E_j = E_i \, \delta_{ij}$). The following decomposition of f is then obtained in [41]:

$$f = \sum_i \frac{1}{z^\beta \ln z + b_i(z)} \; E_i'(z;\Omega,\Omega') \tag{64}$$

where $1/b_i$ and $E_i' = -2i\pi \, z^\beta \, E_i$ are locally analytic at $z = 0$.

Eq. (64) is the analogue of the previous decomposition into partial waves and exhibits in particular, as in this previous case, the non-holonomicity of the sum. Although we have only considered a simplified theory, the analysis suggests that Sato's conjecture [40] on holonomicity of the S matrix has to be modified in general for graphs that include sets of 2r+1 lines, $r \geq 1$, between some vertices, e.g. for three-particle thresholds in two-body processes.

The result (62), which is of the form $\sum_n a_n(p)[z^\beta \ln z]^n$, with analytic coefficients a_n, also supports the conjecture made independently in [34] and described in Sect. 5.1: the Feynman graphs  are precisely equal to $(z^\beta \ln z)^n$ up to locally analytic factors.

Ref. [41] contains a related analysis in field theory in terms of Bethe-Salpeter type equations. It leads to similar conclusions: namely the Green function of the simplified theory should be locally an infinite sum of generalized Feynman integrals

,

where integration is no longer on-mass-shell but with Feynman-like propagators and where G is an irreducible (= locally analytic) off-shell kernel.

APPENDIX - ESSENTIAL SUPPORT THEORY, HYPERFUNCTION THEORY

Note: The mathematical variables of this Appendix will be in the physical context energy-momenta (not space-time) variables, e.g. the components of the energy-momenta, or of the momenta, or local coordinates according to the distributions considered.

1. Essential Support Theory: Basic Notions and Results [24]

Let f be at tempered distribution defined in the real space R^n of an n-dimensional variable $x = (x_1, \ldots, x_n)$.

The essential support $ES_X(f)$ of f at each point X of R^n is a closed cone with apex at the origin in the dual space (= space of Fourier transformed variables) composed of the singular directions along which a generalized Fourier transform of f at X:

$$F(v;\gamma,X) = \int f(x)\ e^{-iv.x-\gamma|v|(x-X)^2}\ dx\ , \qquad (A1)$$

does not fall off exponentially, in a well defined sense, for sufficiently small $\gamma > 0$. More precisely, a direction $\hat{v}_o$ is outside $ES_X(f)$ if there exist a neighborhing cone V of $\hat{v}_o$ with apex at the origin in v-space, $\alpha > 0$, $\gamma_o > 0$, (a polynomial P and $q \geq 0$) such that

$$|F(v;\gamma,X)| < [P(|v|)(\gamma|v|^{-1}]e^{-\alpha\gamma|v|} \qquad (A2)$$

in the region $v \in V$ and for all γ satisfying $0 < \gamma < \gamma_o$. The important factor in (A 2) is the exponential fall-off factor $e^{-\alpha\gamma|v|}$. The definition of $ES_X(f)$ is essentially unchanged if f is replaced by χf where χ is a C^∞ function with compact support around X and is locally analytic and different from zero at X, and if the factor $(x-X)^2$ is replaced by a function that has similar local properties. We note finally that

$$F(v;0,X) \equiv \tilde{f}(v)\ ,$$

where $\tilde{f}$ is the usual Fourier transform of f (and does not depend on X).

By definition, f is microanalytic at X in a direction $\hat{u}$ if $\hat{u} \notin ES_X(f)$.

If f is defined on a real analytic manifold M, $ES_X(f)$ is, at each X, a cone in the cotangent space $T^*_X M$ at X to M(= the dual of the tangent space $T_X M$ at X to M).

The following results are proved: $ES_X(f)$ is empty iff (= if and only if) f is locally analytic at X, f is the boundary value at X of an analytic function $\underline{f}$ from the (imaginary) directions of an open cone Γ^+ if $ES_X(f)$ is contained in the closed (convex salient) dual cone C of Γ. More generally, $ES_X(f)$ is not necessarily convex, or contained in a closed convex cone, and f is not in the latter case the boundary value of an analytic function. The notion of essential support then characterizes in a well defined way all possible decompositions of f, at X, or over real domains D, as a sum of distributions which are boundary values of analytic functions from directions which may depend on X[+)].

Edge-of-the-wedge theorems

The following general theorem follows from the definition of the essential support:

Theorem 1: "If $ES_X(f) \subset C_1$ and $ES_X(f) \subset C_2$ where C_1, C_2 are closed cones with apex at the origin, then

$$ES_X(f) \subset C_1 \cap C_2 \text{"}. \qquad \text{(A3)}$$

In view of the results mentioned above on analyticity properties, this theorem is a powerful generalization of the several-dimensional edge-of-the-wedge theorem which says that, if f is (locally) the boundary of an analytic function $\underline{f}_1$ from the directions of a cone Γ_1 and also the

---

+Let $z = x+iy$, $z=z_1\ldots z_n$, be the complexified variable of x; f is the b.v. at X of $\underline{f}$ from the directions of Γ if, for any Γ' with apex at the origin and closure contained in Γ (apart from the origin), there exists some real neighborhood ω of X and $\rho > 0$ such that f is analytic in $\{z : x \in \omega,\ y \in \Gamma',\ |y| > \rho\}$ and such that

$$f(x) = \lim_{y \in \Gamma',\ y \to 0} f(x+iy) \text{ in } \omega ,$$

in the sense of distributions (i.e. after integrating with C^∞ test functions with support in ω).

boundary value of an analytic function $\underline{f}_2$ from the directions of Γ_2, then $\underline{f}_1$ and $\underline{f}_2$ define the same analytic function $\underline{f}$, f being its boundary value from the directions of the convex envelope of Γ_1 and Γ_2. This result is in fact a particular case of (A3) obtained when C_1, C_2 are both closed convex salient cones.

Theorems on products, integrals, restrictions

Theorem 2: "The product f_1f_2 of two distributions f_1,f_2 is well defined in the neighborhood of a point X if there exists no pair (u_1,u_2), $u_1,u_2 \neq 0$, such that $u_1 \in ES_X(f_1)$, $u_2 \in ES_X(f_2)$, and $u_1+u_2 = 0$. Under this condition, one has moreover

$$ES_X(f) \subset ES_X(f_1) + ES_X(f_2)$$

$$\equiv \{u; u=u_1+u_2, u_1 \in ES_X(f_1), u_2 \in ES_X(f_2)\}." \qquad (A4)$$

By definition, X is a u = 0 point for the product f_1f_2 if it does not satisfy the condition of the theorem. Theorem 2 is generalization of a well known theorem on products of boundary values f_1,f_2 of analytic functions $\underline{f}_1$, $\underline{f}_2$ obtained (at X) from directions of respective cones Γ_1, Γ_2 whose intersection is not empty: f_1f_2 is then the boundary value of the product $\underline{f}_1$ $\underline{f}_2$ from the directions of this intersection. The proof of Theorem 2 is made either by using decompositions of f_1, f_2 into sums of boundary values of analytic functions, or more directly by expressing the Foruier and generalized Fourier transforms of f_1,f_2, or possibly of $\chi_1f_1 \times \chi_2f_2$, in terms of the corresponding transforms of χ_1f_1 and χ_2f_2. The exponential fall-off properties (A3) outside $ES_X(f_1)$ + + $ES_X(f_2)$ are extracted from those of the generalized Fourier transforms at X of χ_1f_1 and χ_2f_2.

Theorem 3: "Let f be a distribution in the space $R^{n+n'}$ of variables x, t, with compact support with respect to t when w varies in a neighborhood of a given point X. The integral $\rho(x) = \int f(x,t)dt$ is a well defined distribution in the neighborhood of X, and

$$ES_X(g) \subset \{u;\ \exists\ T,\ V \text{ such that } (u,V) \in ES_{X,T}(f)\}\ ." \qquad (A5)$$

Theorem 4: "Let f be a distribution defined in R^n and let M be real analytic submanifold of R^n. Then the restriction $f|_M$ of f to M is a well defined distribution in the neighborhood of a point $X \in M$ if $ES_X(f) \cap N_X(M)$ is empty (apart from the origin), where $N_X(M)$ is the conormal space at X to M. Under this condition, one has moreover

$$ES_X(f|_M) \subset ES_X(f)/N_X(M) \qquad (A6)$$

i.e. $ES_X(f|_M)$ is at most the set of points u in $ES_X(f)$ defined modulo addition of vectors in $N_X(M)$".

Theorems 2,3,4 have been used in Sect. 6 in axiomatic field theory.

Theorem 5: "$ES_X(\delta(M)) = N_X(M) \quad \forall X \in M,$ (A7)

where $\delta(M)$ is a product of δ-functions $\prod_j \delta(f_j(x))$ such that the set of equations $f_j(x) = 0$ defines M locally."

The following related result is used to characterize the link between Green functions, S-matrix kernels and scattering functions:

Theorem 6: "Let $F = f \times \delta(M)$ be a distribution in R^n that vanishes outside M, f being defined on M. Then

$$ES_X(f) = ES_X(F)/N_X(M) \quad \forall X \in M\ ." \qquad (A8)$$

Finally, in the application to S matrix theory, one needs results on the essential support of the kernel A(x,y) of a product of bounded operators A_1A_2. (For instance,

is the product of $S^c_{3,3}$ and $(S^{-1})^c_{3,3}$.) If these operators have an appropriate support property (guaranteed in the application by energy-momentum conservation), the following result holds:

<u>Theorem 7:</u> "If (X,Y) is not a u = 0 point for the product A_1A_2, then $ES_{X,Y}(A) \subset \{(u,v) ; \exists\, T,w$ such that

$$(u,w) \in ES_{X,T}(A_1) \quad , \quad (w,v) \in ES_{T,Y}(A_2)\}" \quad . \tag{A9}$$

By definition, (X,Y) is a u = 0 point if $\exists$ T,W such that

$$(0,W) \in ES_{X,T}(A_1)$$

$$(W,0) \in ES_{T,Y}(A2) \quad . \tag{A10}$$

Theorem 7, which provides the $u \neq 0$ structure theorem of Sect. 4, follows from Theorem 2 and 3. It can also be established directly by expressing the generalized Fourier transform of A in terms of the generalized Fourier transforms of A_1 and A_2: see [25].

2. Regularity Property R and u = 0 Results

The origin of the u = 0 problem either in Theorem 2 or in Theorem 7 appears clearly in the direct proofs of these theorems in terms of generalized Fourier transforms. Solutions at u = 0 points are obtained [25,26] if the product itself is well defined (for instance, if f_1, f_2 are locally square integrable functions, the product A_1A_2

of bounded operators being on the other hand always well defined) and if a regularity property R is assumed for f_1, f_2 or A_1, A_2 at the appropriate points and directions of their essential supports from which the problems arise. Precise statements of property R depend on the nature of the distributions considered. The main content is, however, the same. We explain it for a distribution f defined in R^n.

As we have seen, a direction $\hat{v}_o$ is outside $ES_X(f)$ if $F(v;\gamma,X)$ falls off like $e^{-\alpha\gamma|v|}$ in this direction for some $\alpha > 0$ and for sufficiently small γ (see A2 for the precise formulation). It can be shown independently of property R that, being given $\hat{v}_o \notin ES_X(f)$, α can be chosen arbitrarily close to

$$\alpha_o = \text{Max}_{\alpha'} \{\alpha'; \hat{v}_o \notin ES_X(f) \ \forall x \text{ such that} (x-X)^2 < \alpha'\}$$

and can e.g. be chosen equal to some fixed fraction of α_o. The main content of property R is that the maximal value γ_o of γ does not tend to zero faster than linearly with respect to the angle of $\hat{v}$ with the essential support when a direction of the latter is approached. The factor $\gamma_o(\hat{v})|v|$ is then proportional to the distance of v to the essential support. (Note that, in Sect. 5.1, v is replaced by τu and γ is the analogue of the product $\gamma|u|$ of the present Appendix.)

The u = 0 theorems obtained on the basis of this regularity property are then similar to Theorems 2 and 7 except that some limiting procedures that may enlarge the essential support have to be considered in general.

The result obtained is a generalization in the framework of essential support theory of the following more simple u = 0 result on the product f_1, f_2 of two functions that are near X boundary values of analytic functions $\underline{f}_1, \underline{f}_2$ from directions of cones Γ_1, Γ_2. The u = 0 problem arises when $\Gamma_1 \cap \Gamma_2$ is empty. Suppose, however, there is one or more common directions in the intersection of their closures. Property R holds (in the appropriate directions in v-space) for f_1 and f_2 if the shape of the analyticity domains in y-space along these common directions is of the form shown in Fig.15a), and if e.g. $\underline{f}_1, \underline{f}_2$ remain bounded when y tends to the boundary of these domains. Under such conditions $ES_X(f_1,f_2) \subset C_1 + C_2$, as shown in Fig. 15b:

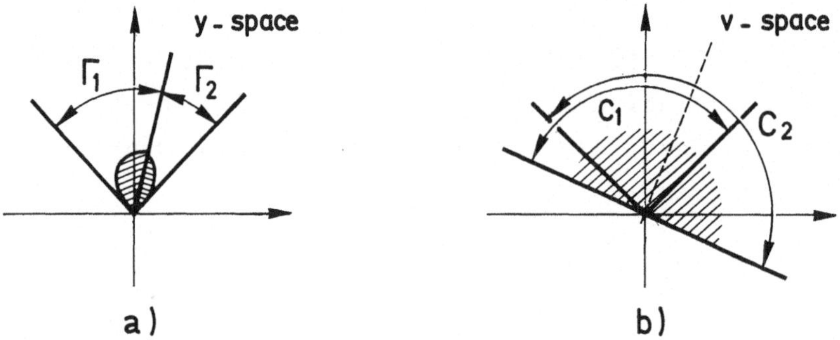

Fig. 15

3. Hyperfunction Theory, Holonomicity

The notion of singular spectrum has been introduced independently by very different methods in hyperfunction theory [27]. (It is also called "singular support" in [27].) General results analogous to those described in Sect. 1 of this Appendix have been obtained in that framework. The essential support and singular spectrum characterize analyticity properties in a very similar way, the only difference being that the quantities f_i appearing in decompositions of f into sums of boundary values of analytic functions may à priori be hyperfunctions in the second case, even if f itself is a distribution. It has, however, been proved in [55] that the two notions do coincide for distributions (and coincide also with Hörmander's "analytic wave front set" [56]).

For completeness, we give some very brief indications on hyperfunctions and microfunctions. A distribution f can always be decomposed as a sum of b.v. f_i of analytic functions $\underline{f}_i$, in a highly non-unique way in general in the multi-dimensional case. Each f_i is here a distribution and $\underline{f}_i$ does not grow faster than an inverse power of the imaginary part y when $y \to 0$. A <u>hyperfunction</u> can be defined as an equivalent class of collections f_i of formal b.v. of analytic functions $\underline{f}_i$, the $\underline{f}_i$ having now no constraint on their behaviour when $y \to 0$.

A <u>microfunction</u> is an equivalent class of hyperfunctions with respect to singular spectrum properties. We only indicate here that by definition:

$$f = g \qquad \text{at} \qquad (X,\hat{u})$$

in the sense of microfunction iff $\hat{u} \in S.S._X(f-g)$ ($\equiv ES_X(f-g)$).

<u>Holonomicity, regular holonomicity [27,29]</u>

A function (or microfunction) is, according to the terminology recently proposed by M. Sato, holonomic if it is solution of a maximally over-determined system of linear micro (= pseudo)-differential equations. Such systems appear to be in some sense the simplest extensions in several variables of ordinary linear differential equations: the space of microfunction solutions of these systems is always finite-dimensional. (This is in some sense an extension of the fact that solutions of ordinary linear differential equations are well determined by the data at one point.)

We explain below the significance of holonomicity in the following simple situation. We consider a distribution f which, in an appropriate local coordinate system

$(x_1,\ldots,x_n)$, is locally analytic in the neighborhood of the origin except on the surface $x_1 = 0$, and which is moreover locally the boundary value of an analytic function $\underline{f}$ of $z = (z_1,\ldots,z_n)$ from the directions $y_1 = \mathrm{Im}\, z_1 > 0$. Then local holonomicity of f near the origin first includes "local maximal analyticity" in the sense that $\underline{f}$ admits a multivalued analytic (or meromorphic) continuation around $z_1 = 0$; more precisely, $\underline{f}$ is analytic (or meromorphic) in a certain covering of $V - \{z_1 = 0\}$, where V is some complex neighborhood of the origin. In the class of distributions f satisfying local maximal analyticity, local holonomicity is moreover characterized by the following "finite-determination property": the vector space generated by all successive determinations F_τ of F in its various sheets is finite-dimensional.

The local holonomicity of f in the case under consideration implies that $\underline{f}$ has the following explicit form:

$$\underline{f}(z) = \sum_{\alpha,j} a_{\alpha j}(z)\, z_1^{\alpha} (\log z_1)^j , \qquad (A11)$$

where α and j run over <u>finite</u> sets, the j's are ≥ 0 integers, the α's are complex numbers and the $a_{\alpha j}$ denote analytic functions defined (i.e. uniform) in $V - \{z_1 = 0\}$.

The distribution f is moreover <u>holonomic with regular singularities</u> (= regular holonomic) locally if and only if the $a_{\alpha,j}$ in (A11) can be chosen <u>analytic</u> locally at $z = 0$.

<u>u = 0 results, second microlocalization</u>

We have already mentioned that u = 0 results were obtained in [52] on products of holonomic functions with regular singularities. Further work has been carried out recently in hyperfunction theory on the u = 0 problem that arises in the study of products, or also in the very closely related problem of restrictions to submanifolds: see [57,58]. The results of [57] are somewhat more general than the earlier ones of [52]. The exact link between these various works and the work of [25,26] is not known to the present author. The approach developed in [58] and probably partly in [57] takes place in the framework of the so-called "second microlocalization" introduced recently [59] in hyperfunction theory. The "second microsupport" might be closely linked with the regularity property R introduced in Sect. 2 of this Appendix.

REFERENCES

1. D. Iagolnitzer, The S-matrix, North-Holland, Amsterdam (1978).
2. D. Iagolnitzer, Complex Analysis, Microlocal Calculus and Relativistic Quantum Theory, Lecture Notes in Physics 126, Springer-Verlag, Heidelberg (1980), p. 263.
3. H.P. Stapp, Structural Analysis of Collision Amplitudes, Ed. by R. Balian and D. Iagolnitzer, North-Holland, Amsterdam (1976) p. 275.
4. A.R. White, Structural Analysis of Collision Amplitudes, Ed. by R. Balian and D. Iagolnitzer, North-Holland, Amsterdam (1976) p. 431.
5. G.F. Chew and V. Poenaru, Phys. Rev. Lett. 45 (1980) 29. V. Poenaru, Univ. d'Orsay, Dept. de Mathematique, preprint no. 81T06 (1981), and references therein.
6. G.F. Chew, The Analytic S matrix, W.A. Benjamin, New-York (1966), and references therein.
7. R.J. Eden, P.V. Landshoff, D.I. Olive, and J.C. Polkinghorne, The Analytic S matrix, Cambridge Univ. Press (1966), and references therein.

8. M.I.W. Bloxham, D.I. Olive, and J.C. Polkinghorne, J. Math. Phys. 10 (1969) 494; 10 (1969) 555; 10 (1969) 553.
9. H.P. Stapp, J. Math. Phys. 9 (1968) 1548.
J. Coster and H.P. Stapp, J. Math. Phys. 10 (1969) 371; 11 (1970) 2743.
10. D. Iagolnitzer, J. Math. Phys. 6 (1965) 1576, Thesis, Paris (1967), and J. Math. Phys. 10 (1969) 1249.
11. S. Coleman and R. Norton, Nuovo Cim. 38 (1965) 438.
12. E.H. Wichmann, and J.H. Crichton, Phys. Rev. 132 (1963) 2788;
R.J. Taylor, Phys. Rev. 142 (1966) 1236.
13. G. Wanders, Helv. Phys. Acta 38 (1965) 142.
14. K. Hepp, Helv. Phys. Acta 37 (1964) 659, and J. Math. Phys. 6 (1965) 2762.
15. R. Omnes, Phys. Rev. 146 (1966) 1123.
16. C. Chandler and H.P. Stapp, J. Math. Phys. 90 (1969) 826.
17. F. Pham, Ann. Inst. Poincaré, Vol. VI, no. 2 (1967) 89.
18. D. Iagolnitzer and H.P. Stapp, Comm. Math. Phys. 14 (1969) 15.
19. D. Iagolnitzer, Lectures in Theoretical Physics, Vol. 11D, ed. by K.T. Mahanthappa and W. E. Brittin, Gordon and Breach, New-York (1969) p. 221.
20. G. Wanders, Nuov. Cim. 14 (1959)168.
21. D. Iagolnitzer, Introduction to S-matrix theory, ADT, Paris (1973), and references 1,2.
22. F. Pham, Hyperfunctions and theoretical physics, Lecture Notes in Mathematics 449, Springer-Verlag, Heidelberg (1975) p. 83.
23. T. Kawai, H.P. Stapp, Publ. R.I.M.S., Kyoto Univ. 12, Suppl. (1977) p. 155.
24. D. Iagolnitzer, Structural Analysis of Collision Amplitudes, ed. by D. Iagolnitzer, North-Holland, Amsterdam (1976) p. 295.
25. D. Iagolnitzer, Comm. Math. Phys. 63 (1978) 49.

26. D. Iagolnitzer, Seminaire Goulaouic-Schwartz, Ecole Polytechnique, Paris 1978-79, p.1, and Complex Analysis, Microlocal Calculus and Relativistic Quantum Theory, Lecture Notes in Physics 126, Springer Verlag, Heidelberg (1980) 1.
27. M. Sato, T. Kawai, M. Kashiwara, Hyperfunctions and Pseudodifferential Equations, Lecture Notes in Mathematics 287, Springer Verlag, Heidelberg (1973), p. 265.
28. M. Kashiwara, T. Kawai, Comm. Math. Phys. 54 (1977) 129, and references therein to previous works by M.Sato and coworkers.
29. M. Kashiwara, and T. Kawai, Tech. Report R.I.M.S. 293, Kyoto Univ. (1979), and Complex Analysis, Microlocal Calculus and Relativistic Quantum Theory, Lecture Notes in Physics 126, Springer Verlag, Heidelberg (1980),p.5.
30. D. Iagolnitzer, Comm. Math. Phys. 41 (1975) 39.
31. H.P. Stapp, Structural Analysis of Collision Amplitudes, ed. by R. Balian and D. Iagolnitzer, North-Holland, Amsterdam (1976),p.191.
32. D. Iagolnitzer, Ch. III of Ref. 1.
33. M. Kashiwara, T. Kawai, and H.P. Stapp, Comm. Math. Phys. 66 (1979) 95.
34. T. Kawai and H.P. Stapp, in preparation.
35. D. Iagolnitzer, H.P. Stapp, Comm. Math. Phys. 57 (1977) 1.
36. D. Iagolnitzer, Comm. Math. Phys. 77 (1980) 251.
37. J. Bros, H. Epstein, and V. Glaser, Helv. Phys. Acta 45 (1972) 149.
38. H. Epstein, V. Glaser and D. Iagolnitzer, to be published in Comm. Math. Phys. (1981).
39. J. Bros, in this volume.
40. M. Sato, Recent developments in Hyperfunction Theory and its Applications to Physics, Lecture Notes in Physics 39, Springer Verlag, Heidelberg (1975), p.13.
41. J. Bros, D. Iagolnitzer, and D. Pesenti, Dph-T-Saclay preprint no. 81/8 (1981).

42. J. Bros, M. Epstein, and V. Glaser, Nuovo Cim. 31 (1964) 1265.

43. W. Zimmermann, Nuovo Cim. 21 (1961) 249.
R. Oehme, Phys. Rev. 121 (1961) 1840.

44. A. Martin, Scattering Theory: Unitarity, Analyticity and Crossing, Springer Verlag, Heidelberg (1970).

45. J. Bros, Analytic Methods in Mathematical Physics, Gordon and Breach, New York (1970), p.85.

46. D. Iagolnitzer, Phys. Rev. D18 (1978) 1275, and Phys. Lett. 76B (1978) 207.

47. R.J. Taylor, op. cit in Ref. 12.

48. D. Ruelle, Helv. Phys. Acta 35 (1962) 147.

49. H.P. Stapp, op. cit in Ref. 9.

50. M. Froissart and R.J. Taylor, Phys. Rev. 153 (1967) 1636.

51. D.I. Olive, Hyperfunctions and Theoretical Physics, Lecture Notes in Mathematics 449, Springer Verlag, Heidelberg (1975), p.133.

52. M. Kashiwara, T. Kawai, Publ. R.I.M.S., Kyoto Univ., 12 Suppl. (1977) 131, Adv. in Math. 34 (1979) 163, and Ref. 29.

53. J. Bros and M. Lasalle, Comm. Math. Phys. 43 (1975) 279.

54. F. Pham, Introduction à l'Etude Topologique des Singularités de Landau, Memorial des Sciences mathématiques, no. 164, Gauthier-Villars, Paris (1967).

55. J.M. Bony, Seminaire Goualouic-Schwartz, Ecole Polytechnique, Paris 1976-77.

56. L. Hörmander, Acta Math. 127 (1972) 79.

57. M. Kashiwara, P. Schapira, Univ. of Paris-Nord, preprint.

58. J.M. Bony and Y. Laurent, in preparation.

59. M. Kashiwara, T. Kawai, Complex Analysis, Microlocal Calculus and Relativistic Quantum Theory, Lecture Notes in Physics 126, Springer Verlag, Heidelberg (1980), p.21.
Y. Laurent, id. p. 77.

Acta Physica Austriaca, Suppl. XXII, 329–400 (1981)

SCATTERING IN QUANTUM FIELD THEORY: THE M.P.S.A. APPROACH IN COMPLEX MOMENTUM SPACE+

by

J. BROS
CEN, Saclay, Boite Postale no.2
91190 Gif-sur-Yvette, France

SUMMARY

In this course, we intend to show how "Many-Particle Structure Analysis" (M.P.S.A.) can be worked out in the standard field-theoretical framework, by using integral relations in complex momentum space involving "ℓ-particle irreducible kernels". The ultimate purpose of this approach is to obtain the best possible knowledge of the singularities (location, nature, type of ramification) and of the ambient holomorphy (or meromorphy) domains of the n-point Green functions and scattering amplitudes, and at the same time to derive analytic structural equations for them which display the global organization of these singularities.

The generation of Landau singularities for integrals and Fredholm resolvents, taken on cycles in complex space, will be explained on the basis of the Picard-Lefschetz formula (presented and used in simple situations).

+Lectures given at XX. Internationale Universitätswochen für Kernphysik, Schladming, Austria, February 17-26, 1981.

Among various results described, we present and analyse a structural equation for the six-point function (and for the 3 → 3 particle scattering function), valid in a domain containing the three-particle normal threshold.

CONTENTS

1. INTRODUCTION, GENERAL IDEAS, AND SOME TYPICAL RESULTS OF M.P.S.A.

In this first lecture, we wish to present some of the ideas of a global consistent approach to the analytic and monodromic structure of Green functions and scattering amplitudes of elementary particles on the basis of general quantum field theory. This approach, which originates in

Symanzik's pioneer program [1] called "Many-Particle Structure Analysis" (in brief M.P.S.A.), has already been presented and investigated by us in its complex momentum space formulation in several articles [2,3].

It integrates in an organized scheme concepts which pertain to three kinds of standard investigations concerning analyticity in particle physics, namely:

(i) Analyticity properties of Green functions $H^{(n)}$ and scattering amplitudes on the basis of the axioms of local fields [4,5];

(ii) Analytic structure of scattering amplitudes in axiomatic S-matrix theory [6];

(iii) Analytic structure of the Feynman amplitudes in the perturbative treatment of Quantum Field Theory [7,8,9,10].

The foundations of the M.P.S.A. approach are the axioms used in (i), supplemented by an extra-postulate called "asymptotic completeness" (A.C.): the latter actually implies the property of "unitarity of the S-matrix" which is one of the basic ingredients of (ii). The method which we use in M.P.S.A. is based on the study of certain integral relations in complex momentum space which have similarities and closed links with the Feynman integrals; the technique which we apply to study the generation of complex singularities through the integration process and the results which we obtain are comparable to those of a standard approach of (iii) [8].

From the framework of (i), the M.P.S.A. approach keeps the benefit of the following results: a) the existence of well defined global analyticity domains D_n in $C^{4(n-1)} = \{(k=(k_1 \dots k_n);\ k_1+\dots+k_n=0\}$ for the Green functions [4] $H^{(n)}(k_1,\dots,k_n)$ $(n \geq 2)$, D_n being called the "primitive domain" of $H^{(n)}$; b) the "reduction formulae" which express the various scattering functions of $m \to n-m$

particle processes as restrictions to the mass shell of appropriate boundary values of the corresponding "amputated" Green functions

$$\hat{H}^{(n)}(k_1,\dots,k_n) = \prod_{i=1}^{n} [H^{(2)}(k_i)]^{-1} H^{(n)}(k_1,\dots,k_n);$$

c) the primitive domains D_n can be enlarged by holo-convex completion (i.e. by computing holomorphy envelopes or parts of the latter); the enlarged domains D'_n, as well as the primitive domains, are "schlicht" (i.e. single-sheeted) and bounded on real momentum space by various sets of the form

$$k_I^2 = (\sum_{i \in I} k_i)^2 \geq M_I^2 \quad (\text{with } I \subset \{1,2,\dots,n\}):$$

they define for each n the so-called "physical-sheet determination" of $H^{(n)}$, and (by restriction to the complex mass shell) of the corresponding scattering functions.

Although these domains contain rather large global regions (which include for n = 4 the domains of dispersion relations [5a,b] on the mass shell), they obey in general the following severe limitations: for n > 4, the restrictions of the known domains D'_n to the corresponding complex mass shell M_n^c ($k_i^2 = m_i^2$; i = 1,2,...,n) are of very small size; in fact, near the physical region of a given (arbitrary) m → n-m particle process, the scattering function has been shown to be the boundary value of an analytic function <u>only</u> from a narrow cone of Im k-space (in M_n^c), which even vanishes when Re k varies in a certain region surrounding the threshold of the considered process (more complicated properties of decomposition in sums of several analytic functions were proved to hold in this region) [5c]. On the other hand, in this axiomatic approach (i), it is impossible to justify and study the properties of analytic continuation of the $H^{(n)}$ <u>across</u> the cuts $k_I^2 \geq M_I^2$ into new "unphysical sheets", without making use of an extra-postulate, such as Asymptotic Completeness (see Sect. 2 for a precise formulation of it).

Exploitation of the A.C. postulate (with appropriate technical assumptions) allows one indeed to investigate the analytic and monodromic structure (i.e. the "type of ramification") of the Green functions and scattering functions around well-defined singular sets (including the thresholds mentioned above), called "Landau varieties". Investigations of this type are in some sense analogous to those which can be carried out in the S-matrix framework (ii) by exploiting the unitarity integral relations. In (ii) however, no primitive global analyticity properties for the scattering functions can be deduced from plausible postulates formulated in Minkowski space-time+ so that the monodromic structure of the scattering functions can only be studied there in the framework of microlocal analysis, the latter having been justified on physical grounds by macrocausality type assumptions [11] and through the method of local Fourier transformation (see [6]).

The integral relations of M.P.S.A. through which the A.C. axiom can be efficiently exploited are exact counterparts of certain identities which relate appropriate formal series of Feynman functions in the framework of perturbation theory (namely equations of the Bethe-Salpeter type). The various terms of these integral relations can be represented by graphs which obey integration prescriptions similar to those of Feynman functions, but whose vertices are associated with "general n-point functions" instead of constants or monomials (we call them "G-convolution functions").

+For the possibility of an alternative global approach which would be purely based on requirements of maximal analyticity for the S-matrix, as proposed in [12], see [7b,13,6] and references therein.

The "general n-point functions" which occur in these relations are not only the Green functions $H^{(n)}$ themselves, but other quantities which are called "ℓ-particle irreducible n-point functions" ($\ell = 1,2,\ldots,\infty$) and which have an intuitive interpretation in perturbation theory [1]. In the axiomatic framework, these functions are expected to be defined through a recursive algorithm starting from the lowest values of the integer ℓ and to enjoy the following important property: for increasing values of ℓ they are analytic in increasingly larger domains of $C^{4(n-1)}$ which all contain the primitive domain of $H^{(n)}$; in the M.P.S.A. approach which we consider, these analyticity properties of the ℓ-particle irreducible functions are expected to be obtained as consequences of the A.C. postulate which supplement the primitive analyticity properties entailed by (i)[+]. At the present stage of our program these properties have only been established for $\ell = 1,2,3$ (see below), but some steps have also been taken in the treatment of the general case.

From considering the integral relations of M.P.S.A., it emerges that the global analytic and monodromic structure of Green functions should be completely governed by the two following operations:

(I) Integration of analytic functions over prescribed classes of cycles is definite complex domains,

(II) Fredholm inversion of analytic kernels depending analytically on parameters over prescribed classes of cycles in definite complex domains.

In the recursive process of M.P.S.A., one can indeed expect that to each Green function $H^{(n)}$ and integer ℓ,

---

[+] In some tractable models of Quantum Fields which have been studied in the constructivist approach (belonging to the class $P(\phi)_2$ at small coupling), the required analyticity properties of ℓ-particle irreducible functions (satisfying a suitable recursive algorithm) have been checked directly [14].

there corresponds a definite domain $D_n^{(\ell)}$ in complex four-momentum space $C^{4(n-1)}$ in which the above operations (I) and (II) should determine exhaustively the Landau singularities and the associated monodromic structure of $H^{(n)}$ implied by the completeness of the $1,2,\ldots,\ell-1$, ℓ-particle asymptotic states of the theory (in brief: the "$1,2,\ldots,\ell-1$, ℓ-particle structure"). The domains $D_n^{(\ell)}$ should be determined by the primitive analyticity domains of the corresponding ℓ-particle irreducible n-point functions; their size, governed by the production threshold of $(\ell+1)$-particle states should increase with ℓ, and to the extent that in the limit $\ell \to \infty$, $D_n^{(\ell)}$ fills up the whole complex four-momentum space $C^{4(n-1)}$, the M.P.S.A. program may have - in principle - the ambition of characterizing at a global level in $C^{4(n-1)}$ the singularities and the organization of the complete monodromic structure of the Green functions $H^{(n)}$.

At the present stage, the obtained results (for $\ell = 1,2,3$) are indications in favour of this wishful description.

Before we describe some typical results which have been obtained for the two-and three-particle structure of the six-point Green function $H^{(6)}$ and of the three-particle scattering function, we shall make a few comments on the way in which the mathematical operations (I) and (II) are acting in the M.P.S.A. approach.

<u>(I) Integration in complex space:</u> Let $t = (t_1,\ldots,t_q) \in C^q$, $z = (z_1,\ldots,z_n) \in C^n$,

$$I(t) = \int_{\Gamma(t)} f(t,z)\, dz_1 \wedge \ldots \wedge dz_n \quad , \tag{1}$$

where f is analytic in a domain $D = \bigcup_{t \in \Delta} (t, D_t)$, and $\Gamma(t) \subset D_t$ is an n-dimensional cycle which has to "vary continuously" with t (see [8]).

A typical example which illustrates the relevance of this operation to M.P.S.A. is the "triangle term" which contributes to the r.h.s. of the structural equation (11) (given below) for $H^{(6)}$, namely:

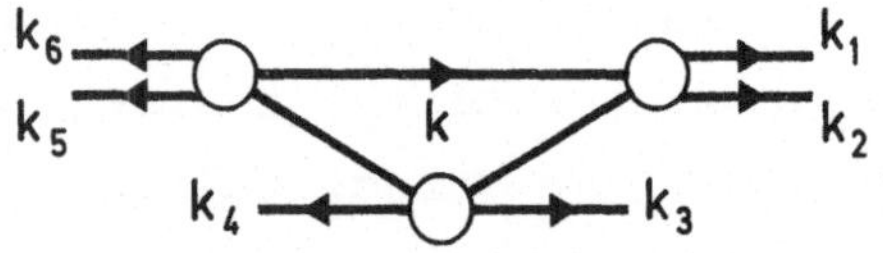

In the latter the bubble vertices represent the Green function+ $H^{(4)}$, and the associated contribution to the six-point function is the (G-convolution) integral:

$$H^{(6)}(k_1,\ldots,k_6)=\int_{\Gamma(k_1,\ldots,k_6)} H^{(4)}(k_5,k_6,k,-k-k_5-k_6)$$

$$H^{(4)}(k_1,k_2,-k,k-k_1-k_2)\ldots H^{(4)}(k_3,k_4,k_5+k_6+k,k_1+k_2-k)$$

$$[H^{(2)}(k)H^{(2)}(k-k_1-k_2)H^{(2)}(k+k_5+k_6)]^{-1}d_4k\ , \qquad (2)$$

$\{\Gamma(k_1,\ldots,k_6)\}$ being a well-defined family of four-dimensional cycles.

At the preliminary step of M.P.S.A. - at which $H^{(4)}$ is only known to be analytic in its primitive domain D_4 - the integral (2) can be shown to define a function which is analytic in the primitive domain D_6: this is in fact the content of the "G-convolution theorem" proved in [3a], in the case when G is the triangle graph, here considered. To proceed further, we now assume that a single mass μ is involved in the spectrum of the theory, so that the integrand of (2) contains the following product of three simple poles: $[k^2-\mu^2]^{-1}\cdot[(k-k_1-k_2)^2-\mu^2]^{-1}\cdot[(k+k_5+k_6)^2-\mu^2]^{-1}$.

+More precisely, a suitable modified form of $H^{(4)}(k_1k_2k_3k_4)$ which is regularized at infinity, e.g. by a factor $\prod_{1\le i\le 4}[(\rho-\mu^2]^{\alpha}[k_i^2-\mu^2]^{-\alpha}$ with $\rho \gg \mu^2$ and $\alpha > 0$ sufficiently large [2a, 3c] (such a factor does not modify the restriction of $H^{(4)}$ to the mass shell).

At the step $\ell = 2$ (namely, the exploitation of the two-particle structure), $H^{(4)}$ can be shown to be analytic in a larger domain which is a two-sheeted Riemann surface with branching singularity $s = (k_1+k_2)^2 = (2\mu)^2$. Correspondingly, the integral (2) can be shown to have an analytic continuation on a certain Riemann surface which contains the so-called "three-particle elastic scattering region on the mass shell" (defined by $k_i^2 = \mu^2$, $i = 1,\dots,6$: $(3\mu)^2 < (k_1+k_2+k_3)^2 < (4\mu)^2$). This Riemann surface has common features with that of the Feynman amplitude corresponding to the Feynman graph 

but it is confined over a certain definite domain: this restriction expresses the fact that only the two-particle structure of the theory has been taken into account at this stage. As in the perturbative case, the Landau singularities of

$H^{(6)}$

are produced by the pinches of the integration cycle $\Gamma(k_1,\dots,k_6)$ by certain "vanishing cells" [8,15] (see Sect. 4) of the singular set **S** of the integrand; here S is the union of the three polar manifolds $k^2 = \mu^2$, $(k_1+k_2-k)^2 = \mu^2$, $(k_5+k_6+k)^2 = \mu^2$, and of the branching manifold $(k_1+k_2+k_3-k)^2 = (2\mu)^2$ (notice that the latter is absent for the corresponding Feynman integrand).

(II) Fredholm inversion in complex space or "Bethe-Salpeter inversion":

One is interested in the analyticity properties (in (t,z,z',λ)) of the Fredholm resolvent F of a given analytic kernel G, satisfying the equation:

$$\forall \lambda \in \mathbb{C}\ ,\ F(t,z,z',\lambda) = G(t,z,z') +$$

$$+\lambda\int_{\Gamma(t)} F(t,z,z'',\lambda)G(t,z'',z')\omega(t,z'')dz''_1\ldots dz''_n \; ; \qquad (3)$$

G is analytic in a domain $D = \bigcup_{t\in\Delta} (t, D_t \times D'_t)$, $\omega(t,z)$ is an analytic weight and $\Gamma(t) \subset D_t \cap D'_t$ is an n-dimensional cycle which has to vary continuously with t (see [16]) for a systematic study of the analytic and monodromic structure of F, under suitable simple assumptions: an account of it is given below in Sect. 5).

A typical example which illustrates the relevance of this operation to M.P.S.A. is the Bethe-Salpeter equation which expresses $H^{(4)}$ as the Fredholm resolvent of the two-particle irreducible four-point function $L^{(4)}$:

$$H^{(4)}(K,z,z') = L^{(4)}(K,z,z') + \frac{1}{2}\int_{\Gamma(K)} H^{(4)}(K,z,z'')$$

$$L^{(4)}(K,z'',z')\omega(K,z'')d_4z'' \quad , \qquad (4)$$

or graphically:

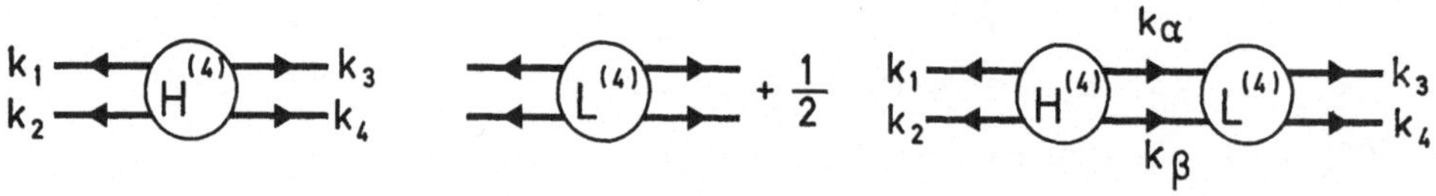

where: $K = k_3+k_4 = -(k_1+k_2) = k_\alpha+k_\beta$,

$$z = \frac{k_2-k_1}{2} \; , \quad z' = \frac{k_3-k_4}{2} \; , \quad z'' = \frac{k_\alpha-k_\beta}{2} \; ,$$

$$\omega(K,z'') = [H^{(2)}(k_\alpha(K,z''))H^{(2)}(k_\beta(K,z''))]^{-1}$$

$$(k_\alpha(K,z'') = \frac{K}{2} + z'' \; , \quad k_\beta(K,z'') = \frac{K}{2} - z'') \; ,$$

and $\Gamma(K)$ coincides with the euclidean subspace⁺ $(iR) \times R^3$ of z-space when K belongs itself to $(iR) \times R^3$.

⁺Here, a Lorentz frame has to be chosen: $K = (K^{(0)}, \vec{K})$, and the Minkowski metric $K^2 = K^{(0)2} - \vec{K}^2$ defines a euclidean metric in $(iR) \times R^3$.

Equation (4) is actually the starting point of the step $\ell = 2$[+] of M.P.S.A. [2], since it accounts for the generation of the two-sheeted Riemann surface of $H^{(4)}$ (quoted in the previous paragraph), once the primitive "schlicht" analyticity domain of $L^{(4)}$ is given. The latter includes the two-particle elastic scattering region on the mass shell ($k_i^2 = \mu^2$, $i = 1,\ldots,4$, $(2\mu)^2 \leq$ $\leq (k_1+k_2)^2 < (4\mu)^2$), and the branching singularity $s = K^2 = (k_1+k_2)^2 = (2\mu)^2$ of $H^{(4)}$ is produced by the pinch of the integration cycle $\Gamma(K)$ by the vanishing cell of the polar set S of $L^{(4)}$, namely:

$$S = \{(K,z''); (\tfrac{K}{2}+z'')^2 = \mu^2\} \cup \{(K,z''); (\tfrac{K}{2}-z'')^2 = \mu^2\} .$$

Let us point out that, if the primitive ("schlicht") domain of $H^{(4)}$ is distinguished (as usual) as the "physical sheet" of $H^{(4)}$, the second sheet or "unphysical sheet" is a domain of meromorphy for $H^{(4)}$, because the solution of (4) can be written in the whole domain of $H^{(4)}$:

$$H^{(4)}(K,z,z') = \frac{N^{(4)}(K,z,z')}{D^{(2)}(K)} ,$$

where $N^{(4)}$ is a general four-point function and where $D^{(2)}$ is a general two-point function which may have zeros in the considered second-sheet domain: if these zeros lie near the physical region, and on the side $\mathrm{Im}\, s < 0$, the corresponding poles of $H^{(4)}$ are interpreted as unstable particles which can distintegrate into a pair of fundamental particles with mass .

At a further step, Bethe-Salpeter equations of the form

[+] Equ.(4), which we quote here, actually corresponds to the case of even quantum field theories: in the latter the step $\ell=1$ (...ℓ odd) does not exist for $H^{(4)}$.

$$\equiv\!\!\!\bigcirc\!\!\!\!F\!\!\!\!\equiv\ =\ \equiv\!\!\!\bigcirc\!\!\!\!G\!\!\!\equiv\ +\ \frac{1}{3!}\ \equiv\!\!\!\bigcirc\!\!\!\!F\!\!\!\!\equiv\!\!\!\bigcirc\!\!\!\!G\!\!\!\equiv$$

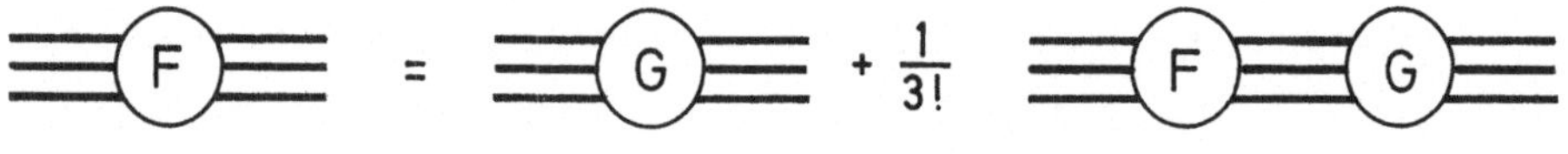

$$(5)$$

have to be considered, and for such equations of type (3) new difficulties arise as far as the analysis of the Landau singularities of F are concerned. These difficulties are linked with the occurrence of singular sets S for G involving "non-separable components" with respect to z- and z'-variables (for instance: poles of the form $1/((z_i-z_j')^2-\mu^2)$. The latter generate infinite sequences of Landau varieties whose elements are singularities for the successive terms of the Neumann series of equation (3).

In the physical case of equations of the type (5), such singularities accumulate to the threshold singularity $(k_1+k_2+k_3)^2 = (3\mu)^2$, and a careful analysis of the various sheets of the resolvent has to be done. The results for $H^{(6)}$ which we shall now state are valid in a domain which is bounded by the cut: $s = (k_1+k_2+k_3)^2 = (3\mu)^2 + \rho$ $(\rho \geq 0)$; in other words, the monodromic structure around the three-particle threshold has not yet been explored there. In order to describe properly these results, we first need to specify some notations.

We consider the case of even Quantum Field Theories with a single fundamental mass μ.

We put $H^{(6)}(k_1,\ldots,k_6) = H^{(6)}(K,z,z')$, where $K = = k_1+k_2+k_3 = -(k_4+k_5+k_6)$, $z = (\underline{k}_4,\underline{k}_5,\underline{k}_6)$ with: $\underline{k}_i = -k_i - \frac{K}{3}$, $i = 4,5,6$; $z' = (\underline{k}_1,\underline{k}_2,\underline{k}_3)$ with $\underline{k}_i = k_i - \frac{K}{3}$, $i = 1,2,3$. K is kept in the manifold $K = (K^{(0)},\vec{0})$, and $K^{(0)}$ varies in a domain $\Delta_M^{cut} = \Delta_M - \{K^o \geq 3\mu\}$, where $\Delta_M = \{K^{(0)}; 0 < \text{Re } K^o < M,$ with $3\mu < M \leq 5\mu\}$ (Fig. 1). Fig. 2 represents the situation in Re $z'^{(0)}$-plane; ω_h is the polar manifold $k_h^2 = \mu^2$, σ_h is the branch manifold $s_{ij} = (k_i+k_j)^2 = 4\mu^2$ $(\{i,j,h\} = \{1,2,3\})$; the dark disk at the center of Fig. 2a) represents in pro-

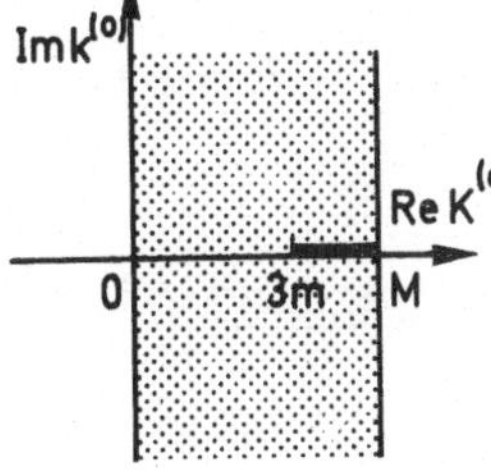

Fig. 1. The domain Δ_M^{cut}.

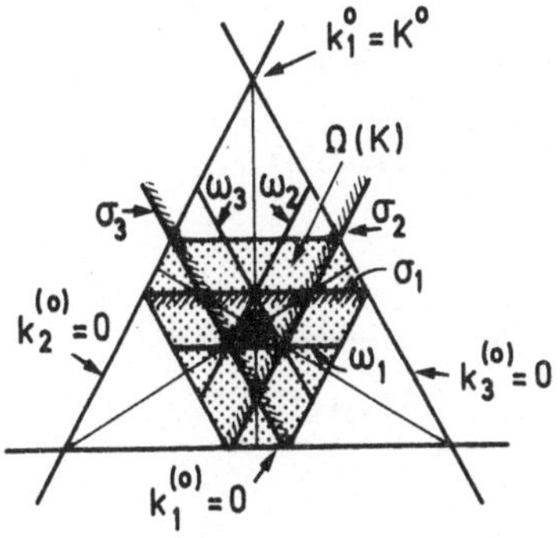

Fig.2a. $K^{(0)} > 3m$.

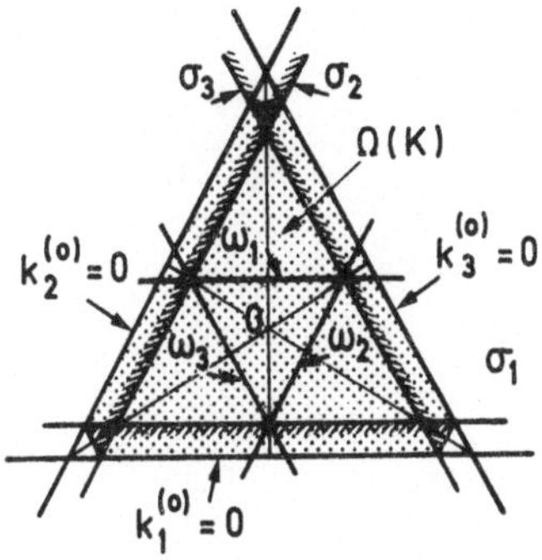

Fig.2b. $K^{(0)} < 3m$.

jection the physical region of the reaction in which the incoming (resp. outgoing) particles are labelled by the set $\{4,5,6\}$ (resp. $\{1,2,3\}$).

For every $K^{(0)}$ in Δ_M, one defines in z-space a domain D_K of the following form:

$$D_K = \{z = (z^{(0)},\vec{z}) \in C^8;\ \mathrm{Im}\ z^{(0)} \in R^2,\ \mathrm{Re}\ z^{(0)} \in \Omega(K),$$

$$\mathrm{Re}\ \vec{z} \in R^6;\ |\mathrm{Im}\vec{z}| < \varepsilon\ \ (\mathrm{Re}\ z^{(0)})\}$$

where the domain $\Omega(K)$ is represented by the shaded area on Fig. 2.

The following domains D_M, D_M^{cut} are used in the results below:

$$D_M(\text{resp.}D_M^{cut}) = \{(K^{(0)},z,z');\ K^{(0)} \in \Delta_M(\text{resp.}\Delta_M^{cut});$$
$$z \in D_K;\ z' \in D_K\} .$$

<u>One- and two-particle dressing of $H^{(6)}$:</u>

The aim of this step is to express+ $H^{(6)}$ in terms of a suitable irreducible part $G^{(6)}$ and of auxiliary kernels which only involve $H^{(4)}$ and $L^{(4)}$. We first define the latter, namely: T_{in} is the tree-graph product

$\sum_{h,n}$ 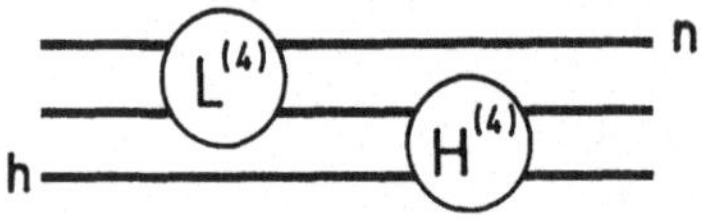

and similarly

$T_{out} = \sum_{h,n}$ 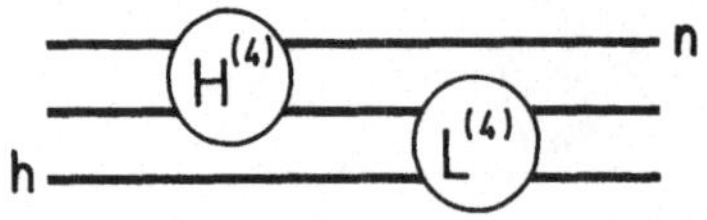

We also introduce the following operators depending on K:

$\Lambda = 1 - \frac{1}{6} \sum_{hn}$ 

and

$V = 1 + \frac{1}{6} \sum_{hn}$ 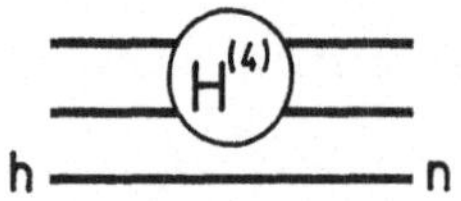

which satisfy:

$$\Lambda V = 1 - \frac{1}{6} T_{in} , \quad V\Lambda = 1 - \frac{1}{6} T_{out} .$$

+More precisely, $H^{(6)}, H^{(4)}, H^{(2)}$ are replaced here by the corresponding regularized forms (see our previous footnote 3).

When $K = (K^{(0)}, \vec{0})$ lies in euclidean space (Re $K^{(0)} = 0$), all these operators are represented by kernels in euclidean z-space (Re $z^{(0)} = 0$, Im $\vec{z} = 0$) equipped with the measure

$$\prod_{i=1}^{3} [H^{(2)}(k_i(K^{(0)}, z))]^{-1} \, dz \; ;$$

when $K^{(0)}$ varies in Δ_M^{cut}, a suitable cycle $\Gamma(K_o)$ in complex z-space (which is a distortion of euclidean z-space) has to be considered. Operators of this type are compact (and the corresponding kernels analytic and properly decreasing at infinity) iff they correspond to <u>connected</u> graphs of six-point functions[+]: this is the case for T_{in}, T_{out} (not for Λ and V). T_{in} and T_{out} thus admit respective Bethe-Salpeter inverses U_{in}, U_{out} (in the sense of (5)); equivalently in the operator notation, one has:

$$(1-\tfrac{1}{6}T_{in})\cdot(1+\tfrac{1}{6}U_{in}) = (1-\tfrac{1}{6}T_{out})\cdot(1+\tfrac{1}{6}U_{out}) = 1 \quad . \tag{6}$$

This allows one to define the inverse of the (bounded but not compact) operator Λ as:

$$\Lambda^{-1} = V\cdot(1 + \tfrac{1}{6}\, U_{in}) = (1 + \tfrac{1}{6}\, U_{out})\cdot V \quad . \tag{7}$$

We now introduce the six-point function $G^{(6)}$ through the equation:

$$G^{(6)} = \Lambda H_1^{(6)} \Lambda - \Lambda T_{out} \tag{8}$$

(note that $\Lambda T_{out} = T_{in}\Lambda$), where:

$H_1^{(6)} = H^{(6)} -$ 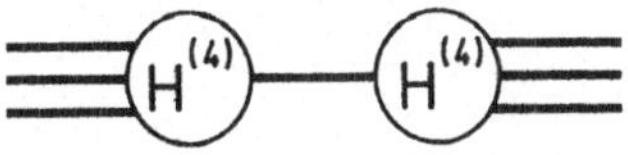 (9)

[+] i.e. if they define G-convolution products in the sense of [3a] (see Sect. 4).

Suitable one- and two-particle irreducibility properties can be proved for $G^{(6)}$, which imply in particular the following analyticity property of

$$\hat{\hat{G}}^{(6)} = \prod_{i=1}^{6} (k_i^2-\mu^2)\, G^{(6)} \;.$$

<u>Lemma:</u> $\hat{\hat{G}}^{(6)}$ is analytic in a domain D_M^{cut}, such that $M = 11\ \mu/3$.

Now the equation (8) which defines $G^{(6)}$ can be inverted and yields the following "two-particle dressing equation":

$$H_1^{(6)} = (1+\tfrac{1}{6}U_{out})VG^{(6)}V(1+\tfrac{1}{6}U_{in}) + U_{out}V \quad . \tag{10}$$

By using the analyticity properties of $H^{(4)}$ and $G^{(6)}$, and by exploiting in complex space some standard combinatorics of Fredholm theory, one can prove the

<u>Theorem:</u> In D_M^{cut} (with $M = 11\ \mu/3$) the following decomposition of $H^{(6)}$ holds:

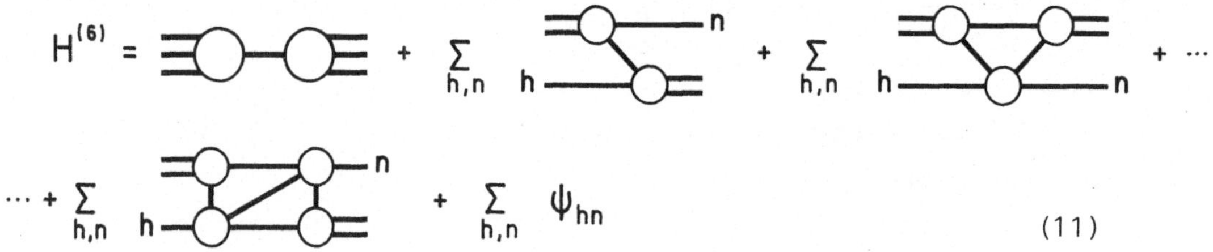

(11)

In the latter, all the "bubbles" of the G-convolution products represent $H^{(4)}$, and each analytic function ψ_{hn} of the residual sum ($h \in \{4,5,6\}$, $n \in \{1,2,3\}$) has the following properties:

i) ψ_{hn} is analytic in a "physical-sheet domain" which is $D_M^{(cut)} - (\underline{\sigma}_h \cup \underline{\sigma}_n)$; here the sets $\underline{\sigma}_h$, $\underline{\sigma}_n$ denote the following "cuts":

$$\underline{\sigma}_h = \{(K,z,z');\ s_{ij}(K,z) = 4\mu^2 + \rho;\ \rho \geq 0\} \text{ and}$$

$$\underline{\sigma}_n = \{(K,z,z');\ s_{\ell m}(K,z') = 4\mu^2 + \rho;\ \rho \geq 0\} \quad .$$

ii) ψ_{hn} admits a local analytic continuation across the sets $\underline{\sigma}_h$ and $\underline{\sigma}_n$, on the Riemann surface associated with the Feynman graph

Besides, each G-convolution term of the decomposition (11) is analytic in D_M^{cut} minus the Landau singular set associated with the corresponding graph G; it moreover admits a local analytic continuation on the Riemann surface associated with G.

Remark: The only singularities of

$$\hat{H}^{(6)} = \prod_{i=1}^{6} [H^{(2)}(k_i)]^{-1} H^{(6)}$$

which are produced in $D_M^{cut} - \bigcup_{h,n} (\underline{\sigma}_h \cap \underline{\sigma}_n)$, considered as the physical sheet defined by the extension of the primitive analyticity domain, are the Landau singularities of the 0-loop, 1-loop, and two-loop "truss-bridge graphs"; the ℓ-loop graphs ($\ell > 2$)

are potentially present in the Neumann expansions of U_{in}, U_{out} and illustrate the two-particle dressing of $H_1^{(6)}$ from both sides of $G^{(6)}$ in formula (10); however, it is a consequence of the previous theorem that their Landau singularities are effective only in other sheets.

By taking the restriction of $\hat{H}^{(6)}$ to the complex mass-shell M_6^c, one obtains the following corollary of the previous theorem:

<u>Corollary:</u> In $D_M^{cut} \cap M_6^c$, the $3 \to 3$ particle scattering function admits the following decomposition as a sum of analytic functions[+]:

$T_{IJ}^{(6)} = \sum_{h,n}$ 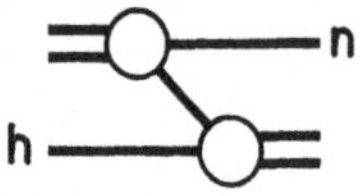 $+ \sum_{h,n}$ 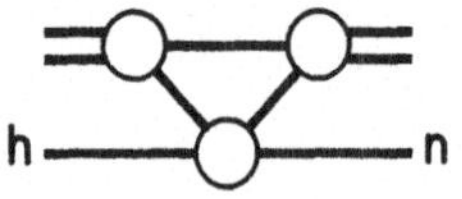 $+ \sum_{h,n} \hat{\psi}_{hn}$, (12)

where each function $\hat{\psi}_{hn}$ is analytic in $D_M^{cut} \cap M_6^c - (\underline{\sigma}_h \cup \underline{\sigma}_n)$; across each cut $\underline{\sigma}_h$, $\underline{\sigma}_n$, $\hat{\psi}_{hn}$ admits a local analytic continuation which spreads in a two-sheeted Riemann surface around the corresponding threshold σ_h, σ_n.

As far as the three-particle structure of $H^{(6)}$ is concerned, we shall only say here that:

i) a three-particle irreducible kernel $L^{(6)}$ has been introduced on the basis of three-particle asymptotic completeness (see Sect. 5).

ii) By using a Bethe-Salpeter type equation which relates $G^{(6)}$ to $L^{(6)}$, it is then possible to show that $G^{(6)}$ and $H^{(6)}$ can be analytically continued at least locally across the cut $s = (3\mu)^2 + \rho$ $(\rho \geq 0)$. The global study of this monodromic structure remains to be done.

<u>Remarks:</u> 1) This result contains as a byproduct the local analyticity properties which are sufficient to ensure macrocausality conditions (in the sense of Stapp-Iagolnitzer [11]) for the $3 \to 3$ scattering process in the corresponding low-energy region.

[+] In the r.h.s. of (12), the G-convolution terms must be considered as amputated (i.e. multiplied by $\prod_{i=1}^{6} [H^{(2)}(k_i)]^{-1}$).

2) A decomposition property for the $3 \to 3$ particle scattering function of the same form as (12) has been derived independently in [17] in the same axiomatic field-theoretical framework (and under slightly weaker technical assumptions) by a method of microlocal analysis[+] and a direct use of A.C. equations.

In the following lectures, we shall describe with more details the basic postulates of our approach (Sect. 2), some useful results on the Green functions (Sect. 3), various developments of the mathematical method involving integration in complex space (Sect. 4 and 5), and finally our main results concerning one-, two-, and three-particle structure analysis (Sect. 6,7,8). In particular, more details about the analysis of the structural equations for $H^{(6)}$ (given above) will be found in Sect. 7, and (as far as results on Fredholm theory in complex space are needed) in Sect. 5. The final Section 8 illustrates how more global analyticity properties can be obtained by combining the results of M.P.S.A. together with standard techniques of analytic completion: the existence of a crossing domain for the $3 \to 3$ particle scattering function (in the equal mass case) is shown there.

2. RECALL ON GENERAL QUANTUM FIELD THEORY [18]

2.I Axioms of relativistic quantum mechanics

- Hilbert space of states $\mathcal{H}$ with the standard interpretation of the scalar product.

- $U(a,\Lambda)$: unitary representation of the Poincaré group ($a \in \mathbb{R}^4$: space-time translation, Λ: Lorentz transformation).

[+] Further refinements are needed, however, to recover local analyticity around the thresholds by this method.

- $U(a,1) = e^{iP_\nu a^\nu}$ admits the four energy-momentum operators P_ν as generators ($P_o = H$ being the total hamiltonian in a given Lorentz frame).

- Spectral condition (= positivity of energy): $Sp\,\{P_\nu\} \in \overline{V^+}$ (the future light-cone).

2.II Axioms of local fields

We just state them for a neutral scalar field Φ:

1) $\Phi(x)$ $(x \in R^4)$ is an operator-valued distribution defined in a certain domain D_Φ of H, namely the set of all operators $\Phi(f) = \int \Phi(x) f(x) d_4x$ are well-defined in D_Φ, for f belonging to an appropriate space of smooth test-functions on R^4.

2) Relativistic covariance of Φ:

$$U(a,\Lambda)\Phi(x)U(a,\Lambda)^{-1} = \Phi(\Lambda x + a) \ . \tag{13}$$

3) The domain D_Φ of Φ contains a vacuum state Ω such that $U(a,\Lambda)\Omega = \Omega$ $(P_\nu\Omega = 0)$, and is spanned by all "vacuum excitation states" of the form $\Phi(f_1)\ldots\Phi(f_n)\Omega$.

4) Locality property of Φ: For every state ψ in the domain D_Φ:

$$[\Phi(x),\Phi(y)]\psi = 0 \qquad \text{if} \qquad (x-y)^2 < 0 \tag{14}$$

(for $x = (x^{(0)},x^{(1)},x^{(2)},x^{(3)}) = (x^{(0)},\vec{x})$, we have put: $x^2 = x^{(0)2} - x^{(1)2} - x^{(2)2} - x^{(3)2} = x^{(0)2} - \vec{x}^2$).
Condition (14) can be weakened (for example for a field derived from a quasilocal observable in the sense of [19]).

2.III The asymptotic completeness postulate for a field Φ

Let us call $H_\Phi \subset H$ the subspace of the field, i.e.

the closure of D_Φ in H. On the basis of the previous axioms (I and II) and of an additional postulate concerning the existence of a subspace of H_Φ associated with a positive discrete mass spectrum, one proves the existence of two subspaces of asymptotic states of the field, $(H_{in})_\Phi$ and $(H_{out})_\Phi$, contained in H_Φ (see in [18b]: Haag-Ruelle theory). There moreover exists a natural isometry S_Φ from $(H_{out})_\Phi$ onto $(H_{in})_\Phi$ which is the scattering operator associated with Φ.

We say that Φ has the property of asymptotic completeness if:

$$(H_{in})_\Phi = (H_{out})_\Phi = H_\Phi \quad . \qquad (15)$$

This implies the unitarity of the operator S_Φ in H_Φ.

Note however that the condition $(H_{in})_\Phi = (H_{out})_\Phi \subset H_\Phi$ is sufficient to ensure the unitarity of S_Φ as an operator in $(H_{out})_\Phi$.

Remark: By stating "Asymptotic Completeness" as a property which is relative to a certain local field Φ and which holds in the corresponding subspace H_Φ (and this is what is actually relevant for working out the M.P.S.A. approach), we do not put any restriction on the possible occurrence of other (more fundamental...) concepts in the theory, such as basic gauge fields acting in a larger Hilbert space (as in Q.C.D.), other types of particle states (in the spirit of the soliton approach...) etc...: in such cases, the local field Φ (of traditional type) might very well exist as a derived quantity (i.e. a certain functional of the fundamental fields), and the M.P.S.A. formalism relative to Φ would then still be valid.

2.IV Green functions, scattering functions, and reduction formulae

One can associate with the field Φ its n-point τfunctions, which are the Fourier transforms in energy-momentum space of the "truncated" or "connected" V.E.V. of the time-ordered products of n field operators:

$$\tau^{(n)}(p_1,\ldots,p_n) = \int e^{i\sum p_i\cdot x_i} \langle\Omega, T(\Phi(x_1)\ldots\Phi(x_n))\Omega\rangle_{tr}\, dx_1\ldots dx_n. \tag{16}$$

The $\tau^{(n)}$ are (up to a global δ-function of energy-momentum conservation) special boundary values of the analytic n-point Green functions $H^{(n)}$ (see Sect. 3), namely:

$$\tau^{(n)}(p_1^{(0)},\vec{p}_1,\ldots p_n^{(0)},\vec{p}_n) = \delta_4(p_1+\ldots+p_n) \lim_{\substack{\varepsilon\to 0\\ \varepsilon>0}} H^{(n)}(p_1^{(0)}(1+i\varepsilon), \vec{p}_1,\ldots,p_n^{(0)}(1+i\varepsilon),\vec{p}_n) . \tag{17}$$

For simplicity (here and in most of the following) we assume that one-particle states are vacuum excitation states created by single field operators (namely $\Phi(f)\Omega$).

The complex mass shell manifold M_n^c of particles with mass μ is the submanifold of $C^{4(n-1)} = \{(k_1,\ldots,k_n) \in C^{4n}; k_1+\ldots+k_n=0\}$ defined by the n equations: $k_1^2=\mu^2$, $k_2^2=\mu^2,\ldots,k_n^2=\mu^2$. In real space, M_n^c has disconnected sheets M_{IJ} associated with all channels (I,J) (i.e. ordered partitions of $\{1,2,\ldots,n\}$), corresponding to the set I of incoming particles and to the set J of outgoing particles; M_{IJ} is defined by the prescription: $-k_i = k_i' \in V^+$ for $i \in I$ and $k_j \in V^+$ for $j \in J$.

For each n and each channel (I,J), the scattering operator S_Φ is represented by a set of n-point kernels $S_{IJ}^{(n)}(\{k_i'\}_{i\in I}; \{k_j\}_{j\in J})$, where $(\{-k_i'\}_{i\in I}, \{k_j\}_{j\in J}) \in \underline{M}_{I,J}$ and $\underline{M}_{IJ} = \{(k_1\ldots k_n); k_i^{(0)} = -\sqrt{\vec{k}_i^2+\mu^2}, \forall i\in I, k_j^{(0)} = +\sqrt{\vec{k}_j^2+\mu^2}, \forall j \in J\}$.

The set of kernels S_{IJ}^n can be replaced (through a standard algorithm [6a]) by an alternative set of kernels

$S_{IJ}^{n\ conn.}$, called "connected scattering kernels".

The theory of reduction formulae [20] yields the following relation between the connected scattering kernels $S_{IJ}^{(n)\,conn.}$ and the n-point functions $\tau^{(n)}$ of the field Φ:

$$S_{IJ}^{(n)\,conn.}(\{-k_i\}_{i\in I};\{k_j\}_{j\in J}) = (\frac{1}{Z})^n \prod_{\ell=1}^{n} (k_\ell^2-\mu^2)\,\tau^{(n)}(k_1\ldots k_n)|M_{IJ} \tag{18}$$

(Z being a normalisation constant of the asymptotic states of mass μ created by the field Φ). Formula (18) has been established in a certain (dense) open set of generic momentum configurations.

By putting

$$S_{IJ}^{(n)\,conn.} = \delta_4\,(\sum_{j\in J} k_j - \sum_{i\in I} k_i')\,T_{IJ}^{(n)}(\{k_i'\}_{i\in I},\{k_j\}_{j\in J}), \tag{18'}$$

one defines the scattering functions $T_{IJ}^{(n)}$, whose analyticity properties on the complex mass shell M_n^c will be obtained by restriction to M_n^c of the corresponding enlarged analyticity domains D_n' of the $H^{(n)}$ (see Sect. 1); in fact (18) will then yield, by analytic continuation (in view of (17) and (18')):

$$T_{IJ}^{(n)}(\{-k_i\}_{i\in I};\{k_j\}_{j\in J}) =$$

$$= b.v.M_{IJ}^{+}[\frac{1}{Z^n}\prod_{\ell=1}^{n}(k_\ell^2-\mu^2)\,H^{(n)}(k_1,\ldots,k_n)]\Big|_{M_n^c} \tag{19}$$

where "b.v.M_{IJ}^{+}" indicates that a suitable boundary value on M_{IJ} from M_n^c has to be taken in the r.h.s. of (19).

In the general case when several masses μ_α occur in the spectrum of the field Φ, formula (19) must be replaced by the following more general formula, in which the notations $T^{(n)}_{IJ,\hat{\alpha}}$ and $M_{IJ,\hat{\alpha}}$ correspond to a given mass prescription: $\ell \to \mu_{\hat{\alpha}(\ell)}$ for each $\ell \in I \cup J$:

$$T^{(n)}_{IJ,\hat{\alpha}}(\{-k_i\}_{i\in I};\{k_j\}_{j\in J}) = \mathrm{b.v.}M^{+}_{IJ,\hat{\alpha}}[[\prod_{\ell=1}^{n} H^{(2)}(k_\ell)]^{-1} \cdot$$

$$\cdot H^{(n)}(k_1,\ldots,k_n)]\Big|_{M_n^c} \quad . \tag{20}$$

3. ANALYTICITY PROPERTIES OF GREEN FUNCTIONS, ABSORPTIVE PARTS, AND OFF-SHELL UNITARITY-TYPE EQUATIONS

3.I The domains D_n; a brief account

For each $n \geq 2$ there exists an analytic Green function $H^{(n)}(k_1,\ldots,k_n)$, defined in a "primitive domain" D_n of $C^{4(n-1)} = \{(k_1,\ldots,k_n);\ k_1+\ldots+k_n = 0,\ k_i \in C^4,\ 1 \leq i \leq n\}$.

D_n is the union of the following sets:

i) a family of tubes $T_n^{(S)}$ (i.e. "generalized half-planes") obtained from locality: in each $T_n^{(S)}$, $H^{(n)}$ admits a boundary value on $R^{4(n-1)}$ which is (the Fourier transform of) a certain "generalized retarded function" $r_n^{(S)}(k_1\ldots k_n) = \langle\Omega, R_n^{(S)}(k_1\ldots k_n)\Omega\rangle$;

ii) a set of regions through which the $T_n^{(S)}$ communicate with each other, and which are limited by spectral conditions of the form: $k_I^2 = (\sum_{i\in I} k_i)^2 < M_I^2, k_I^2 \neq \mu^2$.

Typical examples:

a) $H^{(2)}(k,-k) = H^{(2)}(k);\ k = (k^{(0)},\vec{k})$.

For $\vec{k}$ real, the $k^{(0)}$-section of D_2 is the cut plane of Fig. 3, where:

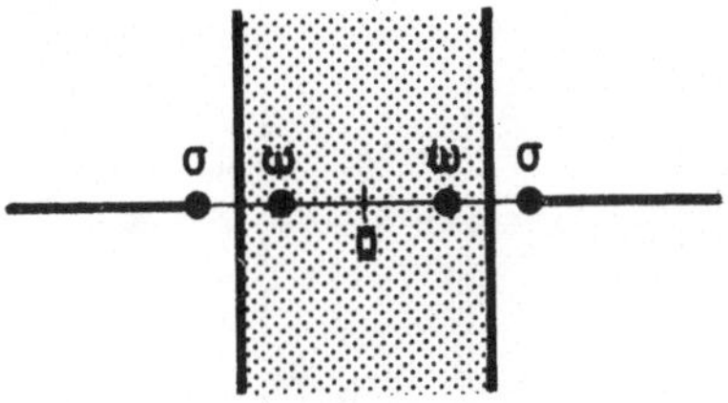

Fig. 3

$$\omega = \{k^{(0)} = \pm\sqrt{\vec{k}^2+\mu^2}\},\quad \sigma = \{k^{(0)} = \pm\sqrt{\vec{k}^2+M^2}\},$$

$M = 2\mu$ (or 3μ in even theories).

b) $H^{(4)}(k_1,k_2,k_3,k_4) = H^{(4)}(K,z,z') =$ 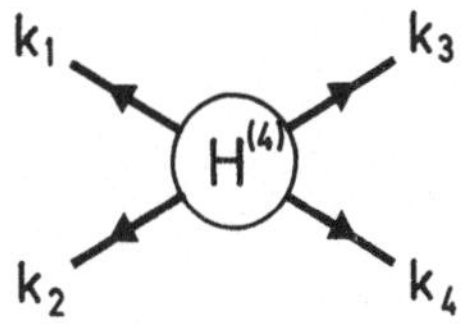

where we have put: $K = k_3+k_4 = -(k_1+k_2)$,

$$z = \frac{k_1-k_2}{2}\,,\qquad z' = \frac{k_3-k_4}{2}\,, \tag{21}$$

or equivalently: $\{k_i = k_i(K,z,z')$: $k_1 = -\frac{K}{2} + z$, $k_2 = -\frac{K}{2} - z$,

$$k_3 = \frac{K}{2} + z',\quad k_4 = \frac{K}{2} - z'\}\,. \tag{22}$$

Section of D_4 in (z,z') space at $K = (K^{(0)},\vec{K})$ fixed, $\vec{K}$ real, $K^{(0)}$ in the cut plane of Fig. 3; this section contains the following set (see Fig. 4): $\{z = (z^{(0)},\vec{z})$, $z' = (z'^{(0)},\vec{z}')$; $\vec{z},\vec{z}'$ real; $(z^{(0)},z'^{(0)}) \in C^2\}$ - {"pole-cut systems": $\underline{\sigma}_i,\omega_i$, $i = 1,2,3,4$, $\underline{\sigma}_{13},\omega_{13},\underline{\sigma}_{23},\omega_{23}\}$ with ω_i: $k_i^2(K,z,z') = k_i^{(0)\,2} - \vec{k}_i^2 = \mu^2$,

$$\underline{\sigma}_i = k_i^{(0)\,2} - \vec{k}_i^2 = M^2 + \rho\,,\quad \forall \rho \geq 0$$

(and similarly for ω_{ij}, $\underline{\sigma}_{ij}$ in terms of $k_{ij}^2 = (k_i+k_j)^2$).

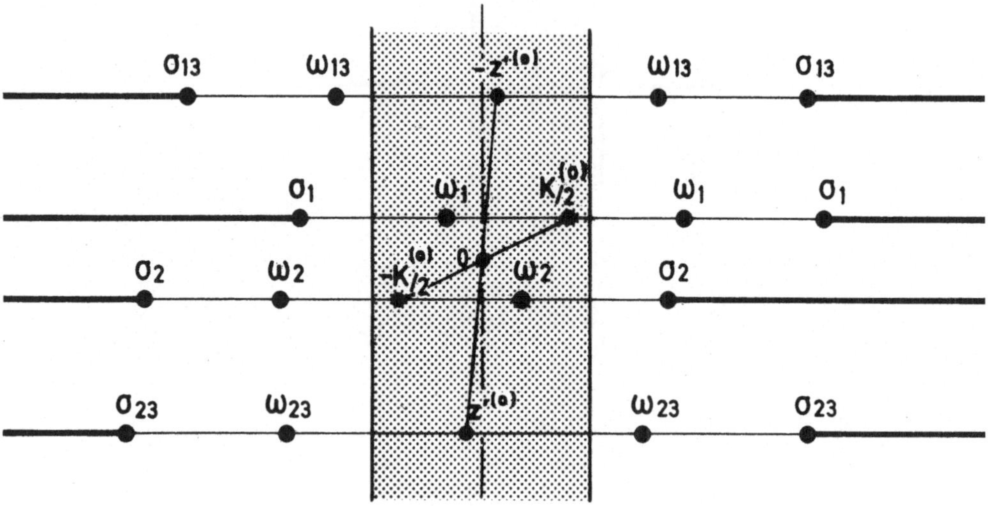

Fig. 4. Section in $z^{(0)}$-plane at K,z' and $\vec{z}$ fixed.

c) A similar description holds for the sections in $\{k_1^{(0)} \ldots k_n^{(0)}\}$-space of the primitive domain D_n of H_n; being given a channel (I,J) ($I \cup J = \{1,2,\ldots n\}$), the associated variables are:

$$K = \sum_{j \in J} k_j = - \sum_{i \in I} k_i , \qquad z = \{z_i, \ i \in I; \ \sum_{i \in I} z_i = 0\},$$

$$z' = \{z'_j, \ j \in J; \ \sum_{j \in J} z'_j = 0\}, \qquad \text{where: } z_i = -k_i - K/|I|$$

$$\text{and } z'_j = k_j - K/|J|$$ ($|X|$ = number of elements in the finite set X).

If all the components of K,z,z' are kept fixed with $(\vec{K},\vec{z},\vec{z}')$ real, except one couple $(z_{i_1}^{(0)}, z_{i_2}^{(0)})$ $(z_{i_2}^{(0)} + z_{i_1}^{(0)} = - \sum_{i \neq i_1, i_2} z_i^{(0)})$, the section of D_n in the complex $z_{i_1}^{(0)}$-plane is the whole plane minus a set of "pole-cut systems" σ_L, ω_L which vary with all the remaining components of (K,z,z') (as in the previous example b)).

Here we shall not describe the full domains D_n in $(k_1 \dots k_n)$-space, nor the corresponding enlarged domains D_n' (obtained by holo-convex completion of the latter) either. The general feature of these domains is that their sections at $(k_1^o,\dots,k_n^o)$ fixed in $(\vec{k}_1,\dots,\vec{k}_n)$-space are, in contrast with those in $(k_1^{(0)},\dots,k_n^{(0)})$-space described above, bounded domains whose size can be very small or even empty for certain sections; the section of D_n' by the mass shell manifold is also of similar type+. Note that the conditions $k_i^2 = \mu^2$, i.e. $\mathrm{Re}\, k_i^2 - \mathrm{Im}\, k_i^2 = \mu^2$, $\mathrm{Re}\, k_i \cdot \mathrm{Im}\, k_i = 0$, imply in fact that $\mathrm{Im}\, k_i$ is always a space-like vector.)

The nature of the enlargement obtained (from D_n to D_n') has precisely to do with analyticity in the variables $(\vec{k}_1,\dots,\vec{k}_n)$; the latter can be characterized (in a very rough way) by the size of the interval

$$0 < \sum_{1 \le i \le n} |\mathrm{Im}\, \vec{k}_i|^2 < \varepsilon(\{k_i^{(0)}\},\{\mathrm{Re}\, \vec{k}_i\})$$

which can be reached.

In the first steps of M.P.S.A., one mainly uses subdomains of D_n' which are tube neighborhoods $B_n(\Omega)$ of the euclidean subspace $E_n = \{(k_1,\dots,k_n);\ k_1+\dots+k_n = 0,\ k_i = (k_i^{(0)},\vec{k}_i),\ \vec{k}_i$ real, $k_i^{(0)}$ purely imaginary$\}$, i.e. domains of the following form:

+However, in certain complex directions of the mass shell manifold, the section of D_n' may have infinite parts; this is the case for $n = 4$, where crossing-type domains can be derived in well-suited complex submanifolds of the mass shell [5b] (in the most favourable configurations, one even obtains cut-planes in which dispersion relations can be written).

$$B_n(\Omega) = \{(k_1 \ldots k_n);\ k_1 + \ldots + k_n = 0;\ (\mathrm{Re}\ \vec{k}_i,\ \mathrm{Im}\ k_i^{(0)};$$

$$1 \leq i \leq n) \in E_n\ ,$$

$$\sum_{i=1}^{n} |\mathrm{Im}\ \vec{k}_i|^2 < \varepsilon(\{\mathrm{Re}\ k_i^{(0)}\});\ (\mathrm{Re}\ k_i^{(0)};\ 1 \leq i \leq n) \in \Omega\},\quad (23)$$

Ω being a suitable region limited by spectral conditions and ε a number whose ideal value is unknown, but which may be estimated in practice and then depends on the completion procedure (used for computing D_n' from D_n). The sections of $B_n(\Omega)$ in arbitrary $k_i^{(0)}$ (or $z_i^{(0)}$)-planes are strips parallel to the imaginary axis (figured out by shaded strips in the examples of Fig. 3 and 4).

<u>3.II The absorptive parts</u>

Let $H^{(n)}(k_1,\ldots,k_n) = H^{(n)}(K,z,z') = I$ 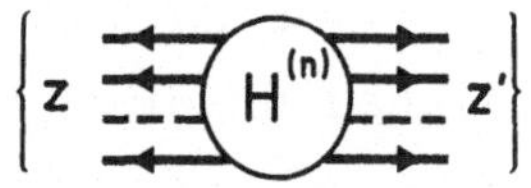

J

with (K,z,z') defined in 3.Ic). The absorptive part of $H^{(n)}$ with respect to the channel (I,J) is the discontinuity function

$$\Delta_I H^{(n)}(K;z,z') = \lim_{\substack{\varepsilon \to 0 \\ \varepsilon > 0}} [H^{(n)}(K^{(0)} + i\varepsilon,\ \vec{K};z,z') -$$

$$- H^{(n)}(K^{(0)} - i\varepsilon,\ \vec{K};z,z')]\ .\quad (24)$$

The support of $\Delta_I\ H^{(n)}$ is the union of the polar manifold ω_I^+:

$$K^{(0)} = \sqrt{\vec{K}^2 + \mu^2},\ \text{and of the cut}\ \underline{\sigma}_I^+:\ K^{(0)} \geq \sqrt{\vec{K}^2 + M^2}$$

$(M = 2\mu$ or $3\mu)$, in the subspace where K is real.

For K fixed in the cut $\underline{\sigma}_I^+$, $\Delta_I\ H^{(n)}$ is analytic with respect to (z,z') in a product domain $D_I(K) \times D_J(K)$.

We put

$$D_n^{(I)} = \{(K,z,z');\ K \text{ real} \in \underline{\sigma}_I^+,\ (z,z') \in D_I(K) \times D_J(K)\} \quad . \tag{25}$$

3.III Asymptotic completeness equations (or off-shell unitarity-type equation)

The following expressions for the absorptive parts $\Delta_I H^{(n)}$ are consequences of the asymptotic completeness property for the field considered [3b]; they hold as identities of analytic functions in $D_n^{(I)}$:

$$\Delta_I H^{(n_1+n_2)}(K,z,z') = \sum_{\ell \geq 1} \left(\frac{1}{z}\right)^{\ell} \frac{1}{\ell!} H_{\varepsilon}^{(n_1+\ell)} *_{\ell} H_{-\varepsilon}^{(n_2+\ell)}(K,z,z') \tag{26}$$

where $n_1 = |I|$, $n_2 = |J|$, and the operation $*_\ell$ denotes the integration on the ℓ-particle mass shell (after an "amputation" by $(k_1^2-\mu^2)\ldots(k_\ell^2-\mu^2)$ on each of the two factors $H^{(n_1+\ell)}$, $H^{(n_2+\ell)}$ has been done):

$$H_{\varepsilon}^{(n_1+\ell)} *_{\ell} H_{-\varepsilon}^{(n_2+\ell)}(K,z,z') = (2i\pi)^{\ell} \int \Big[\prod_{1\leq i\leq \ell} (k_i^2-\mu^2) H_{\varepsilon}^{(n_1+\ell)}(K;z,k_1,\ldots,k_\ell)\Big]\Big[\prod_{1\leq i\leq \ell}(k_i^2-\mu^2) H_{-\varepsilon}^{(n_2+\ell)}(K;k_1,\ldots,k_\ell,z')\Big]$$
$$\times\ \delta_4(K-k_1\ldots k_\ell) \prod_{i=1}^{\ell} \theta(k_i^{(0)})\,\delta(k_i^2-\mu^2)\, d_4 k_i \, . \tag{27}$$

In (26) and (27), the subscript $\varepsilon = +$ or $-$ denotes the half-space ($\mathrm{Im}\, K^{(0)} > 0$ or < 0) from which the corresponding boundary values of $H^{(n_1+\ell)}$ or $H^{(n_2+\ell)}$ must be taken. Both formulae are derived from identities of the following type for the boundary values of $H^{(n)}$ ($\langle R_n^{(S)} \rangle =$ $= H_S^{(n)} \times \delta(k_1+\ldots+k_n)$):

$$<R_n^{(S_+)}> - <R_n^{(S_-)}> = <R_{n_1}^{(S_1)} \cdot R_{n_2}^{(S_2)}>$$

$$= \sum_{\ell \geq 1} <R_{n_1}^{(S_1)} E_{in}^{(\ell)} R_{n_2}^{(S_2)}> = \sum_{\ell \geq 1} <R_{n_1}^{(S_1)} E_{out}^{(\ell)} R_{n_2}^{(S_2)}>, \qquad (28)$$

where S_+, S_-, S_1, S_2 label appropriate boundary values (see 3.Ii)), and $E_{in}^{(\ell)}$ (resp. $E_{out}^{(\ell)}$) denotes the projector onto the space of ℓ-particle incoming (resp. outgoing) asymptotic states of the field. The r.h.s. of (28) can be transformed into the r.h.s. of (26) by using appropriate reduction formulae and an argument of analytic continuation.

It is a direct consequence of Eq.(27) that in the r.h.s. of (26) the term of order ℓ vanishes for $K^2 < (\mu\ell)^2$; thus in any bounded region, the sum occurring in (26) reduces to a finite number of terms. (As for the term $\ell = 1$, its support is contained in the manifold $K^2 = \mu^2$; the factor $\delta(K^2-\mu^2)$ that it contains expresses the fact that $H^{(n)}$ admits a simple pole on this manifold.)

<u>Connection with unitarity equations:</u>

The simplest case to be considered is: n = 4, J = {1,2}, I = {3,4} ($n_1 = n_2 = 2$), the total momentum K varying in the region $4\mu^2 \leq K^2 < 9\mu^2$. In this case, the corresponding equation (26) reduces (for Z = 1) to:

$$\Delta H^{(4)} = H_+^{(4)} - H_-^{(4)} = \frac{1}{2} H_+^{(4)} *_2 H_-^{(4)} \quad . \qquad (29)$$

But in view of the reduction formulae, one has:

$$\prod_{i=1}^{4} (k_i^2-\mu^2) H_+^{(4)}(k_1 \ldots k_4) \Big|_{M_{IJ}} = T_{IJ}^{(4)}(-k_3,-k_4;k_1k_2)$$

and similarly

$$\prod_{i=1}^{4} (k_i^2-\mu^2) H_-^{(4)} (k_1,\ldots,k_4) \Big|_{M_{IJ}} = T_{IJ}^{*(4)} (-k_3,-k_4;k_1k_2),$$

where the function $T_{IJ}^{*(4)}$ is associated with the adjoint scattering operator S^+ (as $T_{IJ}^{(4)}$ is associated with S). Thus, by multiplying both sides of (29) by $\prod_{i=1}^{4} (k_i^2-\mu^2)$, and taking the restriction to the mass shell, one obtains:

$$T_{IJ}^{(4)} - T_{IJ}^{*} = \frac{1}{2} T_{IJ}^{(4)} *_2 T_{IJ}^{*(4)} , \tag{30}$$

which is the unitarity equation for the two-particle scattering function $T_{IJ}^{(4)}$ in the "elastic" region $K^2<9\mu^2$. This shows that (up to the amputation factor $\prod_{i=1}^{4} (k_i^2-\mu^2)$) eq. (29) can be considered as an off-shell analytic extrapolation of the unitarity equation (30).

In the more general cases, the relationship between unitarity equations for the $T_{IJ}^{(n)}$ and relations (29) is not so simple, because several discontinuity formulae of the type (29) must be added (after amputation by $\prod_{i=1}^{n} (k^2-\mu^2)$) and restriction to M_{IJ}) in order to yield the unitarity equation for the corresponding quantity $T_{IJ}^{(n)} - T_{IJ}^{*(n)}$. However since the r.h.s. of (26) involves the operations $*_\ell$ (mass shell integration) as it is the case in the unitarity equations, we shall say that the equations (26), which express the A.C. postulate, are "off-shell unitarity-type equations". (Note that another version of off-shell unitarity type equations based on A.C. has been first written in [21] .)

3.IV Notion of "general n-point function"

We call "general n-point function" any analytic function $F(k_1,\ldots,k_n)$ $(k_1+\ldots+k_n = 0)$ such that:

i) F is analytic in a domain of the type D_n, with a certain specification of the spectral conditions $k_I^2 < M_I^2$.

ii) The various absorptive parts $\Delta_I F$ are analytic in the corresponding domains $D_n^{(I)}$.

The <u>"physical n-point function" (or Green function)</u>: $H^{(n)}(k_1,\ldots,k_n)$ is the one whose appropriate boundary values on real space are the Fourier transforms of V.E.V. of generalized retarded products of the field Φ (see 3.I). A general n-point function is said to be <u>amputated</u> if all the poles $k_i^2 = \mu^2$ are absent. The amputated Green function $\hat{H}^{(n)}$ associated with $H^{(n)}$ is

$$\hat{H}^{(n)}(k_1,\ldots,k_n) = \prod_{i=1}^{n} H^{(2)}(k_i)^{-1} H^{(n)}(k_1,\ldots,k_n).$$

<u>ℓ-particle irreducible n-point functions:</u> A general n-point function F is said to be "ℓ-particle irreducible with respect to the channel (I,J)", if $\Delta_I F$ vanishes for+ $K^2 < (\ell+1)\mu^2$ (i.e. $M_I = (\ell+1)\mu$) $(K = -\sum_{i\in I} k_i = \sum_{j\in J} k_j)$. A basic part of M.P.S.A. consists in constructing "physical" ℓ-particle irreducible n-point functions, linked with the Green functions $H^{(n)}$ of the field Φ via Bethe-Salpeter type equations (see Sect. 5 and 6).

4. INTEGRATION IN COMPLEX SPACE AND G-CONVOLUTION FUNCTIONS

We call "G-convolution functions" analytic functions which generalize the Feynman functions associated with (connected) graphs G: with each vertex is associated a general n-point function instead of a constant (resp. with each internal line, a general two-point function instead of a free propagator).

+ in a theory with a single mass μ in the discrete spectrum associated with Φ.

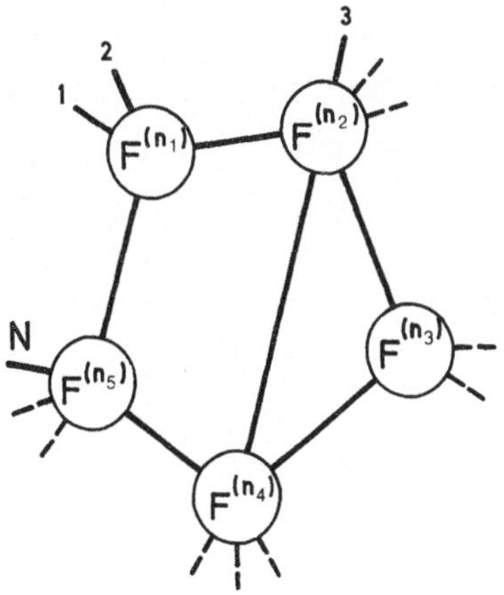

Fig. 5

4.I Definition and primitive domains

Being given: a connected graph G with external lines $\{1,2,\ldots,N\}$ (Fig. 5), a set of amputated n_v-point functions $F^{(n_v)}$ associated with the vertices v of G, a set of two-point functions $F_\alpha^{(2)}$ associated with the internal lines α of G, we define the G-convolution function $F_G^{(N)}$, first in euclidean space $((k_1,\ldots,k_N) \in E_N)$, by the following integral in which the usual Feynman rules (i.e. conservation of energy-momentum at each vertex) have been taken into account:

$$F_G^{(N)}(k_1,\ldots,k_N) =$$

$$= \int_{E_L} [\prod_v F^{(n_v)} \times \prod_\alpha F_\alpha^{(2)}](k_1,\ldots,k_N;z_1,\ldots,z_L)dz_1,\ldots dz_L; \tag{31}$$

in the r.h.s. of (31), $(z_1,\ldots,z_L)$ denotes a set of L loop energy-momentum variables, integrated on the corresponding euclidean space E_L.

The following theorem [3a] ensures the conservation of the primitive analyticity of n-point functions by G-convolution.

Theorem: The integral $F_G^{(N)}(k_1,\ldots,k_N)$ defines by analytic continuation a general N-point function satisfying conditions i) and ii) described in Sect. 3.IV. (A more refined specification of the spectral conditions could be derived; see below: property of additivity of thresholds.)

The proof is based on the following mathematical result; let

$$I(k) = \int_{\Gamma(k)} F(k,z)\, dz_1,\ldots,dz_n \tag{32}$$

where F is analytic in a domain $D = \{(k,z); k \in \Delta, z \in D(k)\}$ and $\Gamma(k)$ is a cycle contained in $D(k)$ which, for $k \in \Delta$, varies continuously with k. Then $I(k)$ is analytic in Δ.

For the integral (31), one can define in each loop four-momentum variable z_i a contour $\Gamma(k_1,\ldots,k_N,z_1,\ldots,z_{i-1})$: in a more tractable way, a recursive argument over the number of loops of G can be done: in each loop four-momentum, Γ can be taken of the form $L \times R^3$, where L is a line in the energy complex plane which goes from $-i\infty$ to $+i\infty$ inside the cut-plane type section of the domain of the integrand (see Sect. 3.I and Fig. 4).

Property of additivity of thresholds:

In any given channel (I,J), the mass threshold $M_I = M_J$ which occurs in the precise definition of the primitive domain of $F_G^{(N)}$ depends on the topology of G: in a very rough way, it should be obtained in general by taking all possible ways of "cutting" G into two connected parts which carry respectively the sets of external lines I and J, determining for each cutting the sum of all mass thresholds (including poles) "encountered" on the cut lines and bubbles, and finally taking the inf. over the "cuttings" of G defined above of the corresponding sums of mass thresholds.

Remark on tree-convolution products:

When G is a tree, the integral (31) is replaced by an ordinary product (with the usual Feynman conservation law at each vertex):

Ex:

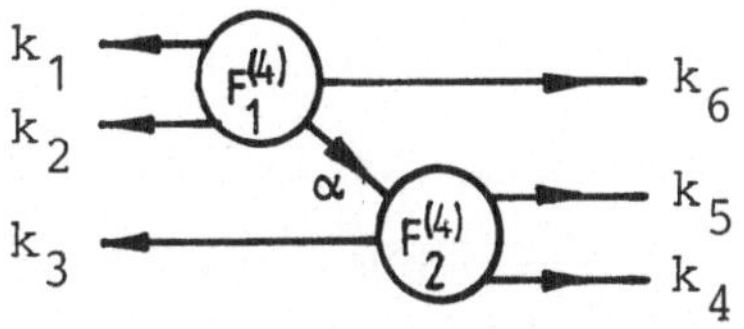

$$k_\alpha = k_3+k_4+k_5 = -(k_1+k_2+k_6) \quad ,$$

$$F_G^{(6)} = F_1^{(4)}(k_1,k_2,k_6,k_\alpha) \times F_2^{(4)}(k_3,k_4,k_5,-k_\alpha) F_\alpha^{(2)}(k_\alpha) \,. \quad (33)$$

Remark on the question of integration at infinity in (3.1)

The functions $F^{(n_v)}$ and $F_\alpha^{(2)}$ are assumed to satisfy appropriate decrease properties at infinity in their respective euclidean spaces, so as to ensure the convergence of (31). (A theory of renormalized G-convolution integrals, similar to that of Zimmermann for ordinary Feynman functions, has also been done [22]; however we do not need it for investigating the analytic structure of Green functions along the line of M.P.S.A..)

Example:

G = 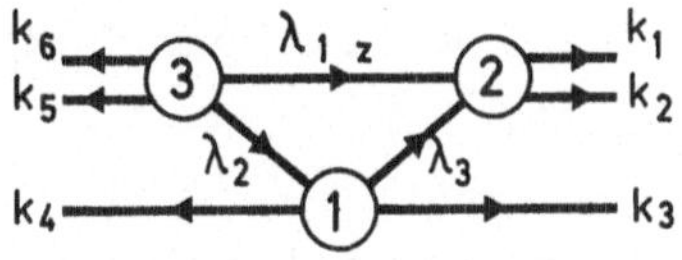

With the bubbles, we associate the function $\hat{H}^{(4)}$ of an <u>even</u> field with single mass μ; with each internal line λ_i, we associate the two-point function $H_{\lambda_i}^{(2)}$ of this field; formula (31) yields in this case:

$$\hat{H}^{(6)}$$ 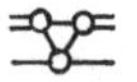 $$(k_1,\dots,k_6) = \int_{\Gamma(k_1,\dots,k_6)} \Big[\prod_{i=1,2,3} \hat{H}_i^{(4)} \cdot$$

$$\cdot \prod_{i=1,2,3} H_{\lambda_i}^{(2)} \Big] (k_1,\dots,k_6;z)\, d_4 z \,, \qquad (34)$$

where z denotes the four-momentum carried by the internal line λ_1, and $\Gamma(k_1,\dots,k_6)$ is a suitable cycle obtained by distortion of $\{z \in (iR) \times R^3\}$. Note that formula (34) is equivalent to the definition of $H^{(6)}$ 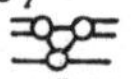 given in Sect. 1 (formula(2)), as it immediately results from the definition of the amputated functions $\hat{H}^{(n)}$ (see Sect. 3.IV). $\hat{H}^{(6)}$ 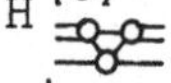 is analytic in a domain $D_6($ 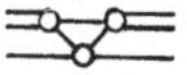 $)$ with the following spectral conditions: $M_{\{12\}} = M_{\{34\}} = M_{\{56\}} = 2\mu$, all other $M_{\{ij\}} = 4\mu$; all $M_{\{ij\ell\}} = 3\mu$.

$\hat{H}^{(6)}$ 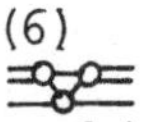 is one of the contributions to the structural equation which is satisfied by $\hat{H}^{(6)}$ near the $3 \to 3$ physical region (see formula (11) in Sect. 1).

4.II Production of Landau singularities for G-convolution functions

As it is the case for Feynman functions, the Landau singularities of the integrals I(k) of the form (32) are produced when the cycle $\Gamma(k)$ is "pinched" by the singular set S of the integrand F(k,z).

The following mathematical pattern applies to a number of cases: $S = S_1 \cup \dots \cup S_M$, where each set S_i is defined by an analytic equation $s_i(k,z) = 0$. Let $A \subset \{1,2,\dots,M\}$, and consider the "stratum" S_A of S, defined as follows:

$$S_A = \bigcap_{i \in A} S_i - \Big(\bigcup_{j \notin A} S_j \Big) \quad (i,j \in \{1,2,\dots M\}).$$

S_A is said to be critical at a point (k_o, z_o) if the projection $(k,z) \overset{\pi_A}{\rightarrow} k$ from S_A to k-space is not of maximal rank at (k_o, z_o). The set L_A of such points (k_o, z_o) is <u>assumed</u> to project isomorphically onto a set L_A in k-space (i.e., there is a single $(k_o, z_o) \in \mathsf{L}_A$ above a given $k_o \in L_A$). L_A (resp. L_A) is called the "Landau set" (resp. "L.s. in projection") associated with S_A.

When k lies near L_A, but not on L_A, $\pi_A^{-1}(k)$ contains a cycle $e_A(k)$ which is holomorphically equivalent to a sphere $S^{(n-m)}$, n being the complex dimension of z-space and $m = |A|$. In appropriate local coordinates, this sphere can be described by the equations:

$$\ell_A(k) - (x_m^2 + \ldots + x_n^2) = 0,$$

$$x_1 = \ldots = x_{m-1} = 0 \quad , \qquad (35)$$

where $\ell_A(k)$ is an analytic function such that on L_A: $\ell_A(k) = 0$, $d\ell_A(k) \neq 0$. When $k \to k_o \in L_A$, this sphere shrinks to zero: $e_A(k)$ is called the "vanishing sphere" of the stratum S_A.

When k turns around L_A (i.e. $t = \ell_A(k)$ turns around the origin in its complex plane, starting from $\underline{t} = \ell_A(\underline{k}) > 0$), the contour $\Gamma(k)$ varies continuously from an initial position $\Gamma^+(\underline{k})$ to a final position $\Gamma^-(\underline{k})$, while staying inside $D(k)$, for each k. This variation of $\Gamma(k)$ is given by the <u>Picard-Lefschetz formula</u> (see [8]):

$$\Gamma^-(\underline{k}) = \Gamma^+(\underline{k}) + Z\, \tilde{e}_A(\underline{k}) \qquad (36)$$

where $\tilde{e}_A(k)$ is a cycle in $D(k)-S$, which lies in a neighbourhood of $e_A(k)$ and is determined canonically from $e_A(k)$ by the "Leray's cobord" operation [15]: the latter can be illustrated as follows in the simplest case $n = m = 1$. Let $e(t)$: $t - x^2 = 0$, then $\tilde{e}(t) = \gamma_+(\sqrt{t}) + \gamma_-(-\sqrt{t})$: see Fig. 6 in z-plane for $t = \underline{t} > 0$. In the general case, $\tilde{e}(k)$ is defined by associating with each point z_o in $e(k)$ a product of n-m circles (in the ambient z-space) which is centered at z_o and depends continuously on z_o.

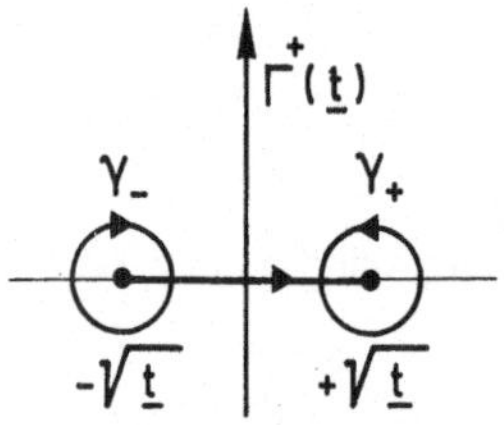

the z-plane for $t = \underline{t} > 0$

Fig. 6

In (36), Z denotes an integer which represents the (algebraic) number of intersection points of $\Gamma(k)$ with the "vanishing cell" $\hat{e}(k)$ supported by $e(k)$ and bounded by the m surfaces S_{i_j} $(A = \{i_1, i_2, \ldots, i_m\})$ (In the example of Fig. 6, $\overset{+}{\Gamma}(\underline{t})$ is the imaginary axis, $e(\underline{t}) = [-\sqrt{\underline{t}}, +\sqrt{\underline{t}}]$, Z = 1; formula (36) can be checked by hand.)

<u>Case</u> when the S_i $(i \in A)$ are simple poles for $F(k,z)$, i.e.:

$$F(k,z) = \frac{\hat{F}(k,z)}{\prod_{i \in A} s_i(k,z)}, \text{ with } \hat{F} \text{ holomorphic on } \bigcup_{i \in A} S_i:$$

Formula (36) yields the following discontinuity formula for $I(k)$ around its <u>Landau singularity</u> L_A:

$$\Delta_{L_A} I(\underline{k}) = I^+(\underline{k}) - I^-(\underline{k}) = \int_{\tilde{e}_A(k)} \frac{\hat{F}(k,z)}{\prod_{i \in A} s_i(k,z)} dz_1 \wedge \ldots \wedge dz_n \tag{37}$$

which also takes the alternative form:

$$\Delta_{L_A} I(\underline{k}) = (2i\pi)^m \int_{e_A(k)} \hat{F}(k,z) \frac{dz_1 \wedge \ldots \wedge dz_n}{ds_{i_1} \wedge \ldots \wedge ds_{i_m}} \, . \tag{38}$$

In (38) the notation

$$\frac{dz_1 \wedge \ldots \wedge dz_n}{ds_{i_1} \wedge \ldots \wedge ds_{i_m}}$$

is a substitute to $\delta(s_{i_1}(k,z)) \ldots \delta(s_{i_m}(k,z))\ dz_1 \ldots dz_n$ which represents a meaningful differential form, defined and analytic for k complex. The derivation of (38) from (37) is obtained by applying a general form of the residue theorem (Leray's residue formula [15]) which amounts in practice to replace the pole factor$(2i\pi)^{-|A|}\ [\prod_{i \in A} s_i(k,z)]^{-1}$ by the corresponding product of δ-functions $\prod_{i \in A} \delta(s_i(k,z))$, and the integration on the vanishing cycle $\tilde{e}_A(k)$ by the integration on the vanishing sphere $e_A(k)$.

<u>Nature of the singularity:</u> near L_A, I(k) has two types of behaviour, according to the parity of dimension:

a) if n-m is even: $I(k) = a(k)\ell_A(k)^{\frac{(n-m-1)}{2}}[1+O(k)]+b(k)$, (39)

b) if n-m is odd: $I(k) = a(k)\ell_A(k)^{\frac{(n-m-1)}{2}} \cdot$

$\cdot \text{Log}\ \ell(k)[1+O(k)] + b(k)$, (40)

where a(k), b(k) and O(k) are holomorphic, and O(k) vanishes on L_A.

This can be proved by using the parametrization (35) of $e_A(k)$ and formula (38) for estimating $\Delta_{L_A} I(\underline{k})$ (the factor $\ell(k)^{(n-m-1)/2}$ comes from the radial jacobian factor in (n-m+1)-dimensional space): the forms (39) and (40) for I(k) then follow from a standard argument on analytic functions.

4.III An example: the Landau singularities of $\hat{H}^6$

The domain D of the integrand of (34) is of the form: $B_6(\Omega) - \Sigma$, where the notation $B_6(\Omega)$ has been defined in (23) and $\Omega = \{(k_1^{(0)},\dots,k_6^{(0)};\ z^{(0)})$ real; $k_1^{(0)}+\dots+k_6^{(0)}=0$, $|k_i^{(0)}+k_j^{(0)}|<4\mu,\ \forall\, i,j;\ \ |k_i^{(0)}|<3\mu,\ \forall\, i;\ |k_1^{(0)}+k_2^{(0)}+k_3^{(0)}|<5\mu;$ $|k_i^{(0)}+k_j^{(0)}+k_\ell^{(0)}|<3\mu,\ \forall\,(ij\ell)\neq(123)$ or (456); $|z^{(0)}|<\zeta(k_1^{(0)},\dots,k_6^{(0)})\}$ ($\zeta(k_1^{(0)},\dots,k_6^{(0)})$ being determined by the various spectral conditions for the three factors $\hat{H}^{(4)}$).

$$\Sigma = \underline{\sigma} \cup \underline{\sigma}_{12} \cup \underline{\sigma}_{56} \cup S \cup S' \ ;$$

$\underline{\sigma}$, $\underline{\sigma}_{12}$, $\underline{\sigma}_{56}$ are respectively the following cuts (in the external variables):

$$K^2 \equiv (k_1+k_2+k_3)^2 = (3\mu)^2+\rho\ (\rho\geq 0)\ ;\ (k_1+k_2)^2 = (2\mu)^2+\rho\ (\rho\geq 0)\ ;$$

$$(k_5+k_6)^2 = (2\mu)^2 + \rho\ (\rho\geq 0)\ ;$$

S is the singular set which gives rise to the various critical strata S_A and Landau sets L_A:

$$S = S_1 \cup S_2 \cup S_3 \cup S_4 \ ,$$

where for $i = 1,2,3$, S_i: $s_i(k_1,\dots,k_6;z) = k_{\lambda_i}^2 - \mu^2 = 0$,

$$S_4:\ s_4(k_1,\dots,k_6;z) = (k_1+k_2+k_3-z)^2-(2\mu)^2 \geq 0;$$

S' is given by other cuts (similar to S_4) which are not critical in D.

The integration contour $\Gamma(k_1,k_2,\dots,k_6) = L(k_1,\dots,k_6)\times R^3$ of (34) is obtained by continuous distortion of $(iR \times R^3)$ inside D. (Near the surfaces L_A, it must be more thoroughly distorted and does not remain of the form $L \times R^3$.)

As a matter of fact, by taking into account the local

analyticity of each factor $\hat{H}^{(4)}$ in the corresponding region $4\mu^2 < k_{ij}^2 = (k_i+k_j)^2 < 16\mu^2$ (see Sect. 6), we can say that the integrand of (34) is analytic even on the cuts $\underline{\sigma}_{12}, \underline{\sigma}_{56}, \underline{S}_4$ which bound the domain $B_6(\Omega)-\Sigma$, except at the points which belong to the set $S_1 \cup S_2 \cup S_3$, S_1, S_2 and S_3 being (simple) polar manifolds for the integrand. Let us then consider the stratum $S_{123} = (S_1 \cap S_2 \cap S_3) - S_4$; S_{123} is critical in the considered domain and admits a Landau set L_{123} defined by the following equations:

$$k_{\lambda_i}(k_1,\dots,k_6,z)^2 = \mu^2;\ i = 1,2,3;\ \alpha_1 k_{\lambda_1} = \alpha_2 k_{\lambda_2} + \alpha_3 k_{\lambda_3}, \tag{41}$$

for any set of complex parameters α_1, α_2, α_3, all $\neq 0$.

The corresponding projection L_{123} in $(k_1,\dots,k_6)$-space (obtained by eliminating z, α_1, α_2, α_3 in Eq. (41)) can be represented by the equation:

$$\ell_{123}(k_1,\dots,k_6) \equiv \mathrm{Det}\,[(k_{\lambda_i}\cdot k_{\lambda_j})] = 0 \tag{42}$$

in which the scalar products $k_{\lambda_i}\cdot k_{\lambda_j}$ are reexpressed in terms of $k_1,\dots,k_6$ through the following formulae:

$$k_{\lambda_i}^2 = \mu^2, \qquad i = 1,2,3,$$

$$k_{\lambda_1}\cdot k_{\lambda_2} = \frac{(k_5+k_6)^2}{2} - \mu^2,$$

$$k_{\lambda_2}\cdot k_{\lambda_3} = -\frac{(k_3+k_4)^2}{2} + \mu^2,$$

$$k_{\lambda_3}\cdot k_{\lambda_1} = \frac{(k_1+k_2)^2}{2} - \mu^2. \tag{43}$$

<u>Remark:</u> In real k-space, only a part of $L_{123}^{(\mathrm{real})}$ is actually

carrying a singularity of $\hat{H}^{(6)}$ (in the physical sheet): this part, called "the positive α branch" is defined by adding the conditions $\alpha_1 > 0$, $\alpha_2 > 0$, $\alpha_3 > 0$ to the Landau equations (41).

In fact, it can be seen by a simple geometrical argument that the contour $\Gamma(k_1,\ldots,k_6)$ of (34) is pinched in the vanishing cell $\hat{e}_{123}(K)$ (see Fig. 7), <u>iff</u> the parameters α_i satisfy these conditions.

On Fig. 8, the "positive α" and "non-positive α" branches of $L_{123}^{(real)}$ have been represented respectively by full and dotted lines.

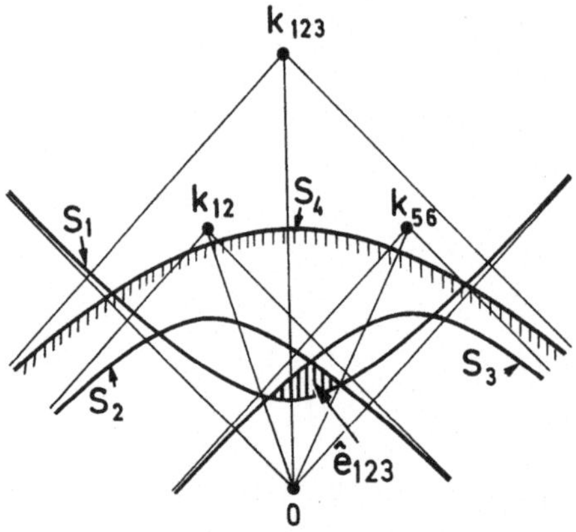

Fig. 7. The sets S_i $(1 \leq i \leq 4)$ and $\hat{e}_{123}$ in real z-space.

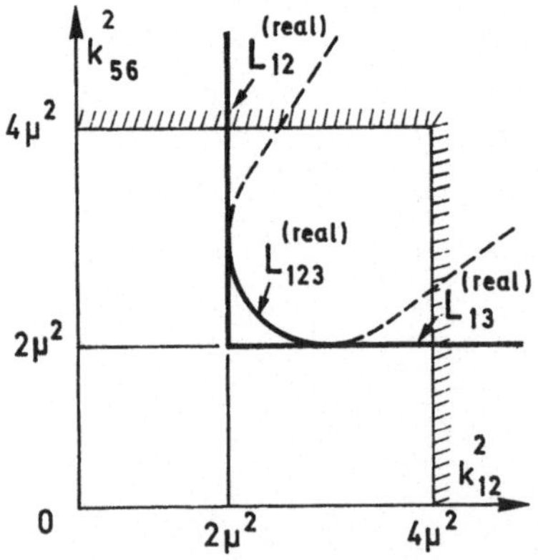

Fig. 8. Section of the Landau set at k_{34}^2 fixed < 0.

The discontinuity formula (38) becomes in the case of L_{123} the following "Cutkosky-type formula":

$$\Delta_{L_{123}} \hat{H}^{(6)}$$ 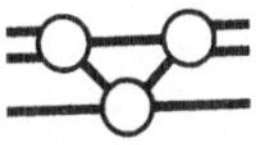 $$(k_1,\ldots,k_6) =$$

$$= \int_{e_{123}(k_1,\ldots,k_6)} [\hat{H}_1^{(4)} \hat{H}_2^{(4)} \hat{H}_3^{(4)}](k_1,\ldots,k_6;z)\cdot$$

$$\delta(s_1)\,\delta(s_2)\,\delta(s_3)\,d_4 z \qquad (44)$$

in which $e_{123}(k_1,\ldots,k_6)$ is the <u>mass shell</u> manifold ($k_{\lambda_i}^2 = \mu^2$; $i = 1,2,3$). <u>Nature of the singularity</u> of $\hat{H}^{(6)}$ 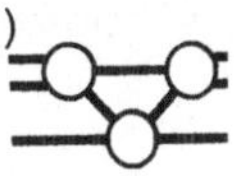 on L_{123}:

By applying formula (40) and putting $k = (k_1,\ldots,k_6)$, one obtains:

$$\Delta_{L_{123}} \hat{H}^{(6)}$$ 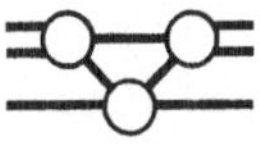 $$(k) = a(k)\ \mathrm{Log}\, l_{123}(k)\,(1+O(k))+b(k)\ . \qquad (45)$$

A similar study can be done for the strata S_{12}, S_{13} and S_{14} whose corresponding Landau sets are respectively $L_{12} = \sigma_{12}$ ($k_{12}^2 = 4\mu^2$), $L_{13} = \sigma_{56}$ ($k_{56}^2 = 4\mu^2$), $L_{14} = \sigma$ ($k_{123}^2 = 9\mu^2$); however for obtaining discontinuity formulae in this case, one must take into account the discontinuities of the relevant factors $\hat{H}^{(4)}$ (i.e. formula (38) is not directly applicable).

5. BETHE-SALPETER-TYPE EQUATIONS IN COMPLEX SPACE AND OFF-SHELL UNITARITY-TYPE EQUATIONS

We consider Fredholm equations of the form (3) of Sect. 1 (with variables t, now called K, interpreted as total energy-momentum variables, and $\omega = 1$: see below, footnote on the next page),

$$F(K,z,z';\lambda) = G(K;z,z') + \lambda \int_{\Gamma(K)} F(K;z,z'',\lambda)G(K;z'',z')dz'' . \tag{46}$$

We call B.S.-type equation an equation of this form in which F denotes a 2m-point Green function

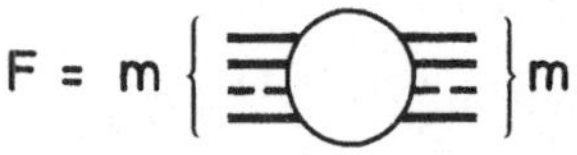

associated with a certain field, and G is supposed to have better analyticity properties than F, namely to be analytic at the threshold $K^2 = (m\mu)^2$ (this is the meaning of m-particle irreducibility). In this section, F and G are moreover considered (by convention) as amputated[+] from the poles $k_i(K,z)^2 = \mu^2$ $(1 \leq i \leq m)$ carried by the incoming (i.e. left-hand side) lines of G. The Fredholm parameter λ is given a fixed numerical value λ_o, determined by field theory (with a certain standard normalisation used in field theory, $\lambda_o = \frac{1}{m!}$). $K = (K^{(0)},0)$ denotes the total energy-momentum of the considered $m \to m$ channel: $K = k_1'+\dots k_m' = -(k_1+\dots+k_m)$; z(resp.z') denotes a set of (m-1) independent relative energy-momentum vectors for the m incoming (resp. outgoing) lines (see Sect. 3.Ic)); eq.(46) can then be graphically represented as follows:

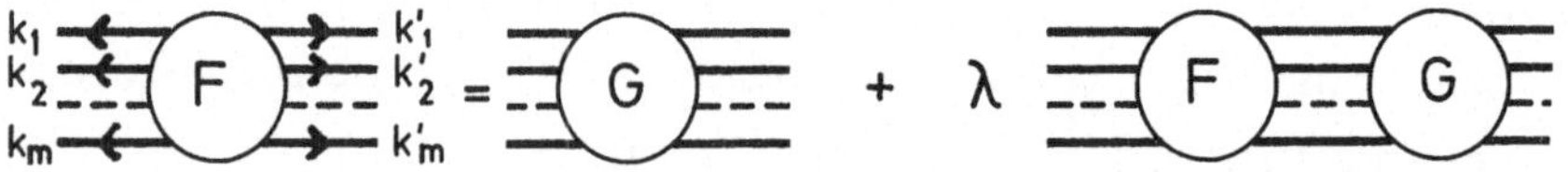

(47)

Equ.(46) (or (47)) is first considered for $K^{(0)}$ varying in a cut domain of the complex plane (see Fig.9); λ being fixed, it defines either F in terms of

---

[+] i.e. deduced from the original Green functions by multiplying the latter by a factor

$$\prod_{i=1}^{m} (k_i^2-\mu^2) \quad \text{or} \quad \prod_{i=1}^{m} [H^{(2)}(k_i)]^{-1} \quad ;$$

this convention accounts for the present choice $\omega(K,z'')=1$ (compare with the choice of ω in Sect.1, for instance in formula (4)).

G, or G in terms of F in a unique way (by Fredholm theory); for instance:

$$F(K;z,z';\lambda) = \frac{N(K;zz';\lambda)}{D(K;\lambda)} ,$$

where N and D are expressed in terms of G through Fredholm series.

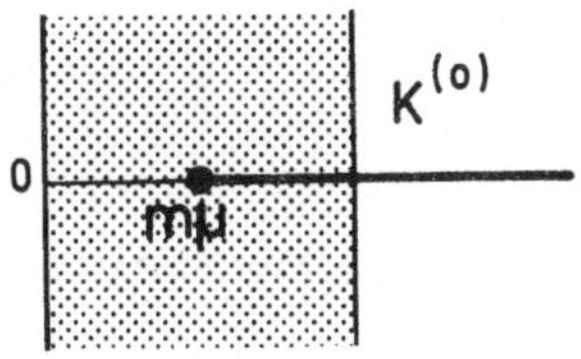

Fig. 9

Starting from its initial position (i.e. the euclidean space), the contour $\Gamma(K)$ varies continuously with K and has to stay inside the analyticity domain of F (and G) in z-space. When $K^{(O)}$ tends to the cut $K^{(O)} \geq m\mu$ from both sides Im $K^{(O)} > 0$ and Im $K^{(O)} < 0$, $\Gamma(K)$ should admit two limiting positions $\Gamma^+(K)$ and $\Gamma^-(K)$ (see however what happens in the case c) below).

If G(K,z,z') is "m-particle irreducible", i.e. analytic on the manifold $K^2 = (m\mu)^2$ (or $K^{(O)} = m\mu$), it is however singular on a certain set S (containing at least the "right-hand side" poles S_i: $k_i'^2 = s_i(K,z') = \mu^2, 1 \leq i \leq m$); the pinching of $\Gamma(K)$ by certain strata of this set S will produce a Landau singularity for F (as for I(k): cf. Sect. 4) on the manifold $K^2 = (m\mu)^2$ ("normal threshold singularity").

In this study of equation (46) (or (47)), two questions can be distinguished:

i) In which way is the irreducibility of G related with off-shell unitarity-type equations (of the kind deduced from asymptotic completeness: see Sect. 3.III)?

ii) If G is irreducible, what can be said about the type of singularity of F at the threshold $K^2 = (m\mu)^2$?

We shall consider three situations (by order of increasing complexity) concerning the structure of the singular set S of **G**:

a) $S = \bigcup_{1\leq i\leq m} S_i$ (only the poles carried by the outgoing lines);

b) $S = (\bigcup_{1\leq i\leq m} S_i) \cup S'$, where S' contains other types of singularities,

such as thresholds of the form $k'^2_{ij} \equiv (k'_i+k'_j)^2 = 4\mu^2$, all of them being singularities in the variables (K,z') (not involving the variables z).

c) Besides components of the type b) above, S also contains singularities in all variables (K,z,z') (or "non-separabl singularities").

a) The purely polar case: In the **exact** theory, this situation occurs only for the two-particle B.S. equation (m = 2); its analogue for general m corresponds to a simplified model of m-particle interaction in which all interactions involving less than m initial and final particles are neglected.

In this case, the two points (i) and (ii) mentioned above can be treated completely.

i) Equivalence between the irreducibility of G and an off-shell unitarity equation for F

Let $F^{\pm}(K,z,z')$, $G^{\pm}(K,z,z')$ be the boundary values of F and G on the cut $K^{(0)} \geq m\mu$ from the respective sides $\mathrm{Im}\, K^{(0)} > 0$, $\mathrm{Im}\, K^{(0)} < 0$, and let: $\Delta F = F^+ - F^-$, $\Delta G = G^+ - G^-$. In the present case when $S = \bigcup_{i=1}^{m} S_i$, the contour $\Gamma(K)$ is pinched in the limit $K^{(0)} \to m\mu$ by the vanishing sphere e(K) of the stratum $\underline{S} = \bigcap_{i=1}^{m} S_i$: for $K^{(0)}$ real $> m\mu$, e(K) is the mass shell manifold $s_i(K,z') \equiv k'^2_i - \mu^2 = 0$ $(1 \leq i \leq m;\ k'_i$ real).

The Picard-Lefschetz formula (36) (Sect. 4) yields here: $\Gamma^+(K) - \Gamma^-(K) = -\tilde{e}(K)$; by taking into account Leray's residue formula (see Sect. 4, formulae (37), (38)), this implies the following algebraic relation:

$$\Theta_+ - \Theta_- = * \tag{48}$$

between the integration operators:

$$\int_{\Gamma^+(K)} \ldots d_m z = \Theta_+ , \int_{\Gamma^-(K)} \ldots d_m z = \Theta_- , -\int_{\tilde{e}(K)} \ldots d_m z =$$

$$= -(2i\pi)^m \int_{e(K)} \ldots d_m z \delta(s_1) \ldots \delta(s_m) = * \quad . \tag{49}$$

We can rewrite equation (46), in the limits $\Gamma \to \Gamma^{\pm}(K)$ (K^o real $> m\mu$) as follows (by using the notations (49)):

$$F^+ = G^+ + \lambda_o F^+ \Theta_+ G^+ \ , \tag{50}$$

$$F^- = G^- + \lambda_o F^- \Theta_- G^- \ , \text{ or also} \tag{51}$$

$$G^+ = F^+ \Theta_+ (1^+ - \lambda_o G^+) \ , \tag{50'}$$

$$G^- = F^- \Theta_- (1^- - \lambda_o G^-) \ , \tag{51'}$$

1^+, 1^- denoting the unit operator respectively on $\Gamma^+(K)$, $\Gamma^-(K)$.

In view of formula (48), we then have:

$$F^+ \Theta_+ G^+ - F^- \Theta_- G^- = F^+ \Theta_+ \Delta G + F^+ \Theta_+ G^-$$

$$+ \Delta F \Theta_- G^- - F^+ \Theta_- G^-$$

$$= F^+ \Theta_+ \Delta G + \Delta F \Theta_- G^- + F^+ * G^- \ . \tag{52}$$

Thus (50) and (51) yield by subtraction (and by taking

(52) into account) the following relation between the discontinuities ΔF and ΔG:

$$\Delta F - \lambda_o \, \Delta F \, \Theta_- \, G^- = \Delta G + \lambda_o \, F^+ \, \Theta_+ \, \Delta G + \lambda_o \, F^+ * G^- \quad . \tag{53}$$

By putting the variables z in $\Gamma^+(K)$ and z' in $\Gamma^-(K)$, we can rewrite eq.(53) as follows:

$$\Delta F \, \Theta_- \, (1^- - \lambda_o G^-) = (1^+ + \lambda_o F^+)\Theta_+ \, \Delta G + \lambda_o F^+ * F^- \, \Theta_-(1^- - \lambda_o G^-)$$

(in the last term, (51') has been taken into account); we thus have:

$$(\Delta F - \lambda_o F^+ * F^-)\Theta_- \, (1^- - \lambda_o G^-) = (1^+ + \lambda_o F^+)\Theta_+ \, \Delta G \; . \tag{54}$$

Due to the invertibility of $(1^+ + \lambda_o F^+)$ and $(1^- - \lambda_o G^-)$, we have the equivalence:

$$\Delta G = 0 \qquad \leftrightarrow \qquad \Delta F - \lambda_o F^+ * F^- = 0 \; ; \tag{55}$$

i.e. irreducibility of G (up to a continuity condition on G which has to be added for $K^{(0)} \to m\mu$).

this is an "off-shell unitarity-type equation" (of the form (29), with $*_2$ replaced by $* = *_m$).

<u>(ii) Nature of the singularity of F at $K^o = m\mu$</u>

We now assume that G is irreducible, i.e. $\Delta G = 0$; the nature of the singularity of F then depends on the parity of the dimension d of the mass shell integration space (i.e. the vanishing sphere) e(K):

$$d = 4(m-1) - m = 3m - 4 \quad . \tag{56}$$

<u>If d is even</u>, i.e. m even: $e(K) \to -e(K)$ when $K^{(0)}$ turns around the threshold $m\mu$. Thus:

$$\Gamma^+(K) \xrightarrow[\text{first loop}]{} \Gamma^-(K) \xrightarrow[\text{second loop}]{} \Gamma^+(K) ,$$

and F is two-sheeted; the behaviour of F at the threshold is of the form: $a(K,z,z')[K^2-(m\mu)^2]^{(3m-5)/2} + b(K,z,z')$, with a and b holomorphic.

<u>If d is odd</u>, i.e. m odd, $e(K) \to e(K)$ when $K^{(0)}$ turns around $m\mu$. Thus

$$\Gamma^+(K) \equiv \Gamma^{(0)}(K) \xrightarrow[\text{first loop}]{} \Gamma^{(1)}(K) \to \ldots \xrightarrow{\tau^{\text{th}}=\text{loop}} \Gamma^{(\tau)}(K),$$

with $\Gamma^{(\tau)}(K) = \Gamma^{(0)}(K) + \tau e(K)$. F is infinite-sheeted and as a matter of fact "non-holonomic"+ (i.e. F cannot be expressed as a finite sum $\sum_{\alpha,r}(K^2-(m\mu)^2)^{\alpha} \times \log^r(K^2-(m\mu)^2)$: singularities of the type

$$\frac{1}{\log(K^2-(m\mu)^2)}$$

occur necessarily).

In both cases, a certain domain is generated in the K^o-plane for the unphysical sheet(s); this domain is smaller than the physical-sheet domain, since it is determined by imposing an additonal constraint, namely that e(K) be contained in the domain D(K) of z-space (for each K).

<u>b) Singularities of separable type; more general cases:</u>

A typical example is the B.S. equation which occurs for the $3 \to 3$ particle process, and relates the two-particle irreducible with the three-particle irreducible six-point function (we limit ourselves to the case of an even field theory and adopt the above "left-amputation" convention: see footnote following (46)).

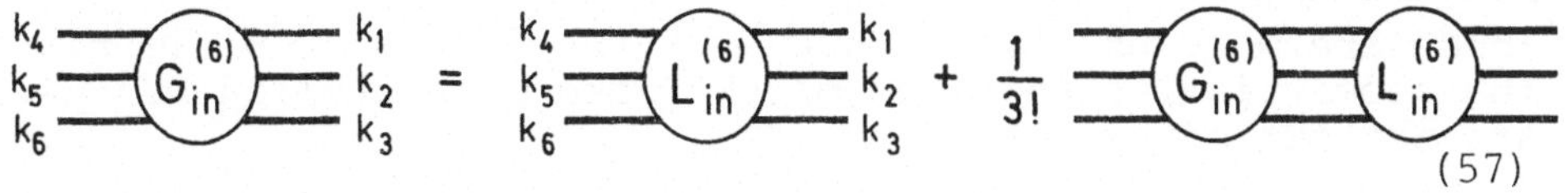

(57)

+In the terminology of M. Sato et al.. In this connection, see our recent work [16].

In the physical sheet domain only considered here (namely $(D_M^{cut} - S)$) the singular set S of $L_{in}^{(6)}$ is composed of the following manifolds:

The three poles $\omega_i : k_i^2 = \mu^2$, $i = 1,2,3$;

the three two-particle thresholds

$$\sigma_i : (k_j + k_\ell)^2 = (2\mu)^2 , \quad (i,j,\ell) = 1,2,3 .$$

The limiting positions $\Gamma^+(K)$, $\Gamma^-(K)$ of $\Gamma(K)$ can still be explicitly defined in this case by a simple geometric rule (A part of the cycle $\Gamma^+(K)$ is "flattened" on a real region of z-space, with a slight $i\varepsilon$ distortion obeying the following prescription: $\mathrm{Im}\, z = -\varepsilon\, \mathrm{Re} z$; indeed, since the point z = 0 lies in all regions $k_i^2 > \mu^2$ (i = 1,2,3), $(k_i + k_j)^2 > (2\mu)^2$ (i,j = 1,2,3), the above prescription ensures that $\Gamma^+(K)$ has no point in common with any manifold ω_i and σ_i, and that it stays in the domain of $L_{in}^{(6)}$.)

At $K^2 = (3\mu)^2$, the singular set S produces several pinches for $\Gamma(K)$: besides the vanishing sphere $e_{123}(K)$ in $\omega_1 \cap \omega_2 \cap \omega_3$ (i.e. the mass shell $k_1^2 = k_2^2 = k_3^2 = \mu^2$), three other relevant "vanishing spheres"+ have to be considered in the intersections $\omega_i \cap \sigma_i$ (i = 1,2,3). Since moreover the functions $G_{in}^{(6)}$ and $L_{in}^{(6)}$ admit σ_i as a branch manifold (with two-sheeted structure) instead of a pole, the P.L. formula (36) is not applicable to computing the discontinuity of $G_{in}^{(6)} \circ L_{in}^{(6)}$ across the cut $K^2 \geq (3\mu)^2$. As a matter of fact the analogue of formula (52) is the

+These are more exactly products of the form $S^2 \times R^4$ so that the geometrical conditions of the case presented in Sect.4.II are not satisfied here.

following (proved in [24]):

$$G_{in}^{(6)+}\,\Theta_{+}\,L_{in}^{(6)+} - G_{in}^{(6)-}\,\Theta_{-}\,L_{in}^{(6)-} = G_{in}^{(6)+}\,\Theta_{+}\,\Delta L_{in}^{(6)} +$$

$$+\,\Delta G_{in}^{(6)}\quad \Theta_{-}\,L_{in}^{(6)-} + G_{in}^{(6)+} * L_{in}^{(6)-} +$$

$$+\,\sum_{\alpha} \Delta_{\alpha}\, G_{in}^{(6)+}(\Theta_{-}*)_{\alpha}\, L_{in}^{(6)} \quad . \tag{58}$$

The last sum at the r.h.s. of (58) takes into account the discontinuities $\Delta_{\alpha} G_{in}^{(6)}$ across the corresponding cuts $\sigma_{\alpha}((k_{\beta} + k_{\gamma})^2 \geq (2\mu)^2)$, and the operation $(\Theta_{-}*)_{\alpha}$ indicates that the internal four-momenta k_{α}, k_{β}, k_{γ} $(k_{\alpha} + k_{\beta} + k_{\gamma} = K)$ are integrated on a set of the form:

$$\{(k_{\alpha}, z_{\alpha} = \frac{k_{\beta}-k_{\gamma}}{2});\ k_{\alpha}^2 = \mu^2, \quad (K-k_{\alpha})^2 \geq (2\mu)^2, \quad z_{\alpha} \in \Gamma_{-}(K, k_{\alpha})\}$$

(graphically:

$$\Delta_{\alpha}\, G_{in}^{(6)+}\,(\Theta_{-}*)_{\alpha}\, L_{in}^{(6)-} =$$ 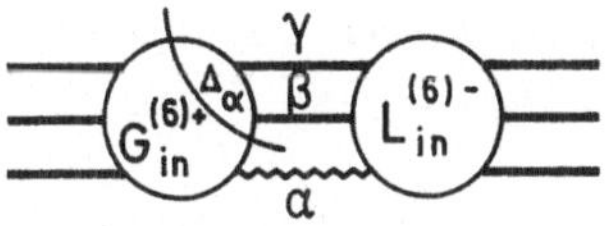

)

By using the discontinuity formula (58), it is possible to treat our point (i) as in the purely polar case a) and to obtain a formula which is the analogue of (54), namely:

$$[\Delta \hat{H}_1^{(6)} - \frac{1}{3!}\,\hat{H}_1^{(6)+} *_3 \hat{H}_1^{(6)-}]\ \Theta_{-}\ \Lambda(1 - \frac{1}{3!}\,L_{out}^{(6)-}) =$$

$$= \Lambda^{-1}(1 + \frac{1}{3!}\,G_{in}^{(6)+})\ \Theta_{+}\ \Delta L^{(6)} \quad , \tag{59}$$

where Λ and $\hat{H}_1^{(6)}$ have been defined in Sect. 1 (formula (9) and above), $G_{in}^{(6)}\Lambda = \Lambda G_{out}^{(6)} = G^{(6)}$ and $L_{in}^{(6)}\Lambda = \Lambda L_{out}^{(6)} = L^{(6)}$ (see Sect. 7). Formula (59) implies the equivalence between the three-particle irreducibility of $L^{(6)}$, namely $\Delta L^{(6)}=0$, and the off-shell unitarity equation for $H_1^{(6)}$ (valid for

$(3\mu)^2 < K^2 < (5\mu)^2)$:

$$\Delta H_1^{(6)} - \frac{1}{3!} H_1^{(6)} *_3 H_1^{(6)} = 0 \quad . \tag{60}$$

The latter can be deduced from the analogous equation for $H^{(6)}$, which is itself a consequence of A.C. (see formula (29)).

Concerning our point (ii), the analysis is much more complicated than in a). Although it is possible to show the existence of a local analytic continuation of $G_{in}^{(6)}$ across the cut $K^2 > (3\mu)^2$ (and for $K^2 < (5\mu)^2$), the complete monodromic structure around the threshold remains to be studied; in particular the occurrence of all Landau singularities associated with graphs of the form

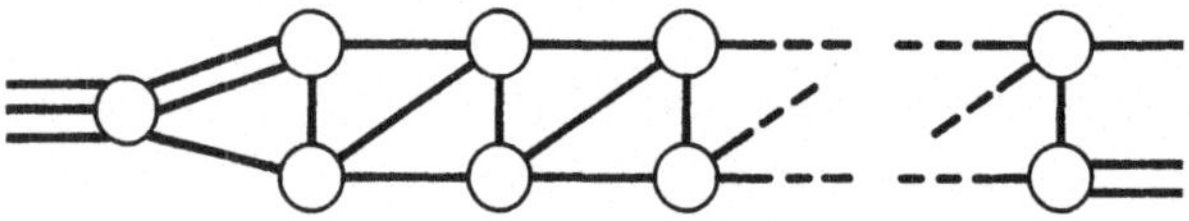

in various unphysical sheets has to be taken into account.

c) Singularities of non-separable type:

A typical example is the auxiliary B.S. equation for the 3 → 3 particle process, which allows one to perform the inversion of the "two-particle undressing kernel"

$$\Lambda = I - \frac{1}{6}\sum_{h,n}$$ 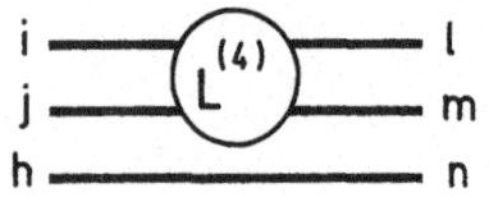 ;

this equation is (see Sect. 1, formula (6))

$$U_{in}(K;z,z') = T_{in}(K;z,z') + \frac{1}{3!}\int_{\Gamma(K)} U_{in}(K;z,z")\, T_{in}(K;z",z')\, dz" \tag{61}$$

where :

$T_{in} = \sum_{h,n}$ 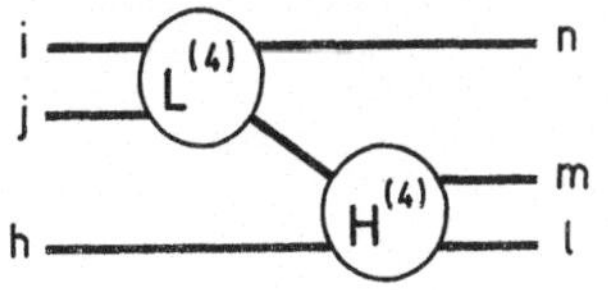

In T_{in}, the transverse poles ω_{hn}: $(k_i+k_j+k_n)^2 = \mu^2$ have to be taken into account, since they cross the relevant part of the primitive domain$^+$ in which $\Gamma(K)$ should stay. Note that in the (z,z')-variables ω_{hn} is represented by:

$$(z'_n - z_i - z_j - \frac{K}{3})^2 = \mu^2 \quad (\text{with } z_i = -k_i - \frac{K}{3},\ z_j = -k_j - \frac{K}{3},$$

$$z'_n = k_n - \frac{K}{3})$$

and is therefore a singularity of non-separable type.

The difficulty consists in the impossibility of finding a contour $\Gamma(K)$ (for Re $K \geq 3\mu$) which can be repeated in all the iterated integrals of Fredholm theory, since the singularities ω_{hn} are moving with the external variables z,z'. In fact, one can write:

$$T_{in}^{(p)}(K;z,z') = \int_{\Gamma(K,z,z')} T_{in}(K;z,z_1)\ldots T_{in}(K;z_{p-1},z')dz_1\ldots dz_{p-1}$$

for a suitable contour $\Gamma(K,z,z')$ in $C^{4(p-1)}$; but in general this contour is not of the form $(\Gamma(K))^{p-1})$. A way of avoiding this difficulty in the present case is the following. For a certain value p_o of p, it turns out that the iterated kernel $T_{in}^{(p_o)}$ of T_{in} (primitively defined in euclidean space) can be analytically continued in the relevant part D_M^{cut} of the primitive domain (see Sect. 1) in such a way that it only has there a singular set of separable type (namely the same set S as in the example considered in b). One can then use a well-known trick in Fredholm theory, namely express U_{in} in terms of the Fredholm resolvent $(U_{in})^{(p_o)}$ of $T_{in}^{(p_o)}$ for the

$^+$as soon as Re $K^{(0)} \geq 3\mu$.

value $\lambda_o^{p_o} = (\frac{1}{3!})^{p_o}$ of the Fredholm parameter; this expression is the following:

$$U_{in} = \sum_{1\leq r<p_o} \lambda_o^r \, T_{in}^{(r)} + \lambda_o^{p_o} \sum_{0\leq r<p_o} \lambda_o^r \int_{\Gamma_r(K,z,z')} (U_{in})^{(p_o)} T_{in}^{(r)} . \tag{62}$$

In view of b), $(U_{in})^{(p_o)}$ can be shown to be analytic in the domain $D_M^{cut} - S$ (as $T_{in}^{(p_o)}$). Moreover by an appropriate choice of the cycles $\Gamma_r(K)$, each term of the summation in the last term of (62) can be shown to be also analytic in the same domain $(D_M^{cut}-S)$. Therefore in the physical sheet, the singular set of U_{in} is the union of the (separable-type) singular set S and of the (non-separable type) singular sets $S_{(r)}$ of the first terms $T_{in}^{(r)}$ (for $1 \leq r < p_o$).

Investigating the complete monodromic structure of U_{in} around the threshold $K^2 = (3\mu)^2$ would involve all the (non-separable type) singular sets $S_{(p)}$ associated with all iterated kernels $T_{in}^{(p)}$: these singularities which are relevant in the unphysical sheets indeed accumulate to the threshold $K^2 = (3\mu)^2$; these are the Landau singularities of all the "truss-bridge" graphs:

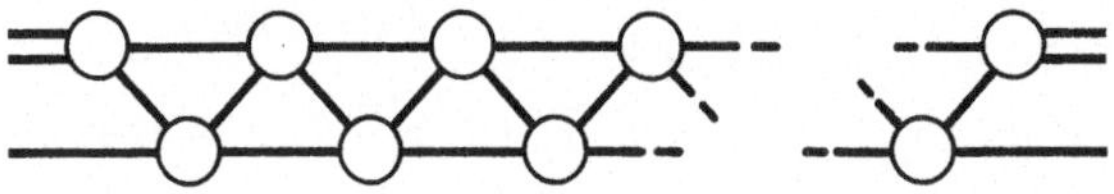

6. THE FIRST STEPS OF M.P.S.A.: ONE-AND TWO-PARTICLE STRUCTURE

6.I One-particle structure

The one-particle irreducible n-point functions with respect to a certain channel (I,J) are defined as follows: ($n_1 = |I|$ and $n_2 = |J|$ being $\geq$ 2):

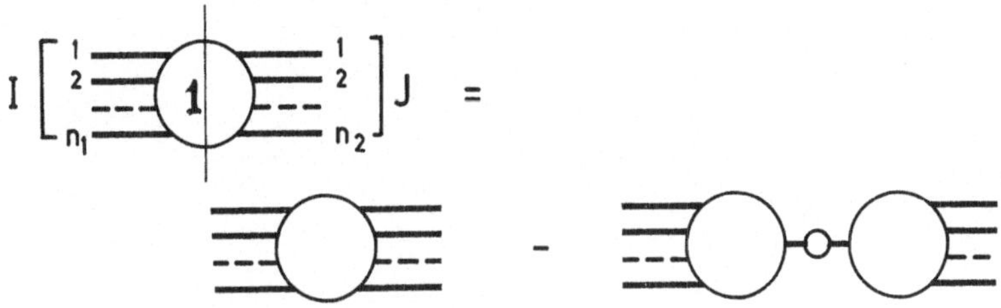

(63)

or:

$$\hat{H}_1^{(n_1,n_2)}(k_J;z,z') = \hat{H}^{(n)}(k_J;z,z') - \hat{H}^{(n_1+1)}(k_J;z)\cdot$$

$$H^{(2)}(k_J)\,\hat{H}^{(n_2+1)}(k_J;z') \qquad (63')$$

($k_J = -k_I = \sum_{j\in J} k_j$; z,z' being defined as in Sect. 3.Ic)). The one-particle irreducibility of $\hat{H}_I^{(n_1,n_2)}$, i.e. the absence of the poles+ $k_I^2 = \mu_\alpha^2$ can be proved by comparing the discontinuity of

$$\hat{H}^{(n_1+1)} \times H^{(2)} \times \hat{H}^{(n_2+1)}$$

with that of $\hat{H}^{(n)}$ expressed through formula (26) (or its generalisation in the case of several masses μ_α in the spectrum of the field); the term $\ell = 1$ in the r.h.s. of (26) is easily seen to reproduce the discontinuity of

$$\hat{H}^{(n_1+1)} \times H^{(2)} \times \hat{H}^{(n_2+1)}$$

near the considered pole.

Formula (63) implies the following <u>factorisation property</u> for the residues $R_\alpha^{(n_1,n_2)}$ of the $\hat{H}^{(n)}$, at the corresponding poles $k_J^2 = \mu_\alpha^2$:

$$R_\alpha^{(n_1,n_2)}(z,z') \equiv (k_J^2-\mu_\alpha^2)H^{(n)}(k_J;z,z')\Big|_{k_J^2=\mu_\alpha^2} =$$

+Note that other poles coming from the two-particle structure (i.e. bound states) may still be present in the $\hat{H}_1^{(n)}$: such poles occur in $H^{(4)}$, but not in $H^{(2)}$.

$$= \frac{R_\alpha^{(n_1,1)}(z) \times R_\alpha^{(n_2,1)}(z')}{R_\alpha^{(1,1)}} \quad , \qquad (64)$$

$R_\alpha^{(1,1)} = Z_\alpha$ being the normalization constant of the one-particle states of mass μ_α created by the field.

Off-shell unitarity-type equations for the $\hat{H}_1^{(n_1,n_2)}$: they follow from the off-shell unitarity type equations for the $\hat{H}^{(n)}$ (see Sect. 3.III, formula (26)); for every channel (I,J), one has:

$$\Delta_I \, \hat{H}_1^{(n_1,n_2)} = \sum_{\ell \geq 2} \frac{1}{\ell!} H_{1,\varepsilon}^{(n_1,\ell)} *_\ell H_{1,-\varepsilon}^{(\ell,n_2)} \qquad (65)$$

or :

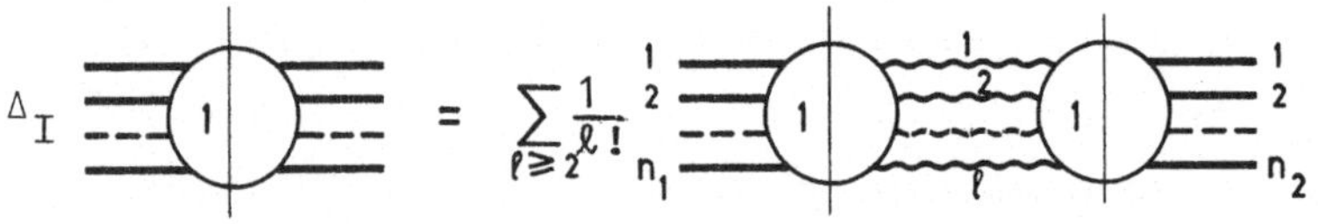

(65')

The one-particle irreducible functions with respect to all channels $\hat{H}_{[1]}^{(n)}$ can be introduced by the recursive algorithm:

$$\hat{H}^{(n)} = \hat{H}_{[1]}^{(n)} + \sum_{\text{trees}} \qquad (66)$$

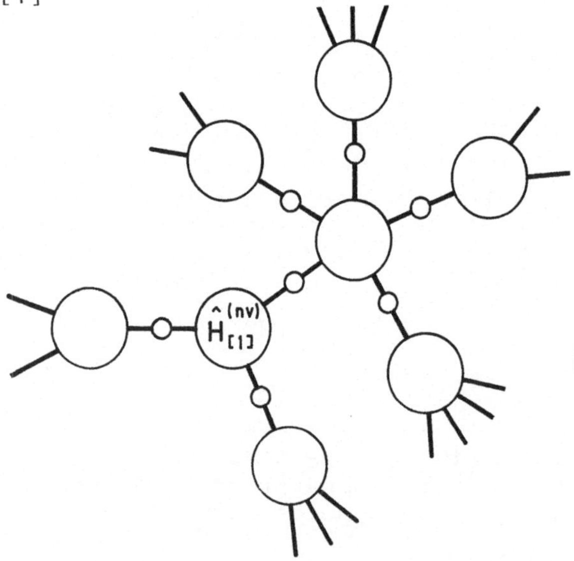

where each tree-graph represents a G-convolution term in which the vertex factors are the 1-p.i. functions $\hat{H}^{(n_v)}_{[1]}$ (n_v being always < n) and the line factors are functions $H^{(2)}$ of the corresponding four-momenta.

The one-particle irreducibility of $\hat{H}^{(n)}_1$, i.e. the absence of poles in all channels, can be proved on the basis of formula (26) (by an argument similar to that given above for $\hat{H}^{(n_1,n_2)}_1$.

6.II Two-particle structure in a given channel

a) Two-particle structure of $H^{(4)}_1$:

The starting point is the B.S. equation for $H^{(4)}_1$:

$$H^{(4)}_1(K,z,z') = L^{(4)}(K,z,z') + \frac{1}{2}\int_{\Gamma(K)} H^{(4)}_1(K,z,z'')\cdot$$

$$L^{(4)}(K,z'',z')\,\omega(K,z'')\,dz'' \qquad (67)$$

or:

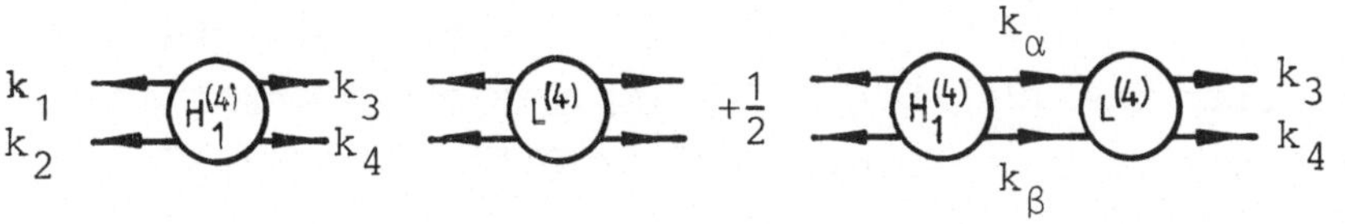

(67')

where the variables K,z,z',z" and ω(K,z) have been defined above (see Sect. 1, after formulae (4), (4')). $H^{(4)}_1$ denotes the one p.i. kernel with respect to the channel (12,34)(i.e.

$$\hat{H}^{(4)}_1 = H^{(4)}_1 \times \prod_{1\leq i\leq 4} [H^{(2)}(k_i)]^{-1} = \hat{H}^{(4)}(K,z,z') -$$

$$- \hat{H}^{(3)}(K,z)H^{(2)}(K) \times \hat{H}^{(3)}(K,z')).$$

In an even theory, one has: $H^{(4)}_1 = H^{(4)}$, so that eq.(67) (resp. (67')) coincides with eq.(4) (resp. (4')). The contour Γ(K) is obtained by distortion of the euclidean z"-space (the initial position for K being itself euclidean). Integrability at infinity on $\Gamma(K)$ is ensured by modifying (if necessary) $H_1^{(4)}$ by a suitable analytic factor of the form:

$$\prod_{1\leq i\leq 4} \left(\frac{m^2-\rho}{k_i^2-\rho}\right)^{\alpha} \quad (\rho >> m,\ \alpha > 1)$$

which is equal to 1 on the mass shell. (Of course, the 2 p.i. kernel $L^{(4)}$ thus defined depends on this cut-off, but the analyticity properties and threshold behaviour of $H^{(4)}$ at $K^2 = (2\mu)^2$ are derived from (67), independently of this ambiguity at infinity.)

The analysis of (67) makes uses of the techniques presented in Sect. 5 (case a)); in the field-theoretical axiomatic framework, the following steps are taken (in their logical order):

1) Being given $H_1^{(4)}$ analytic in the primitive domain D_4, (67) defines $L^{(4)}$ in the same domain.

2) Under a suitable regularity assumption for $H^{(4)}$ on the cut $(2\mu)^2 < K^2 < (3\mu)^2$ (existence of a pointwise limit), the off-shell unitarity equation (65) for $H_1^{(4)}$ (consequence of the similar one (29) for $H^{(4)}$) implies the two-particle irreducibility of $L^{(4)}$, i.e.: $\Delta L^{(4)} = 0$ for $K^2 < (3\mu)^2$ (see (55)).

3) By applying to (67) the results of Fredholm theory in complex space, one obtains the square-root nature[+] of the threshold singularity $K^2 = (2\mu)^2$ of $H_1^{(4)}$, and the meromorphic character of $H_1^{(4)}$ in the second (or "unphysical") sheet, with fixed poles in the variable K^2 (see 6.III).

[+]Note that this two-sheeted structure holds because the space-time dimension is even. In odd-dimensional theory, singularities of the type

$$\frac{1}{\log(K^2-4\mu^2)}$$

would occur (see Sect.5a) (ii)).

b) <u>Two-particle structure</u> of the Green functions $H^{(n)}$ $(n \neq 4)$:

The two-particle irreducible n-point functions $H_2^{(n_1,n_2)}(k_I;z,z')$ with respect to a certain channel (I,J) $(n_1 = |I|, n_2 = |J|)$ are defined as follows:

$$H_2^{(2,2)} = L^{(4)} ; \tag{68}$$

$$\forall n_1 \neq 2, \; H_1^{(n_1,2)} = H_2^{(n_1,2)} + \frac{1}{2} \int_{\Gamma(k_J)} H_1^{(n_1,2)}(k_J;z,z'') \cdot L^{(4)}(k_J;z'',z')\,\omega(k_J,z'')\,dz'' , \tag{69}$$

$\forall n_1, n_2 \neq 2$,

$$H_1^{(n_1,n_2)} = H_2^{(n_1,n_2)} + \frac{1}{2} \int_{\Gamma(k_J)} H_2^{(n_1,2)}(k_J;z,z'') \cdot H_1^{(2,n_2)}(k_J;z'',z')\,\omega(k_J,z'')\,dz'' . \tag{70}$$

We notice that all the functions $H_2^{(n_1,n_2)}$, with $(n_1,n_2) \neq (2,2)$, are defined without solving new Fredholm equations (once $L^{(4)}$ has been defined).

Formulae analogous to (54) show the equivalence [3c] between the following set of off-shell unitarity-type equations, valid in the region $(2\mu)^2 < k_j^2 < (3\mu)^2$:

$$\Delta_I \hat{H}_1^{(n_1,n_2)} = \frac{1}{2} \hat{H}_1^{(n_1,2)} *_2 \hat{H}_1^{(2,n_2)} \tag{71}$$

and the two-particle irreducibility of all functions $H_2^{(n_1,n_2)}$, i.e.

$$\Delta_I H_2^{(n_1,n_2)} = 0 , \qquad \text{for } k_J^2 < (3\mu)^2 . \tag{72}$$

Each general n-point function $H_2^{(n_1,n_2)}$ is thus analytic in a primitive domain whose threshold in the channel (I,J) is at $k_J^2 = (3\mu)^2$: in other words, it is analytic in a neighbourhood of the two-particle region: $(2\mu)^2 \le k_J^2 < (3\mu)^2$ (all other variables k_i being kept in their prescribed domain). Now, by using (67), eq.(69) can also be rewritten

$$H_1^{(n_1,2)} = H_2^{(n_1,2)} + \frac{1}{2} \int_{\Gamma(k_J)} H_2^{(n_1,2)} \cdot H_1^{(4)} \omega(k_J;z")dz". \quad (73)$$

Then the analyticity of $H_2^{(n_1,2)}$ and the two-sheeted analytic structure of $H_1^{(4)}$ around $k_J^2 = (2\mu)^2$ imply the two-sheeted analytic structure of $H_1^{(n_1,2)}$ and (through (70)) of all the $H_1^{(n_1,n_2)}$. From (63), it then follows that in all channels (I,J) each $H^{(n)}$ can be continued across the cut $(2\mu)^2 < k_J^2 < (3\mu)^2$ and has a two-sheeted structure around $k_J^2 = (2\mu)^2$.

6.III Poles in the unphysical sheet

The Fredholm solution of (67) exhibits $H_1^{(4)}$ as a meromorphic function of the form $\frac{N(K;z,z')}{D(K)}$; when $K = (K^{(0)},0)$ lies in the physical sheet, D(K) can only have zeros for $K^{(0)}$ real $< 2\mu$, the latter being interpreted as bound states. When $K^{(0)}$ lies in the unphysical sheet domain (obtained by continuation across the cut $2\mu < K^{(0)} < 3\mu$ from the side $\text{Im}\, K^{(0)} > 0$), the zeros of D(K) are interpreted as unstable particles. They correspond to poles of $H^{(4)}$ but cannot be poles of $H^{(2)}$ and $H^{(3)}$. On the other hand, there may a priori occur poles of $H^{(2)}$ (in the same unphysical-sheet domain of $K^{(0)}$-plane). These poles ζ_β enjoy the same properties as the real poles μ_α^2 extracted in one-particle structure analysis, namely: in each function $\hat{H}^{(n)}$ and each channel (I,J), there will be in general a pole at $k_I^2 = \zeta_\beta$, and the corresponding residues

$R_\beta^{(n_1,n_2)}$ will satisfy factorisation formulae which are similar to (64), namely:

$$R_\beta^{(n_1,n_2)}(z,z') = \frac{R_\beta^{(n_1,1)}(z)\cdot R_\beta^{(n_2,1)}(z')}{R_\beta^{(1,1)}} . \qquad (74)$$

This can be shown by rewriting the corresponding off-shell unitarity-type equation (29) in the region $(2\mu)^2 < k_J^2 < (3\mu)^2$ as follows:

$$\Delta_I \hat{H}^{(n)} = \frac{1}{2}\hat{H}_+^{(n_1+2)} *_2 \hat{H}_-^{(n_2+2)} ,$$

by taking its analytic continuation in the unphysical sheet up to the considered pole ζ_β; one then computes the corresponding residue $R_\beta^{(n_1,n_2)}$ and compares it with those (obtained similarly) of $H^{(n_1+1)}$, $H^{(n_2+1)}$, $H^{(2)}$.

7. COMMENTS ON THE STRUCTURAL EQUATIONS FOR $H^{(6)}$, RELATIVE TO THE 3 → 3 PARTICLE PROCESS+

We refer to Sect. 1 for all the notations introduced, concerning the domains and various six-point kernels. All these kernels are defined (in the primitive domain D_6) by equations involving a finite number of G-convolution terms in which $H^{(6)}$, $H^{(4)}$, $L^{(4)}$ are the only ingredients. The first step consists in proving the relevant irreducibility properties of the following kernels $G_{in}^{(6)}$, $G_{out}^{(6)}$, $G^{(6)}$.

We recall that

$$\hat{H}_1^{(6)} = \hat{H}^{(6)} - \hat{H}^{(4)} \cdot H^{(2)} \cdot \hat{H}^{(4)} \qquad (75)$$

+For simplicity, we considered the case of an even field theory.

or:

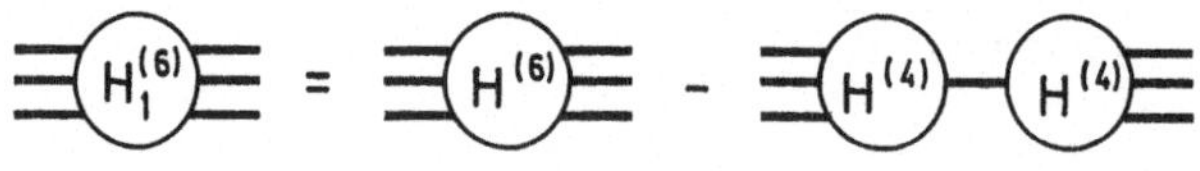

$$\equiv\!\!\left(H_1^{(6)}\right)\!\!\equiv \;=\; \equiv\!\!\left(H^{(6)}\right)\!\!\equiv \;-\; \equiv\!\!\left(H^{(4)}\right)\!\!-\!\!\left(H^{(4)}\right)\!\!\equiv \tag{75'}$$

$$G_{in}^{(6)} = \Lambda \cdot H_1^{(6)} - T_{in} \;, \tag{76}$$

$$G_{out}^{(6)} = H_1^{(6)} \cdot \Lambda - T_{out} \;, \tag{77}$$

$$G^{(6)} = G_{in}^{(6)} \cdot \Lambda = \Lambda \cdot G_{out}^{(6)} = \Lambda H_1^{(6)} \Lambda - T_{in}\Lambda = \Lambda H_1^{(6)}\Lambda - \Lambda T_{out} \;. \tag{78}$$

The operator

$$\Lambda = \mathbb{1} - \frac{1}{6} \sum_{hn} \quad \equiv\!\!\left(L^{(4)}\right)\!\!\equiv \tag{79}$$

applied (resp. on the left or on the right) to $H_1^{(6)}$ has the property of making the latter two-particle irreducible in the three corresponding subchannels (i.e. resp., (ij,h 123), (ijh) = (456), or (456 n,ℓm), (ℓmn) = (123)). For instance, let us show that $\Lambda H_1^{(6)}$ is two p.i. with respect to the channel (45,6123); we have:

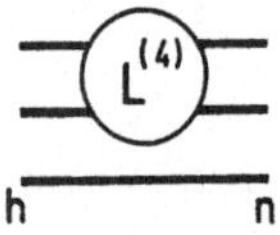

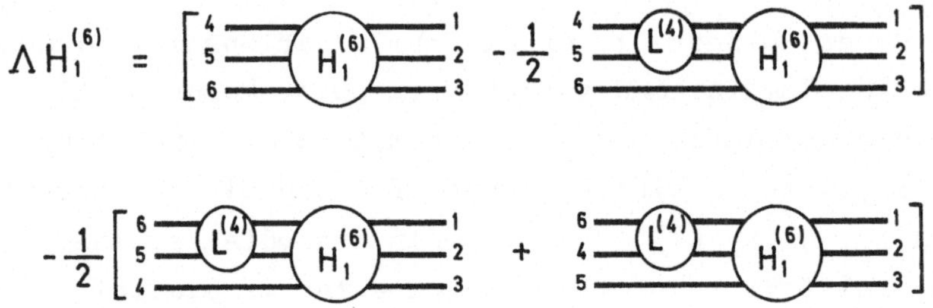

$$\Lambda H_1^{(6)} = \left[\; \equiv\!\!\left(H_1^{(6)}\right)\!\!\equiv \; - \; \frac{1}{2} \; \equiv\!\!\left(L^{(4)}\right)\!\!\left(H_1^{(6)}\right)\!\!\equiv \; \right]$$

$$- \frac{1}{2}\left[\; \equiv\!\!\left(L^{(4)}\right)\!\!\left(H_1^{(6)}\right)\!\!\equiv \; + \; \equiv\!\!\left(L^{(4)}\right)\!\!\left(H_1^{(6)}\right)\!\!\equiv \; \right] \tag{80}$$

In view of equations (69) and (75), the first bracket of (80) can be written:

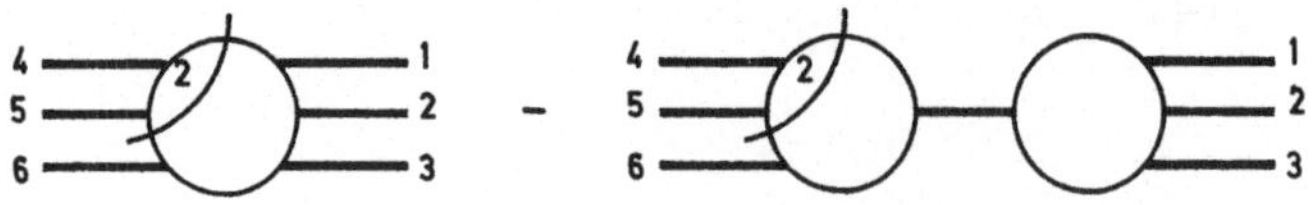

and each term of (81) is two p.i. with respect to the channel (45,6123) (as seen in Sect. 6, this fact being a consequence of the two-particle off-shell unitarity equations).

In the second bracket of (80), each term is two p.i. with respect to the channel (45,1236) because this channel is "transverse" with respect to the integration channel: the corresponding threshold is then $(2\mu + 2\mu)^2$ (see Sect.4.I: property of additivity of thresholds; the latter can be easily checked in the present case).

So Λ can be called a "two-particle undressing operator". The role of the subtracted terms $(-T_{in}, -T_{out})$ in (76), (77) consists in taking off the one-particle structure of resp. $\Lambda H_1^{(6)}$ or $H_1^{(6)}\Lambda$ in all crossed subchannels of the form (ijn, hℓm)

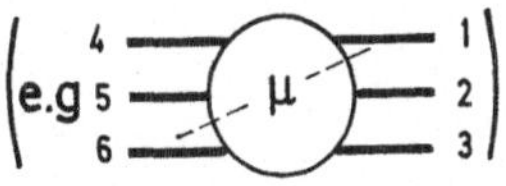

with (hij) = (456), (ℓmn) = (123).

Through similar arguments, one proves that: $G^{(6)}$ is two-particle irreducible with respect to the six channels (ij, h123) and (456n, ℓm), and one p.i. with respect to the nine channels: (ijn, hℓm), (hij) = (456), (ℓmn) = (123) . It is also two p.i. with respect to the main channel (456,123). All these irreducibility properties imply that $\hat{G}^{(6)}$ is analytic in the domain D_M^{cut}, with $M = 11\,\frac{m}{3}$.

<u>Remark:</u> This limitation on M comes from the occurrence of the three-particle thresholds in the crossed channel :

(ijn,hℓm)

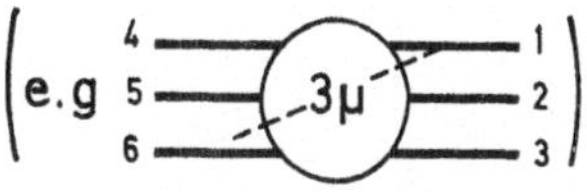

We shall say that a six-point function has normal analyticity properties in D_M^{cut} if its only singularities in this (physical sheet) domain are the poles ω_i, ω_{i+3} $(i = 1,2,3)$ and the two-particle cuts $\underline{\sigma}_i$, $\underline{\sigma}_{i+3}$ $(i = 1,2,3)$: in particular, no non-separable type singularity is admitted.

We shall now give some hint on the derivation of the analytic structural equation (11) (valid in D_M^{cut}) from the two-particle dressing equation (10).

The first step consists in transforming eq. (10) by using the expression (62) of U_{in} (written for a certain choice of p_o) and a similar one for U_{out}. This yields:

$$H_1^{(6)} = \left(\sum_{r=0}^{p} \left(\tfrac{1}{6}\right)^r T_{out}^{(r)}\right) \left[1 + \left(\tfrac{1}{6}\right)^{p+1} (U_{out})^{(p+1)}\right] \cdot$$

$$VG^{(6)} V \left[1 + \left(\tfrac{1}{6}\right)^{p+1} (U_{in})^{(p+1)}\right] \left(\sum_{r=0}^{p} \left(\tfrac{1}{6}\right)^r T_{in}^{(r)}\right) +$$

$$+ \left(\sum_{r=0}^{p} \left(\tfrac{1}{6}\right)^r T_{out}^{(r)}\right) \left[\left(\tfrac{1}{6}\right)^p (U_{out})^{(p+1)} \cdot V\right] +$$

$$+ \left(\sum_{r=1}^{p} \left(\tfrac{1}{6}\right)^{r-1} T_{out}^{(r)}\right) \cdot V \quad . \qquad (82)$$

By using the B.S. equation (67), it can be seen that:

$$\left(\sum_{r=1}^{p} \left(\tfrac{1}{6}\right)^{r-1} T_{out}^{(r)}\right) V = \sum_{r=0}^{p-1}$$ 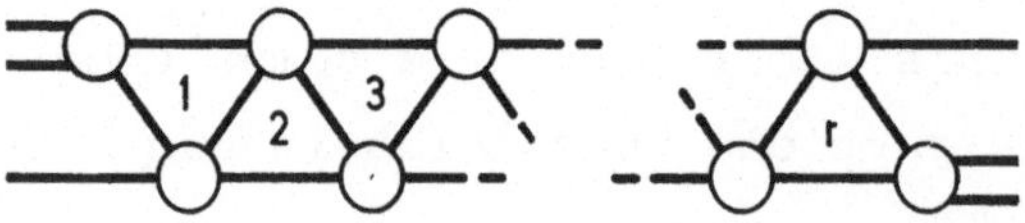

$$+ \; \Phi$$

(83)

where the r-looped G-convolution terms in the sum only contain factors $H^{(4)}$ (and not $L^{(4)}$) associated with each vertex, and Φ is a sum of G-convolution terms of the same form ("trussbridge graphs") whose number of loops is $\geq$ p.

In (82), (83), the choice p = 3 is done and motivated by the fact that for r > 2, any r-looped G-convolution function (of the previous form) has normal analyticity properties in D_M^{cut}. The contribution of the last term of (82) then consists, in view of (83), in:

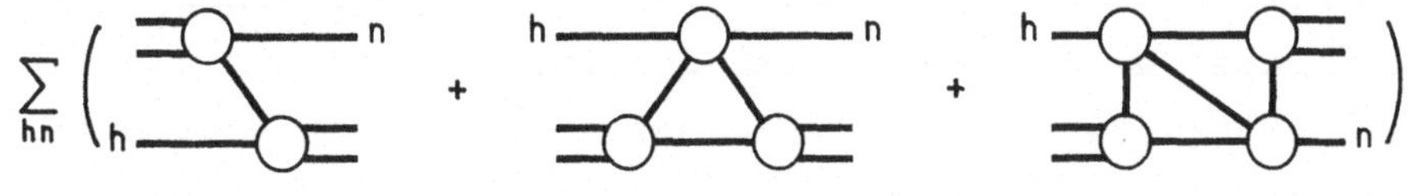

plus a remainder which has normal analyticity properties. The rest of the proof of eq.(11) consists in showing that all the contributions coming from the two first terms at the r.h.s. of (82) also have normal analyticity properties in D_M^{cut}. In fact, this is the case for $VG^{(6)}V$ and for the Fredholm resolvents $(U_{in})^{(3)}$, $(U_{out})^{(3)}$ of $T_{in}^{(3)}$, $T_{out}^{(3)}$; moreover the normal analyticity is stable by integration over $\Gamma(K)$, and G-convolution with T_{in} and T_{out} kernels.

Finally, the exploitation of the three-particle off-shell unitarity equation to obtain a three-particle irreducible kernel $L^{(6)}$, and associated kernels $L_{in}^{(6)}$, $L_{out}^{(6)}$ such that

$$L = L_{in}^{(6)}\Lambda = \Lambda L_{out}^{(6)} \quad , \tag{84}$$

can be done along the line described in Sect. 5 (case b); $L_{in}^{(6)}$ is introduced through the B.S. equation (57), namely:

$$G_{in}^{(6)} = L_{in}^{(6)} + \frac{1}{3!}\, G_{in}^{(6)} \ \Theta \ L_{in}^{(6)} \quad ; \tag{85}$$

the three p.i. property $\Delta L_{in}^{(6)} = 0$ is a consequence of the off-shell unitarity equation $\Delta H^{(6)} - \frac{1}{6} H_+^{(6)} *_3 H_-^{(6)} = 0$ (valid in the region $(3\mu)^2 < K^2 < (5\mu)^2$). The local analytic continuation of $G_{in}^{(6)}$ through the cut $\underline{\sigma}(K^2 > (3\mu)^2)$

then follows from the local analyticity of $L_{in}^{(6)}$ on $K^2 \geq (3\mu)^2$ (implie by $\Delta L_{in}^{(6)} = 0$).

8. GLOBAL ANALYTICITY PROPERTIES IMPLIED BY M.P.S.A.: AN EXAMPLE

The best examples of global-type analyticity properties obtained in general field theory are crossing domains on the complex mass shell. By crossing domain, one means a domain which connects two physical regions of different scattering processes (belonging respectively to M_{IJ} and $M_{I'J'}$, with $|I| + |J| = |I'| + |J'| = n$) and belongs to the n-particle complex mass shell M_n^c. The only crossing domains which have been derived rigorously from the postulates of field theory (i.e. by applying analytic completion techniques to the primitive analyticity domains of the Green functions) are crossing domains for two-particle scattering functions (with arbitrary positive masses) [5b], i.e. domains on M_4^c; in the most favourable mass configurations, these domains contain cut-planes in the energy variable s (at fixed momentum transfer t), and dispersion relations can be written for the scattering functions in the latter.

The limitations which prevent one to generalize the derivation of crossing domains to n > 4 are of the following kind:

a) the spectral conditions which determine the primitive analyticity domains D_n are so low that the completion techniques (using for instance the Jost-Lehmann-Dyson domain) [23] yield "crossing type domains" only on complex manifolds defined by prescribing unphysical values of the squared mass variables $k_i^2 = \zeta_i$ (in general for $\zeta_i < 0$).

b) Extrapolation of such domains to physical values $\zeta_i = \mu^2$ of the squared mass variables is only made possible if sufficient local analyticity properties in several variables (including some ζ_i and energy variable s) around the physical regions on the mass shell can be established.

In principle M.P.S.A. may improve the situation with respect to both points a) and b); it is clear that for suitable ℓ-particle irreducible functions, one may be in better situation with respect to a) (the thresholds of these functions being improved); however one still has to use B.S.- type equations to transfer this "input" into a crossing property for the $H^{(n)}$ (which may be a non-trivial problem). On the other hand, M.P.S.A. may improve in a decisive way our knowledge of local analyticity and yield results of the type described in b). This is precisely the case for $H^{(6)}$ in the 3 → 3 particle region, and in the following we use this result to prove the existence of a crossing domain for the 3 → 3 particle scattering function.

A crossing domain for the $3\pi \to 3\pi$ scattering amplitude

We consider the crossing problem for the couple of channels $(I,J) = (\{456\},\{123\})$ and $(I'J') = (\{156\},\{423\})$, namely we seek an analyticity domain Θ on M_6^c for $T^{(6)} = \hat{H}^{(6)}|_{M_6^c}$ whose boundary contains parts of both physical regions M_6^{IJ} and $M_6^{I'J'}$, and such that it touches M_6^{IJ} from the side $\text{Im } s > 0$ $(s = k_I^2)$ and $M_6^{I'J'}$ from the side $\text{Im } s' > 0$ $(s' = k_{I'}^2)$: the corresponding boundary values of $T^{(6)}$ will then be respectively the scattering amplitudes $T_{IJ}^{(6)}$ and $T_{I'J'}^{(6)}$.

We shall define Θ as a neighbourhood in M_6^c of a certain domain Θ_α in $M_6^c \cap V_\alpha$, V_α being the following one-dimensional submanifold of "forward configurations":

$$V_\alpha = \{(k_1,\dots,k_6);\ k_1+k_4 = 0,\ k_2+k_5 = 0,\ k_3+k_6 = 0;\ k_1^2=\mu^2,$$

$$k_1 = (k_1^{(0)},k_1^{(1)},0,0),\ k_2 = (\mu\text{ch}\alpha,0,\mu\text{sh}\alpha,0),$$

$$k_3 = (\mu\text{ch}\alpha,0,-\mu\text{sh}\alpha,0)\} \quad ;$$

α will be chosen sufficiently small. On this manifold, $T^{(6)}$ is singular since the pole $(k_1+k_4+k_5)^2 = \mu^2$ of $\hat{H}^{(6)}$ reduces to $k_5^2 = \mu^2$ on the mass shell. Therefore on V_α, it is necessary to consider the crossing problem for the forward part of $\hat{H}^{(6)}$ which we define by:

$$\hat{H}_F^{(6)} = \hat{H}^{(6)} - \sum_{hn}{}'$$ 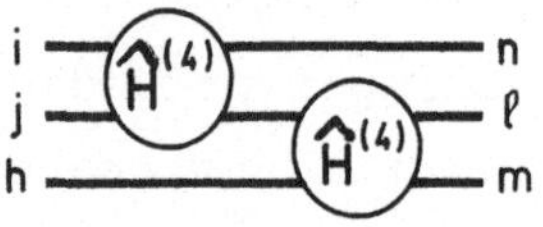

where the sum $\sum'$ runs over the set $\{(h,n);\ h \in \{4,5,6\},\ n \in \{1,2,3\};\ h \neq n+3\}$.

The following lemma can be proved for $\hat{H}_F^{(6)}$.

Lemma: There exists a crossing domain Θ_α on V_α $(0 < \alpha < \alpha_0)$ for the "forward scattering function" $T_F^{(6)} = \hat{H}_F^{(6)}\big|_{M_6^c}$. This domain, whose shape is indicated on Fig. 10, tends to a cut-plane in the variable s when α tends to 0.

We note that on V_α, the physical regions of the channels (I,J), (I',J') are symmetric with respect to the origin if one takes $k_1^{(0)}$ as the variable, and that:

$$k_1^{(0)} = \frac{s-\mu^2(1+4\mathrm{ch}^2\alpha)}{4\mu\,\mathrm{ch}\alpha} = -\frac{s'-\mu^2(1+4\mathrm{ch}^2\alpha)}{4\mu\,\mathrm{ch}\alpha}$$

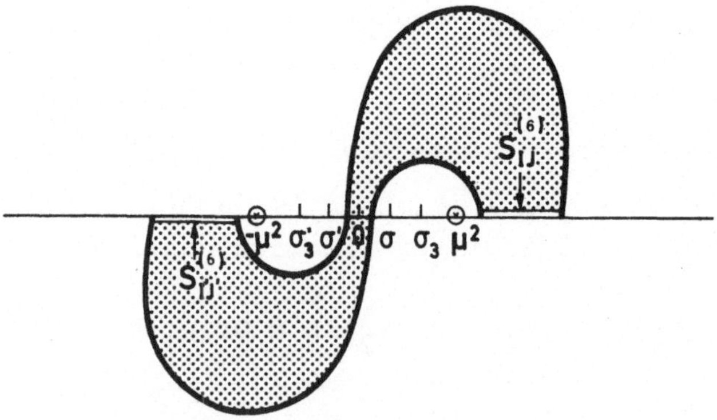

Fig. 10. The crossing domain Θ_α in the $k_1^{(0)}$-plane.

The proof is obtained by performing the analytic completion of the union of the two following regions in the space of two complex variables $k_1^{(0)}$ and $\zeta = k_1^2$.

a) For ζ real $< \zeta_\alpha = \frac{(2-\text{ch}^2\alpha)^2}{\text{ch}^2\alpha}\,\mu^2$, $\hat{H}_F^{(6)}$ is analytic in the **cut**-plane (the cuts being given by $s \geq 9\mu^2$, $s' \geq 9\mu^2$). This results from the Jost-Lehman-Dyson completion [23]in k_1-space. Note that $\zeta_\alpha < \mu^2$ and that $\zeta_\alpha \to \mu^2$ when $\alpha \to 0$.

b) When $k_1^{(0)}$ varies in a neighbourhood of the low-energy physical region considered above (limited by $s < M^2$), $\hat{H}_F^{(6)}$ is analytic in ζ in a neighbourhood N_α of ζ_α whose size is <u>independent</u> of $k_1^{(0)}$ and α, when α tends to zero. This results from the fact that such a domain is contained in the analyticity domain obtained in the decomposition theorem for $\hat{H}^{(6)}$ (see Sect. 1, Eq.(11)).

A standard interpolation technique in two complex variables[+] then yields the crossing domain on Fig. 10, if $\zeta_\alpha - \mu^2$ is small compared with the size of N_α: this is satisfied if α is chosen sufficiently small.

To obtain the crossing property for $\hat{H}^{(6)}$ itself, it is sufficient to notice that $\hat{H}_F^{(6)}$ is analytic in a neighbourhood Θ of Θ_α in M_6^c, and that in Θ, the tree-graph terms

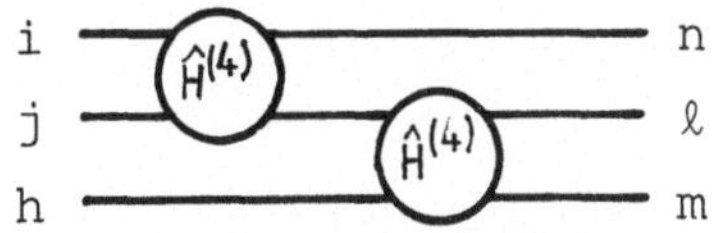

are analytic except on the corresponding poles $(k_i+k_j+k_n)^2 = m^2$ (this follows from the analyticity of the $2 \to 2$ forward scattering).

[+]See for instance [5b], Sect. 5.

REFERENCES

1. K. Symanzik, J. Math. Phys. 1 (1960) 249, and in "Symposium on Theoretical Physics", 3 (1967) 121, New York, Plenum Press.
2. a) J. Bros in "Analytic Methods in Mathematical Physics", p. 85, Gordon and Breach, New York (1970).

   b) J. Bros, M. Lasalle in "Structural Analysis of Collision Amplitudes", p. 97, North-Holland, Amsterdam (1976).

   c) J. Bros, in "Mathematical Problems in Theoretical Physics", Lecture Notes in Physics 116 (1979) 166, Springer Verlag, Berlin, Heidelberg, New York.
3. a) M. Lasalle, Comm. Math. Phys. 36 (1974) 185.

   b) J. Bros, M. Lassalle, Comm. Math. Phys. 43 (1975) 279.

   c) J. Bros, M. Lassalle, Comm. Math. Phys. 54 (1977) 33.

   d) J. Bros, M. Lassalle, Ann. Inst. H. Poincaré 27, 279.
4. H. Epstein, V. Glaser, R. Stora in "Structural Analysis of Collision Amplitudes" p. 7, North Holland, Amsterdam (1976), and references quoted therein.
5. a) R. Omnes in "Relations de dispersion et particules élémentaires", Hermann, Paris (1960), and references quoted therein.

   b) J. Bros, H. Epstein, V. Glaser, Comm. Math. Phys. 1 (1965) 240.

   c) J. Bros, H. Epstein, V. Glaser, Helv. Phys. Acta 45 (1972) 149.
6. a) D. Iagolnitzer, "The S-Matrix", North-Holland, Amsterdam (1978), and references quoted therein.

   b) D. Iagolnitzer, in this volume.
7. a) L.D. Landau, Nucl. Phys. 13 (1959) 181.

   b) R.J. Eden, P.V. Landshoff, D.I. Olive, J.C.Polkingnorne, "The analytic S-matrix", Cambridge Univ. Press (1966), and references quoted therein.
8. D. Fotiadi, M. Froissart, J. Lascoux, F. Pham, Topology (Gr.Brit.) 4 (1965) 159.

F. Pham, "Introduction à l'étude topologique des singularités de Landau", Gauthier-Villars, Mémorial des Sc. Math. 164 (1967).

9. G. Ponzano, T. Regge, E.R. Speer, M.J. Westwater, Comm. Math. Phys. 15 (1969) 83.

10. a) M. Kashiwara, T. Kawai, T. Oshima, Comm. Math. Phys. 60 (1978) 97, and references quoted therein.
b) A. Van den Essen in "Complex Analysis, Microlocal Calculus and Relativistic Quantum Theory", Lecture Notes in Physics 126 (1979) 117, Springer Verlag, Berlin, Heidelberg, New York.

11. D. Iagolnitzer, H.P. Stapp, Comm. Math. Phys. 14 (1969) 15.

12. G.F. Chew, "The Analytic S-matrix", W.A. Benjamin, New York (1966), and references quoted therein.

13. H.P. Stapp and R. White in "Structural Analysis of Collision Amplitudes" p. 275 and p. 431, North-Holland, Amsterdam (1976).

14. a) T. Spencer, F. Zirilli, Comm. Math. Phys. 49 (1976) 1.
b) M. Combescure, F. Dunlop, "n-particle irreducible functions in euclidean Q.F.T." Preprint IHES, Bures-sur-Yvette (1978).
c) H. Koch, Thesis, Genève Univ. (1979).

15. J. Leray, Bull. Soc. Math. de France 87 (1959) 81.

16. a) J. Bros, D. Pesenti, J. Math. Pures et appl. 58 (1980) 375.
b) J. Bros, D. Pesenti, "Fredholm resolvent in complex space: the non-holonomic case", in preparation.
c) J. Bros, D. Iagolnitzer, "Non-holonomic singularities of the S-matrix and Green functions", to be published in Comm. Math. Phys..

17. H. Epstein, V. Glaser, D. Iagolnitzer, to be published in Comm. Math. Phys..

18. a) R.F. Streater and A.S. Wightman, "PCT, Spin & Statistics and All That", Benjamin, New York, 1964.
b) R. Jost, "The General Theory of Quantized Fields", Am. Math. Soc., Providence, 1965.

19. a) H. Araki, "Einführung in die Axiomatische Quantenfeldtheorie", E.T.H. Zürich (1961-1962).
b) R. Haag and B. Schroer, J. Math. Phys. 3 (1962) 248.

20. K. Hepp in "Axiomatic Field Theory", p. 137, Brandeis University, M. Chretien and S. Deser Editors (1965).

21. a) V. Glaser, H. Lehmann, W. Zimmermann, Nuovo Cimento 6 (1957) 1122.
b) O. Steinmann, Comm. Math. Phys. 10 (1968) 245.

22. J. Bros, M. Manolessou-Grammaticou, Comm. Math. Phys. 72 (1980) 175 and 207.

23. R. Jost, H. Lehmann, Nuovo Cimento 5 (1957) 1598.
F.J. Dyson, Phys. Rev. 110 (1958) 1460.
J. Bros, A. Messiah, R. Stora, J. Math. Phys. 2 (1961) 639.

24. A. Katz, "Sur les formules de discontinuité en théorie quantique axiomatique des champs", Thèse de 3ème cycle, Orsay (1979).

Acta Physica Austriaca, Suppl. XXIII, 401–432 (1981)

GEOMETRICAL ASPECTS OF GAUGE CONFIGURATIONS+

by

A. TRAUTMAN

Inst. of Theoretical Physics, Warsaw Univ.

Hoża 69, 00-681 Warsaw, Poland

SUMMARY

These notes contain an informal description of the geometrical foundations of gauge theories. The theory of gravitation is compared to theories of the Yang-Mills type. Space-time symmetries of gauge configurations are defined in terms of automorphisms of principal bundles. Symmetry breaking is related to restricting the structure group of the bundle. The Liénard-Wiechert solution of the Yang-Mills equations is discussed in some detail. An approximate solution of the Yang-Mills equations is shown to allow for the phenomenon of radiation of the colour charge by a classical gluon field.

+Lectures given at the XX. Internationale Universitätswochen für Kernphysik, Schladming, Austria, February 17-26, 1981.

INTRODUCTION

The validity and usefulness of a geometrical view of gauge configurations have been accepted by the physics community [1] and a fair number of surveys written on the description of Yang-Mills theories in terms of infinitesimal connections on principal bundles over spacetime [2-10].

The present notes are based on the series of four lectures given by the author in February, 1981, in Schladming. They contain also a few changes and additions made after a similar series of lectures was delivered in March at the Collège de France. A part of the introductory material is omitted from these notes as it is included in a joint paper by M.E. Mayer and A. Trautman, appearing in this volume.

In my lectures I have tried to emphasize those aspects of the geometry of gauge configurations which I consider to be fundamental and important, but which have not received, so far, as much attention as they deserve. In particular, a considerable amount of time is devoted here to subtleties of symmetry breaking and of spacetime symmetries of gauge configurations. In the theory of gravitation, I like to stress the role of soldering and the rather different nature of translations. In the last lecture, I give a rather detailed analysis of the Liénard-Wiechert solution for the non-Abelian case, intended to show that geometrical methods provide convenient tools for the solution of concrete problems. In the last part of the notes, I discuss also the analogy between gravity and theories of the Yang-Mills type regarding the possibility of secular changes of their sources. At the classical level, one can consider the (hypothetical) phenomenon of radiation of the non-Abelian charge, such as colour, as an analogue to the phenomenon of change of the total energy and momentum due to gravitational waves.

I feel that the most important outcome of the bundle-theoretic approach to gauge fields may be a unification of fundamental interactions and an explanation of their hierarchy and symmetry breaking. This idea, which may be traced back to Weyl, Einstein, Kaluza, Klein, Pauli, Bargmann, Bergmann, and many other authors, was clearly formulated by B.S. DeWitt [11]. The Kaluza-Klein construction for a Yang-Mills field was later considered by several authors [6,12-15]. It was briefly described in my lectures, but is not included in the written text.

The notes should be read in conjunction with the "Brief introduction to the geometry of gauge fields" and other reviews listed in the Bibliography.

PURE GAUGE CONFIGURATIONS

A pure gauge configuration is given by a connection form ω on a principal bundle $\pi : P \to M$ with structure group G. The base space (spacetime) M is usually assumed to be oriented and endowed with a Riemannian (pseudo-Riemannian) metric ds^2. If M is 2n-dimensional, then the Hodge dual $*\alpha$ of an n-form α on M is invariant under the conformal transformations of the metric $ds^2 \to \sigma\, ds^2$. The curvature

$$\Omega = d\omega + \frac{1}{2} [\omega,\omega]$$

is a form defined on P, but since it is horizontal, its dual $*\Omega$ may be computed by reference to the metric and orientation on M. Since Ω is a 2-form, its dual is conformally invariant iff M is 4-dimensional and, in this case, the sourceless Yang-Mills equations

$$D * \Omega = 0$$

are also invariant under conformal changes of ds^2.

In physics, one works most often with local section of π. If U is an open subset of M and

$$s : U \to P , \qquad \pi \circ s = id_U$$

is such a local section, then

$$A = s^* \omega \qquad \text{and} \qquad F = s^* \Omega$$

are the potential and the field strength of the gauge configuration given by ω, relative to s. If s' is another section over the same domain as s, then there exists a map

$$g : U \to G$$

such that

$$s'(x) = s(x)g(x) \quad \text{for any} \quad x \in U .$$

Putting

$$A' = s'^* \omega \qquad \text{and} \qquad F' = s'^* \Omega$$

one obtains the classical transformation formulae

$$A' = g^{-1} A g + g^{-1} dg \quad \text{and} \quad F' = g^{-1} F g$$

where dg is understood as follows: embed G in a group of matrices and compute the derivatives of g entry by entry, i.e. if $g = (g^i_j)$ then $dg = (dg^i_j)$.

INTERACTIONS AND ADDITIONAL STRUCTURE

Clearly, pure gauge fields do not suffice to describe all physics. Moreover, even in such a "pure" case as gravity in empty spacetime, one needs the metric tensor in addition to the linear connection which - in this case - may be identified with the gauge degrees of freedom.

In general, one considers a representation ρ of G in a (finite-dimensional, real or complex) vector space V, i.e. a homomorphism of Lie groups,

$$\rho : G \to GL(V) .$$

This defines a vector bundle $\rho(P) \to M$, associated with P by ρ. One can also form the tensor-fibre product

$$\rho^k(P) = \rho(P) \otimes_M \Lambda^k T^* M, \ k = 1,2,\ldots,\dim M.$$

It is known (cf., for example [16], p.76) that sections of $\rho^k(P) \to M$ are in a bijective and natural correspondence with horizontal k-forms of type ρ on P. For any such k-form ϕ one defines its covariant exterior derivative by

$$D\phi = \text{hor } d\phi = d\phi + \rho'(\omega) \wedge \phi \tag{1}$$

where

$$\rho' : \mathcal{G} \to L(V)$$

is the homomorphism of the Lie algebra $\mathcal{G}$ of G into $L(V)$, derived from ρ,

$$\rho'(A)v = \frac{d}{dt}\, \rho(\exp\, t\, A)v\Big|_{t=0},\ v \in V,\ A \in \mathcal{G} .$$

Clearly, $D\phi$ corresponds to a section of $\rho^{k+1}(P) \to M$ and

$$D^2\phi = \rho'(\Omega) \wedge \phi \quad . \tag{2}$$

If $s : U \to P$ is a local section, then $s^*\phi$ is a local section of $V \otimes \Lambda^k T^*M \to M$, i.e., a V-valued k-form on U.

Quantities of this type occur in both differential geometry and the description of gauge fields interacting with other kinds of matter.

<u>Example 1.</u> Let $G = U(1)$ so that $P \to M$ is an electromagnetic

bundle. If $\rho_n : U(1) \to U(1) \subset GL(1,C)$ is the representation given by

$$\rho_n(z) = z^n , \quad z \in U(1) ,$$

then sections of $\rho_n(P)$ may be identified with wavefunctions of scalar particles of charge equal to n times the elementary charge. If $\phi : P \to C$ is the equivariant map corresponding to such a section,

$$\phi(pz) = z^{-n} \phi(p),$$

then

$$D\phi = d\phi + n \omega \phi$$

corresponds to the "minimal coupling prescription".

Example 2. If $G = SU(N)$ and $\rho = \mathrm{Ad}$ is the adjoint representation of G in its Lie algebra $\mathcal{G}$, then $\phi : P \to \mathcal{G}$ is a standard Higgs field.

Example 3. Let LM be the frame bundle of an n-dimensional manifold. Its structure group is $GL(n,R)$. If ρ is a tensor representation of $GL(n,R)$ in a vector space V, then a zero-form of type ρ, $\phi : P \to V$, corresponds to a tensor field of type ρ on M. In particular, if $V = R^n$ and $\rho = \mathrm{id}$ then ϕ corresponds to a vector field. If $V = L(R^n)$ and $\rho = \mathrm{Ad}$, then ϕ corresponds to a tensor field of valence (1,1). A metric tensor corresponds to the natural representation of $GL(n,R)$ in $L_s^2(R^n,R)$, i.e. in the vector space of symmetric $n \times n$ matrices, etc..

Example 4. Let $P \to M$ be a $GL(n,R)$-bundle over an n-dimensional manifold M. The bundle P is isomorphic (in the category of principal fibre bundles over M) to LM iff it admits a soldering form, i.e. a horizontal 1-form $\theta : TP \to R^n$ of type id [17].

Example 5. If M is n-dimensional, oriented and has a metric

tensor, then the Hodge dual $*$ can be defined on horizontal forms on LM. In particular, to any integer k, $0 \leq k \leq n$, there corresponds the horizontal $\Lambda^k R^n$-valued (n-k)-form with components

$$\eta_{\mu_1\mu_2\ldots\mu_k} = * (\theta_{\mu_1} \wedge \theta_{\mu_2} \wedge \ldots \wedge \theta_{\mu_k})$$

where $\theta = (g^{\mu\nu}\theta_\nu)$ is the soldering form. In particular, $\eta = * 1$ is the volume n-form.

Example 6. The torsion form

$$\Theta = D\theta = d\theta + \omega \wedge \theta$$

corresponding to a linear connection ω on LM is a horizontal 2-form of type id. From (2) there follows the Bianchi identity for torsion

$$D\Theta = \Omega \wedge \theta \ .$$

Example 7. The curvature Ω of a connection ω on any bundle P is a horizontal 2-form of type Ad.

A scalar product h on V is invariant under the action of G on V defined by ρ if, for any $a \in G$ and $u, v \in V$,

$$h(\rho(a)u, \rho(a)v) = h(u,v) \ .$$

By differentiation, this implies

$$h(\rho'(A)u,v) + h(u,\rho'(A)v) = 0 \ , \quad A \in \mathcal{G} \ .$$

In particular, the Killing-Cartan form on $\mathcal{G}$,

$$k(A,B) = \mathrm{Tr}\ \mathrm{Ad}'(A) \circ \mathrm{Ad}'(B) \ , \qquad A,B \in \mathcal{G} \ , \tag{3}$$

is invariant under the adjoint action of G in $\mathcal{G}$, and, since

$$Ad'(A)B = [A,B] \ , \tag{4}$$

the property of invariance implies

$$k([A,B],C) + k(B,[A,C]) = 0 \ , \tag{5}$$

for any $A,B,C \in G$.

Let (e_i) and (e_a) be linear frames (bases) in G and V, respectively. The components of k and h in these frames are, respectively,

$$k_{ij} = k(e_i,e_j) \quad \text{and} \quad h_{ab} = h(e_a,e_b).$$

If

$$[e_i,e_j] = c^k_{ij}\, e_k$$

then, from (3) and (4) there follows the formula

$$k_{ij} = c^k_{i\ell}\, c^\ell_{jk} \quad .$$

It is known that (k_{ij}) is non-singular iff G is semi-simple and, if, moreover, G is compact, then the quadratic form $k(A,A)$ is negative-definite. The connection and curvature forms may be represented by their components relative to (e_i),

$$\omega = \omega^i\, e_i, \ \Omega = \Omega^i\, e_i, \ \Omega^i = d\omega^i + \frac{1}{2}\, c^i_{jk}\omega^j \wedge \omega^k.$$

Similarly, $\phi : P \to V$ is written as

$$\phi = \phi^a\, e_a$$

and

$$D\phi = D\phi^a\, e_a$$

so that eq.(1) becomes

$$D\phi^a = d\phi^a + \rho^a_{bi}\, \omega^i \wedge \phi^b$$

where

$$\rho'(e_i)\, e_b = \rho^a_{bi}\, e_a \; .$$

If U is any function of the invariant

$$\phi^2 = h_{ab}\, \phi^a\, \phi^b$$

and η is a volume element on M, then the form on P given by

$$k_{ij} *\Omega^i \wedge \Omega^j + h_{ab} * D\phi^a \wedge D\phi^b + U(\phi^2)\pi^*\eta \qquad (6)$$

is horizontal and invariant under the action of G. Therefore, it defines a form on M, denoted by L, and used to formulate the principle of least action

$$\delta \int_M L = 0 \; .$$

This is the starting point of <u>classical gauge theories of the Yang-Mills type.</u> The field ϕ is referred to as a (generalized) <u>Higgs field</u> and the property of (6) which consists in the appearance of the derivatives of ϕ only through the form $D\phi$ is a reflection of the <u>principle of minimal coupling</u> between the gauge configuration and the matter fields interacting with it. Incidentally, the scalar product k occurring in (6) need not be given by the Killing-Cartan form (3): it may be any scalar product on G invariant under the adjoint action of G.

BREAKING OF SYMMETRY

The mechanism of spontaneous symmetry breaking has a simple interpretation in terms of restrictions of principal bundles [9,18]. It is discussed in considerable detail in the lectures by Meinhard E. Mayer appearing in this volume. In this short section I wish only to stress the analogy between the role of the metric tensor in general relativity theory and the breaking of symmetry by a normalized Higgs field. As an illustration, the restriction of the SO(3)-bundle over S_2, induced by the 't Hooft-Polyakov solution, is discussed in considerable detail.

Consider a G-bundle $\pi : P \to M$ and a representation ρ of G in a vector space V. Let

$$\phi : P \to W \quad V$$

be a field of type ρ with values in an orbit W of G. In other words, G acts transitively on the manifold W of values of ϕ. Therefore, given a fixed point $w_o \in W$, in each fibre $\pi^{-1}(x)$ of P there is at least one point q such that $\phi(q) = w_o$. The set of all such points,

$$Q = \{q \in P : \phi(q) = w_o\} ,$$

is a submanifold of P, and the restriction of π to Q defines a fibre bundle $Q \to M$. It is a principal bundle: its structure group H is the isotropy (stability) or little group of w_o,

$$H = \{a \in G : \rho(a) w_o = w_o\} .$$

The embedding $Q \to P$, $H \to G$ defines a restriction of P to H in the sense described in the "Brief Introduction".

Conversely, if such a restriction is given, one can define $\phi : P \to G/H$ by putting $\phi(qa) = a^{-1}H$ for $q \in Q$ and $a \in G$. Under rather general assumptions the "non-linear realization" of G in $W = G/H$ can be extended to a linear representation of G in a vector space V containing W.[+]

[+] I am indebted to Profs. L. Michel and A. Sparzani for having drawn my attention to the last problem and explained its subtleties.

Example 8. Let H be a closed subgroup of GL(n,R). A restriction of the bundle LM of linear frames of an n-dimensional manifold M defines an H-structure on M. In particular, a Riemannian geometry on M is an O(n)-structure. In other words, the introduction of a metric tensor of Lorentz signature on spacetime is equivalent to breaking the symmetry from GL(4,R) down to the Lorentz group.

Example 9. Consider the static, spherically symmetric solution of the Yang-Mills equations with G = SO(3) and a source corresponding to a standard Higgs field. The 't Hooft-Polyakov Ansatz assumes a regular potential A; therefore, the corresponding bundle is trivial. Removing the time axis from R^4, one can represent

$$M = \{(t,x,y,z) \in R^4 : x^2 + y^2 + z^2 > 0\}$$

as the product $R^2 \times S_2$. Because of spherical symmetry it suffices to consider the trivial bundle

$$P = S_2 \times SO(3) \to S_2 \quad . \tag{7}$$

The (normalized) Higgs field

$$\phi : S_2 \times SO(3) \to S_2 \quad R^3 = \text{Lie algebra of } SO(3)$$

is given by

$$\phi(r,a) = a^{-1} r \tag{8}$$

where

$$r = (x,y,z) \in S_2 \text{ and } a \in SO(3).$$

Let

$$r_o = (0,0,1) \; ;$$

then

$$H = \{a \in SO(3) : ar_o = r_o\} = SO(2)$$

and

$$Q = \{(r,a) \in P : a^{-1}r = r_o\} \approx SO(3).$$

Therefore ϕ reduces the trivial SO(3)-bundle (7) to a non-trivial SO(2)-bundle

$$\pi : SO(3) \to S_2 , \qquad \pi(a) = ar_o . \qquad (9)$$

The latter bundle is isomorphic to the bundle of oriented dyads of S_2 and admits a canonical SO(2)-connection corresponding to a magnetic pole with a charge equal to twice the lowest (Dirac) value [19]. This connection is obtained by projecting the one on P onto the direction of r_o, considered as an element of the Lie algebra of SO(3).

The bundle (7) admits an obvious section

$$s : S_2 \to P , \qquad s(r) = (r,I)$$

where I is the unit element of SO(3). The pull-back $s^*\phi$,

$$s^*\phi(r) = r ,$$

corresponds to the "hedgehog" representation. On the other hand, if s' is any (local) section of (9), then

$$s'^*\phi(r) = r_o$$

corresponds to the description of the same Higgs field in a gauge exhibiting an alignment of the field along the third axis.

The field ϕ given by (8) is spherically-symmetric. To make this statement meaningful it is necessary to lift the action of SO(3) from the base, S_2, to the bundle,

$P = S_2 \times SO(3)$. This can be done in many ways. For example, if $a,b \in SO(3)$ and $r \in S_2$, then the formula

$$\gamma_a^o(r,b) = (ar,b)$$

defines a lift, but

$$\phi \circ \gamma_a^o \neq \phi \text{ , unless } a = I.$$

However, the action given by

$$\gamma_a^1(r,b) = (ar,ab)$$

leaves ϕ invariant,

$$\phi \circ \gamma_a^1 = \phi \quad .$$

SYMMETRIES OF GAUGE CONFIGURATIONS

The last example leads to the following general question: how to define space time symmetries of gauge configurations, and, in particular, of infinitesimal connections and of Higgs fields?

To appreciate the subtleties of the problem, consider first a simple, local situation. Let A be the G-valued 1-form of potential defined on M and $f : M \to M$ a transformation (diffeomorphism). One says that <u>A is invariant under f</u> if there exists (a gauge transformation) $g : M \to G$ such that

$$f^*A = g^{-1} Ag + g^{-1} dg \quad . \tag{10}$$

Clearly, eq.(10) implies

$$f^*F = g^{-1}\,Fg \ , \text{ where } F = dA + \frac{1}{2}\,[A,A] \ . \tag{11}$$

The converse is true if $H^2(M,R) = 0$ and G is Abelian. Indeed, if G is Abelian, then (11) reads

$$f^*F = F \qquad \text{and} \qquad F = dA \ ,$$

therefore

$$0 = f^*dA - dA = d(f^*A - A)$$

and $f^*A - A$ is exact by virtue of the topological assumption. However, for a non-Abelian G, the implication (11) → (10) is false even in the case $M = R^n$. For example, let G = SO(3) and consider a "plane wave" in Minkowski space [20,21],

$$A(t,x,y,z) = (a(u)x + b(u)y)du \ ,$$

where

$$u = t - z \ , \quad a,b : R \to G \approx R^3 \quad \text{and} \quad [a,b] \neq 0 \ .$$

The field

$$F = (adx + bdy) \wedge du$$

is invariant under the translation $(t,x,y,z) \to (t,x+\lambda,y,z)$, but the potential is not.

The topological condition

$$H^2(M,R) = 0$$

is essential for the implication (11) → (10) to be true in the Abelian case. For example, the field strength of a magnetic pole, $\vec{B} = q\vec{r}/r^3$, exhibits spherical symmetry for any q, but the corresponding potential is spherically symmetric - in the sense of eq. (10) - iff the magnetic charge q satisfies the Dirac quantization condition.

A section $s : M \to P$ defines an isomorphism of

bundles $i : M \times G \to P$ given by

$$i(x,a) = s(x)\, a \qquad \text{where} \quad x \in M \quad \text{and} \quad a \in G.$$

A connection form ω on P, pulled-back by this isomorphism, assumes the form

$$\omega_s = i^*\omega = a^{-1}(da + A(x)a)$$

where

$$A = s^*\omega \quad .$$

An automorphism $h : P \to P$ (cf. the Brief Introduction) covering $f : M \to M$, pulled-back by i to $M \times G$, becomes $h_s = i^{-1} \circ h \circ i$ where

$$h_s(x,a) = (f(x), g(x)^{-1}a) \text{ and } h(s(x))g(x) = s(f(x)).$$

A simple computation gives

$$h_s^* \, \omega_s = a^{-1} \; (da + A'a)$$

where

$$A' = gf^* \, Ag^{-1} - (dg)g^{-1} \; .$$

Therefore, condition (10) is equivalent to

$$h_s^* \, \omega_s = \omega_s$$

and this, in turn, is equivalent to

$$h^*\omega = \omega \quad . \tag{12}$$

To summarize, a gauge configuration described by a connection form ω on a principal bundle $P \to M$ is defined to be invariant under a diffeomorphism $f : M \to M$ if there is a lift h of f to Aut P such that (12) holds. The analogous condition of invariance for a Higgs field $\phi : P \to V$ is

$$\phi \circ h = \phi . \quad (13)$$

It is instructive to list typical situations when diffeomorphisms can be lifted and to give examples showing that this cannot always be done.

Example 10. If $P = M \times G$ then $f : M \to M$ lifts to $h: M\times G \to M\times G$ given by $h(x,a) = (f(x),a)$. Moreover, the group Aut P is a semi-direct product of G^M by Diff M [17].

Example 11. If P is a natural bundle - a bundle $L^r M$ of frames of M of differential order r - then f can be lifted to P by the very definition of a natural bundle.

Example 12. If both M and G are compact, then any one-parameter group (f_t) of diffeomorphisms of M can be lifted to a one-parameter group (h_t) of automorphisms of $\pi : P \to M$ in such a way that $\pi \circ h_t = f_t \circ \pi$. It suffices to take a connection on P and the flow generated by the horizontal lift of the vector field induced on M by (f_t).

However, a diffeomorphism which is not homotopic to the identity need not lift:

Example 13. Complex conjugation on $U(1)$, $f(z) = \bar{z}$, does not lift to the exponential bundle $\pi : R \to U(1)$, $\pi(t) = \exp(2\pi\, it)$, considered as a principal Z-bundle. (It does, however, lift to an automorphism of the bundle structure.)

Example 14. Similarly, the space inversion $f : S_2 \to S_2$, $f(r) = -r$, does not lift to the principal Hopf bundle $S_3 \to S_2$.

From the point of view of physical applications, we are most often interested in lifting a Lie group of transformations of M to a Lie group of automorphisms of $\pi: P \to M$, covering the given action in M. There are subtleties, as shown by the following

Example 15. Any rotation of S_2 lifts to the Hopf bundle $S_3 \to S_2$ (this follows from example 12), but the natural action of SO(3) on S_2 does not lift. However, the action of SU(2) does lift, as is obvious from the identification $S_3 \approx SU(2)$.

Example 16. There is a simple construction of all G-bundles $\pi : P \to M$ admitting a Lie group K of automorphisms transitive on the fibres of π (cf., e.g. [16] p. 105 and [22]). Let J be the subgroup of K, leaving invariant a point $x_o \in M$. If $p_o \in P$ is such that $\pi(p_o) = x_o$, then for any $a \in J$, the point ap_o is in the same fibre as p_o. There is thus an element $\lambda(a)$ of G such that

$$a\, p_o = p_o\, \lambda(a)$$

and

$$\lambda : J \to G \qquad (14)$$

is a homomorphism. Moreover, $K \times G$ acts transitively on P by

$$(a,b)\, p = apb^{-1}$$

and

$$(a,b)\, p_o = p_o \leftrightarrow a \in J \text{ and } b = \lambda(a) \;.$$

Therefore, P is diffeomorphic to the quotient $(K \times G)/J$, where J is assumed to act on $K \times G$ by

$$(a,b)c = (ac,\, b\lambda(c)) \;, \quad c \in J \;.$$

Conversely, given a group K acting transitively on M, and a homomorphism (14) of the stability group J of a point $x_o \in M$, one constructs the bundle by taking

$$P_\lambda = (K \times G)/J$$

and defining the action of $K \times G$ on P_λ in the standard manner.

For example, if $K = SU(2)$, $G = U(1)$ and $M = S_2$, then $J = SO(2) = U(1)$ and all homomorphisms $\lambda : J \to G$ are of the form

$$\lambda_n(z) = z^n \quad \text{for some} \quad n \in Z .$$

It is seen by inspection that P_{λ_n} is isomorphic to the lens space

$$SU(2)/Z_n \quad , \quad n \in Z . \tag{15}$$

If one starts, however, with $K = SO(3)$, then one gets in this manner only the even lens spaces,

$$SO(3)/Z_n \approx SU(2)/Z_{2n} \quad . \tag{16}$$

In terms of magnetic monopoles, the bundles (15) and (16) correspond to the Dirac and Schwinger quantization conditions, respectively.

GRAVITATION+

The similarities and differences between gravitation and theories of the Yang-Mills type have been discussed by many authors (cf., e.g., [9], [23] and the references given there). In this brief section, I intend only to summarize my views and to comment on the "Abelian nature of gravitational waves".

1. Gravitation is a theory based on fibre bundles wich are

+This section has been influenced, in part, by a discussion with R.P.Wallner. I am indebted to Peter Aichelburg and Roman Sexl of the University of Vienna for hospitality and stimulating conversations. I wish also to acknowledge discussions on this matter held at various occassions with Jürgen Ehlers, Friedrich Hehl, J.Nitsch and M. Schweizer.

"concrete" and, as such, have a richer structure (cf. Example 4) than "abstract" bundles underlying gauge theories of the Yang-Mills type.

2. The soldering form on LM leads to the notion of torsion which has no analogue in electromagnetism or the Yang-Mills theory.

3. The metric tensor is somewhat analogous to a Higgs field: it breaks down the full linear symmetry to the Lorentz group (cf. Example 8). There is a complete analogy between the condition of compatibility between the metric tensor ($g_{\mu\nu}$) and a linear connection, expressed by

$$Dg_{\mu\nu} = 0 \tag{17}$$

and a similar equation

$$D\phi^{a} = 0 \tag{18}$$

assumed to be satisfied by the ground state of a gauge configuration.

4. There has been a lot of discussion on the choice of the structure group G for a theory of gravitation. Essentially, there are two (minor) problems to consider:

(i) whether one wishes to incorporate translations - or even take them as the starting point;

(ii) whether the metric is to be introduced ab initio, as part of the definition of the bundle, or rather as an additional structure, restricting the linear or the affine group to its Lorentz or Poincaré subgroup, respectively.

Concerning the first problem, it is convenient to summarize the situation by reference to the following diagram of bundles over an n-dimensional manifold:

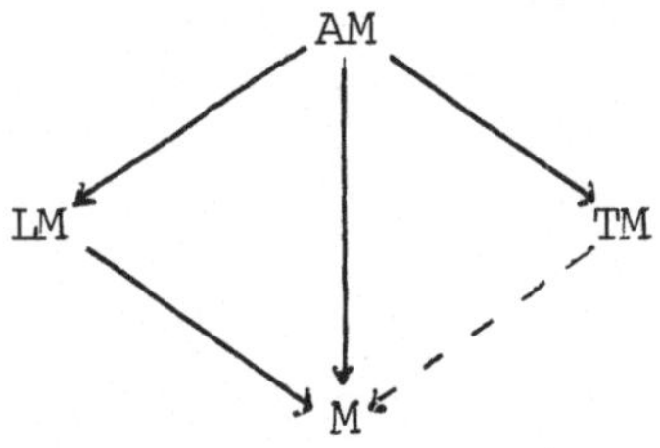

(19)

Here TM is the tangent bundle and

$$AM = LM \underset{M}{\times} TM$$

is the affine bundle. Its structure group is the general affine group,

$$GA(n,R) \quad = \quad GL(n,R) \times R^n$$

which acts on AM as follows: if $e : R^n \to T_xM$ is a linear frame at $x \in M$, considered as an isomorphism of vector spaces, $u \in T_xM$, $a \in GL(n,R)$ and $b \in R^n$, then

$$(e,u)(a,b) \;=\; (ea,\; u + e(b)) \;.$$

All *solid* arrows in (19) denote projections of *principal* bundles with appropriate groups. The tangent bundle, however, does not admit a natural structure of principal bundle. It is easy to see that the introduction of an R^n-action on TM, to make out of it a principal bundle, is equivalent to giving a global section of $LM \to M$, i.e. a teleparallelism structure on M [24]. A connection on $TM \to M$, compatible with such a structure, has torsion but no curvature.

There is no essential difference between considering connections on LM and AM. According to a classical theorem (cf.[16], p. 127), any affine connection is defined by a linear connection and a tensor field of type Ad.

5. An essential difference between theories of the Yang-Mills type and gravitation is in the choice of the Lagrangian leading to the field equations. In the Yang-Mills case, there is only one kind of duality that can be used to construct the left-hand side of the field equations: the Hodge dual operating on Ω (or F) considered as a G-valued 2-form. In the gravitational case, there is the Levi-Civita $\Lambda^2 R^n$-valued (n-2)-form $\eta_{\mu\nu}$ (cf. Example 5) which leads to the Einstein(-Cartan) Lagrangian

$$\frac{1}{2}\,\eta_{\mu\nu} \wedge \Omega^{\mu\nu} \quad .$$

6. The vanishing of torsion, assumed in Einstein's theory, considerably reduces the "degrees of freedom" inherent in a metric connection. In particular, plane gravitational waves are, in a well-defined sense, Abelian. To see this, consider the metric

$$ds^2 = e^1 \otimes e^1 + e^2 \otimes e^2 + e^3 \otimes e^4 + e^4 \otimes e^3 \qquad (20)$$

where

$$e^1 = dx, \quad e^2 = dy, \quad e^3 = du, \quad e^4 = dv + H\,du \qquad (21)$$

and H is a function of the coordinates u = t-z, x and y. The connection 1-form Γ, referred to the coframe (e^μ),

$$\Gamma_{\mu\nu} = (a_{\mu\nu}(u)\,x + b_{\mu\nu}(u)y)\,du \qquad (22)$$

is compatible with (20) iff

$$a_{\mu\nu} + a_{\nu\mu} = 0 \qquad \text{and} \qquad b_{\mu\nu} + b_{\nu\mu} = 0\,, \qquad (23)$$

i.e. iff the polarization "vectors" $a = (a_{\mu\nu})$ and $b = (b_{\mu\nu})$ have values in the Lie algebra SO(1,3) of the Lorentz group. The vanishing of torsion,

$$de^\mu + \Gamma^\mu_{\ \nu} \wedge e^\nu = 0\,, \qquad (24)$$

restricts the values of a,b to the commutative Lie subalgebra n of SO(1,3), isomorphic to R^2 [25]+. Therefore

$$[a,b] = 0$$

and the gravitational wave, described completely by eqs. (20) - (24), is really plane, in contradistinction to a plane-fronted Yang-Mills wave with $[a,b] \neq 0$.

GROUPS OF GAUGE TRANSFORMATIONS [21]

Gauge transformations may be defined as automorphisms of a principal bundle preserving the absolute elements of a gauge theory. Putting it in a slightly different way, a gauge theory is based on a category C, which is a subcategory of the category of principal G-bundles over M. Gauge transformations are simply isomorphisms in C.

For any principal bundle $P \to M$, there is the exact sequence

$$I \to \mathrm{Aut}_o P \to \mathrm{Aut}\, P \to \mathrm{Diff}\, M \qquad (25)$$

where $\mathrm{Aut}_o P$ is the group of vertical (based) automorphisms of P. If $\mathcal{G}$ is the subgroup of Aut P preserving the absolute elements of $P \to M$ and

$$\mathcal{G}_o = \mathcal{G} \cap \mathrm{Aut}_o P \,,$$

then one can form the exact sequence

$$I \to \mathcal{G}_o \to \mathcal{G} \to \mathcal{G}/\mathcal{G}_o \to I \,.$$

+Moshe Flato pointed out that n corresponds to the nilpotent part of the Iwasawa decomposition of the Lorentz group. I gratefully acknowledge the hospitality extended to me by M. Flato during my visit to Dijon.

The elements of $\mathcal{G}_o$ are <u>pure gauge transformations,</u> whereas the elements of $\mathcal{G}$ can be referred to as <u>gauge transformations.</u> In the case of a Yang-Mills theory over S_4, the group $\mathcal{G}/\mathcal{G}_o$ coincides with O(5) whereas $\mathcal{G}_o$ is "large". In Einstein's theory, $\mathcal{G}_o = \{id\}$, but $\mathcal{G} = \text{Diff } M$ is "large". The sequence (25) splits if (i) $P \to M$ is trivial [17] or (ii) P is natural.+ I do not know whether it splits in any case not covered by (i) or (ii).

TIME-DEPENDENT GAUGE CONFIGURATIONS

There are not many time-dependent exact solutions of the Yang-Mills equations. Coleman's plane-fronted waves have been already briefly discussed here. Waves with spherically-symmetric wave fronts have wire singularities [25,26]. In this section, I analyze in some detail the Liénard-Wiechert solution, adapted by Arodź [27] to the Yang-Mills case. A method used by Roskies [28] to study the asymptotic properties of Yang-Mills configurations leads to a simple estimate of the rate of radiation of the colour charge carried by a classical gluon field. There is an analogy between the energy-momentum vector in Einstein's theory and the colour charge in chromodynamics. Physically, the analogy is related to a presumed similarity in the self-interaction of gravitons and gluons. To appreciate the analogy formally, it is convenient to write both the Einstein and Yang-Mills equations in terms of differential forms,

$$dU - 4\pi i = 4\pi j$$

where j is a vector-valued 3-form describing the sources. In the Yang-Mills case, U is the Hodge dual, $*F$, of the field strength, i.e. a G-valued 2-form. The 3-form

+I am indebted to Ivan Kolàr for having pointed out this to me.

$$i = -\frac{1}{4\pi}[A,*F]$$

is the gluon contribution to the conserved total current $i + j$.

Similarly, in the case of gravitation, $U = (U^\mu)$ is the 2-form of Freud's "superpotential" [29]

$$4U^\mu = *(dx^\mu \wedge dx^\nu \wedge dx^\rho) \wedge \omega_{\nu\rho} \quad ,$$

where $\omega^\nu_\rho = \Gamma^\nu_{\rho\sigma} dx^\sigma$ are the connection 1-forms. The currents $i = (i_\mu)$ and $j = (j_\mu)$ correspond, respectively, to the energy-momentum densities of the gravitational field and of its source,

$$i_\mu = *dx^\nu \, t_{\mu\nu} \quad , \qquad j_\mu = *dx^\nu \, T_{\mu\nu} \quad ,$$

where $(t_{\mu\nu})$ is the Einstein "pseudotensor" and $(T_{\mu\nu})$ is the stress tensor of the source [10].

In both cases the field contribution i to the total current is highly gauge-dependent: no physical meaning can be attached to the notion of a local distribution of either gravitational energy or colour charge of the gluon field. If the fields satisfy suitable boundary conditions at large distances, say F or $\Gamma = O(r^{-2})$, one can compute the total charge q (energy-momentum or colour) from the Gauss law

$$4\pi q = 4\pi \int_{B_R} (i + j) = \oint_{S_R} U \quad (R \to \infty) \quad , \tag{26}$$

where S_R is the surface (boundary) of a ball B_R of radius R. The surface integral converges for $R \to \infty$ even if F or $\Gamma = O(r^{-1})$, provided that the "electric" component of the $1/r$ part of the field is tangent to S_R. Such is the case of pure outgoing waves. The boundary conditions are presumed to fix the gauge at large distances so that q is well-defined, up to the transformation $q \to g^{-1} q g$, $g \in G$ (up to a Lorentz transformation in the case of gravity). It should be noted, however, that imposing suitable boundary conditions is a subtle matter.

To construct the Liénard-Wiechert solution of the Yang-Mills equations with group G, consider a point particle of colour charge q, whose history is represented by a time-like world-line z in Minkowski space. A priori, the charge may depend on time. It is, therefore, a (dimensionless) function $q : R \to G$.

Let $z^\mu(s)$ be the Cartesian coordinates of z. The world-line is parametrized by its proper time, $g_{\mu\nu} \dot{z}^\mu \dot{z}^\nu = 1$, $\dot{z}^\mu = dz^\mu/ds$, $\dot{z}^0 > 0$. One associates with z a system of comoving spherical coordinates (t,r,θ,ϕ) by writing [30]

$$x^\mu = z^\mu(u) + r\ell^\mu(\theta,\phi)/p(u,\theta,\phi) \ ,$$

where $u = t-r$ is a retarded time,

$$r = g_{\mu\nu}(x^\mu - z^\mu(u))\dot{z}^\nu(u) \geq 0$$

is a radial distance measured in the rest frame of an observer moving along z, and $\ell = (\ell^\mu)$ is the null vector field,

$$\ell = (1, \sin\theta \cos\phi, \sin\theta \sin\phi, \cos\theta) \ .$$

These definitions imply

$$g_{\mu\nu}\, dx^\mu\, dx^\nu = (1-rp^{-1}\dot{p})\, du^2 + 2\, du\, dr - p^{-2}\, r^2(d\theta^2 + \sin^2\theta d\phi^2)$$

where a dot denotes differentiation with respect to u. The function $p = g_{\mu\nu} \dot{z}^\mu \ell^\nu$ has a simple physical interpretation: if ω is the frequency of a beam of light moving in the (θ,ϕ)-

direction, as seen from rest in the coordinate system (x^μ), then $\omega p(u,\theta,\phi)$ is the frequency of the same beam measured at $z^\mu(u)$ by the observer moving along z. The world-line z is straight, $\ddot{z}^\mu = 0$, if and only if $\dot{p} = 0$.

The Liénard-Wiechert potential may be written as

$$A = q(u)\ r^{-1}\ \dot{z}_\mu(u)\ dx^\mu \quad . \tag{27}$$

This form of the potential fixes the gauge almost completely; the only remaining freedom is of constant gauge transformations, $q \to g^{-1}\ qg,\ g \in G$. Such global gauge transformations are not enough to align a time-dependent colour along a fixed direction.

The field strengths corresponding to (27),

$$F = qr^{-2}\ du \wedge dr + r^{-1}\ du \wedge \{\dot{q}dr + grd(p^{-1}\dot{p})\}\ , \tag{28}$$

satisfy the Yang-Mills equation $D * F = 0$ $(r \neq 0)$ if and only if

$$\dot{q} + [q,\dot{q}] = 0 \tag{29}$$

and

$$\frac{\partial}{\partial u}\ (p^{-1}\ \dot{q}) = 0 \quad . \tag{30}$$

If G is either (i) Abelian, or (ii) compact and semi-simple, then eq.(29) implies $\dot{q} = 0$. It is worth noting that, in the important case (ii), strict conservation of colour follows from the Yang-Mills equation alone. Moreover, the field (28) has the same structure as in electrodynamics; it contains a Coulomb-like r^{-1} term and a radiative r^{-2} term, linear in $(\ddot{z}^\mu)$. Clearly, the latter term gives rise to outgoing radiation of energy. The expressions for the Poynting vector and the total intensity may be obtained from the corresponding formulae derived in electromagnetism by replacing the square of the electric charge by $-\operatorname{Tr} q^2$. Moreover, it is easy to see that, if $\dot{q} = 0$, then the colour current j corresponding to (27) is a distribution with support on the world-line z.

If $\dot{q} \neq 0$, then eq. (30) leads to

$$\dot{p} = 0 \qquad \text{and} \qquad \ddot{q} = 0 \; .$$

The particle is thus unaccelerated and its colour changes linearly with time,

$$q(t) = at + b \; , \tag{31}$$

where, by virtue of eq. (29),

$$[a,b] = a \neq 0 \; .$$

The x-coordinates may now be adjusted so that p = 1 and the solution assumes a manifestly spherically-symmetric form,

$$A = r^{-1} \; (au + b) \; dt \; , \tag{32}$$

$$F = r^{-2} \; (at + b) \; dt \wedge dr \; . \tag{33}$$

In this case, colour appears to change, but there is no transfer of energy by the gluon field. Incidentally, the field strengths (33) may also be derived from the potential r^{-1} (at + b) dt which is not gauge equivalent to (32).

Gauss's law applied to the field strengths (33) gives the time-dependent charge (31). A closer analysis shows, however, that the gauge configuration described by (32) and (33) is, in fact, time-independent. This is easily seen in the simple, but typical, case when G = SL(2,R) and

$$a = \begin{bmatrix} 0 & 1 \\ 0 & 0 \end{bmatrix}, \quad b = \begin{bmatrix} -1/2 & 0 \\ 0 & 1/2 \end{bmatrix}, \quad I = \begin{bmatrix} 1 & 0 \\ 0 & 1 \end{bmatrix}.$$

If $S = I + au$, then $S^{-1} = I - au$ and

$$A' = S^{-1} AS + S^{-1} dS = r^{-1} b\, dt + a\, du,$$

$$F' = S^{-1} FS = r^{-2} (ar + b)\, dt \wedge dr \quad . \tag{34}$$

The gauge-transformed field (34) is explicitly time-independent, but contains an unexpected r^{-1} term which makes it impossible to apply Gauss's law.

Returning to the analogy between the Yang-Mills and Einstein theories, one should bear in mind that a single, spherically symmetric body cannot radiate gravitational waves, but a system of bodies, moving under their mutual attraction, is believed to lose energy due to gravitational radiation. One is thus led to consider the asymptotic behaviour of a bounded, time-dependent source of a Yang-Mills field [28]. An approximate computation, described below, shows that the total colour charge of such a source may indeed change as a result radiation of colour, in analogy to the gravitational case.

Consider a classical Yang-Mills field in Minkowski space. Introduce a system of spherical coordinates (r,θ,ϕ) and put $u = t - r$. Assume that there is a gauge such that

$$j = O(r^{-3}) \text{ and } A = A_o + O(r^{-3}), \text{ for large } r,$$

where

$$r^2 A_o = (Kr+P)\,du + (Lr+Q)\,dr + (Mr+R)\,r\,d\theta + (Nr+S)\,r \sin\theta\, d\phi, \tag{35}$$

and the Lie algebra-valued functions $K, L, \ldots, S$ depend on

u, θ, and ϕ only. It follows from these assumptions that the field strength is

$$F = F_o + O\,(r^{-3})$$

where

$$F_o = dA_o + \frac{1}{2}\,[A_o, A_o] = O(r^{-1}) \ .$$

The Yang-Mills equation

$$d * F + [A, * F] = 4\pi j$$

is seen to be equivalent, to order r^{-3}, to

$$d * F_o + [A_o, * F_o] = O(r^{-3}) .$$

The last equation reduces to the system

$$\ddot{L} = 0 \quad , \qquad \dot{L} + [L, \dot{L}] = 0 \quad , \tag{36}$$

$$\frac{\partial}{\partial u}\left(\frac{\partial L}{\partial \theta} - [L, M]\right) = [L, \dot{M}] \quad , \tag{37}$$

$$\frac{\partial}{\partial u}\left(\frac{1}{\sin\theta}\frac{\partial L}{\partial \phi} - [L, N]\right) = [L, \dot{N}] \ , \tag{38}$$

$$\frac{\partial}{\partial u}\,(K + [K, L] + \dot{Q}) + [K, \dot{L}] = [M, \dot{M}] + [N, \dot{N}] + $$

$$+ \frac{1}{\sin\theta}\frac{\partial}{\partial u}\left(\frac{\partial}{\partial\theta} M \sin\theta + \frac{\partial N}{\partial\phi}\right) \quad , \tag{39}$$

where dots denote again derivatives with respect to u. Any solution to this system gives an approximate solution (35) of the Yang-Mills equation. Asymptotically, the solution is an outgoing (retarded) wave. Eqs.(36) can be solved, $L = au + b$, where a and b depend on θ and ϕ, and

$$[a, b] = a \quad . \tag{40}$$

For G semi-simple and compact, the only solution to (40) is a = 0. Therefore $\dot{L} = 0$, and if L = 0 is assumed, then the system of equations reduces to (39) [28]. The field strength is now

$$F = du \wedge (\dot{M} d\theta + \dot{N} \sin\theta \, d\phi) + r^{-2}(K + \dot{Q}) \, du \wedge dr + \ldots \quad (41)$$

where dots stand for terms which do not contribute to the surface integral (26). The first term in (41) is of order 1/r and represents the radiative, purely transverse part of the field. Total colour charge can be now evaluated by computing (26) for u = const. and $r = R \to \infty$. Using $* \, (du \wedge dr) = r^2 (d\theta \wedge \sin\theta \, d\phi)$ and equ. (39) one obtains

$$4\pi q = \oint (K + \dot{Q}) \, d\theta \wedge \sin\theta \, d\phi \quad ,$$

$$4\pi \dot{q} = \oint ([M, \dot{M}] + [N, \dot{N}]) \, d\theta \wedge \sin\theta \, d\phi \quad , \quad (42)$$

where the integrals are taken over a unit sphere. It is clear from (42) that radiation of colour is a truly non-linear and non-Abelian phenomenon requiring at least two particles with non-commuting charges. Radiated energy is computed from the Yang-Mills Poynting vector,

$$4\pi \dot{E} = \oint \mathrm{Tr}(\dot{M}^2 + \dot{N}^2) \, d\theta \wedge \sin\theta \, d\phi \quad . \quad (43)$$

Here E is the total energy of the system and Tr denotes the scalar product in the Lie algebra of G, defined by its (negative-definite) Killing form k.

ACKNOWLEDGEMENT

These notes are based on lectures given at the XX. Internationale Universitätswochen für Kernphysik, Schladming, Austria, February 17-26, 1981. I am grateful to Professor

H. Mitter for the hospitality in Schladming. The notes have been written in March, 1981, during my visit to the Collège de France. I thank Professors André Lichnerowicz and Yvonne Choquet for stimulating discussions and warm hospitality in Paris. I am also indebted to Mme M.-P.Serot Almeras for her help in editing my manuscript.

REFERENCES

1. T.T. Wu and C.N. Yang, Phys. Rev. D12 (1975) 3845.
2. A. Trautman, Rep. Math. Phys. (Toruń) 1 (1970) 29 and 10 (1976) 297.
3. W. Drechsler and M.E. Mayer, "Fiber Bundle Techniques in Gauge Theories", Lecture Notes in Physics No. 67, Springer-Verlag, Berlin (1977).
4. M.F. Atiyah, Geometrical Aspects of Gauge Theories, Proc. Intern. Congress Math., vol.II, pp. 881-885, Helsinki (1978).
5. A. Jaffe, Introduction to Gauge Theories, ibid. pp. 905-916.
6. R. Hermann, "Yang-Mills, Kaluza-Klein and the Einstein Program", Math. Sci. Press, Brookline, Massachussetts (1978).
7. M. Daniel and C.M. Viallet, Rev. Mod. Phys. 52 (1980) 175.
8. G.H. Thomas, Rev. Nuovo Cimento 3:4 (1980) 1-119.
9. A. Trautman, Fiber Bundles, Gauge Fields, and Gravitation, in "General Relativity and Gravitation", vol. I, pp.287-308, ed. by A. Held, Plenum Press, New York (1980).
10. W. Thirring, "A Course in Mathematical Physics", vol.II, Springer-Verlag, Wien-New York (1979).
11. B.S. DeWitt in "Relativité, Groupes et Topologie", p.725, edited by C. DeWitt and B.S. DeWitt, Gordon and Breach, New York (1964).
12. R. Kerner, Ann. Inst. Henri Poincaré 9 (1968) 143.

13. Y.M. Cho, J. Math. Phys. 16 (1975) 2029; Y.M. Cho and P.G.O. Freund, Phys. Rev. D12 (1975) 1711.
14. W. Kopczyński, Acta Phys. Polon. B10 (1979) 365 and in "Differential Geometric Methods in Mathematical Physics" p. 462, ed. by P. Garcia et al., Lecture Notes in Mathematics No. 836, Springer-Verlag, Berlin (1980).
15. A. Trautman, Lecture at Convegno di Relativita, Rome, 1980, to be published by the Accademia dei Lincei.
16. S. Kobayashi and K. Nomizu, Foundations of Differential Geometry, vol.I, Interscience-Wiley, New York (1963).
17. A. Trautman, in "Geometrical and Topological Methods in Gauge Theories", edited by J. Harnad and S. Shnider, Lecture Notes in Physics No. 129, Springer-Verlag, Berlin (1980).
18. J. Madore, Commun. Math. Phys. 56 (1977) 115.
19. A. Trautman, Intern. J. Theor. Phys. 16 (1977) 561.
20. S. Coleman, Phys. Lett. B70 (1977) 59.
21. A. Trautman, Bull. Acad. Polon. Sci., ser. sci. phys. et astron. 27 (1979) 7.
22. J.P. Harnad et al., Phys. Lett. B76 (1978) 589 and J. Math. Phys. 20 (1979) 931.
23. W. Thirring, Acta Phys. Austriaca, Suppl. XIX (1978) 439.
24. R.P. Wallner, Notes on Gauge Theory and Gravitation, UW Th Ph-81-3 preprint.
25. A. Trautman, J. Phys. A13 (1980) L1.
26. I. Robinson and A. Trautman, Phys. Rev. Lett. 4 (1960) 431.
27. H. Arodź, Phys. Lett. 78B (1978) 129.
28. R. Roskies, Phys. Rev. D15 (1977) 722.
29. P. Von Freud, Ann. Math. 40 (1939) 417.
30. A. Held, E.T. Newman and R. Posadas, J. Math. Phys. 11 (1970) 3145.

Acta Physica Austriaca, Suppl. XXIII, 433–476 (1981)

A BRIEF INTRODUCTION TO THE GEOMETRY OF GAUGE FIELDS+

by

M. E. MAYER
Dept. of Physics, Univ. of California
Irvine, California, 92717, USA

and

A. TRAUTMAN
Inst. of Theoretical Physics, Warsaw Univ.
Hoża 69, 00-681 Warsaw, Poland

1. INTRODUCTION

In view of the common background required for the understanding of the lectures of both authors, and in order to avoid unnecessary duplications, we have decided to present jointly this brief introduction to the language and properties of fiber bundles. By now the advantages of the fiber-bundle formulation of gauge field theories have led to a widespread acceptance of this language, and a number of reviews of the subject have appeared or are in course of publication. These, together with a number of standard textbooks are listed in the references to this

+Lectures given at the XX. Internationale Universitätswochen für Kernphysik, Schladming, Austria, February 17 - 26, 1981.

introduction. Nevertheless, we felt that it would be convenient for the reader of these proceedings to have at his disposal a summary of the basic facts. We also tried to clarify a number of concepts and propose an acceptable terminology wherever a standard has not been established in the literature. This refers, in particular, to the terms gauge transformation , pure gauge transformation, and the related (infinite-dimensional) groups as well as to the concepts of extension, prolongation, restriction, and reduction of bundles, which are used with slightly varying meaning in different texts.

In the oral presentation most of the general background material was presented by Andrzej Trautman, and the material related to reduction and symmetry of connections was given in Meinhard Mayer's lectures. Little, if anything, in this introduction is original. The actual text has been written in California by the first author and slightly revised by the second during his stay in France after the Schladming meeting.

No detailed proofs are given here, but wherever possible illustrations and examples are used to make the concepts plausible to physicists. Many proofs are straightforward and can be carried out by introducting local coordinates and bases. However, we recommend to the reader who wants to become familiar with the spirit of modern, coordinate-free, differential geometry to try to stay away from bases and indices as much as possible.

2. EXAMPLES OF BUNDLES. FUNDAMENTAL DEFINITIONS

We assume that the reader is familiar with such fundamental notions as differential manifold (including the concept of charts, atlases, diffeomorphism, etc.) and with the calculus of exterior differential forms, as well as with the fundamental concepts related to Lie groups and Lie algebras. In this section we list a few examples, illustrated by pictures, which will provide the intuitive background for understanding the more formal definitions and statements of the remainder of this lecture.

Of course, the most important examples of bundle structures appearing in contemporary physics are furnished by abelian and nonabelian gauge theories, and by the (pseudo-) Riemannian manifolds of general relativity. In the most familiar abelian gauge theory - electrodynamics of a charged field - the field ψ defined on Minkowski space and with values in some complex vector space (for simplicity, consider a complex scalar, or a Dirac or Pauli spinor) is subjected to the "point-dependent phase transformation", $\psi(x) \to \exp\{i\alpha(x)\}\psi(x)$, and in order to reestablish invariance of the equations of motion, one replaces in them (or in the Lagrangian) the ordinary space-time derivatives ∂_μ by the "covariant" derivative $\nabla_\mu = \partial_\mu + iA_\mu$ (we set $e = c = 1$) where the new field A_μ subject to the "gauge transformation of the second kind" $iA_\mu \to i(A_\mu + \partial_\mu\alpha) = iA_\mu + g(x)^{-1}\partial_\mu g(x)$, where $g(x) = \exp(i\alpha(x))$ is a smooth function on spacetime with values in the group U(1). We see here the appearance of a function on spacetime with values in a Lie group, or more generally, a copy of the group U(1) attached to each point of spacetime - the trivial principal bundle $M \times U(1)$. A simple analysis of magnetic monopoles shows that this picture is not adequate, and that in some situations a more intimate mixture of spacetime and the gauge group U(1) becomes necessary, where a product representation is valid only locally.

Similarly, if one considers a field which transforms under a representation of a nonabelian compact Lie group G (e.g., G = SU(2), in the original work of Yang and Mills), one is led to a structure in which a copy of G is attached to each point of space-time, and locally, in a neighborhood

U of a point, this can be represented as the product U × G. The place of the electromagnetic vector potential is taken by the Lie-algebra-valued (matrix-valued) one-form A = $= \sum A_\mu^a(x)\, e_a dx^\mu$, where (e_a) is a basis of the Lie algebra G of G (in the case of SU(2), $e_a = i\sigma_a$, a = 1,2,3, σ_a are the Pauli matrices), and as the field $\phi(x)$ is subjected to the "local gauge transformation" $\phi(x) \to g(x)\phi(x)$ the equations of motion are preserved if A is subjected to the affine transformation $A(x) \to g(x)^{-1}A(x)g(x) + g(x)^{-1}dg(x)$ provided ordinary differentials are replaced by "covariant differentials" $d\phi \to D\phi = d\phi + [A,\phi]$. These differentials do not commute and their "commutator" is related to the Yang-Mills field strength two-form

$$F = dA + \frac{1}{2}[A,A] \quad . \tag{2.1}$$

This two-form is subject to the "Bianchi identity" (just as the electromagnetic field strength two-form satisfies dF = 0)

$$DF = dF + [A,F] = 0 \quad . \tag{2.2}$$

In distinction from the abelian case, F is not gauge-independent, but transforms as

$$F(x) \to g(x)^{-1}F(x)g(x) \quad ,$$

and, if one wishes to write down the inhomogeneous Yang-Mills equation, one may generalize the Maxwell equations to

$$D{*}F(x) = -{*}J(x) \quad . \tag{2.3}$$

Here $*$ means the Hodge-duality operator which associates to a p-form in n-space the dual (n-p)-form (provided the manifold is oriented and has a Riemannian metric). In particular, in Minkowski space $*$ can be defined through its action on the basis coordinate forms:

$$*dx^o = dx^1 \wedge dx^2 \wedge dx^3 \text{ (cycl.perm.)},$$

$$*(dx^1 \wedge dx^2) = dx^o \wedge dx^3 \text{ , etc..} \qquad (2.4)$$

Alternatively, $*F$ can be thought of as the two-form with components given by $*F_{\mu\nu} = \frac{1}{2}\varepsilon_{\mu\nu\rho\sigma} F^{\rho\sigma}$. $*J$ denotes the current three-form of the matter field which satisfies the "covariant divergence" equation $D*J = 0$.

In order to see that the concept of "a group labeled by a point in a manifold". i.e., a principal bundle, or a vector space labeled by a point in a manifold, i.e., a vector bundle, appear naturally in geometry, and that matrix-valued forms, such as A and F, are usually associated with such objects, we consider some more elementary geometric examples.

In looking at these examples it is important to remember that the concept of fiber bundle generalizes the notion of direct (or cartesian) product of two spaces, and that the concept of section generalizes the graph of a function. Thus, the simplest example (and the one easiest to picture) is the cartesian product of two sets X (the domain space of the function $f:X \to Y$) and Y (the range space). The cartesian product $X \times Y$ can be viewed as formed by copies of Y (fibers) attached to each point of X, called the base space. In the usual treatment of cartesian products X and Y are treated on an equal footing. In fiber-bundle theory the base space X plays the role of a label space, whereas the typical fiber Y is quite distinct (and may be of a different nature, e.g., a group G, a vector space V, a homogeneous (coset) space G/H, such as a sphere, etc.). A distinguished role is also played by the projection $\pi : X \times Y \to X : (x,y) \to x$. Another distinction to be kept in mind is the fact that in a cartesian product we automatically identify points in different fibers "which lie on the same horizontal". In a fiber bundle there is no such automatic identification; it must be introduced as extra structure, by defining a section (or a basis of sections, in a vector bundle), i.e., picking a distinguished point or basis in each fiber. As we shall see this cannot always be done (if it can, and the bundle is principal, then it may be identified with a product, and is called trivial). Some of these notions are illustrated in Fig.1. The reader should keep in mind that most of our illustrations are for products (i.e., trivial bundles, since the nontrivial cases are difficult to draw; the reader might think of the Möbius band or the Klein bottle as examples of nontrivial bundles).

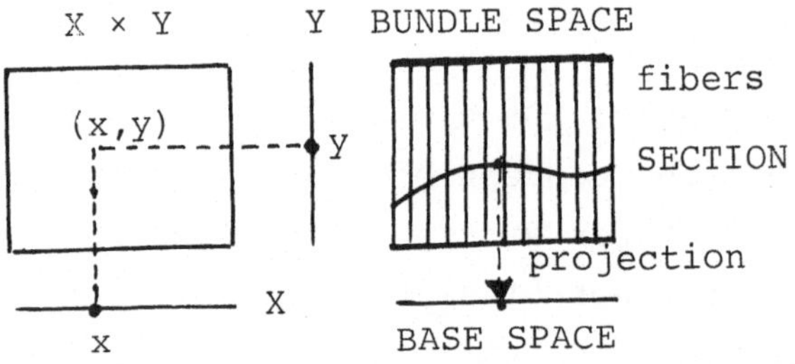

Fig. 1

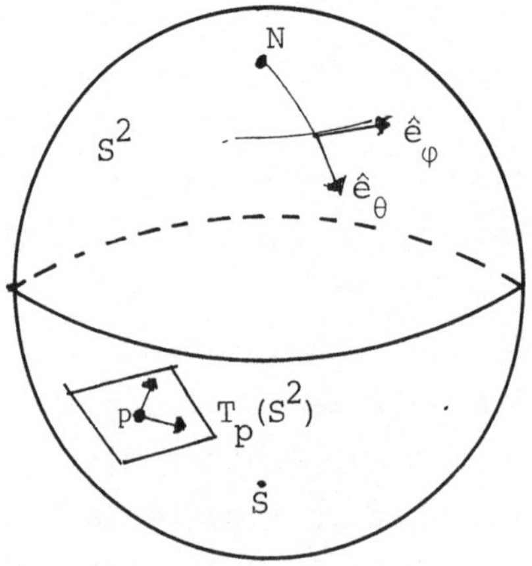

Fig. 2

An example which illustrates nontrivial bundles and some of the general concepts to be defined later is the tangent bundle of the sphere S^2 and the associated frame bundle of dyads on an oriented sphere (Fig. 2). Consider the unit sphere S^2 in R^3 as the base space X of our bundle. This sphere is a differentiable manifold, and an atlas of charts consists, e.g., of two open sets obtained by removing caps around the north and south poles, and the mappings of the remaining portions of the sphere onto R^2 realized by stereographic projections from these two poles. To each of the points on S^2 we attach the two-dimensional tangent space spanned, e.g., by the two tangent vectors $\hat{e}_\theta$, $\hat{e}_\phi$. We denote this tangent space at the point p by $T_p(S^2)$. The union of all these two-dimensional vector spaces, as the point p ranges over the sphere, is the tangent bundle TS^2. It is a four-dimensional manifold, since each point in it is labelled by the two coordinates of p and the two coordinates of the tangent vector at p. Moreover, since the "sphere cannot be combed", we cannot represent this bundle as a cartesian product of S^2 and R^2 globally (although in any chart of S^2 this is possible), and therefore we are dealing with a vector bundle on S^2 which is nontrivial, but locally trivial. Associated with this vector bundle is a bundle - the frame bundle - the fiber of which is isomorphic to a group. Indeed, consider the bundle of oriented dyads (pairs of unit vectors tangent to S^2 at each point). In an embedding into R^3 it is clear that the total space of this bundle is isomorphic to SO(3), since any dyad can be taken into any other dyad by an orientation-preserving rotation of R^3. Locally, such a rotation can be viewed as consisting of the two parameters labelling the point p on the sphere and an angle ψ which takes a "standard dyad" (e.g., the east-north dyad) into the given dyad, i.e., locally, SO(3) decomposes into the product of an open set U in S^2 and a copy of the group SO(2) - the typical fiber, the projection π being the smooth map associating the point p to the frame at that point.

The bundle SO(3) → S^2 is an example of a principal bundle, in which the typical fiber is a Lie group G (just as for a gauge theory), which also acts on the bundle space on the right (since we like to write frame transformations as right multiplication by matrices, reserving the left multiplication for actions of the group on vector components), and the tangent bundle is a vector bundle associated with this principal bundle by the fundamental representation of SO(2) by rotation matrices in two dimensions.

A generalization of this example to four dimensions is the bundle of tetrads (Vierbeins) often used in general relativity. Here the typical fiber is the Lorentz group which takes the standard tetrad $e_a^o(x)$ at a given point into an arbitrary tetrad $e_b(x)$. In distinction from the previous example, if the base space is homeomorphic to R^4, such a bundle will be trivial, i.e., will admit a product representation M × SO(1,3).

The principal bundles of dyads, tetrads, and, more generally, frame bundles (which may be of higher differential order than one) are more special or "richer" than the principal bundles occurring in gauge theories of the Yang-Mills type. The elements of the former bundles are "concrete": they can be defined and easily visualized in terms of geometrical constructions referring only to the base space. In other words, frame bundles are soldered to the base. This important concept distinguishes the theory of gravitation among all gauge theories; it is discussed in some detail in the lectures by the second author.

To illustrate the concept of a nontrivial principal bundle and of its local triviality, consider a last example (Fig. 3), where the base-space is the circle S^1 and the fiber is the discrete group of all integers Z. We easily obtain two bundles: the trivial product S^1 × Z (left) which is just the union of a countable set of circles "stacked"

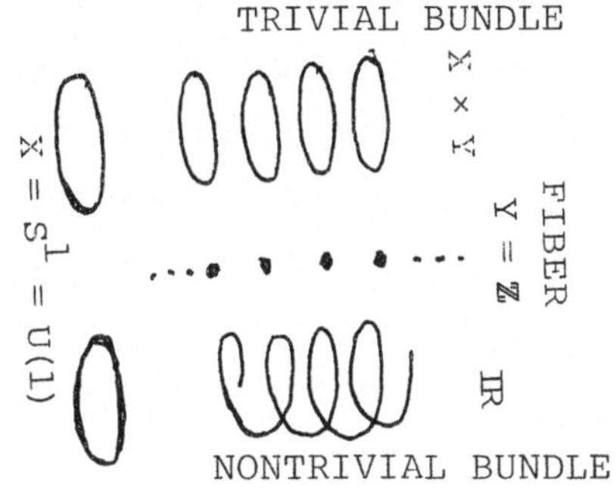

Fig. 3

over S^1, and the nontrivial helix (right) homeomorphic to R with projection map $\pi : R \rightarrow S^1$ given by $\pi(t) = \exp(2\pi it)$. It is obvious that acts on the bundle space by "translation" $(t,n) \rightarrow t + n$, where $t \in R$, $n \in Z$. The attentive reader will recognize in this example an essential ingredient of the Riemann surface of the logarithm. The nontriviality of this bundle reflects the impossibility of defining the logarithm as a smooth function on the pointed complex plane $C - \{0\}$.

We are now ready to summarize the precise definitions to be used in the sequel. Further examples will be discussed in Section 5.

A gauge theory consists of various mathematical structures associated with a <u>principal bundle</u> $P(M,G) = (P,M,G,\pi)$, where M, the <u>base space,</u> is spacetime (Minkowski space, its imaginary-time version R^4, or the compactification S^4 of the latter, or one of the (pseudo)-Riemannian manifolds of general relativity), G is the <u>structure group</u> of the bundle (the gauge group, in the physics literature), P is a smooth manifold which locally, i.e., over a covering of M by open sets U_i, has a product structure: $P|U_i \simeq U_i \times G$, and π is the projection map

$\pi : P \to M$, a smooth surjection of P onto M, such that the inverse image $\pi^{-1}(x) = P_x$, the fiber over $x \in M$, is diffeomorphic to G. The bundle is called trivial if it is isomorphic to $M \times G$ (globally). The above mentioned isomorphism of the restriction of P to each open set U_i to a product is called a local trivialization. The group G acts on P to the right in such a manner that the equivalence induced by this action is the same as that induced by the projection π, i.e., all points in the same orbit of G project onto the same point in M.

A section (or cross section) of a principal bundle is a (smooth) mapping $s : M \to P$ such that $\pi(s(x)) = x$. Global sections may not exist in principal bundles; in fact if such a section exists, the bundle is trivial, and vice versa. However, in view of the local triviality, local sections, i.e., sections over properly chosen open sets U_i always exist and can be used to describe the local trivialization. Indeed, let us consider the diffeomorphism between $P|_U$ and $U \times G$ given by $y \to (\pi(y), \phi(y))$, $y \in P$, $\phi(y) \in G$, such that for any $g \in G$, $\phi(yg) = \phi(y)\cdot g$. Then the local section s_U is defined by $s_U(x) = y\cdot\phi(y)^{-1}$ for y in $\pi^{-1}(x)$, obviously independent of the point y, because one can be taken into another by the action of G. The local section corresponds to an identification of the identity in the group with the submanifold of $\pi^{-1}(U)$ corresponding to U. It is sometimes convenient to describe the principal bundle in terms of charts, and the change of coordinates from one chart to another by means of transition functions (G-valued cocycles), i.e., maps $g_{UV} : U \cap V \to G$ satisfying the cocycle identity on $U \cap V \cap W$: $g_{UW}(x) = g_{UV}(x)\cdot g_{VW}(x)$. A change of local trivialization subjects the sections to a local gauge transformation $s'_U(x) = s_U(x)\cdot g_U(x)$ where g_U is a G-valued function on U, and the transition functions to $g'_{UV} = g_U^{-1} g_{UV} g_V$.

The principal bundle P(M,G) is the kinematic background of the gauge theory, since it specifies both the spacetime manifold M and the gauge group G. It will carry some of the dynamics of the field theory, since the connections and curvatures (see Section 3) will be identified as gauge potentials and gauge fields, and the metric structure of the base space may be identified with the gravitational field.

However, in order to accommodate the various matter fields describing particles, we have to associate to the bundle P(M,G) various vector bundles (or bundles with homogeneous spaces G/H) as fibers), in which the particle fields can be considered as sections. Alternatively, we may consider the particle fields as smooth maps from the bundle space P into a vector space V on which G acts (on the left) by means of a representation r, and which are equivariant under this action. More precisely, let $\phi: P \to V$ be such a map. Then equivariance means that for any $g \in G$ and $p \in P$

$$\phi(p \cdot g) = r(g^{-1})\phi(p) \quad . \tag{2.5}$$

In other words, the field ϕ defined on the bundle space P with values in the vector space V in fact depends only on the projection $x = \pi(p)$, i.e., is a field defined on spacetime in the usual sense. When composed with, or pulled-back by, a section s_U, the field will be $\phi_U = s_U^* \phi = \phi \circ s_U$, where ϕ_U is a V-valued function of $x \in U$, and if the trivialization is changed by a local gauge transformation , $s_U(x) \to$ $\to s_U(x) g_U(x) = s'_U(x)$, the field will be subjected to the "gauge transformation of the second kind" $\phi'_U(x) = r(g_U(x)^{-1}\phi_U(x)$. The same method works if V is replaced by an arbitrary manifold F on which G acts on the left.

Alternatively, one can first define a fiber bundle with typical fiber F (or V, in the case of vector bundles) <u>associated with P(M,G)</u> by the representation (action) r of G in the following manner: consider the cartesian product

$P \times F$. The group G acts naturally on the right on this product in the following way. Let $p \in P$, $f \in F$, $g \in G$. Then $(p,f)\cdot g = (p\cdot g,\ r(g^{-1})f)$. The orbit space of this action, i.e., the set of equivalence classes under the equivalence $(p,f) \sim (p\cdot g, r(g^{-1})f)$ is denoted by $P \times_G F$ and is called the fiber bundle associated with P by the action of G by r on the typical fiber F. It is denoted by E(M,F,G) and has a natural projection $\pi_E : E \to M$, obtained by factoring the composition of the projection of $P \times F$ on P with the bundle map $P \to M$ through the quotient map $P \times F \to E$. If $F = V$ is a vector space we obtain a vector bundle, and if $F = G/H$ is a coset space we obtain a bundle with homogeneous spaces G/H as fibers.

It is an easy exercise to show that a section of E(M,F,G) defines an equivariant function on P with values in F, and vice versa. Therefore, matter fields can also be regarded as sections of the associated (vector) bundle E, which is convenient in some constructions.

Before going on to the definition of connections, curvature and holonomy, we recall several general constructions involving fiber bundles which will be used in the physical applications.

3. MORPHISMS OF BUNDLES. FIBERED PRODUCTS AND PULLBACKS. EXTENSION AND RESTRICTION OF THE STRUCTURE GROUP

We limit our definitions to the case of principal bundles, although some are valid for more general bundles, and all transpose easily to associated fiber bundles. We list only the most important definitions, referring the reader for more details to the literature.

Let $P(M,G,\pi)$ and $P'(M',G',\pi')$ be two principal bundles. A morphism m of P into P' is a pair $m = (u,h)$, where $u : P \to P'$ is a C^∞-map and $h : G \to G'$ is a Lie-group homomorphism such that $u(p\cdot g) = u(p)\cdot h(g)$ for all p in P and g in G. It is clear that u takes fibers of P into fibers of P' (remembering that a fiber in a principal bundle is the orbit under the right group action), and therefore induces a C^{∞}-map $v : M \to M'$ such that the diagram

$$\begin{array}{ccc} P & \xrightarrow{\;u\;} & P' \\ \downarrow{\scriptstyle \pi} & & \downarrow{\scriptstyle \pi'} \\ M & \xrightarrow{\;v\;} & M' \end{array}$$

is commutative. This can be taken as the definition of a morphism of bundles which are not necessarily principal. If h is an isomorphism (we then identify G with G') and v is a diffeomorphism then the map u is also a diffeomorphism, and the morphism m is called a bundle isomorphism. In particular, an isomorphism of P onto itself is called a _bundle automorphism._ The group of all bundle automorphisms of a given principal bundle is an infinite-dimensional group, denoted by Aut P. Usually, in gauge theories, we are interested in bundle automorphisms which reduce to the identity map on the base space, i.e., for which $v = Id_M$. We call such bundle automorphisms _vertical_ or _based._ Automorphisms which leave some absolute elements invariant will be called _gauge transformations._ A gauge transformation which is vertical is called a _pure gauge transformation_ (for a more detailed discussion of this distinction which is important in general-relativistic contexts, cf. Section 6).

If we consider a local trivialization of the bundle P over the open set U of M, defined by a section s_U (the reader may think of the local section s_U as a way of identifying the group identity e_x over each point x of the base M), then it is easy to see that the pure gauge transformation $u : P \to P$ is implemented by a function

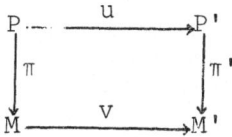

$g_U : U \to G$ which is smooth, so that the section s_U is taken into the section $s'_U(x) = s_U(x)g_U(x)$ (i.e., it reduces to a gauge transformation in the sense employed by physicists). Thus, the group of pure gauge transformations $\mathcal{G}_o$ may be viewed as the infinite-dimensional group of all smooth G-valued functions on M (cf. Section 6).

Another concept which is useful and where the group structure of the principal bundle does not intervene is the concept of fibered product and pullback.

The <u>fibered product</u> of two bundles (here it does not matter whether we are dealing with principal bundles, vector bundles or other fiber bundles) (T_1, M, P_1), (T_2, M, P_2), over the same base manifold M, with total spaces T_i, and projections P_i is defined as the bundle

$$(T_1 \times_M T_2, M, p_1 \times_M p_2) \tag{3.1}$$

which has as total space the submanifold of $T_1 \times T_2$ consisting of all pairs (t_1, t_2) such that $p_1(t_1) = p_2(t_2)$ (i.e., points in the fibers project onto the same point of the base), and as projection, the restriction of the product-projection to that subspace. The local triviality of the fibered product is easily established. This concept is also useful for maps.

An important special case of the fibered product is obtained if one considers instead of one of the bundles above a (smooth) mapping of a manifold M' into the base manifold M of a bundle. On obtains by the same construction the <u>pullback</u> or <u>induced bundle</u> over M with the same fiber. More precisely, let $\lambda = (E, M, p)$ be a bundle over M with fiber F (a group G in the case of principal bundles, a vector space in the case of vector bundles), and let $f : M' \to M$ be a (smooth) map of M' into M. Then the triple $f^*(\lambda) = (M' \times_M E, M', p')$ is a bundle, called the pullback of λ by f (or the induced bundle, or reciprocal image bundle) and denoted by $f^*(\lambda) = M' \times_M \lambda$. Here p' is the restriction of the projection onto the first factor of $M' \times E$ to the fibered product $M' \times_M E$. In particular, if M' is a submanifold and f is the injection, the pullback is the same as the induced bundle on the submanifold.

The fiber of the pullback bundle is the same as the fiber of the original bundle, but that the base space has been replaced by M'. In particular, this will be one way of defining gauge theories over extensions of the usual space-time manifold. A section $s : M \to E$ of the bundle ξ is taken into a section $s' : M' \to f^*(\xi)$ of the pullback bundle by means of the relation $s'(x') = (x', s(f(x')))$, where x' is a point in M'. This also defines a map $f' : M' \times_M E \to E$ such that $f' \circ s' = s \circ f$. The map f' is the restriction of the second projection of the product $M' \times E$ to the fibered product.

In physical and geometrical applications of fiber bundles it is often necessary to change the structure group of the bundle. Thus, if one deals with a gauge theory with symmetry breaking the original Lagrangian is defined on a principal bundle P with structure group G, and the "vacuum" (or classical critical field) may have the lower symmetry group H, a closed subgroup of G. There arises the question of constructing a principal bundle Q with structure group H, and the relation between P and Q. Similarly, in general-relativistic contexts, one considers the frame bundle with structure group $GL(4,R)$, whereas physics often imposes either restrictions of $GL(4,R)$ to one of its subgroups, such as $SO(1,3)$, if the metric is to be contemplated on the same footing as a Higgs field, or extensions to larger groups, such as the affine group $GL(4,R) \times R^4$ (semidirect product). In general, one can consider four distinct operations related to changes of the structure group. The terminology employed in the literature in this context is not uniform (and there are sometimes subtle differences in definitions). Here we adopt the terminology proposed by one of the authors (A.T., 1976) which is at variance with that used in some of the literature and in some of the lectures by the other author (although the distinction plays no important role in the Yang-Mills context).

Consider two principal bundles $\xi = (Q, M, H, p)$ and $\eta = (P, M, G, \pi)$ over the same base space M, and a morphism $m = (f,v)$ of Q into P, where f is a smooth map of the bundle space Q into the bundle space P, v is a diffeomorphism of M onto itself (not necessarily the identity), and the corresponding homomorphism of the structure groups is denoted, as before, by $h : H \to G$.

If both f and h are <u>injective immersions</u> (i.e., one-to-one into, and such that the tangent map is injective at each point of the respective manifold), then ξ is called a <u>restriction</u> of η relative to the morphism $m = (f,v,h)$, or, simpler, Q is a <u>restriction</u> of P to the subgroup H of G, and η is called an <u>extension</u> of ξ relative to the morphism m, or simpler, P is called an <u>extension</u> of Q to the structure group G.

If both f and h are <u>surjective submersions</u> (i.e., their tangent maps are surjective at each point), then η is called a <u>reduction</u> of ξ relative to m, or simpler, P is a reduction of Q to the structure group G, and ξ is called a <u>prolongation</u> of η relative to m, or Q is a prolongation of P to the group H.

In particular, we shall assume in the sequel that $v = \mathrm{Id}_M$ and that H is a closed Lie subgroup of G, such that the coset space G/H is a differentiable manifold. If $f : Q \to f(Q) \subset P$ is a diffeomorphism of Q onto a closed submanifold of P, such that $f(q \cdot h) = f(q) \cdot h$ for all $h \in H$, $q \in Q$, $f(q) \in P$, then we are in the restriction-extension situation, which we discuss in more detail.

Note. The terminology here follows that of Trautman (1976) and is closest to Bourbaki and Dieudonné, whereas Kobayasi-Nomizu (as well as Mayer) call our restriction reduction. In the Yang-Mills context there is usually no risk of confusion, but in the discussion of G-structures and spin structures some care is indicated.

There is no difficulty in obtaining an extension of the bundle Q with structure group H to a larger group G. Indeed, construct the associated bundle $Q \times_H G$ with fiber G based on the left action of H on G. We now let G act on the right on this space (consisting of H-orbits) by $p \cdot g = ((q,g')H) \cdot g = (q,g' \cdot g)H$ with $p \in P$, $q \in Q$, $g,g' \in G$ (recall that $Q \times_H G$ consists of equivalence classes of pairs $(q,g) \sim (q \cdot h, h^{-1} \cdot g)$, where we have not written out the homomorphism of H into G which is understood), and the action fibers this manifold over M, resulting in the bundle P. The morphism $f : Q \to P$ is given by $f(q) = (q,e)H$, where e is the identity in G (and H). It is easy to see that the projection of the associated bundle is the projection in P, and that the local triviality of Q induces local triviality of P.

On the other hand, restriction of P to a subgroup H of G is not always possible. This can be easily seen if one notices that a necessary and sufficient condition for a restriction to exist is that there should be a covering of M by open sets such that the corresponding transition functions take values in the subgroup H of G. Globally, one obtains a more interesting condition:

The bundle P(M,G) has a restriction to the group H iff the associated bundle $E(M,G/H,G,P) = P \times_G G/H = P/H$ admits a cross section $\sigma : M \to E$.

It is easy to see that the orbit space of P under the action of the subgroup H of G, P/H, can be identified with the associated bundle E. Denoting by γ the canonical projection of G onto G/H, we can set for $p \in P, \delta(p) = p\cdot\gamma(e)$, where e is the identity of G. The mapping $\delta: P \to E$ is a projection for the **new** principal bundle (P, H, E, δ) over the larger base $E = P \times_G G/H$ which is canonically identified with the orbit space P/H (this is illustrated in Fig. 4, in the middle).

Let now $\sigma : M \to E$ denote a section of E and σ^* : (P, H, E, δ) = (Q, H, M, ρ) the pullback (induced bundle) of this map. It is obvious (cf. Fig. 4, right) that this is now a principal bundle with structure group H over M, and its extension to G is isomorphic to the original bundle P. Two different sections σ_1 and σ_2 of E will define isomorphic restrictions iff they are mapped into each other by a pure gauge transformation (G-M-automorphism) of P. Otherwise different sections of E determine different (nonisomorphic) restrictions.

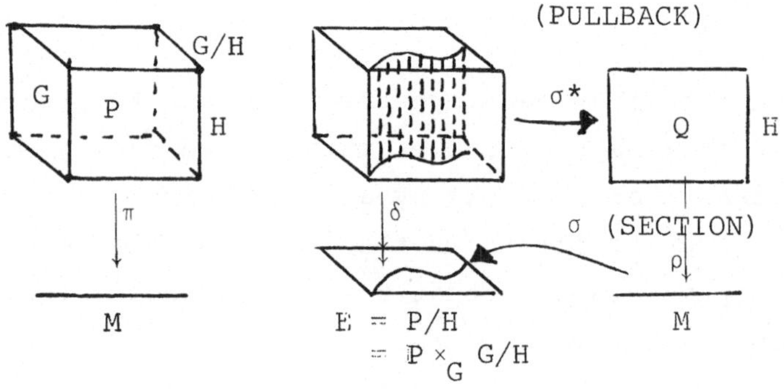

Fig. 4

4. CONNECTIONS, CURVATURE, AND HOLONOMY

We have seen in Sec. 2 that gauge potentials and fields are described by differential forms with values in the Lie algebra of the gauge group and act on fields through covariant differentiation. We have also made it plausible that fields should be considered as equivariant functions on a principal bundle with values in a vector space V (or a homogeneous space G/H) on which the gauge group G acts on the left. In this section we discuss the concept of a connection and its curvature on a principal bundle and show how these concepts are related to the familiar Yang-Mills potential and its field-strength (considered as matrix-valued one- and two-forms, respectively, on spacetime) and to the Levi-Civita connection (represented by the familiar Christoffel symbols) and the Riemann-Christoffel curvature tensor in the four-dimensional pseudo-Riemannian manifolds describing spacetime in general relativity.

The modern concept of a connection in a principal fiber bundle, and the associated covariant differentiations in associated bundles, has evolved during the first half of this century through the work of many geometers. There exist several equivalent definitions of a connection. We restrict our attention to one which is most useful in the context of gauge theories.

The fibered structure of a principal bundle, in which each fiber is isomorphic to the Lie group G, the structure group of the bundle, suggests that the tangent space $T_p(P)$ to the bundle space P at the point p contains a distinguished subspace of <u>vertical vectors,</u> Ver_pP, vectors which are tangent to the fiber P_x over a point x of the base M. As is well known, the tangent vectors to a Lie group G at a point g form a vector space $T_g(G)$ which is linearly isomorphic to $T_e(G) = G$, the Lie algebra of G. Thus, there exists an isomorphism between vertical vectors in T(P), the tangent bundle of P, and elements of the Lie algebra G, allowing us to identify these (sometimes it is convenient to introduce a field A^* of vertical vectors corresponding to an element A in G; A^* is called the fundamental vector field, and the right action of G on P intertwines with the adjoint action on G, i.e., $R_{g*}A^*$ is the fundamental vector field corresponding to $Ad(g^{-1})A)$. Since locally the bundle space P is isomorphic to the product between a subset of M and the group G, the tangent space $T_p(P)$ is isomorphic to a direct sum of $Ver_p(P)$ and a vector space $Hor_p(P)$ which must be isomorphic to $T_{\pi(p)}(M)$. Unfortunately, since the bundle P has no intrinsic "orthogonality" structure, there is no canonical way of identifying the horizontal subspace, and defining a connection in P means <u>defining</u> at each point p such a horizontal subspace $Hor_p(P)$, smoothly over p, and equivariantly under the action of G, i.e., $R_{g*}Hor_p(P) = Hor_{p \cdot g}(P)$ for $g \in G$. Giving such a distribution of horizontal subspaces allows us to compare tangent vectors in different fibers, i.e., leads to a notion of <u>parallel transport.</u>

This definition of a connection in terms of horizontal subspaces of T(P) is easy to explain, but hard to compute with. Therefore it is more convenient to introduce a dual definition, in terms of differential forms on P, which pick out the vertical component of tangent vectors and vanish on horizontal vectors (we remind the reader that this is the only way of defining a direct sum decomposition in a vector space without metric). More precisely, a <u>connection</u> on P is defined (globally) by a one-form ω on P with values in the Lie-algebra G such that for a vector $X \in T_p(P)$ we have

$$\omega(X) = A \in G \text{ where } A^*(p) = Ver_p X \,, \tag{4.1}$$

i.e., the value of the one-form on the vector X equals the

element of the Lie algebra which is isomorphic to the vertical part of X (the part tangent to the fiber). The horizontal subspace $Hor_p(P)$ of $T_p(P)$ is then the kernel of the one-form ω, i.e. $\omega(Hor_p(P)) = 0$. The invariance of the horizontal space under the right action of G implies that for any vector $X \in T_p(P)$

$$\omega(R_{g*} X) = Ad(g^{-1})\,\omega(X), \qquad \text{or} \qquad R_g^*\omega = Ad(g^{-1})\,\omega \; . \tag{4.2}$$

We remind the reader that R_g means the right action of g on a manifold, R_{g*} denotes the tangent (derivative) linear map of this action on vector fields, whereas R_g^* denotes the pullback of this action to differential forms. $Ad(g^{-1})\omega = g^{-1}\omega g$ if both g and ω are interpreted as matrices.

The existence of a connection establishes an isomorphism between $Hor_p(P)$ and $T_x(M)$, where $x = \pi p$, and thus the connection defines a lift to T(P) of any vector field X on M, denoted by X^* and called the horizontal lift of X. Similarly, any curve in the base space M can be lifted into a horizontal curve in P, i.e., a curve which has the horizontal lifts of the tangent vectors as tangents. This allows one to establish a correspondence between points in different fibers along a curve, correspondence which is called parallel transport of fibers. Indeed, starting from a point p_o in the bundle, the horizontal lift of a curve $\gamma = x_t$, $0 \leqq t \leqq 1$, (assumed smooth, or piecewise smooth and continuous) in M defines a curve γ^* in P, with end point p_1. As p_o varies over the fiber P_{x_o}, p_1 will vary over the fiber P_{x_1}, and the horizontal curves establish an isomorphism $\gamma : P_{x_o} \to P_{x_1}$ between fibers, obviously commuting with the right action of G on fibers.

The reader familiar with the theory of Lie groups will have noticed a certain similarity between the connection form ω and the left-invariant 1-form θ on a Lie group; in particular, the left-invariant form θ on G transforms according to (4.2) under the right action of G on itself. Moreover, the form θ satisfies the Maurer-Cartan structure equation

$$d\theta + \frac{1}{2}[\theta,\theta] = 0 . \tag{4.3}$$

The left-hand side of this equation measures the deviation obtained by parallel transport around an infinitesimal parallelogram (spanned, e.g., by two vectors X, Y tangent to G), i.e., it shows that a Lie group is <u>flat</u> with respect to this parallel transport. Since the bundle consists of fibers isomorphic to G, it makes sense to calculate an expression of the form (4.3) for the connection form, expression which will measure by how much a connection "differs from being a Maurer-Cartan form". This leads to the definition of the <u>curvature two-form on P:</u>

$$\Omega \underset{\text{def}}{=} d\omega + \frac{1}{2}[\omega,\omega] = D\omega \quad . \tag{4.4}$$

Here D denotes covariant <u>exterior differentiation of vector-valued forms</u> defined by

$$D\alpha = \text{hor } d\alpha \tag{4.5}$$

where, for any k-form β on P, one has

$$\text{hor } \beta\,(u_1,\ldots,u_k) = \beta(\text{hor } u_1,\ldots,\text{hor } u_k)$$

for any vectors $u_1,\ldots,u_k \in T_pP$. In particular, since Ω is a <u>horizontal form,</u> the Bianchi identity

$$D\Omega = d\Omega + [\omega,\Omega] = 0 \tag{4.6}$$

follows easily. Both the structure equation (4.4) and the Bianchi identity can be verified by evaluating the right- and left-hand sides on horizontal and vertical vectors. Under the right action of G the two-form Ω transforms according to the adjoint representation of G on G:

$$R_g^*\Omega = Ad(g^{-1})\Omega \ . \tag{4.7}$$

Although there is a superficial similarity between these concepts and the Yang-Mills fields introduced in Section 2, the most striking difference is the fact that the latter are defined on spacetime, i.e., on the base space of the bundle. A connection one-form ω and its curvature two-form Ω can be pulled down to the base space M of the principal bundle P locally, i.e., over an open set U in M, where a section σ_U defines a local trivialization of the bundle. More precisely, let $\{U_i\}$ be an open covering of M and let $\phi_i : \pi^{-1}(U_i) \to U_i \times G$ be the local trivializations, $\psi_{ij} = \phi_i \phi_j^{-1}$, and $\sigma_i(x) = \phi_i^{-1}(x,e)$ the corresponding local sections. Let θ denote the left-invariant (Maurer-Cartan) one-form on G. Then the transition functions ψ_{ij} (which are G-valued on the intersection $U_i \cap U_j$) define a G-valued one-form θ_{ij} on $U_i \cap U_j$ by pullback:

$$\theta_{ij} = \psi_{ij}^*\theta \quad . \tag{4.8}$$

In each open set U_i the section σ_i pulls back the connection form ω to U_i thus defining a family of G-valued one-forms on M

$$A_i = \sigma_i^*\omega \tag{4.9}$$

which are subject to the following "overlap conditions" in $U_i \cap U_j$:

$$A_j = \mathrm{Ad}(\psi_{ij}^{-1})A_i + \theta_{ij} \quad . \tag{4.10}$$

In particular, ψ_{ij} may be thought of as the transition function describing the transition from one local trivialization to another $\sigma_i = \sigma_j \cdot \psi_{ij}$; then the forms θ_{ij} are obviously given by the expression $\theta_{ij} = \psi_{ij}^{-1} d\psi_{ij}$, and leaving out the indices i, j, if we operate in a fixed open set, we obtain the law of gauge transformation:

$$A \to A' = \mathrm{Ad}(\psi^{-1})A + \psi^{-1}d\psi \quad , \tag{4.11}$$

which agrees with the form given in Sec. 2 for the Yang-Mills potential.

Similarly, the curvature two-form Ω pulls back to a family of G-valued two-forms defined on the open covering $\{U_i\}$ of M:

$$F_i = \sigma_i^* \Omega \quad , \tag{4.12}$$

which under a change of trivialization (or in the overlap of two open sets) transforms under the adjoint action of the transition function:

$$F_j = \mathrm{Ad}(\psi_{ij}^{-1})F_i \quad . \tag{4.13}$$

It is easy to verify that the structure equation (4.4) and the Bianchi identity (4.6) "pull down" to the local forms F_i, A_i in the form:

$$F_i = dA_i + \frac{1}{2}[A_i, A_i] \quad , \tag{4.14}$$

$$DF_i = dF_i + [A_i, F_i] = 0 \quad , \tag{4.15}$$

coinciding, respectively, with the definition of the (local)

Yang-Mills field strength and the "homogeneous" Yang-Mills equation (2.1), (2.2).

It should be noted, that if the principal bundle P admits a global section, i.e., is trivializable (which is the case, e.g., if the base space is all of Minkowski space, or any space homeomorphic to an R^n), then the open cover consists of one set, and the forms A, F are globally defined. Conversely, it can be shown that if P is non-trivial, then A and F cannot be globally defined; the reader should keep in mind the example of the Dirac monopole, where A cannot be globally defined, but a connection form can be defined in terms of two pull-backs, to two overlapping sets on S^2, with a gauge transformation on the overlap, or the situation encountered in the Bohm-Aharonov effect.

Finally, it should be noted that parallel transport and covariant differentiation are defined in an obvious manner in any associated vector bundle.

The curvature two-form measures the "nonintegrability" of the connection (or parallel transport) locally, i.e., for transport around an infinitesimal parallelogram in the base space. Parallel transport around an arbitrary loop in the base space leads to the concept of holonomy group of a connection in P.

We consider continuous curves in M which are piece-wise differentiable, and call closed curves starting and ending at a point x loops. The loops based at a point x form a group under the obvious composition, with the zero loop playing the role of identity, and the loop with opposite orientation playing the role of inverse. We denote the group of loops based on the point x in M by L_x, and the subgroup of contractible loops by L_x^o (we assume that M is a connected manifold). Parallel transport of fibers associated with a connection ω in the principal bundle P(M,G) leads to a representation of the group L_x by automorphisms of the fiber P_x, called the holonomy group of P with reference point x, and denoted by H(x). If we restrict the loops to contractible loops, we obtain a subgroup of H(x), denoted by $H^o(x)$ and called the restricted holonomy group at x. Both these groups can be realized as subgroups of the structure group G of P, since the automorphism of the fiber P_x associated to parallel transport around a loop γ can also be realized by the right action of a group element g, with the obvious composition property for successive transport around two loops, the inverse, or the trivial loop, provided one chooses a fixed point p in the fiber above x, where the parallel transport starts. Thus the choice of p and the group of loops L_x (or contractible loops L_x^o) determines a subgroup H(p) (or $H^o(p)$) of G, called holonomy group (or restricted holonomy group) at $p \in P$. Another way of defining H(p) is as that subgroup of elements $h \in G$ such that p and $p \cdot h$ can be joined by a horizontal curve. It can be shown that the holonomy groups H(p) and $H^o(p)$ are actually Lie subgroups of G, that $H(p)/H^o(p)$ is discrete (countable), and that the holonomy groups based at different points x or p are conjugate of each other.

Let P(M,G) be a principal bundle (with M, as always, connected, paracompact) with a connection ω, and p_o an arbitrary point of P. Then the set of points in P which can be joined to p_o by horizontal curves coincides with the restriction of P(M,G) to the subgroup $H(p_o)$ of G; this set is denoted by $P(p_o)$ and is called the holonomy bundle at p_o. Moreover, the connection G is reducible to a connection in $P(p_o)$, i.e., the Lie algebra G of G admits a decomposition into a direct sum G = H + M, where H is the Lie algebra of $H(p_o)$, and the H-component of ω restricted to $P(p_o)$ is a connection form on $P(p_o)$. In other words, the horizontal subspace of $T_q(P)$ is tangent to

$P(p_o)$ for every $q \in P(p_o)$. It is clear that the points in P lie either in the same holonomy bundle, or else their holonomy bundles are disjoint, i.e., the bundle space P is decomposed into a disjoint union of its holonomy bundles, and these are taken into each other by those elements of G which are not in $H(p_o)$. For this reason we may always consider that a gauge group which survives a symmetry breaking is a holonomy group of the bundle (or by abuse of language, a holonomy group of the vacuum).

It is clear that, since the curvature Ω of the connection measures the "infinitesimal holonomy", there must be a relation between curvature and holonomy. There are various theorems establishing such relations, the most important one being the theorem of Ambrose and Singer. This theorem states that in a principal bundle P(M,G) over a connected manifold M, with connection ω and curvature Ω, the Lie algebra of the holonomy group H(p) with reference point p is the subspace of G spanned by all elements of the form $\Omega_v(X,Y)$ where v is a point in the holonomy bundle P(p) and X,Y are arbitrary horizontal vectors at v.

Let E(M,V,P,G) be a vector bundle associated with P and $s : M \to E$ a section of E, which can be represented by an equivariant function $f : P \to V$. Then parallel transport of sections can be defined by the parallel transport in P. In particular, each element of the holonomy group at p is represented by a linear transformation on V, and in terms of the vector-valued function f one can represent this action by a matrix defined on the horizontal lifts of loops in M, acting on the function f, or alternatively, introducing local bases of sections, as product-integrals of matrices around loops (or path-ordered exponentials).

5. MORE EXAMPLES. UNIVERSAL BUNDLES AND UNIVERSAL CONNECTIONS

Before proceeding to a discussion of symmetries of connections we give in this section several examples of principal bundles which seem to be less familiar to the physics community. Some of these examples (Stiefel manifolds fibered over Grassmann manifolds) have a "universal" character, meaning that any bundle with an orthogonal or unitary structure group can be obtained from such a bundle as a pullback of a mapping of the base space into the appropriate manifold (more precisely, a homotopy class of such mappings, since homotopically equivalent maps lead to isomorphic bundles). We also mention briefly the existence of "universal connections", a construction which will certainly find many physical applications in the near future.

One of the simplest examples of principal bundles can be obtained if one considers a Lie group G with two nested closed subgroups H, K ,

$$K \subset H \subset G , \tag{5.1}$$

where K is an invariant (normal) subgroup of H, and such that G/K and G/H are differentiable manifolds. Then it is easy to see that the projection

$$\pi : G/K \to G/H \tag{5.2}$$

yields a principal bundle over G/H with structure group H/K. The projection is the mapping which associates to each left coset of K in G the left coset of H in G which contains it (if K is not a normal subgroup, (5.2) is still a bundle, with fibers the homogeneous space H/K).

In this case, if G is semisimple and the pair G, H

is reductive, i.e., $G = H \dotplus M$ (direct sum) and $[H,M] \subset M$, then the canonical left-invariant form θ on G has a H-component which projects to an H/K connection on G/K (this result is a special case of a theorem by Narasimhan and Ramanan, which was used in a theorem by Harnad on invariant gauge fields, cf. also Nowakowski and Trautman, 1978).

This construction becomes more transparent if we specialize the groups G, H, K, to be orthogonal, unitary or symplectic groups, which act, respectively, on R^n, C^n, or H^n (H denotes the quaternions). For simplicity, we discuss only the real case, but a simple change of notation (replacing R by C or H, SO(n) by U(n), or Sp(n)) yields the results in the complex or quaternionic cases. It should be noted that we could have started from the general linear groups, but there are theorems stating that the bundles defined by the maximal compact subgroups suffice, i.e., the restriction of GL(n,R) to O(n) (or SL(n) to SO(n)) is always possible.

The real <u>Stiefel manifold</u> $S_{n,k}(R)$ (we omit the R in the sequel) is defined as the manifold of all orthonormal k-frames ($0 \leqq k \leqq n$) in R^n, i.e., the set of all linear isometric mappings of R^k into R^n, defined by the set of $n \times k$ matrices with orthonormal rows. To see that this is a manifold, embed the matrices in R^{nk} and verify that the gradients of the orthonormality conditions are nonzero and mutually orthogonal, thus defining a compact manifold without boundary of dimension $nk - k(k + 1)/2$. $S_{n,n}$ is clearly identical to the manifold O(n), and $S_{n,0}$ is a point, whereas $S_{n,1}$ is S^{n-1} and $S_{n,2}$ is the submanifold of TS^{n-1} consisting of unit tangent vectors. It is somewhat harder to show that SO(n) is diffeomorphic to $S_{n,n-1}$ (the diffeomorphism adds to the $n \times (n-1)$ matrix in $S_{n,n-1}$ a column which makes it into an SO(n)-matrix, i.e., completes the orthonormal n-1-frame to an n-frame).

The subgroup of O(n) which stabilizes the orthonormal k-frame is isomorphic to O(n-k), hence $S_{n,k}$ can be interpreted as the coset space $S_{n,k} = O(n)/O(n-k)$. We could have started from oriented orthonormal frames, and then we would have obtained $S_{n,k} = SO(n)/SO(n-k)$. (The reader should be warned that the complex Stiefel manifold $S_{n,k}(C) = U(n)/U(n-k)$ is a real $2nk - k^2$ dimensional compact manifold without boundary which is not a complex manifold in the technical sense; the quaternionic Stiefel manifold $S_{n,k}(H) = Sp(n)/Sp(n-k)$ has real dimension $4nk-(2k^2-k)$ and is also compact without boundary).

The real <u>Grassmann manifold</u> $G_{n,k}(R)$ $(0 \leqq k \leqq n)$ is the set of all k-planes (k-dimensional subspaces) through the origin of R^n. By considering an orthonormal frame in such a subspace, and its complementary (n-k)-frame one can see that the Grassmann manifold admits charts which map its open sets into the space $R^{k(n-k)}$ of real $k \times (n-k)$ matrices. In fact, it is clear that $G_{n,k}$ is a homogeneous space: each k-plane can be considered as a coset in the group O(n) with respect to the subgroup $O(k) \times O(n-k)$,

$$G_{n,k}(R) = O(n)/(O(k) \times O(n-k)) .$$

If we replace k-planes by oriented k-planes in the above definition, we obtain the oriented Grassmann manifolds $G^+_{n,k}(R) = SO(n)/(SO(k) \times SO(n-k)) = O(n)/(O(k) \times SO(n-k))$. Important special cases are: $G_{n,0} = G_{n,n}$ = point; $G_{n,1}(R) = RP^{n-1}$, the n-1 dimensional real projective space through the origin (which is conveniently parametrized by means of the homogeneous coordinates $(x_1/x_j, \ldots, x_{j-1}/x_j, x_{j+1}/x_j \ldots x_n/x_j))$.

The generalizations to the complex and quaternionic cases are obvious:

$$G_{n,k}(C) = U(n)/(U(k)\times U(n-k)), G_{n,k}(H) = Sp(n)/(Sp(k)\times Sp(n-k)). \tag{5.3}$$

The construction given at the beginning of this section now can be applied to nested groups $O(n) \supset O(k) \times \times O(n-k) \supset O(k)$ or their unitary or symplectic counterparts. This leads to a particularly simple principal bundle with structure group $O(k)$ (respectively, $U(k)$, $Sp(k)$):

$$\xi : S_{n,k} \xrightarrow{\pi} G_{n,k} , \qquad (5.4)$$

where the projection associates to each k-dimensional frame the k-plane spanned by it. Rewriting this in the form

$$O(n)/O(n-k) \xrightarrow{\pi} O(n)/(O(k) \times O(n-k)) \qquad (5.5)$$

it is clear that the fiber of this bundle is $O(k)$, and that $O(k)$ acts on the homogeneous space on the right by acting on $O(n)$ (remember that $S_{n,k}$ consists of left cosets, and that $O(k)$ is isomorphic to the factor group $O(k) \times O(n-k)/O(n-k)$). The principal bundle ξ is in a certain sense universal, in that for a given $O(k)$ and large enough n any $O(k)$-principal bundle over a compact manifold can be obtained as a pullback of a mapping of that manifold into $G_{n,k}$ (up to homotopy, which leads to an isomorphic bundle). The same construction holds in the complex and quaternionic cases.

The bundle (5.4) also has a canonical connection - called a universal connection .(The existence of universal connections was proved by Narasimhan and Ramanan, and a recent improvement was found by Roger Schlafly; since they involve embeddings in spaces of high dimension, they may be of use in the discussion of the large N limit of SU(N) gauge theories, and may have other physical applications.) Starting from the $n \times n$ matrix defined by an orthonormal frame in R^n, an element of $O(n)$ which we denote by X, we have the left-invariant Maurer-Cartan form on $O(n)$:

$$\theta = X^{-1}\, dX = -{}^{t}\theta \quad . \tag{5.6}$$

If we restrict the skew-symmetric matrix θ to the $k \times k$-matrix corresponding to the k-frames of the Stiefel manifold $S_{n,k}$, we obtain a connection on the bundle (5.4). Whereas the connection θ is flat, on account of the Maurer-Cartan structure equation (4.3), its restriction ω to the first k rows and columns will have a nonvanishing curvature, as can be seen explicitly by rewriting (4.3) and (5.6) in a basis adapted to $S_{n,k}$ (replacing X^{-1} by ${}^{t}X$).

The Narasimhan-Ramanan theorem considers an arbitrary compact Lie group G embedded in O(k). Then the canonical connection on $S_{n,k}$ can be pulled back into a connection on the principal bundle with structure group G:

$$O(n)/O(n-k) \rightarrow O(n)/(G \times O(n-k)) \quad . \tag{5.7}$$

This yields a canonical connection on this canonical G-bundle. Now for an arbitrary manifold M of dimension $\dim M \leqq m$, and for $n \geqq \frac{1}{2}((k+m)^2 + 7(k+m) + 10)$ any principal G-bundle with connection over M may be obtained as the pullback of the canonical G-bundle (5.7) and its canonical connection by some smooth map $f : M \rightarrow O(n)/(G \times O(n-k))$. Schlafly has extended this theorem to the case when the base space is a compact Riemannian manifold with an isometric action of a compact Lie group H, which also acts on the principal bundle space P and preserves the connection (see next section for the meaning of the last statement). He has shown that in this situation there always exists a representation $r : H \rightarrow O(n)$ and an H-map $f: M \rightarrow O(n)/G \times O(n-k)$, such that the bundle is isomorphic to the bundle induced by the pair of maps f, r.

We now turn to the promised discussion of symmetries an invariance properties of connections on principal bundles.

6. INVARIANCE AND SYMMETRIES OF CONNECTIONS

In spite of the fact that conditions for the invariance of a connection have been discussed in the mathematical literature over twenty years ago, and Wang's theorem can be found in textbooks, physicists rediscovered them only in 1978-79. This section contains a brief survey of this topic, which has been discussed from a more physical point of view by Jackiw in last year's Schladming lectures.

The problem is quite simple when viewed globally, on the principal bundle; complications arise only when one tries to express the invariance conditions for the connection forms on local trivializations of P.

Before discussing connections we summarize the definitions of gauge transformations to be used. An isomorphism of a principal bundle onto itself is called an automorphism of the bundle. Such an automorphism consists of a pair of diffeomorphisms (u,v) of P and M such that $\pi \circ u = v \circ \pi$ (Eq.(3.1)), and $u(p\cdot g) = u(p)\cdot g$ for all $p \in P$, $g \in G$. An automorphism is called vertical if $v = \mathrm{Id}_M$. If we denote the group of all automorphisms (an infinite-dimensional group) by Aut P, the subgroup of all vertical automorphisms $\mathrm{Aut}_M P$ is a normal subgroup, the quotient being the group of all diffeomorphisms of M onto itself, i.e., we have the exact sequence of homomorphisms:

$$I \to \mathrm{Aut}_M P \xrightarrow{\ i\ } \mathrm{Aut}\, P \xrightarrow{\ j\ } \mathrm{Diff}\, M \to I\ , \tag{6.1}$$

where i is the canonical injection and $v = j(u)$. If $u \in \mathrm{Aut}_M P$, its action is in the fiber and therefore can be implemented by an element U(p) of G such that for any p in P and g in G

$$u(p) = p\cdot U(p),\ U(p\cdot g) = g^{-1}U(p)g\ . \tag{6.2}$$

Thus, there is a natural isomorphism of $\mathrm{Aut}_M P$ onto the multiplicative group of (smooth) maps $U : P \to G$, subject to the equivariance condition (6.2), or equivalently, to sections of the associated bundle $P \times_{AdG} G$ with fibers G, but the right action replaced by the adjoint action.

The group Aut P (as well as $\mathrm{Aut}_M P$) acts on (local) sections of P in the following manner: if $s : V \to P$ (V an open subset of M), then its transform is $s' = u \circ s \circ v^{-1}$. If $u \in \mathrm{Aut}_M P$, the subset V of M is left invariant and the section is subject to what a physicist would call a gauge transformation:

$$s'(x) = s(x) \cdot U(s(x)), \quad x \in V \subset M \quad . \tag{6.3}$$

If one deals only with Yang-Mills fields over a flat spacetime (or a Euclidean, compact version thereof) one is thus entitled to identify $\mathrm{Aut}_M P$ with the group of gauge transformations (this is the definition adopted by Atiyah, Singer, and many other mathematicians). However, in theories involving gravity, or other structures on spacetime, it is convenient to introduce a further differentiation.

Definition. The gauge group of a theory in which the bundle has some absolute elements, such as the metric tensor of special relativity, or some other structure element of P or M, is the subgroup $\mathcal{G}$ of Aut P such that the diffeomorphism v and the projection preserve the absolute elements of M. The group of pure gauge transformations consists of the vertical automorphisms in $\mathcal{G}$; this group will be denoted by $\mathcal{G}_o = \mathcal{G} \cap \mathrm{Aut}_M P$, it is a normal subgroup of $\mathcal{G}$, and the quotient $\mathcal{G}/\mathcal{G}_o$ in the exact sequence

$$I \to \mathcal{G}_o \xrightarrow{i} \mathcal{G} \xrightarrow{j} \mathcal{G}/\mathcal{G}_o \to I \tag{6.4}$$

is the subgroup of Diff M leaving the absolute elements invariant (e.g., if M is Minkowski space, $\mathcal{G}/\mathcal{G}_o$ is the Poincaré group; this corresponds to the necessity of sometimes combining a gauge transformation with a change of Lorentz frame in some calculations).

Invariance of connections under automorphisms of the bundle P is simply expressed as the fact that the pullback of the connection form ω on P by the mapping $u \in \mathrm{Aut}\, P$, $\omega' = u^*\omega$ is again a connection form on P. If u is a vertical automorphism (in particular, a pure gauge transformation), then

$$\omega' = \mathrm{Ad}(U^{-1}(p))\omega + U^{-1}(p)dU(p) \quad , \tag{6.5}$$

where U(p) is the map defined in Eq.(6.2). We see that the form ω is subject to the usual gauge transformation of a gauge potential (albeit, on P rather than on M). The curvature form Ω' of the pullback $u^*\omega$ is given by the adjoint action of U(p) on the original curvature form:

$$\Omega' = \mathrm{Ad}(U^{-1}(p))\Omega \quad . \tag{6.6}$$

The equations (6.5), (6.6) can easily be pulled down to the forms A, F on the base space given by a locally trivializing section s. Here one can either pull ω back to M by the transformed section, or pull ω' back by the original section, obtaining the usual gauge transformation formulas for A and F:

$$A' = \mathrm{Ad}(S^{-1})A + S^{-1}dS \;, \quad F' = \mathrm{Ad}(S^{-1})F \;, \tag{6.7}$$

where $S = U \circ s$.

Among the automorphisms of the principal bundle P with a connection ω and the associated bundles carrying the particle fields, symmetries are distinguished by the fact that they preserve the connection ω and the absolute elements of the theory (e.g., they preserve the action, or they modify the Lagrangian density by a divergence). In particular, a symmetry of a gauge theory is a gauge transformation (in the wider sense defined above) which leaves the connection form ω invariant (in addition to the other absolute elements):

$$u^*\omega = \omega, \; u^*\Omega = \Omega \quad ; \qquad (6.8)$$

since a nonabelian gauge theory is not completely determined by the curvature, it is not sufficient to require invariance only of the curvature form.

When this condition is pulled back by a local trivialization to the base space, it will usually be formulated as the requirement that the one-form A be unchanged up to a pure gauge transformation, or in other words, a gauge field is invariant under a symmetry, if the symmetry transformation can be compensated by a gauge transformation of the locally trivializing section (this is the formulation given by Bergmann and Flaherty, Trautman, Jackiw, and other authors).

To write the invariance condition (6.8) for the physical fields A, F, we consider first a one-parameter group $u_t : R \to \mathrm{Aut}\, P$ of automorphisms of P. Let Y denote the corresponding vector field on P, and X the projection of Y onto M:

$$X = \pi_* Y \quad . \qquad (6.9)$$

The vector field X generates a one-parameter group $v_t = j(u_t)$ of transformations on M. Let ω be a u_t-invariant connection on P,

$$u_t^*\omega = \omega, \quad u_t^*\Omega = \Omega \quad . \qquad (6.10)$$

For an arbitrary point p_o in P the groups u_t, v_t define curves in P, M, respectively:

$$p_t = u_t(p_o) \ , \quad x_t = v_t(\pi p_o) = \pi(p_t) \ . \tag{6.11}$$

The connection defines a horizontal lift of x_t which we denote by h_t. Then it is obvious that $p_t = h_t g_t$ for a suitable element g_t of G, and g_t is a one-parameter Lie subgroup of G, generated by the Lie algebra element $T = \omega_{p_o}(Y)$. The invariance of the connection and its curvature on P can be expressed infinitesimally as the vanishing of their Lie derivatives with respect to Y:

$$L_Y\omega = 0 \ , \qquad L_Y\Omega = 0 \ . \tag{6.12}$$

(Recall that for forms the Lie derivative is defined by $L_Y = d \circ Y \lrcorner + Y \lrcorner \circ d$, where denotes the interior product of Y with the differential form following the sign.)

The expressions (6.12) for the invariance of connections are identical to the usual conditions for the invariance of fields encountered in physics, but hidden behind the simple form is the gauge freedom inherent in the theory, particularly if one works in terms of the pullbacks A, F, to the base space. If we denote the value of the one-form ω_p (at the point p in P) on the vector field Y at p by $Z = \omega_p(Y)$, we obtain an equivariant map of P into the Lie algebra $Z : P \to G$, $Z \circ R_g = \mathrm{Ad}(g^{-1}) \circ Z$. Its covariant exterior differential

$$DZ = dZ + [\omega, Z] \tag{6.13}$$

is a horizontal one-form (with values in G) of type Ad, and the definition of the Lie derivative and Eq. (6.14) yield the detailed form of the invariance condition:

$$L_Y\omega = Y \lrcorner \Omega + DZ = 0, \quad L_Y\Omega = D(Y \lrcorner \Omega) + [\Omega, Z] = 0 \tag{6.14}$$

(Trautman, 1979). If we use a local section s to pull back the connection and curvature to the gauge potential A, and the field strength F on M, the vector field Y is to be replaced by the generator X of the transformations in M, and the Lie-algebra-valued function on P, Z, defines a function on M, $\Phi = Z \circ s : M \to G$. Then the invariance conditions for A and F under the symmetry induced on M by the vector field X (such a vector field always has a horizontal lift under the given connection; adding an arbitrary vertical vector field of the type of Z to it, will give a field on P) can be written in the form

$$X \lrcorner F + D\Phi = 0 \tag{6.15}$$

where $D\Phi = d\Phi + [A,\Phi]$, and

$$D(X \lrcorner F) + [F,\Phi] = 0 \,. \tag{6.16}$$

In terms of the potential one-form A the invariance condition can be rewritten as $L_X A = DW(X)$, where $W(X)$ differs from Φ by the zero-form $-X \lrcorner A$. The right-hand side of the last equation has the infinitesimal form of a gauge transformation, and under a change of chart (gauge transformation) with transition functions g_{ij} the function W is subject to the transformation

$$W_j = \mathrm{Ad}(g_{ij}^{-1})\, W_i + g_{ij}^{-1}\, X \lrcorner dg_{ij} \quad . \tag{6.17}$$

If X_1 and X_2 denote two vector fields on M inducing symmetries of the connection A, then consistency requires that

$$2F(X_1,X_2) = \Phi([X_1,X_2]) - [\Phi(X_1),\Phi(X_2)] \quad , \tag{6.18}$$

where the left-hand side denotes the value of the two-form

F on the two vector fields X_1, X_2, and the right-hand side expresses the dependence of the G-valued 0-form on the vector field X_i (and implicitly, on the trivializing section s). The infinitesimal forms of the invariance conditions have been independently discovered by Forgacs and Manton, Harnad, Shnider and Vinet, and Jackiw (cf. the bibliography to Mayer's contribution for references), and the usefulness of Eqs.(6.15), (6.18) (with a difference in sign) has been discussed in Jackiw's 1980 Schladming lectures.

To end this section we give, for the convenience of the reader, a brief statement of Wang's theorems on invariant connections, in a notation which is close to the one used by Kobayashi and Nomizu, where the detailed proofs can be found.

Consider, as before, a principal bundle P(M,G), with a connection ω which is invariant with respect to a group of automorphisms K of P(M,G), assumed to be a connected Lie group with fiber-transitive action, i.e., for any two fibers there is an element of K which maps one into the other, hence K acts transitively on the base space M. We denote by u_o a reference point in P, chosen once and for all, and by x_o its projection in M, $x_o = \pi(u_o)$. Furthermore we denote by J the isotropy subgroup of K at x_o, i.e., the subgroup of all transformations in K which leave x_o invariant (it is clear that M can then be viewed as the homogeneous space K/J). We denote the Lie algebras of the groups G, K, J by $\mathfrak{g}$, $\mathfrak{k}$, $\mathfrak{j}$, respectively and, when it exists, the subspace of $\mathfrak{k}$ complementary to $\mathfrak{j}$ by $\mathfrak{m}$: $\mathfrak{k} = \mathfrak{j} \dot{+} \mathfrak{m}$ (direct sum). Then we define a linear mapping $\Lambda : \mathfrak{k} \to \mathfrak{g}$ by $\Lambda(X) = \omega_{u_o}(\tilde{X})$, where $X \in \mathfrak{k}$ and $\tilde{X}$ is the vector field on P induced by X, which has the properties

(i) $\Lambda(X) = \lambda_*(X)$ for $X \in \mathfrak{j}$; here λ_* is the homomorphism $\lambda_* : \mathfrak{j} \to \mathfrak{g}$ defined as the differential of the homomorphism

$\lambda : J \to G$, which assigns an element $g \in G$ taking the point u_o into the same point as the left action of $j \in J$: $ju_o = u_o g : g = \lambda(j)$;

(ii) for $j \in J$ and $X \in \kappa$, $\Lambda(Ad(j)(X)) = Ad(\lambda(j))(\Lambda(X))$, where $Ad(j)$ is the adjoint action of J on κ and $Ad(\lambda(j))$ is that of G on $\mathfrak{g}$. The geometric meaning of these homomorphisms should be clear from our discussion of the lifting of the horizontal projection of any one-parameter group of automorphisms given by Eq.(6.11) and the discussion following it. Note that u_o denotes our previous p_o (and <u>not</u> the value of the automorphism at t = 0), and the vertical action $\lambda(j)$ is the same as the previous g_t.

It is easy to verify, by using the definition of curvature (the structure equation), that the curvature form Ω satisfies the condition (from which Eq.(6.18) follows by pullback to M):

$$2\Omega_{u_o}(\tilde{X},\tilde{Y}) = [\Lambda(X),\Lambda(Y)] - \Lambda([X,Y]), \text{ for } X, Y \in \kappa . \quad (6.19)$$

What Wang's theorem asserts is the existence of a bijection between the set of K-invariant connections in P and the set of linear mappings $\Lambda : \kappa \to \mathfrak{g}$ satisfying the conditions listed above, bijection which is given by

$$\Lambda(X) = \omega_{u_o}(X), \text{ for } X \in \kappa \quad . \quad (6.20)$$

The proof is straightforward and can be found, e.g., in Kobayashi and Nomizu (p.107, with the same notations as here).

It also follows immediately that a K-invariant connection is flat (i.e., has vanishing curvature) iff $\Lambda : \kappa \to \mathfrak{g}$ is a Lie algebra homomorphism (since then the right-hand side of Eq.(23) vanishes, and hence so does the left-hand side).

Moreover, if in addition the Lie subalgebra $\mathfrak{j}$ admits a complementary subspace $\mathfrak{m}$ in $\mathfrak{k}$ such that $Ad(J)(\mathfrak{m}) = (\mathfrak{m})$, then there is a bijection between the set of K-invariant connections in P and the set of linear mappings $\Lambda_{\mathfrak{m}} : \mathfrak{m} \to \mathfrak{g}$, such that for $X \in \mathfrak{m}$, $j \in J$ we have $\Lambda_{\mathfrak{m}}(Ad(j)(X)) = Ad(\lambda(j))(\Lambda_{\mathfrak{m}}(X))$, with the bijection given in terms of the Λ defined above by $\Lambda(X) = \lambda(X)$ if $X \in \mathfrak{j}$, and $\Lambda(X) = \Lambda_{\mathfrak{m}}(X)$ if $X \in \mathfrak{m}$. The curvature form of the K-invariant connection defined by the linear mapping $\Lambda_{\mathfrak{m}}$ satisfies the following condition:

$$2\Omega_{u_0}(X,Y) = [\Lambda_{\mathfrak{m}}(X),\Lambda_{\mathfrak{m}}(Y)] - \Lambda_{\mathfrak{m}}([X,Y]_{\mathfrak{m}}) - \lambda([X,Y]_{\mathfrak{j}}), \qquad X,Y \in \mathfrak{m},$$

where the subscripts on the brackets denote components in the corresponding subspaces of the algebra $\mathfrak{k}$ where the bracket is originally defined. If $\Lambda_{\mathfrak{m}} = 0$ then the corresponding invariant connection is called the <u>canonical invariant connection</u> with respect to the decomposition $\mathfrak{k} = \mathfrak{j} \dotplus \mathfrak{m}$. Physically, this corresponds to choosing the gauge functions Z and the connection A in eqs.(6.13) - (6.18) so that the components of Φ in the subspace $\mathfrak{m}$, corresponding to the given decomposition, should vanish.

It is to be noted that the existence of a complementary subspace $\mathfrak{m}$ invariant under the adjoint action of J is equivalent to the reductivity of the homogeneous space K/J = M, a rather restrictive condition on the base space M.

Finally, it should be noted that the Lie algebra of the holonomy group of a K-invariant connection at u_0 is defined by a sum of iterated brackets of $\Lambda(\mathfrak{k})$ with the subspace $\mathfrak{m}_0$ of $\mathfrak{g}$ spanned by the right-hand side of eq. (6.19) (for details we refer the reader again to Kobayashi-Nomizu, p.110-111).

ACKNOWLEDGEMENT

The authors would like to thank Professor H. Mitter and the other organizers of the XX. Internationale Universitätswochen for the warm hospitality in Schladming, and many of the participants, both lecturers and auditors, for stimulating discussions.

REFERENCES

Textbooks and Lecture Note Volumes

M.F. Atiyah, Geometry of Yang-Mills Fields, Academia Nazionale dei Lincei, Pisa, 1979.

R.L. Bishop, R.J. Crittenden, Geometry of Manifolds, Academic Press, NY 1964.

D.D. Bleecker, Gauge Theory and Variational Principles, Addison-Wesley, to appear.

N. Bourbaki, Variétes différentielles et analytiques, El. de math. XXXIII and XXXIV, Hermann, Paris, 1967, 1971.

J. Dieudonné, Treaties on Analysis, vols. III and IV, Acdemic Press, NY 1972, 1974 (French original: Traite d'Analyse, Gauthier-Villars 1974, 1971).

Y. Choquet-Bruhat, C. De Witt-Morette, and M. Dillard-Bleick, Analysis, Manifolds, and Physics, North-Holland, 1977.

W. Drechsler, M.E. Mayer, Fiber-Bundle Techniques in Gauge Theory, Lect. Notes in Physics, vol. 67, Springer Verlag, Berlin-Heidelberg-New York, 1977.

F. Hirzebruch, Topological Methods in Algebraic Geometry, Springer, 1966.

D. Husemoller, Fibre Bundles, 2nd Edition, Springer Verlag, 1975.

S. Kobayashi, K. Nomizu, Foundations of Differential Geometry, 2 vols. Wiley-Interscience, NY, 1963, 1969.

A. Lichnerowicz, Global Theory of Connections and Holonomy Groups, Noordhoff, 1976.

M.E. Mayer, Gauge Theory, Vectur Bundles, and the Index Theorem, Springer, to appear.

I.M. Singer, J.A. Thorpe, Lecture Notes on Elementary Topology and Geometry, Springer Verlag, New York, Heidelberg, Berlin, 1976.

M. Spivak, A Comprehensive Introduction to Differential Geometry, Publish or Perish, 1971.

S. Sternberg, Lectures on Differential Geometry, Prentice-Hall, Englewood Cliffs, 1964.

R. Sulanke, P. Wintgen, Differentialgeometrie und Faserbündel, Birkhäuser, Basel,1972.

Various summer and winter school lecture notes (Cargese, Erice, Schladming, etc.).

Articles

P.G. Bergmann, E.J. Jr Flaherty, J. Math. Phys. 19 (1978) 212.

M. Daniel, C.M. Viallet, Rev. Mod. Phys. 52 (1980) 175.

T. Eguchi, P.B. Gilkey, and A.J. Hanson, Phys. Reports 66 (1980) 213-393.

M.E. Mayer, Ann. Israel Phys. Soc. 3 (1979) 80-99; Hadronic J. 4 (1981) 108-152.

A. Trautman, Rep. Math. Phys. (Toruń) 1 (1970) 29; 10 (1976) 297;
Czech. J. Phys. B29 (1979) 107-116;
Bull. Acad. Polon. Sci., ser. sci. phys. et astron. 27 (1979) 7;
Fiber Bundles, Gauge Fields, and Gravitation, in General Relativity and Gravitation, vol. I, edited by A. Held, Plenum Press, New York 1980;
J. Phys. A13 (1980) L1.

Additional references can be found in the lectures of the two authors in this volume.

S.S. Chern, Geometry of Characteristic Classes, Proc. 13th Biennial Seminar, Canadian Mathematical Congress, vol. I, 1972, p.1-40.

M.S. Narasimhan, S. Ramanan, Existence of Universal Connections, Amer. J. Math. 83 (1961) 563-572.

J. Nowakowski, A. Trautman, Natural Connections on Stiefel Bundles are Sourceless Gauge Fields, J. Math. Phys. 19 (1978) 1100.

R. Schlafly, Universal Connections, Inventiones math. 59 (1980) 59-65.

V.A. Rokhlin, D.B. Fuks, Nachal'nyi kurs topologii (Introductory course in topology), Nauka, 1977.

Acta Physica Austriaca, Suppl. XXIII, 477–490 (1981)

THE GEOMETRY OF SYMMETRY BREAKING IN GAUGE THEORIES+

by

M. E. MAYER
Department of Physics, Univ. of California
Irvine, California, 92717, USA

ABSTRACT

This together with Sections 3 and 6 of the joint contribution with A. Trautman (this volume, pp. 433 to be referred to as Mayer-Trautman) constitutes a summary of the first two lectures. Much of the material is available elsewhere [1], so only results and some open questions are discussed. The subject matter of the second two lectures is treated in the following contribution (pp. 491).

1. INTRODUCTION

The motivation of these lectures is a search for an alternative to the traditional Brout-Englert-Higgs-Kibble (BEHK) method of symmetry breaking in gauge theories, based on the geometry of principal bundles with connections. In the BEHK approach the action of the classical theory of a Dirac or Weyl field interacting with a Klein-Yang-Mills field A, F is

---

+Lectures given at the XX. Internationale Universitätswochen für Kernphysik, Schladming, Austria, February 17-26, 1981.

$$S_{DYM} = \frac{1}{4} \int_M \mathrm{Tr} F \wedge {*F} + \int_M {*\{\psi \not{D} \psi\}} \quad , \tag{1.1}$$

where M denotes the spacetime base manifold of the bundles, F is the Yang-Mills field strength two-form on M (the pull-down of the curvature two-form to M, cf. Mayer-Trautman, Eq. (4.12)), $*$ is the Hodge-dual on M, ψ denotes a Dirac or Weyl spinor which transforms under a representation of the structure group G (a local representation of a section in a tensor product of a spin bundle and a vector bundle associated to the gauge principal bundle P by that representation). $\not{D}$ denotes the "gauge-covariant Dirac-Weyl operator" (in coordinates, with e_a a basis for the representation of the Lie algebra G, in which I denotes the unit matrix, and γ^μ the usual Dirac matrices) $\not{D} = I\not{\partial} + A^a_\mu \gamma^\mu e_a$. In order to produce the symmetry breaking, leading to a restriction of the original bundle P to a subbundle Q(M,H), the BEHK model introduces "by hand" a scalar field ϕ which represents a section of an associated bundle (or a smooth function on P with values in some representation space V of G, cf. Mayer-Trautman, Eq. (2.5)), which is supposed to be an extremal of the Ginzburg-Landau action:

$$S_{GL} = \int D\phi \wedge {*D} + V[\phi], \; V[\phi] = -\mu\|\phi\|^2 + \lambda\|\phi\|^4 \quad , \tag{1.2}$$

where the norms in the Ginzburg-Landau functional V are to be understood as the result of integration over M of the hermitian norm in V. For positive λ,μ the functional V has nontrivial critical points ϕ_o and the stabilizer subgroup of these is H, the symmetry group of the "vacuum" to which the bundle is then restricted. Reducing the connection form A and the curvature form F to the corresponding Lie algebra H, one can choose a gauge in which the terms involving the nonvanishing ϕ_o^2 in (1.2) appear like "mass terms" for the components of A in the complement $M = G - H$ of the Lie algebra of H in that of G, thus leading to a loss of conformal invariance for the appropriate Yang-Mills equations. The surviving Higgs fields and the fermions are also **aquiring** "masses" by this mechanism. For details of this and other aspects of traditional symmetry breaking the reader is referred to the review by O'Raifeartaigh [2] where further references can be found.

A closer look at the BEHK mechanism (and this will partially be true of the geometric models discussed in this lecture too) shows that the presence of the Ginzburg-Landau potential is not really at the heart of the matter. The scalar field and the quartic interaction were chosen because they lead to a renormalizable quantum theory and are the simplest combination which does the job. In effect, the reason why they do the job is revealed by a careful analysis of group actions on manifolds, particularly of pairs of groups such as G and H, and their homogeneous spaces G/H. Such an analysis was carried out in other contexts of symmetry breaking by Michel and Radicati [3] over a decade ago, and a good summary can be found in Ref.[2]. It turns out that if one is given G and H, there are relatively few invariants which lead to the desired physical results, among them the Ginzburg-Landau action.

The same remarks apply, mutatis mutandis, to the orbit structure of the associated bundles E(M,P,G/H,p) discussed below in symmetry breaking models. In fact, the symmetry breaking sections should appear from a detailed analysis,in the spirit of Michel and Radicati, of the orbits and strata of group actions in these bundles. There does not seem to exist in the literature an explicit discussion of this topic, and it was hoped that such a discussion could be included here. However, time pressure forced me to defer this to a future publication.

Until recently, Higgs fields were considered almost sacrosanct, but the view that they really exist (and the appropriate particles should indeed manifest themselves experimentally) is becoming less widely held. Many of us who were not satisfied by the artificial introduction of the Higgs fields have been searching for alternatives to the introduction of the Higgs bosons, yielding the same results which made this model so appealing and successful in the electroweak unification, and in all grand unified gauge theories. I will not discuss theories in which the Higgs scalars are treated as bound states of more elementary fermions, or metric theories of the Kaluza-Klein type (for an interesting recent attempt to obtain the putative $SU(3) \times SU(2) \times U(1)$ symmetry of strong and electroweak interactions, I refer the reader to a recent paper by Witten [4] but will describe briefly two "geometric" approaches to the problem which can be described in the language of bundle restrictions (cf. Mayer-Trautman, Secs. 3 and 6). The first, which I call symmetry breaking by "harmonic sections" [5,6] is related to nonlinear sigma-models and leads to a quantum theory which is not obviously renormalizable (in the usual sense of linear quantum field theories). The second is based on the introduction of hidden dimensions followed by a "dimensional reduction" [7,8,9] and makes use of symmetries of connections and curvatures discussed in Sec. 6 of Mayer-Trautman, and should be considered a direct application of the methods discussed there.

For some details we refer the reader to my recent lectures [1]. Additional references, not mentioned in the text, are given at the end of this lecture.

2. SYMMETRY BREAKING BY HARMONIC SECTIONS

We consider, as before, a gauge theory described by a principal bundle $P(M,G)$, with the group G describing a hidden gauge symmetry, and a subgroup H - the holonomy group of the vacuum, to which the bundle P is ultimately to be restricted. The theory of bundle restrictions (Sec.3 of M-T) showed that the orbit space of the action of H on P, which is identical to the associated bundle E(M,G/H,P,p) = $= P \times_G (G/H)$ with fiber the coset space G/H, plays a distinguished role, since each restriction is given by a section $\sigma : M \rightarrow E$ of this bundle, and gauge-inequivalent sections lead to different restrictions, i.e., inequivalent symmetry breakings. It is only natural to let these sections take over the role of the Higgs scalars in the theory. The question then arises how to determine the section which gives rise to the appropriate restriction. In principle, this is again a problem of invariant theory and group actions on manifolds and could be atacked by the Michel-Radicati approach. A simpler point of view was proposed independently by C. Misner [5] and by the author [6] (a closely related perturbative calculation was recently published by Aref'eva and Slavnov [10]) and is based on determining the symmetry-breaking section σ from an action principle.

If we assume that M is compact, Riemannian (e.g., the S^4 compactification used in many applications of gauge theory), then σ is a smooth map from the Riemannian manifold M to the Riemannian manifold E (or alternatively, an equivariant map from M to G/H, the latter being equipped with a Riemannian metric, which it inherits from the Killing-Cartan metric on G). Variational principles for such maps have been extensively studied by mathematicians [11] under the name harmonic maps (since they generalize the Dirichlet integral in Euclidean space which leads to harmonic functions as extremals). The harmonic map action is a positive quadratic functional in "derivatives of the section". The last term requires some clarification. Since the bundle E is not a vector bundle, covariant derivatives or differentials of sections are not immediately defined, and it is not obvious how to form an action four-form

(Lagrangian density) whose critical elements are the sections we are looking for.

However, similar objects occur as fields in non-linear sigma-models: the field quantity is a map from spacetime into the sphere $S^n = SO(n+1)/SO(n)$, i.e., in bundle language, a section of the trivial bundle $M \times S^n$. The action is formed by taking the square of the (co-variant) gradient of the sigma-field and adding the constraint that the vector be of length one by means of a Lagrange multiplier.

Similarly, in the case considered here, if the bundle is trivial, the section σ can be viewed as the graph of a G/H-valued function on M. If we denote the fiber metric (the metric of the compact homogeneous space, i.e., a quadratic form on the tangent space at each point) by h and denote the tangent map of σ by $\sigma':T(M) \to T(E)$, we use the connection on P to define at each point $y \in E$ the horizontal subspace of T_yE (the vertical subspace is tangent to the fiber). Denoting by $D_\omega \sigma$ the horizontal projection of σ' we are thus led to the harmonic map action

$$S_{HM} = \int_M h\ (D_\omega \sigma, D_\omega \sigma)\, dVol \quad , \tag{2.1}$$

where dVol is the integration element on M. In a local trivialization and in terms of local coordinates the action (2.1) will look like the action of a nonlinear sigma-model (e.g., on $S^2 = SO(3)/SO(2)$, $\sigma = \sigma(\theta,\phi)$ is a function of the ordinary spherical coordinates and the action involves derivatives with respect to these). The terms of the harmonic map action (2.1) containing the components of the gauge potential A lying in the sub-space $M = G - H$ of the Lie algebra will appear squared, and thus simulate a mass term in the Yang-Mills action for these components, with coefficients determined by the symmetry-breaking harmonic section σ. Thus, this section takes over the role of the Higgs bosons.

The quantum theory of such a "homogeneous-space-valued field" presents considerable difficulties (it is not renormalizable by ordinary standards) and the whole area needs to be explored further. There has been considerable activity in the theory of nonlinear sigma-models, CP^{n-1}-models, and models where the field takes values in a Grassmann manifold, but only in two spacetime dimensions. Misner (cf. Ref.[5], and unpublished work) has considered models in which the nonlinear field is given by harmonic projection maps . A recent article by Aref'eva and Slavnov [10] discusses just such a model from a perturbative point of view, and they show that the results are to a large extent equivalent to those obtained in the BEHK method.

3. HIDDEN DIMENSIONS AND SYMMETRY BREAKING

Another approach to symmetry breaking (more correctly, a whole class of approaches) is based on the introduction of "hidden dimensions" into the principal bundle on which a gauge theory is based, with the result that certain components of the connection take over the role of the Higgs bosons and produce the required violation of conformal invariance of the Yang-Mills equations. There are essentially two ways of introducing hidden dimensions into a principal bundle, which I will call the Kaluza-Klein, and Weyl methods, respectively. In a Kaluza-Klein approach one starts from a Riemannian or pseudo-Riemannian manifold of dimension 4 + k, writes down the Einstein-Hilbert action (linear in the curvature) for the metric in this space, and treats the non-block-diagonal terms as a Yang-Mills connection. The appropriate terms in the action then yield, among others, terms which are quadratic in the Yang-Mills curvature and can thus be interpreted as a Yang-Mills action. The general theory of such models has been discussed in many places; cf., e.g., Trautman's lecture in this volume, the forthcoming book by Bleecker [12], and Witten's recent atempt [4] to obtain an $SU(3)\times SU(2)\times U(1)$ gauge theory from an eleven-dimensional Kaluza-Klein model.

The Weyl approach, also known as "dimensional reduction" or "fiberflipping", has been particularly popular among supersymmetrists (I will not discuss supersymmetric gauge models here), and has been successfully used by Manton [8] in a model which derives many of the features of the standard electroweak unification from a G_2-principal bundle by this method.

The general theory of such symmetry breaking can be formulated as follows. We start out from a principal bundle P(M,G) over four-dimensional M, with hidden symmetry group G as the structure group. The symmetry group of the vacuum H, a closed subgroup of G, is assumed known. A given symmetry breaking is then described, as already discussed, by a section of the associated bundle E(M,P,G/H,p). We now take E as the base space of a new bundle R(E,G) with structure group G (not H, as was the case for the restriction Q(M,H)) obtained as the pullback of P under the projection p of the associated bundle E: $R = p^*P$. (Pictorially, one can think of R as the bundle obtained from P by "reattaching, or flipping" part of the fiber, G/H, so that it appears both in the base space and in the fiber.)

The result is a larger principal bundle, where the "hidden dimensions" of the manifold G/H appear both in the base space E and in the fiber G. The group G acts both on the base space and the bundle space, and therefore the results obtained in Sec. 6 of the Mayer-Trautman article in this volume apply. The connection on P is pulled back into a connection on R, and both this connection and its curvature acquire extra components, since they are now defined on a larger base space. Let us denote by script letters the pullbacks to E of the connection and curvature on R in a local trivialization determined by a section $s:U \subset E \to R$:

$$\mathcal{A} = s^* p^* \omega \quad , \quad \mathcal{F} = s^* p^* \Omega \quad , \tag{3.1}$$

where ω and Ω are the connection and curvature on P, and $s^* p^*$ denote their pullback to R pulled down to E by s (this symbolic notation can be interpreted easily in terms of local bases, which we leave as an exercise for the reader). If G/H is k-dimensional, then $\mathcal{A}$ is a $\mathcal{G}$-valued one-form on the 4 + k-dimensional manifold E, and $\mathcal{F}$ is a $\mathcal{G}$-valued two-form. Thus $\mathcal{A}$ can be thought of as a set of 4 + k matrices, and $\mathcal{F}$ as a set of $(4 + k)(3 + k)/2$ matrices. Moreover, one can choose in the Lie algebra $\mathcal{G}$ a basis adapted to the splitting $\mathcal{G} = \mathcal{H} + \mathcal{M}$, $\dim \mathcal{G} = n$, $\dim \mathcal{H} = m$, $\dim \mathcal{M} = n - m = k$. Clearly, the physical surviving components of $\mathcal{A}$ and $\mathcal{F}$, which we will denote by A and F, respectively, are a one-form and two form on M with values in $\mathcal{H}$, and the remaining components will be subjected to symmetry and gauge transformations, thus reducing the Yang-Mills action on E to a Yang-Mills-Ginzburg-Landau action on M!

Consider the Yang-Mills action on R (numerical factors are omitted)

$$S_{YM} = \int \mathrm{Tr}\,(\mathcal{F} \wedge {}^*\mathcal{F}) \quad , \tag{3.2}$$

where the trace is the Killing-Cartan trace on $\mathcal{G}$, and the Hodge-dual is taken on the oriented Riemannian manifold E. (This presupposes a fiber metric on G/H, which is the same as the h of the preceding section; it should be recalled that $\mathcal{F}$ is a two-form, hence ${}^*\mathcal{F}$ is a (2 + k)-form, and the integrand is a (4 + k)-form, as it should be.) The connection and its curvature are clearly invariant under the action of G on the base space E, hence we can apply Eqs.(6.15) and (6.18) of Mayer-Trautman (p.433 this volume). We can obviously split the curvature F into components along M (spacetime) and those along directions tangent to G/H. We denote the former components by $F_{!!}$ and the latter by $F_{??}$, whereas the mixed components (one along M, the other along G/H) will be denoted by $F_{!?}$; the Hodge-dual can be reexpressed in terms of the corresponding contravariant components. Then the integrand of (3.2) becomes

$$\mathrm{Tr}\,(F_{!!}F^{!!} + 2F_{!?}F^{!?} + F_{??}F^{??}) \ . \tag{3.3}$$

Exploiting the invariance of the connection with respect to transformations in the ?-directions, i.e., assuming the vector fields X, Y in Eqs.(6.15) and (6.18) in Mayer-Trautman to be along G/H, the components $F_{!?}$ can be expressed as the $D_!\Phi(?)$, where $\Phi(?)$ is the Lie-algebra-valued 0-form corresponding to the invariance of A with respect to the vector field ?, in the G/H direction of E. Thus, the middle term in Eq.(3.3) becomes, symbolically,

$$\mathrm{Tr} \sum D_!\Phi(?)\, D^!\Phi(?) \ , \tag{3.4}$$

where the summation is over the repeated symbols !, ?. The first term in (3.3), after integration over the homogeneous spaces G/H and reduction to the Lie-algebra H, becomes the Yang-Mills action for the reduced Yang-Mills theory on Q. Finally, in order to handle the third term, which involves the contraction $F_{??}$ of F with two vector fields lying along G/H, we make use of the equation (6.18) in Mayer-Trautman, which becomes:

$$2F_{??} = [\Phi(?),\Phi(?)] - \Phi([?,?]) \ , \tag{3.5}$$

with the obvious meaning for the bracket of two?. Thus, the third term in Eq.(3.3) reduces to what is essentially a Ginzburg-Landau polynomial in the components of Φ:

$$\mathrm{Tr}F_{??}F^{??} = \frac{1}{4}\mathrm{Tr}([\Phi,\Phi]-\Phi)^2 , \qquad (3.6)$$

where the square means contraction in the appropriate vector field directions with the metric h on G/H. As was pointed out by Professor O'Raifeartaigh, it is necessary to analyze the expression (3.6) more carefully, since the presence of the brackets may in some cases lead to instability problems. However, special cases which were considered show that Eq.(3.6) has indeed the properties required of a Ginzburg-Landau-Higgs potential, and moreover the relative signs of the quartic and quadratic terms are correct, and only one overall normalization constant (rather than the two which are usual in the expression (1.2)) is needed.

There remains, of course, the problem of how to introduce spinors into a model of this type, and how to couple the spinors to the new fields Φ which have been introduced. There are two obvious ways in which spinors can be handled in this context, neither of which leads to satisfactory results in physical contexts. The first is to treat the spinors as tensor products of four-dimensional spinors with objects behaving trivially on G/H. The second is to introduce spinors on E (i.e., objects transforming under the group Spin (4 + k)), and then carry out the reduction.

4. CONCLUDING REMARKS

In this brief outline (for details we refer to Ref.[1]) we have made it clear that symmetry breaking, at least at the classical level, which was the only one considered here, has a deep underlying geometric meaning, and that there are many substitutes for the BEHK mechanism.

In fact, if it should turn out that quarks are not elementary, but have an underlying gauge theory (so that each fiber of our principal bundle is itself a bundle, or some other "foliated" manifold), then the connections and curvatures at this higher level will be constructed out of more elementary objects, and it may well be that the fields Φ of Section 3, or the Sections σ of Section 2 are "vacuum expectation values of products of more elementary spinors" (called "quinks" or "quints", in the physics literature). Thus the symmetry breaking becomes somewhat analogous to what is encountered in superconductivity: the Ginzburg-Landau model is supplanted by the correlated pairs of the BCS theory.

It should also be emphasized once again that many of the aspects of the symmetry breaking, whether by BEHK, Harmonic sections, Kaluza-Klein, or dimensional reduction, are to a certain degree independent of the precise model used, thus pointing to an invariant-theoretic origin of the results. A careful analysis of group actions, orbit structure, and strata, in the spirit of Michel and Radicati, for principal and associated bundles with connections and invariance groups should be carried out and will certainly yield fruitful results which will shed light on this subject.

Finally, it should not be forgotten that quantum phenomena are bound to play a most important role in this subject, phenomena which have not been discussed at all in these lectures. Unfortunately, we will be unable to discuss the quantum aspects of symmetry breaking in the following lectures on quantization of gauge theories, but hope to return to this subject in the near future.

REFERENCES

1. M.E. Mayer, in Proc. of the Workshop on Gauge Theories, Mexico City, 1980, Hadronic J. 4 (1981) 108-152.
2. L. O'Raifeartaigh, Hidden Gauge Symmetry, Rep. Prog. Phys. 42 (1979) 159-223.
3. L. Michel and L. Radicati, Ann. Phys. NY 66 (1971) 758; Ann. Inst. H. Poincare 18 (1973) 185, and further references there.
4. E. Witten, Search for a Realistic Kaluza-Klein Theory, Princeton Preprint, 1981.
5. C.W. Misner, Phys. Rev. D18 (1978) 4510.
6. M.E. Mayer, Ann. Israel Phys. Soc. 3 (1979) 80-99.
7. E.Witten, Phys. Rev. Lett. 38 (1977) 121.
8. N. Manton, Nucl. Phys. B158 (1979) 141-153; W. Mecklenburg, J. Phys. G6 (1980) 1049.
9. M.E. Mayer, in Math. Probl. Theor. Phys. (Lausanne 1979), Lect. Notes Phys. Springer, vol. 116 (1980) p. 291.
10. I. Ya Aref'eva and A.A. Slavnov, Theor. Mathem. Phys. 44 (1980) 3-16 (Engl. Transl. 1981).
11. J. Eells Jr. and M. Lemaire, A Report on Harmonic Maps, Bull. London Math. Soc. 10 (1978) 1, and references to earlier work there.
12. D.D. Bleecker, Gauge Theories and Variational Principles, Addison-Wesley, to be published.

Additional references on invariant connections:

P.G. Bergmann and E. Flaherty, J. Math. Phys. 19 (1978) 212.

P. Forgàcs and N.S. Manton, Commun. Math. Phys. 72 (1980) 15.

J. Harnad, S. Shnider and L. Vinet, J. Math. Phys. 21 (1980) 2719.

R. Jackiw and N.S. Manton, Ann. Phys. (NY), to appear.

R. Jackiw, 1980 Schladming Lectures (Acta Phys. Austriaca, Suppl.XXII)

J. Madore, Commun. Math. Phys. 56 (1977) 115.

A.S. Schwarz, Commun. Math. Phys. 56 (1977) 79-86.

A. Trautman, Bull. Acad. Polon.Sci.Ser.Phys.Astron.27 (1979) 7, and this volume.

ADDITIONAL NOTE (APRIL 25, 1981)

After my return from Schladming I received an interesting preprint by Christopher Isham: "Spontaneous Symmetry Breaking and Topological Charge", ICTP/80/81-14, Imperial College, London, February 1981, where bundle restrictions and symmetry reductions are considered from a topological point of view.

I was also pleased to learn from a seminar given by Richard Slansky at U.C. Irvine this month that he and Murray Gell-Mann are engaged in a detailed investigation of the hierarchy of symmetry breakings in grand-unified models in terms of the Michel-Radicati approach to group actions, orbits, and strata on homogeneous spaces. As already mentioned in the text, I feel that such an analysis should also be carried out in the fiber-bundle approach to symmetry breaking.

Acta Physica Austriaca, Suppl. XXIII, 491–524 (1981)

GEOMETRIC ASPECTS OF QUANTIZED GAUGE THEORIES+

by

M.E. MAYER
Dept. of Physics, Univ. of California
Irvine, California, 92717, USA

1. INTRODUCTION

These notes represent the second two lectures delivered at the school. If they differ somewhat from the preliminary notes handed out before the lectures and from the material presented in the lectures themselves, this is the result of constructive feedback I have received during the lectures from several participants, to whom I express my indebtedness here. The lectures covered essentially two major topics: the geometric aspects of canonical quantization with indefinite metric, in particular, the meaning of gauge-fixing and the relation of "ghosts" to geometry, and quantization in terms of holonomy operators, an outgrowth of Mandelstam's formulation of gauge theory without potentials, but involving path-ordered integrals.

Before entering into the subject properly, one might ask the question: why quantize a gauge theory, if one considers it a theory of connections and curvatures in a

+Lectures given at XX. Internationale Universitätswochen für Kernphysik, Schladming, Austria, February 17-26, 1981.

principal bundle? To this question one can give a number of answers which may not satisfy everybody, but will serve as our main motivation in the discussion which follows:

1. The most obvious answer is the success of quantum electrodynamics - the earliest and best-known gauge theory - as well as the successes of the perturbative electroweak theory, and maybe even perturbative QCD. Here by success I mean the ability to carry out calculations to high order of perturbation theory, the results of which are not in disagreement with the experimental data (and in the case of QED, the agreement between calculations and experiment is really astounding). Although we know how to calculate we do not always understand what we are doing and why perturbation theory works as well as it does.

2. Because Yang-Mills fields are coupled to particle fields which are described by quantized fields, the gauge fields themselves should in some sense be quantized.

3. As a "laboratory" for quantum gravity. Indeed, if one considers gravitation as a gauge field, lessons learned from quantum gauge theory might be applicable to the considerably more complex case of gravity. (This was indeed the motivation for Feynman [1] to treat the quantum theory of the Yang-Mills field, which led to the discovery of "ghosts" and the difficulties of gauge-fixing as well as Mandelstam's [2] motivation for his path-dependent approach, and led to the work of Bialynicki-Birula [3], De Witt [4], and Faddeev-Popov [5].)

4. As an intellectual exercise - because it is difficult and nonlinear, but "the least nonlinear" of known theories.

5. To understand confinement and the relation to lattice gauge theories, and the results of recent computer experiments related to them.

6. As a problem in geometry: one might learn how to combine differential geometry and functional analysis in a "non-elliptic" context. Much of what is known in the geometry of vector bundles (index theory, structure of spaces of connections, even path integral methods) refers to "elliptic" problems on compact base spaces. A true quantum gauge theory would have to take hyperbolicity into account.

Before giving a brief outline of the subject matter covered in these lectures, I will try to remind the reader of the evolution of quantized gauge theory. This brief history is of necessity incomplete and probably not completely accurate since it is highly subjective. I apologize in advance to all those authors whose work has not been properly quoted or omitted altogether because of the haste with which these notes have been written up. No omission was intentional, and the fact that some authors were quoted does not mean that there does not exist equally valuable work which I have overlooked. No references to classics are given.

In QED gauge conditions in quantization were first systematically discussed by Fermi (1932) who noticed that the Lorentz gauge condition of classical electrodynamics, $d^{*}A = 0$, cannot be imposed as an operator equation in quantum theory, but must be considered as a supplementary condition on physical states $|\psi_{phys}\rangle$ in Hilbert space: $d^{*}A|\psi_{phys}\rangle = 0$, and that Gauss' law, for instance, is only valid as an expectation value in physical states. This theory was extended by Stueckelberg in 1938 to general neutral vector fields, and we shall use a version of the Stueckelberg approach in our discussion of gauge-fixing. After Pauli had analyzed gauge invariance in general in his 1939 Solvay report (which was published in Rev. Mod. Phys. in 1941 and served as the training manual for coming generations of physicists), gauge invariance was extensively used in the heroic days of QED, but some authors, wary of the supplementary conditions, preferred to work in the Coulomb gauge, or in terms of field strengths only, at the cost of losing locality and Lorentz invariance.

The next major step forward came in 1950, when Bleuler and Gupta independently recognized that the Fermi supplementary condition cannot be imposed in an ordinary (positive-metric) Hilbert space and that one needs a Hilbert space with indefinite metric (previously discussed for other reasons by Dirac and Pauli), with the physical vectors being selected by the criterion: $d^{*}A^{(+)}|\psi_{phys}\rangle = 0$, where (+) denotes the annihilators in a second-quantized formalism. They showed that this condition insures cancellation of the contribution of longitudinal and temporal "photons" in expectation values, that gauge transformations affect only the occupation numbers of these unphysical (negative, or zero-norm) states, and,in spite of the reservations often encountered towards indefinite metrics, was and is used in most local covariant gauge-invariant treatments of QED, and has been put on a rigorous footing within axiomatic quantum field theory by Strocchi and Wightman [6], to which we refer the reader for details.

After Yang and Mills rediscovered nonabelian gauge theory in 1954 (we remind the reader of the little-known fact that it had been developed by Oscar Klein in 1938, but apparently forgotten until recently), they quantized the theory right away, using the Coulomb gauge. Coulomb and Lorentz gauge quantization was used also in subsequent work on nonabelian theories by Schwinger, Arnowitt and Fickler, Loos and Treat, Goto and Utiyama, and others, but it was not until nonabelian theory was successfully applied to the electroweak unification that closer attention was paid to the problems of gauge-fixing and the necessity of introducing ghosts (which had been discovered by Feynman [1], De Witt [4], and Faddeev and Popov [5]). For a good introduction particularly from the viewpoint of Feynman path integrals and perturbation theory, which will not be discussed here, cf. the review of Abers and Lee and the stimulating "Diagrammar" of 't Hooft and Veltman [7] as well as the recent book by Faddeev and Slavnov [8].

The perturbative calculations in gauge theories which became necessary in connection with the successes of electroweak theory led to an intensive investigation of Ward-Takahashi (Slavnov-Taylor) identities (which had been proposed earlier, among others by Veltman and independently the author), and the role of ghosts in perturbation theory. Probably the most interesting results to come out of this line of investigation are the Becchi-Rouet-Stora transformations [9] involving the gauge potentials, Feynman-De-Witt-Faddeev-Popov ghosts, and gauge-fixing Lagrange multipliers which we will discuss in detail later. Here I only want to point out that in his Erice lectures of 1975 [9], Raymond Stora proposed an interpretation of the ghost fields as the left-invariant forms on the group of gauge transformations, interpretation which was taken up in a simplified form in a series of papers by Thierry-Mieg and Ne'eman [10]. I will try to convince you that Stora's original proposal is the most attractive from a differential-geometric viewpoint.

The role of indefinite metric in the quantization of nonabelian gauge theories was stressed in several papers, most definitively in the already mentioned paper of Strocchi and Wightman [6], and the role of Gauss' law and the "Maxwell" equations was discussed by Strocchi [11] (cf. also the recent review of Jackiw [12]). But a definitive treatment of canonical quantization with indefinite metric including many important consequences is due to Kugo and Ojima [13] (we recommend their review paper as an excellent introduction to the subject, and for a source of further references).

The other line of inquiry into quantized gauge theories - which I will call the holonomy approach - was started by Mandelstam [2] in 1962 when he proposed a quantum electrodynamics (and quantum gravidynamics) without potentials and without indefinite metric, at the cost of having the particle fields (and in the nonabelian case, the Yang-Mills fields) defined with the help of a "holonomy operator", i.e., each such field is the holonomy transform along a path of a corresponding field defined at a point at "spacelike infinity". Mandelstam was able to show the equivalence of his approach to Coulomb-gauge and Lorentz-gauge treatments of QED and showed that one can obtain the Feynman-Dyson perturbation series from it. The Mandelstam approach was extended in 1963 to the Yang-Mills case by Białynicki-Birula [3], and was shown in 1968 by Mandelstam [14] to be equivalent to the Feynman-De Witt-Faddeev-Popov approach, yielding the same perturbative expansions. In 1975 the author [15] proposed to consider the holonomy group as the object to be quantized in a gauge theory, but unfortunately this program has never been completed; however, much of the success of lattice gauge theories [16] is due to the fact that the main object of study is the Wilson holonomy operator (loop product) $W[C] = P[\exp(\int A)]$, where P means "path-ordering" (this is a symbolic notation for a product integral of elements of the holonomy group along the loop C). There has been much recent activity related to reformulating gauge quantum field theory in terms of quantities defined on loops only [17], an approach which is still awaiting a complete and rigorous mathematical formulation, but which allows an easier comparison with lattice gauge theories.

There are many areas of quantum gauge theory which have not been mentioned in this introduction, not because they are not deemed important, but because they are not directly related to the remainder of this lecture. This refers in particular to lattice gauge theories [18] and the important results of Buchholz and Fredenhagen [19] on a Haag-Kastler quantum electrodynamics.

These notes are organized as follows: in Section 2 I give a quick review of indefinite-metric quantum field theory, following essentially Strocchi and Wightman [6] and Kugo and Ojima [14]; this is followed in Section 3 by a translation into geometric language of the indefinite metric approach in nonabelian theory [14], including ghosts and BRS transformations. Section 4 is devoted to a description (again in geometric terms) of the path-dependent approach to gauge quantum field theory. Section 5 contains a discussion and lists problems which need to be investigated further.

2. QUANTUM FIELD THEORIES WITH INDEFINITE METRIC

In the mentioned paper of Stroc hi and Wightman [6] the reader can find a detailed proof showing that a treatment of the field strength two-form F and the potential one-form A in abelian gauge theories as operator-valued distributions on Minkowski space, with a Hilbert space with positive-definite inner product, is incompatible with the requirements of locality, the existence of a unique vacuum and invariance with respect to the Poincaré group, i.e., a theory satisfying all these requirements is of necessity trivial. Hence the free electromagnetic field cannot be treated as a quantum field á la Gårding-Wightman in ordinary Hilbert space. One must take into account the fact first recognized by Bleuler and Gupta that Maxwell's equations, in particular, Gauss' law div E = 0 (for the free field) is not an operator equation, i.e., the equation $d^*F = -d(d^*A)$ is valid only on the subspace H' of states satisfying the Bleuler-Gupta condition

$$d^*A^{(+)}|\psi_{phys}\rangle = 0 \ , \tag{2.1}$$

where the superscript (+) denotes the negative-frequency (annihilation) part of the expansion of A in a complete set of orthonormal modes. This leads to the result that the Hilbert space H is equipped with a nondegenerate sesquilinear form (inner product) <,> which is positive-semidefinite on the subspace H' of vectors satisfying the condition (2.1) (i.e., states which have equal numbers of "longitudinal" and "temporal" photons), and the Hilbert space of physical states can be defined as the quotient $H_{phys} = \overline{H'/H_o}$, H_o being the subspace of states of zero norm, the only one affected by gauge transformations. Maxwell's equations on H' take on the form ($^*J$ is the current three-form which vanishes in the free-field case):

$$\langle\Phi,(d^*F + {}^*J)\psi\rangle = 0 \quad \text{for} \quad \Phi,\Psi \in H' \ . \tag{2.2}$$

Consistency of the formalism requires that the subspace H' should be stable under the action of the Poincare group and the operators $d^*F + {}^*J$, which are to be considered as operator-valued distributions defined on a suitable space of test functions (cf. Refs.[6,13], for details).

For a Hamiltonian formulation which lends itself easily to nonabelian generalization, it is convenient to supplement the Lagrange density of the Maxwell field (in appropriate units, e = 1)

$$L_M = \frac{1}{4} F \wedge {}^*F \tag{2.3}$$

(note that this four-form already includes the integration measure d^4x of the conventional notation) with a gauge-fixing term of the Nakanishi-Lautrup [20] form (when no ambiguity arises we set $\alpha = 1$)

$$L_{GF} = B(x)d^*A + (\alpha/2)B(x)^*B(x) \ , \tag{2.4}$$

where B(x) is a 0-form (smooth function) and $^{*}B(x)$ is the dual 4-form (i.e. $^{*}B = Bd^4x$ in the usual notation; I have omitted the wedge product in multiplying by a zero-form) which acts as a Lagrange multiplier to the Lorentz-condition 4-form $d^{*}A$, and the addition of the term (2.4) to the action makes B into the "momentum conjugate to $A_o(x)$". The equations of motion derived from $L_M + L_{GF}$ are:

$$-^{*}d^{*}A + \alpha B = 0 \ , \tag{2.5}$$

$$-^{*}d^{*}dB = \Box B = 0 \quad . \tag{2.6}$$

The field B(x) is reminiscent of the scalar field B introduced by Stueckelberg in his 1938 treatment of the massive neutral spin-one field; indeed, Stueckelberg's action can be obtained from ours by adding the appropriate mass-terms.

A quantum theory of the fields A, B is obtained by imposing the commutation relations:

$$[B(x),B(x')] = 0 \ , \tag{2.7}$$

$$[A(x),\ B(x')] = -idD(x-x') \ , \tag{2.8}$$

$$\eta \lrcorner \lrcorner [A(x),\ A(x')] = iD(x-x')ds^2, \tag{2.9}$$

where D(x) is the Pauli-Jordan commutation function for the massless field (the function whose Fourier transform is $\delta(k^2)\varepsilon(k_o)$ in the usual notation). The left-hand side of (2.9) is a symbolic double contraction of the one-forms A(x), A(x') with the Minkowski metric tensor η; $ds^2 = \eta_{\mu\nu}dx^\mu dx^\nu$. It is an attempt to write in approximately coordinate-free form the usual relation between components

$$[A_\mu(x),\ A_\nu(x')] = i\eta_{\mu\nu}D(x-x') \quad . \tag{2.9a}$$

Since B(x) satisfies the wave equation (and is almost classical on account of Eq.(2.7)), one can replace the condition (2.1) by the condition

$$B^{(+)}|\psi_{phys}\rangle = 0 \, , \qquad (2.10)$$

which is equivalent to it on account of Eq.(2.5) as long as $\alpha \neq 0$ (the latter condition is violated in the Landau gauge). The remainder of the construction is the usual textbook Bleuler-Gupta formalism, and it can be shown that in the physical Hilbert space one can construct a self-adjoint Hamiltonian, respectively a unitary S-matrix, and that the only physically observable particles are the transverse photons (cf., e.g., Ref.[13,20]).

The abelian gauge theory, being governed by linear equations in the free-field case thus leads easily to a consistent theory. Possible modifications of the subsidiary conditions in superselection sectors with nonzero total charge can easily be taken into account in this formalism.

In the nonabelian case the inherent nonlinearity of the source-free Yang-Mills equations makes the imposition of a condition of the type (2.1) or (2.5) inappropriate: even if it is true in one-particle states, the nonlinearity will violate it.

3. NONABELIAN GAUGE THEORY; GHOSTS AND BRS

As was mentioned in the introduction, it is possible to give a consistent Hamiltonian formulation of nonabelian gauge theory by fixing the gauge (Coulomb gauge) and the reference frame (cf., e.g., the discussion in Ref.[8] and a recent very explicit proof by Cronström [21] who points out that it is convenient to combine gauge transformations and Poincaré transformations). However, the elimination of "unphysical" degrees of freedom achieved in this manner destroys two of the most attractive features of quantum field theory: locality and relativistic invariance which play such an important role in renormalization theory. If one wishes to maintain locality, relativistic, and gauge invariance one must pay the price of using indefinite-metric Hilbert spaces, and moreover if one wants to avoid the discussion of constraints, Dirac brackets, etc. (cf. again Ref.[8] for a brief introduction to constrained Hamiltonians), one must introduce additional fields, such as the Feynman-De Witt-Faddeev-Popov (FDWFP) ghosts, the analog of the gauge-fixing Lagrange multiplier field B of the previous section, and, in order to restore gauge invariance of the whole procedure, the Becchi-Rouet-Stora transformations among the arbitrary quantities. This has been done in a complete and clear manner in the papers by Kugo and Ojima [13], and this section is an attempt to translate their formalism, as far as possible, into the geometric language. (Thus, one may consider this section as a renewed attempt to give geometric meaning to the FDWFP-ghost fields, which have been previously interpreted by Stora [9] and Thierry-Mieg and Ne'eman [10] as some sort of point-dependent Maurer-Cartan one-form along the fiber, thus being naturally anticommutative.)

Historically, the ghosts made their appearance in Feynman's reasoning [1] used to reestablish unitarity in perturbation theory, compensating, by their negative sign in loops, the undesirable terms introduced by the gauge-field propagators. They reappeared both in De Witt's quantization of gauge fields [4] and in Mandelstam's path-dependent approach [5], although most practitioners of gauge field theory became aware of the need to introduce ghosts into the action of a Yang-Mills theory through the Feynman path integral formulation of Faddeev and Popov [5]. In this approach the "ghosts" make their appearance in an indirect way and seem to be an artifact of the functional integration method: in trying to eliminate the integration over the orbits of the group of gauge transformations by choosing a section, an additional determinant appears in the path integral, which Faddeev and Popov raised into the exponent by considering it as the result of a Berezin integration [22] over a fictitious Grassmann variable c (and its conjugate $\bar{c}$) coupled to the nonabelian gauge potential A. This Grassmann variable c(x) being a space-time scalar, one often sees the statement that its Fermi-Dirac quantization violates the spin-statistics theorem. We will argue that its anticommutativity is natural, since it is a one-form on P in some sense supplementary to the one-form $A = A_\mu dx^\mu$.

Very qualitatively, one can advance the following reason for considering ghost-forms outside perturbation theory or the Berezin-integration trick. If one treats the connection form as a quantized (unobservable) field one should quantize its fluctuations around a "background connection" (in the same spirit in which gravity is quantized). Although in the background connection one can (in principle) eliminate the unphysical components (those which obey constraint equations rather than equations of motion), e.g., by gauge-fixing and choosing a special chart in the base space (locally), one should allow the fluctuations of the connection to be in "arbitrary directions in $T_p(P)$", forgetting for the moment the decomposition of the tangent space into horizontal and vertical vectors given by the background connection. The vertical fluctuations, being fluctuations of the Maurer-Cartan form on the fiber of the principal bundle P, are naturally one-forms which cannot be expanded in terms of the one-forms of a local basis in $T_x^*(M)$, and are thus "one-form fields" naturally graded by the anticommutation of one-forms. It is also natural to treat these "vertical fluctuations" as varying smoothly from point to point (more correctly, in a quantized version, as some sort of operator-valued De-Rham currents on such smooth forms). But then they are no longer the Maurer-Cartan forms of the structure group G of the bundle (made artificially x-dependent in Thierry-Mieg's treatment), but more correctly, they will be left-invariant one forms of the group of gauge transformations (or the group of pure gauge transformations, if it is distinct from the latter). One must introduce these "ghost forms", multiplied by an appropriate Lagrange multiplier - the "antighost", into the action of the Yang-Mills field, at the same time that one introduces the gauge-fixing term analogous to (2.4).

Any variation of the redundant components of the Yang-Mills potential, the gauge-fixing field $B(x)$, and the ghost fields $c(x)$ and $\bar{c}(x)$, must be subject to the condition that the resulting change should in fact be a vertical bundle morphism, i.e., a pure gauge transformation, and this results in the Becchi-Rouet-Stora symmetry of the theory. Thus the BRS transformation, and the associated conserved currents and charges in the graded algebra obtained, are a detailed expression of the gauge invariance of the whole theory. That is why the Kugo-Ojima treatment where all these entities are treated as operators (or operator-valued distributions and currents) involves the BRS and ghost "charges" in the formulation of the supplementary condition which selects the physical Hilbert space, rather than the straightforward analog of the Lautrup-Nakanishi condition (2.10) which was sufficient in the abelian case, where the "ghosts decouple from A and B".

We are now ready to construct the Kugo-Ojima version of the Lagrange 4-form for a Yang-Mills field coupled to a matter field. As far as will be possible, I adhere to the notations of the "Brief Introduction", considering a principal bundle P over spacetime M, with a connection ω and curvature Ω; if M is Minkowski space, assume P trivialized by a global section $s(x)$, and define the pull-back forms $A = s^*\omega$, $F = s^*\Omega$ - otherwise these are local trivializations; a gauge transformation is a change of trivialization induced by a change of (local) section $s'(x) = g(x)s(x)$, where $g(x)$ is a smooth G-valued function on M. The "vertical freedom" remaining after the extremization of the Lagrangian of the matter field, Yang-Mills-field, and gauge-fixing field, will be expressed by the introduction of the "ghost-one-forms" $c(x)$ and its multiplier field $\bar{c}(x)$, with the hermiticity assignment of Kugo and Ojima. In order not to complicate matters the various "fields" will be written as operator-valued fields, rather than as operator-valued distributions or currents (partly because of the problem of how to define the product and bracket of such currents, which I have not yet solved in an acceptable way [15]). The trace in the adjoint representation of G or an equivalent hermitean inner product will be denoted by $<,>$, and all expressions are valid up to possible changes in signs or numerical normalization factors, to which little attention is given.

Thus, the action of the Yang-Mills field is given, as usual, by the integral over M of the 4-form

$$L_{YM} = -\frac{1}{4} <F \wedge {*F}> \quad , \tag{3.1}$$

where $*$ is the Hodge-dual. For the matter field we can take either the Dirac or the Duffin-Kemmer Lagrangian:

$$L_D = <\bar{\psi}\gamma \lrcorner {*D\psi}> + m<\bar{\psi}{*\psi}> \text{ or } L_{DK} = <\Phi\beta \lrcorner {*D\Phi}> + \mu^2<\Phi\beta^o{*\Phi}>, \tag{3.2}$$

(or the equivalent Klein-Gordon, or Landau-Ginzburg-Higgs-Kibble Lagrangian); here D denotes the covariant derivative appropriate for the field under consideration, contracted with the Dirac or Duffin-Kemmer matrices. The Hodge stars on the spinors symbolize the operation needed to produce the 4-forms, just as for the scalars; no explicit use will be made of either of these terms, so they are denoted just by $L_m(A,f)$, where f stands collectively for all possible matter fields.

The Lautrup-Nakanishi gauge-fixing 4-form is conveniently written in a form dual to that of the abelian theory:

$$L_{GF} = \langle dB(x) \wedge {*}A \rangle + \frac{\alpha}{2} \langle B(x)\, {*}B(x) \rangle \ , \tag{3.3}$$

making the matrix $B(x)$ into a "field coordinate" and $A_o(x)$ into its conjugate momentum. The "ghost one-form" is a Lie-algebra valued one-form not on M but on the fiber P_x, denoted by $c(x)$. It enters the action through its covariant exterior differential $Dc(x)$ (i.e., the horizontal projection of its exterior differential with respect to its x-dependence), and its Lagrange multiplier is denoted by $d\bar{c}(x)$ - the exterior differential of the antighost $\bar{c}$. Without entering into a discussion of the uniqueness of choice of the "ghost" action which is suggested, for example by the Faddeev-Popov trick involving Berezin-integration, or reasoning based on a "vertical"completion of the term (3.1) requiring "linearity in A", it is simplest to postulate the ghost Lagrangian:

$$L_{Ghost} = -i \langle d\bar{c} \wedge {*}Dc \rangle \ , \tag{3.4}$$

where the wedge product is to be interpreted as usual for Lie-algebra valued forms (i.e., as a bracket, the grading of the product being taken into account since c and $\bar{c}$ are already 1-forms, albeit not on M), and the covariant exterior differential Dc being defined as usual since c is in the adjoint representation:

$$Dc = dc + [A,c] \ . \tag{3.5}$$

The factor -i was introduced by Kugo and Ojima in order to make c and $\bar{c}$ formally into hermitean operators

$$c^+ = c, \quad \bar{c}^+ = \bar{c} \ , \tag{3.6}$$

differing from the BRS assignment trivially, but leading

to a hermitean Lagrangian, a formally self-adjoint Hamiltonian, and, provided it exists, a unitary S-matrix.

The variation of the total action, given by the Lagrange 4-form

$$L = L_{YM} + L_m(A,f) + L_{GF} + L_{Gh} , \tag{3.7}$$

leads to the equations of motion (I write the Duffin-Kemmer version of the matter field equations, although the Dirac or Klein-Gordon-Higgs version can replace them in a manner inessential for the rest of our discussion):

$$F = DA, \; DF = 0 , \tag{3.8}$$

(this is either the definition of F in terms of A, or, an equation of motion in the first-order formalism, the Bianchi identity being in any case an integrability condition)

$$D{*}F = {*}dB + {*}J_m - i[{*}d\bar{c},c] , \tag{3.9}$$

(J_m is the matter current, e.g. in the Duffin-Kemmer case $J_m = \langle \Phi\beta \lrcorner D\Phi \rangle$)

$$d{*}A + \alpha{*}B = 0 ,$$

(the $*$ is necessary to have equality of rank of the forms)

$$d{*}Dc = 0 , \tag{3.10}$$

$$D{*}d\bar{c} = 0 . \tag{3.11}$$

Finally, the matter field equations are

$$\beta \lrcorner D\Phi + \mu\beta^{o}\Phi = 0 , \tag{3.12}$$

where β is the Duffin-Kemmer matrix four-vector, contracted with the one-form $D\Phi$ (a 5-dimensional column vector in the scalar case); the covariant differential $D = d + [R(A), \;]$ involves the representation of the Lie-algebra-valued one-form A in the representation R to which the field Φ belongs; the same symbols are used in J_m, following Eq.(3.9). Since the Duffin-Kemmer treatment of gauge couplings to scalar fields does not seem to be familiar to many physicists, I would like to make a parenthetical remark regarding it. Superficially, the minimal coupling of a Duffin-Kemmer field to a gauge field looks different from the Klein-Gordon version since the former involves the gauge potential linearly, just as the Dirac field does, whereas the latter contains terms quadratic in the potential; it would thus seem that different diagrams appear in the corresponding perturbation theories. However, a more careful analysis shows that the propagator of the Duffin-Kemmer version of scalar theory contains a contact-term involving a delta function, thus yielding effectively the same diagrams as the quadratic coupling $A^2\phi^2$ of the Klein-Gordon theory. The same will be true in a path-integral approach, although I am not aware of an explicit proof in the literature. The Duffin-Kemmer version is clearly advantageous in formal developments, and it would be worthwhile carrying out an analysis of the Brout-Englert-Higgs-Kibble model in this formalism.

As a preliminary to writing down commutation relations (i.e., treating the various fields as operator-valued distributions) one must select the canonical variables (or use the Peierls method). It is advantageous to treat the time-component $A_o(x)$ (in a chosen Lorentz frame; A_o is still a matrix in the adjoint representation, as is the field B(x) which becomes the canonical coordinate) as the canonical momentum conjugate to the field B(x). The canonical momenta conjugate to the spatial components A_k are, as usual, the "electric" components of the curvature, $F_{ok} = E_k$. The momentum conjugate to the "ghost" form c will be $i\dot{\bar{c}}$ (the time-derivative of the antighost), and the momentum conjugate to $\bar{c}$ is the "covariant" time-derivative $-i(\dot{c} + [A_o, c])$.

This choice leads to the equal-time commutation-anti-commutation relations. (The "fields" c, $\bar{c}$ being one-forms should be taken as anticommutative, i.e., they are formally fermions, although they have no spin; the correct anti-commutation also follows if one uses Schwinger's (or Peierls') method of splitting the kinematic matrix into its hermitian and antihermitian parts.) I will not write out the equal-time commutation-anticommutation relations, since they will not be used.

It is clear that the additional degrees of freedom which have been introduced to take care of the constraints, together with the redundant degrees of freedom which the constraints are supposed to eliminate, are to a large arbitrary, and therefore are subject to transformations which in the end must reduce to vertical bundle automorphisms (pure gauge transformations) or bundle automorphisms which project into Poincaré automorphisms of the base space (gauge transformations combined with inhomogeneous Lorentz transformations). These transformations were first introduced in general infinitesimal form by Becchi, Rouet, and Stora (they designate part of these transformations as Slavnov transformations, since Slavnov used them earlier), and are now known as the BRS-transformations. In infinitesimal form, in terms of the notation I have used so far, the BRS transformations can be written as (δ means variation, not codifferentiation, which has been avoided so far and replaced by ⁚d⁚):

$$\delta A(x) = \lambda \wedge Dc(x) , \tag{3.13}$$

$$\delta \Phi(x) = iR(\lambda \wedge c(x))\Phi(x) , \tag{3.14}$$

$$\delta c(x) = -\lambda \wedge [c(x), c(x)]/2 , \tag{3.15}$$

$$\delta \bar{c}(x) = i\lambda B(x) , \tag{3.16}$$

$$\delta B(x) = 0 \ , \tag{3.17}$$

also

$$\delta(Dc) = 0 \ , \quad \delta[c,c] = 0 \ ; \tag{3.18}$$

here $\lambda_\wedge$ is a constant (x-independent) Grassman element (vertical) acting on the vertical forms c, $\bar{c}$ in such a manner that the left-hand sides have the correct properties (one may think of it, alternatively, as a contraction or inner product, with a tangent vector along the fiber, except in Eq.(3.16), where it is the constant left-invariant form dual to that vector); R denotes the representation of the Lie-algebra element corresponding to the representation of Φ. The conditions (3.18) together with the gauge invariance of the original action easily lead to invariance of the new action under the BRS transformations (3.13)-(3.17). The brackets in (3.15), (3.16) are to be interpreted as the Lie-algebra bracket of the c-s, rather than their quantum-mechanical commutator which I will denote as $[\,,]_\mp$ (the lower sign is the anticommutator, or graded commutator, in general.

Alternatively, the BRS-invariance can be formulated in terms of a conserved Noether current 3-form

$$*J_{BRS} = \langle B*Dc - *dBc + (i/2)*d\bar{c}[c,c] - d(Fc)\rangle \tag{3.19}$$

which is conserved:

$$d*J_{BRS} = 0 \ , \tag{3.20}$$

and hence has an associated conserved charge (for a choice of time-direction):

$$Q_{BRS} = \int d^3x \langle B\partial_0 c - \partial_0 Bc + B[A_0,c] + (i/2)\partial_0\bar{c}[c,c]\rangle \ , \tag{3.21}$$

so that the right-hand sides of Eqs.(3.13)-(3.17) can be

written as commutators with $i\lambda \wedge Q_{BRS}$ (cf. Ref.[13] for details in standard notation). This leads to the "graded" algebra:

$$[iQ_{BRS}, A]_- = Dc \ , \tag{3.22a}$$

$$[iQ_{BRS}, B]_- = 0 \ , \tag{3.22b}$$

$$[iQ_{BRS}, \Phi]_- = iR(c)\Phi \ , \tag{3.23}$$

$$[iQ_{BRS}, c]_+ = -[c,c]/2 \ , \tag{3.24}$$

$$[iQ_{BRS}, \bar{c}]_+ = iB \ , \tag{3.25}$$

$$[Q_{BRS}, Q_{BRS}]_+ = 2Q^2_{BRS} = 0 \ . \tag{3.26}$$

The nilpotency of Q_{BRS} shows that one can form a cochain complex with it and define standard cohomological operations.

It is convenient to introduce another conserved charge, the Faddeev-Popov charge, associated to the ghost fields

$$Q_{\Phi\Pi} = i\int d^3x \langle \bar{c}\partial_o c - \partial_o \bar{c}\ c + \bar{c}[A_o, c] \rangle \ , \tag{3.27}$$

associating a "ghost number" + 1 to c, -1 to $\bar{c}$, and 0 to all other fields. The two charges (3.21), (3.27) form the graded algebra:

$$[Q_{BRS}, Q_{BRS}]_+ = 0 \ ,$$

$$[iQ_{\Phi\Pi}, Q_{BRS}]_- = Q_{BRS} \ ,$$

$$[Q_{\Phi\Pi}, Q_{\Phi\Pi}]_- = 0 \ , \tag{3.28}$$

showing that the BRS-charge carries one unit of "ghost charge". The algebraic properties (3.28) have been used by Kugo and Ojima to show that the unphysical states determined by the subsidiary condition

$$Q_{BRS}|phys\rangle = 0 \ , \tag{3.29}$$

requiring that the physical state be BRS-invariant (hence gauge-invariant), always appear in quartets, with effects which cancel mutually and lead to "confinement" of unphysical quanta. Moreover, under certain assumptions, they also claim that the condition (3.29) and the algebra (3.28) can be responsible for color confinement.

There are numerous other consequences of this formalism which the reader can find in the review of Kugo and Ojima, and in some later work published by these authors, jointly or separately. Of these I mention only the result that the "Maxwell equation" is valid as an expectation value equation in any physical state:

$$\langle d{*}F + {*}J_{YM} + {*}J_{matter}\rangle = 0 \ . \tag{3.30}$$

Kugo and Ojima discuss the unitarity of the S-matrix (assuming asymptotic completeness, the meaning of which for the Yang-Mills field - or even the existence of asymptotic states-is not at all clear and needs to be further investigated), the superselection structure of the resulting model, the derivation of the Ward-Takahashi identities, and many other topics which I cannot review here. It is certainly a formalism which should be taken seriously and may help allay the uneasiness many mathematical physicists show when confronted with theories involving indefinite metric.

In conclusion of this section, I would like to express my indebtedness to Arthur Wightman who brought the work of Kugo and Ojima to my attention, and who, together with Franco Strocchi, was responsible for putting indefinite-metric quantum gauge field theory within reach of the mathematical physics community.

4. THE HOLONOMY QUANTIZATION OF GAUGE THEORIES

In this section I start with an overview of Mandelstam's [2] approach to quantization of gauge theories, which was the precursor of all schemes involving loops, rather than points, in spacetime as the fundamental geometric element. The role of the holonomy group in classical gauge theory was, to my knowledge, first stressed by Loos [23]; it was used extensively in lattice gauge theories [16] and has been actively discussed in the recent literature [17]. An early attempt by the author to treat quantum gauge theory in terms of mappings from the holonomy group of a connection to morphisms of field operators was, unfortunately, never carried far enough to be included in this lecture [15]. I will try only to give a feeling for the flavor of the subject, hoping that it will help renew interest in this area, which many mathematicians and mathematical physicists think deserves more attention.

In order to understand the geometric meaning of Mandelstam's approach, let us start with the abelian case, defined by a U(1)-principal bundle on Minkowski space. In order to define elements of the "holonomy groupoid" (which is to the holonomy group what the fundamental groupoid of a topological space is to the fundamental group; more precisely, the holonomy groupoid consists of all parallel transports along paths; only those paths where the end of the first coincides with the beginning of the second can be composed, leading to a new element; the inverse is parallel transport along the opposite path) it is convenient to think of a "spacelike one-point compactification of M" which is achieved by assuming the existence of a single point at spacelike infinity. Alternatively, one may think first of an Euclidean version of the theory, with the conformal compactification S^4 of R^4; in this case one will have to distinguish between spacelike and nonspacelike paths by means of some "reflection-positivity" requirement, or another prescription designed to take into account the fact that in the real world the Yang-Mills system is hyperbolic, and that paths used for holonomy operations should avoid the characteristic conoids (or cones, in flat space) which propagate the connection.

All "fields", i.e., both the matter fields and later on the components of the curvature in the nonabelian case are the holonomy transforms along a spacelike path of "reference fields" defined at the distinguished point (spacelike infinity, in Mandelstam's terminology). Thus, one starts from conventional scalar electrodynamics, described by the scalar field $\phi(x,t)$ (here I use the Klein-Gordon formalism, rather than the Duffin-Kemmer approach, mainly for typographical convenience) satisfying the equations of motion

$$[(d - ieA)*(d - ieA) - m^2]\phi = 0 ,$$

$$[(d + ieA)*(d + ieA) - m^2]\phi^+ = 0 , \qquad (4.1)$$

with Maxwell's equations for A, F (more precisely, for A):

$$F = dA , \quad dF = 0 ,$$

$$d*F = -*J, (d*dA + *J = 0) \qquad (4.2)$$

where $*J$ is the Klein-Gordon current (including the term $*(\phi^+ A\phi)$, coming from the $A^2\phi^2$ term in the Lagrangian) three-form, and the usual equal-time commutation relations (I write only the nonvanishing ones, all other quantities commute at $t = 0$):

$$[\dot{\phi}(x,0), \phi^+(y,0)]_- = [\dot{\phi}^+(x,0), \phi(y,0)]_- = -i\delta^3(x-y) ,$$

$$[\dot{A}_\mu(x,0), A_\nu(y,0)]_- = i\eta_{\mu\nu}\,\delta^3(x-y) , \qquad (4.3)$$

where x and y are the space-coordinates of two points at $t = 0$, and the remainder of the notation is self-explanatory (including the Minkowski metric tensor and the three-dimensional delta-function).

In terms of the distinguished point chosen "at spatial infinity" and denoted here by x_o (Mandelstam calls it $-\infty$) and a particular spacelike path P joining x_o to x, Mandelstam defines what I will call the holonomy-transformed fields (note that the scalar fields $\phi(x)$ which are gauge-dependent and hence are to be eliminated depend on x and not on x_o, only the holonomy operator which achieves their "parallel transport to infinity" depends on x_o and P):

$$\Phi(x;P]=\exp[-ie\int_P A]\cdot\phi(x),\Phi^+(x;P]=\phi^+(x)\cdot\exp[ie\int_P A], \qquad (4.4)$$

where the integral of the one-form A is along P and path-dependent, the square bracket on the right of the newly defined fields is to remind us that we are dealing with functionals of the path P which also depend on the end-point x; the holonomy operators $\exp[\pm ie\int A]$ are written on the left and right of the fields, respectively, in order to have expressions which can be transcribed directly in the nonabelian case, or if one chooses to consider Duffin-Kemmer 5-vectors for the fields.

The response of the functionals $\Phi(x;P]$ to a variation of the path P consisting of attaching an infinitesimal loop spanning the area bivector σ attached to the path at the point z (such a "bubble" looks like the Greek letter Ω and thus is sometimes suggestively called an Ω-variation of P at z) is given by the expressions:

$$\delta_\sigma\Phi(x;P]=-ie[\int_\sigma F(z)]\cdot\Phi(x;P],\delta_\sigma\Phi^+(x;P]=ie\Phi^+(x;P]\cdot[\int_\sigma F(z)] \,, \qquad (4.5)$$

where the integral is to be understood in the limit as the surface σ contracts to the point z; alternatively, by writing out the area element $dx^\mu \wedge dx^\nu$ one can rewrite these equations in terms of functional derivatives with respect to $\sigma^{\mu\nu}$, which will make the field components $F_{\mu\nu}$ appear in the right-hand sides. The integrability conditions in this case are easily established since the F-s commute along different points of a spacelike path (in the non-abelian case there is intrinsic noncommutativity which requires the derivation of a "nonabelian Stokes theorem"; such theorems have been discussed, among others, by Aref'eva and by Gu). The integrability condition turns out to be the Bianchi identity $dF = 0$, which comes as no surprise, since it was put in by assuming $F = dA$. The differentials of the new fields $\Phi(x;P]$ for a displacement of the endpoint x of the path P by dx (this is a "directional differential") will be denoted by ∂:

$$\partial\Phi(x;P] = \exp[-ie\textstyle\int A]\cdot D\phi(x), \partial\Phi^+(x;P] = \bar{D}\phi^+(x)\cdot\exp[ie\textstyle\int A], \quad (4.6)$$

where $D = d - ieA$, $\bar{D} = d + ieA$ are the usual gauge-covariant differentials. The differentials ∂ do not commute and we get, obviously,

$$[\partial,\partial]\Phi(x;P] = ieF\cdot\Phi(x;P] , \quad (4.7)$$

where the field two-form is taken at the endpoint x (and the noncommutation of path-extensions in different directions leads to the appearance of the field F).

One can derive equations of motion for the fields Φ, Φ^+, which look like the free Klein-Gordon equation with the derivatives replaced by the noncommuting derivatives ∂ (they are equivalent to the gauge-dependent equations (4.1)) and are a consequence of varying a Lagrangian which looks like the free Klein-Gordon Lagrangian plus the free Maxwell Lagrangian (again with derivatives replaced). The current three-form of the fields Φ, Φ^+ which makes its appearance in Maxwell's equations is

$$*J = -ie*(\Phi^+\partial\Phi - \Phi\partial\Phi^+) \quad , \tag{4.8}$$

and also looks formally like the Klein-Gordon current, the F-dependence being hidden in the differentials. The fact that Φ^+ and Φ always appear as products in the Lagrangian and the current makes the path-independence of these quantities obvious (and this is equivalent to gauge-independence of observables in the original theory). The equal-time commutation relations have to be obtained either by use of the Dirac theory of constraints, or by using the Peierls method; the calculation can be found in Mandelstam's paper, and the only nonvanishing commutators (at equal time t = 0, x, y are three-dimensional coordinates) are

$$[\partial_o\Phi^+(x,0;P],\ \Phi(y,0;P]]_- = [\partial_o\Phi(x,0;P],\Phi^+(y,0;P]]_- =$$

$$= -i\delta^3(x-y),[\dot{\Phi}^{(+)}(x,0;P],F_{oi}(y,0)]_-=(\mp)\int dz_i\,\delta^3(y-\mathbf{z})\,\dot{\Phi}^{(+)}(x;P], \tag{4.9}$$

where the following notational shortcuts have been used: the dot denotes either ordinary or covariant time-derivative (in both sides of the equation), the superscript (+) means that the equation is valid with or without the $^+$ on both sides, with the upper sign (-) in front of the integral corresponding to the absence of the conjugation (I have avoided the use of * for conjugates, since it is reserved for Hodge-duality; in the Dirac or Duffin-Kemmer cases a bar is more appropriate but involves the conjugation matrix, leaving $^+$ as the only choice for conjugate or adjoint), and ∂_o denotes the covariant time-derivative in the sense of (4.6). The commutation relations between the field strengths F are the same as for the free-field case and will not be written down.

Mandelstam has proved that one can get all the results of the Coulomb gauge formulation from this approach, as well as the perturbation series of the Feynman (Lorentz) gauge with indefinite metric.

The nonabelian case was treated in 1963 by I. Białynicki-Birula [3] and was then shown in 1968 by Mandelstam to lead to the perturbative series of non-abelian theory, including ghost loops. In the nonabelian case the curvature form pullback F is not gauge-independent, and therefore in a holonomy approach it will lead to a path-dependent field $F(x;P]$, in distinction from QED:

$$F(x;P] = P\exp[-\int A]F(x)\exp[\int A] , \tag{4.10}$$

where I have switched to the antihermitean matrices A (with values in the Lie algebra, just as F is valued) and P means path-ordering (analogous to Dyson's time-ordered product in the interaction picture). In fact the integrals in (4.10) are "product integrals", i.e., the limits of ordered products of elements of the holonomy group taken along the path P; for a detailed definition, cf. the recent book [24], a concept which should be better known among physicists.

Since the field F is path dependent, even the source-free Yang-Mills theory will have some features which only the matter fields exhibit in QED. In particular, the analog of Eq.(4.5) for $F(x;P]$ takes the form (I set the coupling $g = 1$):

$$\delta_\sigma F(x;P] = \int_\sigma [F(x;P],F(z;P']] \tag{4.11}$$

where [,] is the Lie-algebra bracket, and P' denotes the portion of the path P leading from x_o to z where the loop spanning the area σ is attached; the integration is over z only. The Bianchi identity and the Yang-Mills equation look simpler, since instead of the covariant differentiation they will contain the differential ∂:

$$\partial F(x;P] = 0 , \quad \partial^* F(x;P] = 0 ; \tag{4.12}$$

however, one should remember that when one goes to path-independent quantities, these differentials are in fact covariant differentials. In other words, the differentiation involved in Eq.(4.12) is along the horizontal lift of the path P into the bundle, F being a horizontal form, and this expresses the gauge-dependence of this operation.

From these equations Mandelstam has succeeded in deriving sets of equations for the path-dependent Green's functions of the model, which when expressed in terms of "auxiliary, path-independent quantities", i.e., the gauge-dependent components of the connection and the curvature, lead to the same perturbation theory as the Feynman-De Witt approach. The path-dependent, gauge-independent operators F(x;P] are related to the gauge-dependent, but path-independent quantities F(x) by the Eq.(4.10) which involves the product-integrals (path-ordered exponentials) first introduced by Białynicki-Birula [3], and for which various identities which are technically quite useful were derived by Mandelstam [14] and recently rediscovered and used in many papers involving loops [17]. I will not pursue this topic further, except to say that it is a geometrical approach which should be taken seriously and developed further.

In most of the more recent holonomy approaches to quantum gauge theory the Wilson loop holonomy operator of Eq.(4.10) plays a fundamental role. This operator can be defined for a path or a loop (it is the "nonintegrable phase factor" studied by Wu and Yang):

$$W[P] = \exp \int_P A = \lim \prod_P \exp(A(x,\Delta x)) = \lim \prod_P U(x,\Delta x), \tag{4.13}$$

where the product is along the path P, and the limit of products of elements of the holonomy group U(x,Δx) (parallel transport along a polygonal approximation to the path) defines the product integral in (4.13) (cf. Ref.[24] for details of these definitions).

In basing the gauge field theory on the holonomy operator one has a choice of approaches, and at the moment it is not clear which of them is the better guess. Polyakov, for instance, starts out from a "chiral theory" with fields taking their values in a group, and from there derives heuristically equations for loop quantities. Migdal and Makeenko, on the other hand, derive the renormalized perturbation theory for the large N (number of colors) from a set of loop equations. Fröhlich, in his lectures, stresses the importance of a "random geometry" approach by introducing probability measures on the space of loops (on a lattice model).

5. SPECULATIONS ON THE SHAPE OF A FUTURE THEORY

In this section I will ask the forbearance of the reader and will try to speculate on what the form of a gauge field theory based on holonomy transformations ought to be. I realize the risks of such speculations, but consider it important to open up for discussions this important subject. Some of the concepts I mention can be defined easily, at least if one places the proper restrictions on the base space of the bundle (such as compactness, etc.) and on the fibers. Others have as yet to be defined properly and may lead to interesting purely mathematical problems.

I assume that the "kinematical framework" is given by a principal bundle P over spacetime M (appropriately compactified if necessary; meaning that in the associated vector bundles the cross sections fall off at infinity rapidly enough), with structure group G. To this bundle are associated the vector bundles E carrying the "matter fields". The latter are defined as operator-valued distributions on a topological vector space of "test-cross-sections" such as the Schwartz space S or other spaces in the following way.

The first step consists in introducing a Hilbert space and unitary representations of the symmetry group of M (e.g., the Poincaré group) and of the structure group G, or of the infinite group G_o of pure gauge transformations. Next, one defines equivariant (under the group of gauge transformations) operator-valued distributions on the above TVS. So far, the approach is simply modeled on the Gårding-Wightman axioms.

The holonomy groupoid is defined as implementing the Wilson-loop or path integral, e.g., by unitary path-dependent operators of the type introduced in Eqs. (4.10), (4.13):

$$U[P]\Phi(s)U^{-1}[P] = \Phi(W[P]s) \tag{5.1}$$

where $s:M \to E$ is a "test-section" of the associated vector bundle E, Φ is the operator-valued field distribution defined on s, W[P] is the holonomy along the path P acting on the section s, and U[P] is the putative unitary implementation of this action.

If it is impossible to implement (5.1) by unitary operator-functionals of the path, one has to consider more general mappings of the operators which give a representation of the holonomy groupoid by automorphisms of the operator algebra. I remind the reader that the product integrals W[P] from a groupoid, since they can be composed only if the endpoint of the first factor coincides with the starting point of the first; it may be convenient to pick a fixed reference point (such as the spacelike infinity of the Mandelstam approach) and express all path-dependent quantities as differences of Mandelstam-Birula type operators along the same path to the reference point.

The next step would consist in forming Wightman functions out of products of quantities appearing in (5.1) and derive functional differential equations for Ω-differentiations of the paths.

Such a theory would have to satisfy a number of self-consistency requirements and should also be a limit, as the spacing goes to zero, of a good lattice theory.

If a full-fledged theory cannot be developed, one should try various model theories, as suggested by exactly soluble classical gauge theories. One should, of course, analyze the symmetry-breaking models discussed in my first two lectures in the framework of this quantum theory.

Another interesting line of research which should be pursued is the use of universal bundles and connections in quantization schemes (cf. "Brief Introduction",Sec. 5, for the appropriate definitions). Since the universal connections in vector bundles are defined by "projection" from a high-enough dimension of a canonical "flat" connection, one could start a quantum theory by quantizing the latter. As a preliminary exercise, one should develop a theory of quantized Maurer-Cartan forms (or matrices) which is essentially "chiral field theory". It thus becomes less surprising that results derived in various ways in the "Large N limit", where N is the order of the SU(N) color group considered, seem simpler than those obtained for small N. The importance of universal connections for studying the large-N limit of gauge field theories has been emphasized in seminars by I.M. Singer.

Finally, one might want to try to put the Feynman path integral approach in vector bundles on a more rigorous footing. Attempts in this direction are under way; what is needed is a "probability theory" in bundles with connections, in fact, a full-fledged "stochastic geometry". Let us hope that this subject will be covered in one of the Schladming Schools of the near future.

ACKNOWLEDGEMENTS

I would like to thank the organizers of the "Universitätswochen", particularly Heinz Mitter and his colleagues from Graz, for making it possible for me to participate in this 20-th get-together, and to renew friendships started at the first Schladming meeting in 1962, where we hardly suspected how important gauge theories would become twenty years later. I would like to acknowledge stimulating discussions and remarks made during and between the lectures by many participants. Particular thanks go to Andrzej Trautman, the collaboration with whom on the "Brief Introduction" helped clarify and make precise many ideas.

REFERENCES

1. R.P. Feynman, Acta Physica Polonica 24 (1963) 697.
2. S. Mandelstam, Ann. Phys. (NY) 19 (1962) 1, 25; Phys. Rev. 175 (1968) 1580,1604.
3. I. Białynicki-Birula, Bull. Acad. Polon. Sci. (Phys. Astr.) 11 (1963) 135.
4. B.S. DeWitt, Phys. Rev. 160 (1967) 1113, 162 (1967) 1195, 1239.
5. L.D. Faddeev and V.S. Popov, Phys. Lett. 25B (1967) 29, and ITP Preprint Kiev, 1967.
6. F. Strocchi and A.S. Wightman, J. Math. Phys. 15 (1974) 2198; F. Strocchi, numerous preprints and articles.
7. E. Abers and B.W. Lee, Phys. Reports 9 (1974) 1; G. 't Hooft and M. Veltman, Diagrammar, CERN Report 73-9, Geneva, 1973.
8. A.A. Slavnov and L.D. Faddeev, Vvedenie v kvantovuyu teoriyu kalibrovochnykh polei (Introduction to the quantum theory of gauge fields), "Nauka", Moscow, 1978; English Transl. Addison-Wesley, 1980.

9. C. Becchi, A. Rouet, and R. Stora, Commun. Math. Phys. 42 (1975) 127; Ann. Phys. (NY) 98 (1976) 287; lectures in "Renormalization Theory", G. Velo and A.S. Wightman, Eds. (Erice, 1975), Reidel, Amsterdam-Boston, 1976.
10. J. Thierry-Mieg, These de Doctorat, Orsay, 1978; J. Math. Phys. 21 (1980) 2834; Y. Ne'eman and J.Thierry-Mieg, Ann. Phys. (NY) 123 (1979) 247; Ann. Isr. Phys. Soc. vol. 3, 1979.
11. F. Strocchi, Proc. of Nijmegen Group Theory Conference, Springer Lect. Notes in Physics, 1977; Commun. Math. Phys. 56 (1977) 57; Phys. Lett. 62B (1976) 60; Phys. Rev. D17 (1978) 2010.
12. R. Jackiw, Rev. Mod. Phys. 52 (1980) 661.
13. T. Kugo and I.Ojima, Suppl. Prog. Theor. Phys. 66 1979, contains a complete bibliography to their previous work, as well as the papers of Nakanishi, Lautrup, Ferrari, Picasso and Strocchi (and many others) relevant to this section. I. Ojima, Z. Physik C 5 (1980) 227.
14. S. Mandelstam, 1968 papers, Ref.[2].
15. M.E. Mayer, in Lect. Notes in Mathematics, vol. 570, Springer Verlag, 1977.
16. K. Wilson, Phys. Rev. D10 (1974) 2445. K. Osterwalder and E. Seiler, Ann. Phys. (NY) 110 (1978) 440. J.Fröhlich (with D. Brydges and E. Seiler), Lectures on Gauge Theory, IHES Preprint, 1979. See also J. Kogut, Rev. Mod. Phys. 51 (1979) 659, with an extensive bibliography on lattice gauge theory at a more "physical" level.
17. J.Gervais and A.Neveu, Phys. Lett. 88B (1979) 255; Nucl. Phys. B153 (1979) 445; E. Corrigan and B.Hasslacher, Phys. Lett. 81B (1979) 181 (and private communication by E. Corrigan). Y. Nambu, Phys. Lett. 80B (1979) 372. C.N. Yang, Phys. Rev. Lett. 33 (1974) 445. A.M.Polyakov, Aspen Preprint 1979 (Nucl.Phys. 1980), Phys. Lett. 82B; A.A. Migdal, Ann. Phys. 126 (1980) 279. I. Ya Aref'eva, Lett. Math. Phys. 3 (1979) 241; Theor. Math. Phys. 43 (1980) 111. Yu.M. Makeenko and A.A. Migdal, Yad. Fiz. 32 (1980) (Sov.J.Nucl. Phys. 32 (1981)).

18. Reviews by Fröhlich and Kogut, Ref.[16].
19. D. Buchholz, Lecture at Lausanne Conf. 1979, Lect. Notes in Physics, vol. 116, 1979; D. Buchholz and K. Fredenhagen, Nucl. Phys. B154 (1979) 226-238. Oral accounts of Buchholz' results by R. Haag (1979) and D. Kastler (1988/81).
20. B. Lautrup, Mat. Fys. Medd. Dan. Vid. Selsk. 35 (1967) 11.
N. Nakanishi, Prog. Theor. Phys. 35 (1966) 1111, 49 (1973) 640. Suppl. no. 51 (1972).
21. C. Cronström, Quantization of Nonabelian Gauge Theory I. Gauge Invariant Hamiltonian Formulation, Preprint HU-TFT 81-1, Helsinki, January, 1981.
22. F. Berezin, Metod vtorichnogo kvantovaniya, Nauka, Moscow 1965 (The Method of Second Quantization, Academic Press, NY, 1968). Integration over Grassmann variables has been discussed a long time ago by Schwinger.
23. H. Loos, J. Math. Phys. 8 (1967) 2114; Phys. Rev. 188 (1969) 2342.
24. J. D. Dollard and C. N. Friedman, Product Integration, see also Appendix II by P. Masani; Encyclopedia of Mathematics, vol. 10, Addison-Wesley, Reading, MA, 1979.

Acta Physica Austriaca, Suppl. XXIII, 525–575 (1981)

RECENT DEVELOPMENTS IN FINITE ENERGY (TOPOLOGICAL) MONOPOLE THEORY+

by

L. O'RAIFEARTAIGH and S. ROUHANI
Dublin Institute for Advanced Studies
10 Burlington Road, Dublin 4, Ireland

ABSTRACT

Recent activity in the study of static, finite-energy, topologically charged monopole systems is reviewed. The main developments have been the study of axially symmetric systems, the proof of existence of static multimonopole configurations for large monopole separations, and the explicit construction of single monopoles of arbitrary charge and of multimonopole configurations for small monopole separations.

1.INTRODUCTION

According to current views the fundamenal physical interactions are described by unified gauge theory. Since

---

+Lectures given at the XX. Internationale Universitätswochen für Kernphysik, Schladming, Austria, February 17-26, 1981.

this theory differs from Maxwell-Lorentz theory in that it may be non-abelian and spontaneously broken,it is natural to look for properties of the theory that are consequences of these new features. The two most striking properties that have been discovered so far are asymptotic freedom [1] and the existence of stable finite energy (or finite action) field configurations [2]. These properties have turned out to be of interest not only in their own right but also in connexion with quark confinement [3].

In the present lectures we shall consider the stable, static, finite energy configurations in three space dimensions. The identification of such configurations with magnetic monopoles, their relevance for confinement, and the explicit construction of spherically symmetric solutions have all been discussed some time ago [3], but until recently no progress was made in the construction, or even proof of existence, of configurations which are not spherically symmetric. In the past year or so, however, some dramatic progress has been made in this direction and it is this progress which will be reviewed.

As mentioned in the abstract, the main developments have concerned the questions of existence of static monopole configurations, the construction of non-spherically symmetric monopoles (of magnetic charge greater than unity) and the study of axisymmetric systems. All these developments are based on the reduction of the usual second-order field equations to a first-order system known as the Bogomolny (B) equations [4].

Accordingly, the review will commence with a sketch of the earlier developments and the introduction of the B-equations. The sketch will include the simplification of the equations which takes place in Yang's R-gauge [5], and their subsequent linearization by means of Bäcklund transformations [6]. The review then goes on to sketch one of the major recent developments, namely, an existence proof for static separated monopole configurations due to Taubes [7]. The axisymmetric configurations are then discussed. The most striking results here are (i) the fact that axisymmetric systems can not describe separated monopoles of the Taubes kind, but only single monopoles of arbitrary charge [8],(ii) the existence of a master-potential for all the invariants [9],and (iii) the equivalence of axisymmetric B-equations to the Ernst equation of General relativity [10]. We then come to the second major development of recent times, namely, the explicit construction of solutions which describe single (axisymmetric) monopoles of arbitrary strength [11,12]. The properties of such solutions and the problem of establishing their regularity for higher monopole strengths is discussed. In one of these constructions (due to Ward) use is made of the vector-bundle formalism of Atiyah and Ward [13],and because of its wider importance this formalism is described in some detail. Finally, a recent construction by Ward of a solution which describes two slightly separated monopoles, and a proposal to extend this construction to the n-monopole case are described.

2. EARLY DEVELOPMENTS

Let $(\vec{A},\Phi)$ be a static magnetic SU(2) Yang-Mills-Higgs system, with the Higgs field in the adjoint representation, and Hamiltonian

$$H = \frac{1}{2} \int d^3x \, \{B^2 + (D\Phi)^2 + 2V(\Phi)\} \quad , \tag{2.1}$$

where

$$\vec{B} = \vec{\nabla} \times \vec{A} + \frac{1}{2} \vec{A} \hat{\times} \vec{A} \, , \quad \vec{D}\Phi = \vec{\nabla}\Phi + \vec{A} \wedge \Phi \, , \; V = \lambda(\Phi^2 - c^2)^2 \geq 0, \tag{2.2}$$

and wedge denotes SU(2) outer product. The corresponding field equations are

$$\vec{D}\cdot\vec{B} = 0, \quad \vec{D} \times \vec{B} = \Phi \wedge \vec{D}\Phi, \quad D^2\Phi = \partial\hat{V}/\partial\Phi \quad , \tag{2.3}$$

and by finite-energy configurations are meant solutions of (2.3) which make the integral (2.1) converge. The reason that the system admits such solutions is that the spontaneously broken potential admits (indeed requires) the boundary condition

$$\mathop{\mathrm{Lt}}_{r\to\infty} \Phi(r,\theta,\varphi) \to c\phi(\theta,\varphi) \quad , \qquad \phi^2 = 1, \tag{2.4}$$

and the topology of SU(2) is such that $\phi(\theta,\varphi)$ may be non-trivial. More precisely, $\phi(\theta,\varphi)$ is a mapping from S_2 (space) to S_2 (isospace) and such mappings fall into discrete homotopy classes labelled by an integer n. The boundary functions $\phi(\theta,\varphi)$ which belong to the non-trivial homotopy classes $n \neq 0$ (e.g. $\phi = (\cos\theta, \sin\theta\cos\varphi, \sin\theta\sin\varphi)$) can not be gauge-transformed to the trivial function $\phi = 1$, and generate the non-trivial finite energy solutions.

An important operator in this respect is the topological charge, defined as

$$Q = \frac{1}{4\pi}\int d\Omega(\phi,\partial_\psi\phi \wedge \partial_\theta\phi) \, . \tag{2.5}$$

Q is gauge-invariant and takes the value n when $\phi(\theta,\varphi)$ belongs to the nth homotopy class [14,15]. Thus it acts as a Casimir label for the homotopy classes. Q is also a superselection operator (commutes with all the fields), and hence it guarantees the stability of different homotopy sectors. The generalization of (2.5) for arbitrary volumes R is

$$Q = \int_R d^3x (\vec{\nabla}\cdot\vec{j}) \quad \text{where} \quad j_r = \varepsilon_{rst}(\phi, \partial_s\phi \wedge \partial_t\phi), \quad \phi = \Phi/|\Phi|, \tag{2.6}$$

and since the divergence is identically zero when ϕ is regular we see that the charge is actually located only at those points where the Higgs field Φ is zero.

By completing the square in (2.1) and carrying out some partial integrations [4] one may write H in the form

$$H = \frac{1}{2} \int d^3x \{(B - D\,\Phi)^2 + 2V\} + Q \tag{2.7}$$

which shows that Q also provides a lower bound for the energy.

The first non-trivial solution of the system (2.2) was found by t'Hooft [3] who assumed spherical symmetry and thus reduced the field equations to two nonlinear, coupled, but ordinary equations for the norms of the Higgs and gauge fields h(r) and k(r). He then used numerical methods to establish that these equations almost certainly had a solution and to approximate it. Later a rigorous proof for the existence of these solutions was given [16] and it was shown [17] that they must be real analytic functions of r. Finally, in the special case in which the potential V is set equal to zero (but the boundary condition (2.4) is retained) it was shown [18] that the solutions took the elementary form

$$h(r) = c\left(\frac{\cosh cr}{\sinh cr} - \frac{1}{cr}\right), \qquad k(r) = \frac{cr}{\sinh cr} \quad . \tag{2.8}$$

The spherically symmetric solutions so obtained turned out to have unit topological charge ($n = \pm 1$) and it was soon shown [19] that this was the only possible charge which could be obtained from spherically symmetric configurations.

(Conversely, systems with unit topological charge had to be spherically symmetric [20]).

3. THE BOGOMOLNY EQUATION AND ITS LINEARIZATION

According to the results just mentioned, monopoles or monopole systems of charge greater than unity can only be obtained by dropping the assumption of spherical symmetry. But then the field equations (2.2) become extremely complicated, and the problem would probably have remained intractable had it not been noticed [4] that in the limit when the potential V is zero, the field equations reduce to the much simpler first-order set

$$\vec{B} = \vec{D}\Phi \quad . \tag{3.1}$$

This follows from the fact that, for fixed Q, equ. (3.1) manifestly minimizes the energy in (2.6). The equations (3.1) are called the Bogomolny equations, and they also have a geometrical significance, because if we identify Φ with the 4th component A_4 of a 4-dimensional Euclidean gauge-potential they can be regarded as the static version of the 4-dimensional self-dual equations

$$F_{\mu\nu} = \frac{1}{2}\,\varepsilon_{\mu\nu\lambda\sigma}\,F_{\lambda\sigma}\,, \quad \text{where } F_{ij} = \frac{1}{2}\,\varepsilon_{ijk}\,B_k \text{ and } F_{i4} = D_i\Phi \;. \tag{3.2}$$

The Bogomolny system (3.1) not only has the mathematical advantages just discussed, but it has also the advantage that it is the only case in which one can hope to describe a system of separated monopoles in static equilibrium, as well as single monopoles of arbitrary strength. In fact, unless the Higgs field is in the adjoint representation, the potential is zero and the monopole charges have the same sign, there are even long-range forces between the monopoles. On the other hand, if these three conditions and the Bogomolny equation are satisfied, the components of the stress-tensor, and hence the forces between the monopoles vanish everywhere [21].

The Bogomolny system (3.1) contains nine equations for the twelve functions $(\vec{A},\Phi)$, three functions remaining undetermined due to the gauge freedom. However, by using a construction of Yang for the self-dual form (3.2) they can be reduced to three equations for three unknown functions [5]. In fact, by choosing a suitable gauge (the R-gauge) and partially solving the self-dual equations, Yang reduced them to

$$\Box \ln f = \frac{1}{f^2} (\partial e \cdot \partial g) \tag{3.3}$$

and

$$\partial_{\bar{u}} \left(\frac{\partial_u e}{f^2}\right) + \partial_v \left(\frac{\partial_{\bar{v}} e}{f^2}\right) = 0$$

$$\partial_u \left(\frac{\partial_{\bar{u}} g}{f^2}\right) + \partial_{\bar{v}} \left(\frac{\partial_v g}{f^2}\right) = 0, \quad \text{where } u = x + iy, \quad v = 3 + ix_4 \,, \tag{3.4}$$

$$A_\mu = -\frac{1}{2f} \begin{bmatrix} \eta^3_{\mu\nu}\partial_\nu f & \eta^-_{\mu\nu}\partial_\nu e \\ \eta^+_{\mu\nu}\partial_\nu g & -\eta^3_{\mu\nu}\partial_\nu f \end{bmatrix}, \text{ and } \quad \eta^a_{\mu\nu} = \varepsilon_{4a\mu\nu} + \delta_{a\mu}\delta_{\nu 4} - \delta_{a\nu}\delta_{\mu 4}, \quad \eta^\pm = \eta_1 \pm i\eta_2 \,. \tag{3.5}$$

Later Corrigan [6] et al. (CFGY) observed that the Yang equations (3.4) were invariant with respect to the involutive transformations

$$I : \begin{bmatrix} e_I & f_I \\ f_I & g_I \end{bmatrix} = \begin{bmatrix} e & f \\ f & g \end{bmatrix}^{-1} , \tag{3.6}$$

and

$$B : \quad f^b = \frac{1}{f} , \quad \begin{array}{l} (e^b_u, g^b_v, e^b_{\bar{v}}, g^b_{\bar{u}})/f^b \\ =(-g_v, e_u, g_{\bar{u}}, -e_{\bar{v}})/f \end{array} \quad \text{where } e_u = \partial_u e \text{ etc..} \tag{3.7}$$

In fact equs.(3.4) are just the integrability conditions for (3.7). The transformation (3.6) is actually a gauge-transformation, but the transformation (3.7) is not. In fact, (3.7) is a Bäcklund transformation i.e. is a first order differential transformation connecting different solutions of a second order differential equation.

Using the (non-involutive) product IB, Corrigan et al.[6] were able to obtain at least one class of solutions of the Yang equations by linearizing them as follows:

First let $\Delta_o(x)$ be any solution of the d'Alembertian equation. Then

$$f = e = g = \Delta_o^{-1} \quad \text{where} \quad \square \Delta_o = 0 \tag{3.8}$$

is a solution of the Yang equations. Next, let $\Delta_r(x)$ for $r = 0, \pm 1, \pm 2, \ldots$ be any set of functions satisfying the Cauchy-Riemann-like equations

$$\frac{\partial \Delta_r}{\partial \bar{u}} = -\frac{\partial \Delta_{r+1}}{\partial \bar{v}} , \quad \frac{\partial \Delta_r}{\partial v} = \frac{\partial \Delta_{r+1}}{\partial u} , \tag{3.9}$$

(which imply the d'Alembertian equation) and for any integer $k \geq 1$ form the matrix

$$D(k) = \begin{bmatrix} \Delta_{k-1} & \Delta_{k-2} & \cdot & \Delta_1 & \Delta_o \\ \Delta_{k-2} & \Delta_{k-3} & \cdot & \Delta_o & \Delta_{-1} \\ - & - & - & - \\ \Delta_1 & \Delta_o & \cdot & \Delta_{3-k} & \Delta_{2-k} \\ \Delta_o & \Delta_{-1} & \cdot & \Delta_{2-k} & \Delta_{1-k} \end{bmatrix} . \qquad (3.10)$$

Then, provided that the matrix D(k) is invertible, the corner elements of the inverse matrix

$$D^{-1}(k) = \begin{bmatrix} e & . & . & . & . & f \\ . & . & . & . & . & . \\ . & . & . & . & . & . \\ f & . & . & . & . & g \end{bmatrix} \qquad (3.11)$$

form a solution of the Yang system. Furthermore, the Bäcklund transformations IB and BI connect the k with the k ± 1 solutions. In particular, each family of solutions is generated from the (k = 1) solution Δ_o of the d'Alembertian equation. However, the reality and singularity properties are not preserved by the Bäcklund transformations, so the choice of Δ_o is not trivial.

We shall refer to a set of functions Δ_r satisfying (3.9) as a Durham string, and it is perhaps worth mentioning at this point that each Durham string can be generated by a potential function as follows: Let

$$\Delta = \sum_{-\infty}^{\infty} \Delta_r \, e^{ir\psi} \qquad (3.12)$$

be the Fourier transform of the Δ_r. Then the Durham string equation (3.9) is just the condition that Δ should depend on only three of the five variables (x,ψ), namely,

$$\Delta = \Delta(e^{i\psi}, \mu, \nu) \tag{3.13}$$

where (in cylindrical coordinates)

$$\mu = ix_4 + z + \rho e^{(i\varphi + \psi)} \quad , \quad \nu = ix_4 - z\, e^{-i(\varphi + \psi)} \quad . \tag{3.14}$$

Thus any function of $(e^{i\psi},\mu,\nu)$ (with a Fourier transform) may be used to generate a Durham string and hence a solution of the self-dual and Bogomolny equations. The deeper reason for this result will be seen when we consider the AW construction in section 7.

An interesting result, due to Prasad [22], is that the norm of the Higgs field is directly related to the determinant of the Durham matrix D(k) by the formula

$$\Phi^2 = c^2 - \Delta \ln (\det D(k)) \ . \tag{3.15}$$

The proof of this formula, which uses the Bäcklund transformations IB, is given in Appendix A.

All of the recent advances in monopole theory have been made using the Bogomolny system (3.1) and hence, from now on only this system will be considered. We shall see that it admits solutions describing both single monopoles of arbitrary strength and separated monopoles. It should be noted, however, that in the more realistic case when the potential is not set equal to zero, the separated monopole configuration becomes unstable. Thus for the separated solutions the Bogomolny condition is crucial, whereas for the solutions describing single monopoles of arbitrary strength it is probably only a technical device that allows us to obtain the solutions in closed form.

4. THE EXISTENCE OF STATIC MULTI-MONOPOLE SYSTEMS

The first important recent development in the theory of monopoles, and perhaps the most fundamental development, is a proof, due to Taubes [7], that the Bogomolny equations (3.1) do indeed admit static, separated, monopole configurations. Here the monopoles are supposed to be separated by distances rather larger than the monopole "core", the latter being defined as the region outside of which the two "massive" gauge fields fall-off exponentially, leaving only the magnetic and Higgs fields with long-range (1/r) components. (One recalls that for a Hamiltonian of the form (2.1) two of the three gauge fields acquire masses because of the Higgs-Kibble mechanism.)

The idea in Taubes'proof is to start from an initial approximate configuration $(\vec{A}_o, \Phi_o)$ and, by successive iteration of the Bogomolny equation, to construct a sequence $(\vec{A}_s, \Phi_s)$ which converges to an exact solution as $s \to \infty$. It is clear that the basic problems are

(i) to construct a function space for $(\vec{A}, \Phi)$ in which convergence makes sense and is reasonably likely;

(ii) to choose the initial configuration sufficiently close to the expected solution for the sequence to have a reasonable chance of converging.

The function space chosen by Taubes was the Sobelev space

$$\int_R d^3x\{(\ {}_iA_j,\ {}_iA_j) + (\ {}_i\Phi, \nabla_i\Phi) + (A_iA_i) + (\Phi,\Phi)\} < \infty \quad (4.1)$$

(for every finite volume R). A comparison of (4.1) with the Hamiltonian (2.1) shows that there is a close connection between the configurations $(\vec{A}, \Phi)$ which lie in (4.1) and the configurations with finite-energy, and this is what makes (4.1) a natural choice. However, (in contrast to the case of vortices and spherically symmetric monopoles) the exact equivalence of (4.1) with finite energy has not been established. (Note that (4.1) is actually gauge-dependent.)

In order to construct an initial configuration Taubes divided the Euclidean 3-space E(3) into three regions - the expected monopole cores of radius c^{-1}, shells of thickness c^{-1} surrounding the cores, and an 'exterior' region consisting of the rest of E(3). He then chose as initial configuration

(a) the exact 1-monopole solutions of section 2 inside the cores,
(b) an exact 'exterior' solution of the Bogomolny equations in the exterior region,
(c) some C^{∞} transition functions, to connect these two sets of solutions smoothly, in the shells.

The exact 'exterior' solutions are actually solutions of the Maxwell-Higgs U(1) subsystem of SU(2), up to some topological factors which are chosen to produce unit charge within each core [21].(Taubes actually trivialized these factors by using a singular (Dirac) gauge.)

Using the properties of the Bogomolny system, in particular the fact that it is an elliptic system, Taubes was able to show that the iteration of this initial configuration does indeed converge in the norm (4.1). Furthermore, he showed that the exact solutions had to be real analytic. The actual proof is far from trivial (it runs to fifty typed pages) and constitutes a major tour de force.

5. AXIALLY SYMMETRIC CONFIGURATIONS

With the question of existence of multi-monopole solutions settled by Taubes result the emphasis shifts to the question of single monopoles of arbitrary strength and to the explicit construction of solutions. It is not obvious, of course, that such solutions exist or can be constructed in terms of elementary functions, but experience with the spherically symmetric solution (2.7) and the geometrical nature of the Bogomolny equations suggests these possibilities. A natural first step towards such a construction is to consider axially symmetric configurations, since these are the next simplest to spherically symmetric ones. Axisymmetric systems are characterized [24,8] by the fact that they admit a smooth isovector $\omega(x)$ which implements rotations around the axis of symmetry,

$$D_{\varphi} \Phi(x) = \omega(x) \wedge \Phi(x) , \tag{5.1}$$

where φ is the azimuthal angle. Further, since (5.1) should hold not only for $\Phi(x)$ but also for all of its covariant derivatives, it is easy to see that $\omega(x)$ must satisfy the integrability condition

$$D_i \omega(x) = \varepsilon_{i \varphi j} B_j(x) \quad . \tag{5.2}$$

Eqs. (5.1) and (5.2) characterize the axial symmetry.

At first sight one should expect the axisymmetric solutions to describe both single monopoles of arbitrary strength and colinear multi-monopole systems, in particular to describe two monopoles. But here one encounters a major surprise. It turns out [8] that axisymmetric systems can describe only single monopoles!

The proof of this rather surprising result runs as follows: First, because Φ is subharmonic and $|\Phi| \to c \neq 0$ at infinity, the zeros of Φ and hence the topological charge can only lie at points or on one-dimensional curves. Next, from the definition (2.5) of the topological charge (restricted to an arbitrary volume V that intersects the z-axis at two points z_1 and z_2) and from (5.1), it follows after some algebra that the charge contained in the volume V is just

$$\Delta Q = b(z_1) - b(z_2) \text{ where } b = (\omega,\phi) \text{ and } \phi = \Phi/|\Phi| \quad . \tag{5.3}$$

By letting z_1 and z_2 coincide we can see first of all that there is no charge located <u>off</u> the axis. Next from (5.1) and (5.2) we have (for smooth $(\vec{A},\Phi)$)

$$\omega \wedge \Phi = 0 \qquad \text{and} \qquad \omega^2 = \text{constant} \tag{5.4}$$

on the z-axis. Combining (5.3) and (5.4) we then have

$$b^2 = \text{constant, and hence } \Delta Q = 0, \pm \text{ constant,} \tag{5.5}$$

on the axis. For like charges (which as we have seen, is the only stable case) equation (5.5) implies that all the charge must be concentrated at a single point, as required.

Although the result just obtained restricts axial symmetry to single monopoles, such systems are still of interest, so we consider now some further properties of axial symmetry. The first such result is that axisymmetric systems admit a master-potential W(x) from which the gauge-invariants Φ^2, ω^2 and (ω,Φ) may be obtained by differentiation [9]. For if we take the inner-product of (5.1) with Φ and ω respectively, we obtain the Cauchy-Riemann equations

$$\partial_i \omega^2 - \rho^2 \partial_i \Phi^2 = 2\,\varepsilon_{ij}\,\partial_j(\omega,\Phi) \quad , \tag{5.6}$$

and it is easy to see that the content of these equations is precisely that there should exist a scalar W(x) such that

$$\frac{\partial W}{\partial z} = (\omega,\Phi), \; 2\rho\frac{\partial W}{\partial \rho} = \omega^2 - \rho^2(\Phi^2 - c^2), \; \Delta W = \Phi^2 - c^2 \; , \tag{5.7}$$

where z and ρ are the usual cylindrical coordinates.

By comparing (5.7) with (3.15) we obtain also a remarkable connection between the master-potential and the Durham determinant det D(k), namely,

$$W(x) = -\ln[\det D(k)] \quad . \tag{5.8}$$

Equ. (5.8) actually follows from (3.15) only up to a harmonic function. But by a proof analogous to that used to establish (3.15), the harmonic function can be shown to be at most a constant (see Appendix A).

Another interesting property of the axisymmetric case is the simplification obtained for the Bogomolny equations. However, it should be noted that since axial symmetry is abelian, it cannot reduce the number of functions in the Bogomolny equations but only determine their φ-dependence. On the other hand, axial symmetry can be supplemented by mirror-symmetry (symmetry under reflexions in any plane through the axis), and then the number of functions in the Bogomolny equations is reduced from twelve to six, including one gauge function. The six gauge fields take the form

$$\Phi = (\phi_1, \phi_2, 0) \qquad A_\rho = (0, 0, W_\rho)$$

$$A_\varphi = (\eta_1, \eta_2, 0) \qquad A_z = (0, 0, W_z) \quad , \tag{5.9}$$

and since all the fields are φ-independent, the system has then effectively been reduced to a Maxwell, or U(1), system, with one gauge-potential $\vec{W}$, and two real two-component Higgs fields η and ϕ. It is then not surprising that the Bogomolny equations reduce to the U(1)-covariant system

$$D_\rho \phi = \rho^{-1} D_z \eta$$

$$D_z \phi = - \rho^{-1} D_\rho \eta$$

$$\nabla \times W = \rho^{-1} (\eta \wedge \phi) \quad , \qquad (5.10)$$

where D denotes U(1) covariant derivative, and wedge the 2-dimensional outer product. (The factor ρ^{-1} occurs because the coordinates (ρ,z) are curvilinear.) The system (5.10) was first proposed (without using mirror symmetry) by Manton [9] and by Jang, Park and Wali [23] and is sometimes known as the Manton Ansatz. The system consists of five equations for the six unknown functions (5.9), and it can be partially solved in two different ways, as follows:

I. The four equations (a) (b) can be solved, and the solution turns out to be just the master-potential solution (5.6). This shows that <u>all</u> the gauge-invariants can be derived from the master-potential in the mirror symmetric case. The remaining equation (c) can then be written [9] in the form

$$\Delta h = (D\phi)^2 h \qquad \text{where} \qquad h = |\Phi|, \qquad \phi = \Phi/h, \qquad (5.11)$$

as an implicit (4th order) equation for the master-potential

II. The three equations (a), (c) or (b), (c) can be solved, and in a suitable gauge the solution is

$$(\phi_1, \phi_2) = f^{-1}(g_{,z}, -f_{,z}) \; , \qquad w_\rho = \phi_1 \; ,$$

$$(\eta_1, \eta_2) = \rho f^{-1}(g_{,\rho}, -f_{,\rho}) \; , \qquad w_z = \rho^{-1}\eta_1 \; , \qquad (5.12)$$

f and g being unknown functions. When the solution (5.12) is inserted in the remaining two equations, the latter reduce to an equation of the form [10]

$$(\mathrm{Re}\ \varepsilon)\Delta\varepsilon = (\vec{\nabla}\varepsilon)^2 \qquad \text{where} \qquad \varepsilon = f + ig \quad . \qquad (5.13)$$

This is a remarkable result because, as those who are familiar with General Relativity will recognize, (5.10) is just the Ernst equation for axially symmetric gravitational fields! Thus the axisymmetric Bogomolny equations for monopoles and the Ernst equations for gravitation are the same!

In order to see whether the axisymmetric system actually admits solutions of monopole charge greater than unity, a numerical analysis for n = 2,3 was carried out independently by Adler [25] and by Rebbi and Rossi [26]. The numerical analysis suggested strongly that solutions exist and, rather surprisingly, they indicated that the highest concentration of energy would not be at the centre of the monopole. This result is confirmed by the beautiful graphs for n = 2,3,4,5 shown by Palla at this conference. Palla's result shows that the maximum energy density is situated in a torus which coincides roughly with the edge of the monopole core.

The most remarkable result for the axisymmetric case, however, is that exact solutions of the axisymmetric field equations (5.10) have now been found. They have been found independently by Palla et al. using the Ernst equation (5.13) and by Ward. As the Palla construction will be described by Dr. Palla in his seminar I shall consider here only the Ward construction. Although this construction yields the same solution as Palla's, it is quite different, and, as we shall see, it can be generalized to the non-axisymmetric case.

6. THE ATIYAH-WARD CONSTRUCTION: MINI-MODEL

The Ward construction [11] of the n = 2 monopole solutions is based on the use of vector-bundles to construct self-dual fields. Originally the vector-bundles were intended [13] for the case of instantons, but the method has turned out to be more useful for monopoles. To illustrate the basic AW idea we consider first a mini-model in real 2-dimensional Euclidean space E(2).

The straight lines on E(2) are of the conventional form

$$\alpha x + \beta y + \gamma = 0 \;, \tag{6.1}$$

and hence are parametrized by the variables (α,β,γ) modulo a common factor δ where δ is any real number. Thus they form the 2-dimensional (compact) projective space P(2). It is not possible to cover P(2) with a single system of 2-dimensional coordinates, but nevertheless it will be useful to use the conventional 2-dimensional system (m,c) obtained by neglecting the vertical lines and writing (6.1) in the form

$$y = mx + c \quad . \tag{6.2}$$

The AW idea is a generalization of the fact that the scalar fields $\phi(m,c)$ on P(2) and the abelian gauge-fields B(x) on E(2) can be considered as transforms of one another in the following sense: Let B(x) be any abelian gauge-field on E(2) and (m,c) any line in P(2). Then the transform $\phi(m,c)$ of F(x) is defined to be

$$\phi(m,c) = \int_\Delta dx\; dy\; B(x,y) \;, \tag{6.3}$$

where Δ is the triangle enclosed by the line y(m,c) and the coordinate axes in E(2). Conversely, if $\phi(m,c)$ has the correct analyticity properties for $\phi(m, y-mx)$ to have a Laurent expansion in m, then the inverse of (6.3) is

$$B(x,y) = \frac{1}{2\pi i}\frac{\partial}{\partial x}\frac{\partial}{\partial y}\oint \frac{dm}{m}\,\phi(m,y-mx) \tag{6.4}$$

where the integration is around the unit circle in complex m-space. Of course, since E(2) is 2-dimensional B(xy) has only one component so that (6.3) and (6.4) is a transform between scalars on P(2) and E(2) (rather similar to, but not identical with, the Radon transform). However, it is convenient to think of B(xy) as a gauge-field, because of the later generalization and because, as we shall see, the gauge-potential $\vec{A}(xy)$ for B(xy) enters naturally in the proof of (6.4) from (6.3).

To establish (6.4) we first write $B(x) = \vec{\nabla} \times \vec{A}$ and then convert (6.3) into an integral around the perimeter of the triangle to obtain

$$\phi(m,c) = \oint A_k \, dx^k \quad , \qquad k = 1,2. \tag{6.5}$$

Next we let (xy) be any point on the line (m,c) (between the vertices of the triangle) and split the integral into

$$\phi = \phi_+ + \phi_- \quad \text{where} \quad \phi_+ = \oint_{(0,0)}^{(x,y)} \quad , \quad \phi_- = \oint_{(x,y)}^{(0,0)} \tag{6.6}$$

each integral being taken anti-clockwise along the perimeter. Differentiating along the line (m,c) we then have

$$A_x + mA_y = (\partial_x + m\partial_y)\phi_+ = -(\partial_x + m\partial_y)\phi_- \quad . \tag{6.7}$$

But since the vertices of the triangle (c/m,0) and (0,c) we see that if $\phi(m,y-mx)$ is analytic for $m \neq 0,\infty$ then ϕ_+ is analytic for $m \neq \infty$ and ϕ_- for $m \neq 0$. Hence A_x and A_y are the leading terms in the expansions of $\phi_\pm$ respectively, and we have

$$A_x = \frac{1}{2\pi i} \int \frac{dm}{m} \partial_x \phi_+ \quad \text{and} \quad A_y = \frac{-1}{2\pi i} \int \frac{dm}{m} \partial_y \phi_- \quad . \tag{6.8}$$

Equation (6.4) then follows immediately from the definition

$\vec{B} = \vec{\nabla} \times \vec{A}$. Note that the gauge freedom $A_k \to A_k + \partial_k \Lambda$ corresponds exactly to the freedom of assigning the m-independent term in the expansion of $\phi(m,c)$ to ϕ_+ or ϕ_-, indeed $\Lambda = \phi_+^o - \phi_-^o$ where superscript zero denotes the zero-order term in the Taylor expansions of ϕ_+ and ϕ_- respectively.

One might now ask the question: what is the connection between the transformation $\phi(m,c) \leftrightarrow F(xy)$ and fibre-bundles? The point is that since P(2) is not completely covered by the coordinates (m,c) it is natural to think of $\phi(m,c)$ as section of a fibre-bundle with P(2) as base space and fibres of one dimension. Then the mini-model may be interpreted as a relationship between sections of fibre bundles over P(2) and gauge fields on E(2).

7. THE ATIYAH-WARD CONSTRUCTION: MAXI-MODEL

The actual AW construction (maxi-model) consists in lifting the relationship between functions over P(2) and E(2) of the mini-model to a similar relation between functions over CP(3) and C(4), where CP(3) denotes complex projective 3-space and C(4) denotes complex Euclidean space. However, in the maxi case the constraints on the functions are much tighter - the functions on C(4) must be self-dual and those on CP(3) (which are SL(2,c) - valued) must possess a generalized Laurent expansion (to be defined below).

The space E(2) is replaced by C(4) by introducing 4 complex coordinates x_μ. The 'lines' in this space (the analogoues of (6.1)) are then the complex 2-planes

$$\begin{bmatrix} \omega_1 \\ \omega_2 \end{bmatrix} = \begin{bmatrix} ix_4 + x_3 & x_1 + ix_2 \\ x_1 - ix_2 & ix_4 - x_3 \end{bmatrix} \begin{bmatrix} \pi_1 \\ \pi_2 \end{bmatrix} , \qquad (7.1)$$

where (ω_1,ω_2) and (π_1,π_2) are complex 2-spinors. It is clear that the complex 2-planes are parametrized by the four complex numbers $(\omega_1,\omega_2,\pi_1,\pi_2)$ modulo λ, where λ is any complex number. Hence the complex 2-plane form the (3-complex-dimensional) space CP(3). Although CP(3) is compact and cannot be completely spanned by three complex coordinates, it is convenient, nevertheless, to introduce a 3-dimensional coordinate system

$$(\mu,\nu,\xi) = (\omega_1/\pi_1,\ \omega_2/\pi_2,\ \pi_2/\pi_1) \ . \tag{7.2}$$

Such a coordinate system omits the planes $\pi_1 = 0$ and $\pi_2 = 0$ and reduce the planes (7.1) to

$$\mu = i\,x_4 + x_3 + (x_1 + i\,x_2)\xi \ ,$$

$$\nu = i\,x_4 - x_3 + (x_1 - i\,x_2)\xi^{-1} \ . \tag{7.3}$$

Note that if we identify ix_4 with t and ξ with $e^{i\psi}$, the functions μ,ν in (7.3) agree with those in (3.14). Note also that the variables (μ,ν,ξ) are just the well-known 'twistor' variables introduced by Penrose in the context of General Relativity.

To obtain the analogue of equation (6.3) of the mini-model we first note that the "coordinate-axes" are

$$x_1 - ix_2 = x_3 + ix_4 = 0 \quad \text{and} \quad x_1 + ix_2 = x_3 - ix_4 = 0 \ , \tag{7.4}$$

respectively, and that for fixed (μ,ν,ξ) the plane (7.3) intersects them at

$$R(\mu,\nu,\xi) = \tfrac{1}{2}(\nu\xi, i\nu\xi, \mu, -i\mu) \quad \text{and}$$

$$Q(\mu,\nu,\xi^{-1}) = \tfrac{1}{2}(\mu\xi^{-1}, -\mu\xi^{-1}, -\nu, -i\nu) \ , \tag{7.5}$$

respectively. Then for any gauge-potential $A_\mu(x)$ over C(4) one forms the function

$$g(\mu,\nu,\xi) = P\, e^{\int_R^Q A_\mu(x)\,dx^\mu} , \tag{7.6}$$

where the integral is to be taken along a path from R to Q within the plane (7.3), and P denotes path-ordering. Equation (7.6) is the required analogue of (6.3) (the integration along the axes being omitted for later convenience). Note that $g(\mu,\nu,\xi)$ depends only on (μ,ν,ξ) and is SL(2,C)-valued.

It is precisely in (7.6) that the self-duality enters. The point is that since (7.6) can be along any path in the plane the function $g(\mu,\nu,\xi)$ will be uniquely defined if and only if the integral is path-independent, and this will be so if and only if the field $F_{\mu\nu}(x)$ for $A_\mu(x)$ is <u>self-dual</u>. The reason is simple: the planes (7.3) are <u>anti-self-dual</u> in the sense that

$$dx\,\pi = dx'\pi = 0 \leftrightarrow dx_{(\mu}dx'_{\nu)} + \varepsilon_{\mu\nu\lambda\sigma}\, dx_\lambda dx'_\sigma = 0 , \tag{7.7}$$

(see Appendix B) and hence the projections $F_{\mu\nu}\, dx_\mu\, dx'_\nu$ will be zero if and only if $F_{\mu\nu}$ is self-dual. Thus for C(4) only self-dual fields admit the transformation into functions over CP(3).

The converse is also true - not every SL(2,c)-valued function over CP(3) admits the transformation into self-dual fields. To see this, choose any point x_μ in the plane (7.3) and make the decomposition

$$g(\mu,\nu,\xi) = g_+(\xi,x)\; g_-(\xi^{-1},x) \tag{7.8}$$

where

$$g_+(\xi,x) = P\, e^{\int_Q^x A_k dx^k} \quad \text{and} \quad g_-(\xi^{-1},x) = P\, e^{\int_x^R A_k dx^k} . \tag{7.9}$$

From the expressions for Q and R in (7.5) we see that g_+ and g_- are regular for $\xi \neq \infty$ and $\xi \neq 0$ respectively, and so (7.8) is a generalized 'Laurent' decomposition. But not every SL(2,c)-valued function over CP(3) admits such a generalized Laurent decomposition. Thus finally the transformations, $F_{\mu\nu}$ on C(4) to $g(\mu,\nu,\xi)$ on CP(3) and back, are seen to be limited both ways - to self-dual $F_{\mu\nu}$ and to Laurent-decomposable $g(\mu,\nu,\xi)$.

We have anticipated here the result that every Laurent decomposable $g(\mu,\nu,\xi)$ does indeed admit a transformation to a self-dual $F_{\mu\nu}$. To see that it does, and to obtain the $F_{\mu\nu}$ field, one proceeds as in the mini-model: differentiating the decomposition (7.8) in the plane (7.3) and using the Liouville theorem one finds that

$$P_\mu\, g_+^{-1}\partial_\mu g_+ = P_\mu g_-\partial_\mu g_-^{-1} = P_\mu A_\mu , \text{ where } P_\mu = \sigma_\mu(\xi\ ), \tag{7.10}$$

the σ_μ are the Pauli matrices $(i,\vec{\sigma})$, and the $A_\mu(x)$ are independent of ξ. The $A_\pm(x)$ are then the potentials for the self-dual fields.

In fibre bundle language the functions $g(\mu,\nu,\xi)$ may be regarded as sections of an SL(2,c)-fibred bundle with CP(3) as base-space. The condition that $g(\mu,\nu,\xi)$ is Laurent-decomposable may be regarded as the condition that $g(\mu,\nu,\xi)$ trivializes on a line in CP(3) (i.e. for fixed x in (7.1)).

8. THE ATIYAH-WARD ANSATZ

The AW construction translates the problem of finding solutions of the Bogomolny equation into finding functions $g(\mu,\nu,\xi)$ which are Laurent-decomposable and which lead to the correct reality, regularity and boundary conditions for the fields $(\vec{A},\Phi)$.

The full class of Laurent-decomposable $g(\mu,\nu,\xi)$ is not yet known, but Atiyah and Ward have pointed out that a sufficient condition for decomposability is that $g(\mu,\nu,\xi)$ be of the form

$$g(\mu,\nu,\xi) = \begin{bmatrix} \rho(\mu,\nu,\xi) & \zeta^{\ell} \\ -\zeta^{-\ell} & 0 \end{bmatrix} , \qquad (8.1)$$

where ℓ is an integer and ρ is an arbitrary function. Equation (8.1) is known as the AW Ansatz. In the case of instantons (which are algebraic) the Ansatz is sufficient to describe all the self-dual solutions, but it is not clear that such will be the case for monopoles. However, as we shall see, the Ansatz yields at least two important classes of solutions.

Since the AW Ansatz reduces the choice of $g(\mu,\nu,\xi)$ to the choice of a single function $\rho(\mu,\nu,\xi)$ one might expect to find a relation between the Ansatz and the single potential $\Delta(\mu,\nu,e^{i\psi})$ of the Durham string, and indeed there is a remarkably simple connection, namely,

$$\Delta(\mu,\nu,e^{i\psi}) = \rho(\mu,\nu,\xi) \quad \text{for} \quad \xi = e^{i\psi} \quad . \qquad (8.2)$$

This is the deeper reason for the existence of the Durham potential Δ.

In terms of the AW-Ansatz the Durham string may be then written as

$$\Delta_r(x) = \frac{1}{2\pi i}\int \frac{d\xi}{\xi}\, \xi^r\, \rho(\mu,\nu,\xi) \quad , \tag{8.3}$$

where μ and ν are given by (7.3). The great advantage of (8.3) is that instead of having to derive the gauge-potentials from $\rho(\mu,\nu,\xi)$ by means cumbersome decomposition formula (7.10), one can derive them from (8.3) and the algebraic formulae (3.11) and (3.5). The whole problem then reduces to finding a $\rho(\mu,\nu,\xi)$ which leads to the correct reality, regularity and boundary conditions for the gauge fields $(\vec{A},\Phi)$.

9. THE WARD SOLUTION AND ITS GENERALIZATION

Although the AW Ansatz (8.1) was originally intended for instantons, it has turned out to be more useful for monopoles, where by a judicious choice of $\rho(\mu,\nu,\xi)$, Ward [11] was able to construct the same single-monopole of charge-two solution as Forgacs et al.[12].Ward's form of the solution has a gauge which is not manifestly real or time-independent, but has the advantage that the solution can easily be generalized to monopoles of arbitrary strength and even to separated monopole solutions.

Ward's starting point was to note that for the known spherically symmetric charge-one monopole the unknown function $\rho(\mu,\nu,\xi)$ in the AW ansatz turned out to be just the entire function

$$\rho_1(\mu,\nu) = \frac{e^{\mu} - e^{\nu}}{\gamma} \quad , \qquad \gamma = \frac{\mu-\nu}{2} \quad , \tag{9.1}$$

where, from (7.3),

$$\mu = (t+z)+\rho e^{i\varphi}\xi, \quad \nu = (t-z)+\rho e^{-i\varphi}\xi^{-1}, \quad \gamma = z+\tfrac{1}{2}\rho(e^{i\varphi}\xi-e^{-i\varphi}\xi^{-1}),$$

and, in order to simplify the notation, the Higgs constant

c has been normalized to 1. For reasons which will become clearer below Ward then guessed that the generalization of (9.1) to the charge-two case would be the entire function

$$\rho_2(\mu,\nu) = \frac{e^{\mu} + e^{\nu}}{\gamma^2+(\frac{\pi}{2})^2} \quad . \tag{9.2}$$

For the same reasons the generalization of (9.1) to the case of arbitrary ℓ is guessed to be the entire function

$$\rho_\ell(\mu,\nu) = \frac{e^{\mu} + (-1)^{\ell} e^{\nu}}{H_\ell(\gamma)} \quad , \tag{9.3}$$

where

$$H_\ell(\gamma) = \gamma \prod_{m=1}^{(\ell-1)/2} (\gamma^2+m^2\pi^2) \quad \text{and} \quad H_\ell(\gamma) = \prod_{m=1}^{\ell/2} (\gamma^2+(m-\tfrac{1}{2})^2\pi^2) , \tag{9.4}$$

for ℓ odd and even respectively.

It is easy to see from (9.1) and (9.3) that the components of the Durham string generated by the ρ-functions in (9.3) are just

$$\Delta_r(x) = \frac{e^{2t}}{2\pi} \int_0^{2\pi} d\psi \, e^{-\rho\cos\psi} \frac{\sin(\rho\sin\psi + iz)}{H_\ell(\rho\sin\psi + iz)} e^{ir\psi} \tag{9.5}$$

for odd ℓ, and a similar expression with sin → cos for even ℓ. In particular the central function Δ_o takes the form

$$\Delta_o(x) = \sum_{-\frac{(\ell-1)}{2}}^{\frac{\ell-1}{2}} \frac{\sin h\, r(k)}{r(k)} \quad \text{where } r^2(k) = \rho^2+(z+i\pi k)^2 \quad . \tag{9.6}$$

Because of the trivial time-dependence of the Δ_r, one sees also that the d'Alembertian equation reduces to the Yukawa equation

$$\Box\Delta_r(x) = 0 \rightarrow (\Delta - c^2)\Delta_r(x) = 0 \qquad (9.7)$$

where the Higgs constant c has been recalled in order to emphasize its role.

Since the AW construction guarantees that the fields derived from the $\rho_\ell(\mu,\nu)$ in (9.3) satisfy the self-dual equations (3.2), what has to be checked is that the solutions satisfy the other requirements for monopole solutions of charge n. These are

(i) reality and time-independence ,
(ii) the boundary condition $|\Phi| \rightarrow c - \frac{n}{r}$ as $r \rightarrow \infty$ $(c = 1)$,
(iii) regularity.

The boundary condition (ii) is necessary and sufficient to describe a monopole of charge n because, when the Bogomolny equations are satisfied, we have from (2.1) and (2.7)

$$4\pi c\ Q = H = \int d^3x\,(D\Phi)^2 = \tfrac{1}{2}\int d^3x\Delta\Phi^2 = \int r^2 d\Omega|\Phi|\frac{\partial|\Phi|}{\partial r}\ . \qquad (9.8)$$

In making the Ansatz (9.3) the ρ-functions have, of course, been chosen in order to satisfy the conditions (i)-(iii), and we now consider these conditions in turn. First, condition (i) requires that the AW-matrix $g(\mu,\nu,\xi)$ be time-independent and hermitian, or be gauge-transformable to such a $g(\mu,\nu,\xi)$. Here, hermitian means that $g^+(\mu,\nu,\xi) = g(\mu,\nu,-\bar{\xi}^{-1})$ where dagger denotes hermitian conjugate, and the existence of a gauge transformation means that there should exist a pair of invertible 2 x 2 matrices $X(\xi^{-1})$ and $Y(\xi)$, regular in ξ^{-1} and ξ respectively, such that

$$g(\mu,\nu,\xi) \rightarrow \tilde{g}(\mu,\nu,\xi) = X(\xi^{-1})g(\mu,\nu,\xi)Y(\xi) \tag{9.9}$$

where $\tilde{g}$ is time-independent and hermitian. It is clear that the $g(\mu,\nu)$ constructed with the ρ-functions in (9.3) are not themselves time-independent or real, but they admit a transformation of the form (9.9), namely

$$\begin{bmatrix} \rho & -\xi^{\ell} \\ \xi^{-\ell} & 0 \end{bmatrix} = \begin{bmatrix} e^{\frac{\nu}{2}} & 0 \\ 0 & e^{-\frac{\nu}{2}} \end{bmatrix} \begin{bmatrix} \frac{e^{\gamma}+(-1)^{\ell}e^{-\gamma}}{H_{\ell}(\gamma)} & e^{-\gamma}\xi^{\ell}(-1)^{\ell} \\ e^{-\gamma}\xi^{-\ell} & H_{\ell}(\gamma)e^{-\gamma} \end{bmatrix} \begin{bmatrix} e^{\frac{\mu}{2}} - H_{\ell}(\gamma)\xi^{\ell}e^{-\frac{\mu}{2}} \\ 0 \quad e^{-\frac{\mu}{2}} \end{bmatrix}, \tag{9.10}$$

and the transformed matrix is manifestly time-independent and real.

Actually, the matrices X and Y in (9.10) will be regular in ξ and ξ^{-1} only if the denominator function $H_{\ell}(\gamma)$ is a polynomial of degree at most ℓ, and this is why the denominators in (9.3) are chosen to be polynomials. The zeros of the polynomial are chosen partly in order to guarantee that ρ is non-singular in x for $|\xi| = 1$ (in fact it turns out to be non-singular for $\xi \neq 0,\infty$) and partly to satisfy condition (ii) which we now discuss.

For condition (ii) the first point to establish is that $\Phi^2 \rightarrow c^2$ as $r \rightarrow \infty$. For this we first note from (9.7) that if the Δ_r become spherically symmetric as $r \rightarrow \infty$, then the $\Delta_r \rightarrow \exp cr$ and hence, from (3.5) and (3.11), $\Phi^2 \rightarrow c^2$ as $r \rightarrow \infty$, just as in the spherically symmetric case. It turns out that if the zeros of the polynomial $H_{\ell}(\gamma,\xi)$ are chosen to be the first ℓ zeros of the numerator, then the relevant Δ_r, namely those for $|r| \leq \ell$, do indeed become spherically symmetric as $r \rightarrow \infty$. Thus $\Phi^2 \rightarrow c^2$ as required. The second problem is to obtain the second term in the expression (ii). For this we use the fact that, from quite general considerations it can be shown that if $\Phi^2 \rightarrow c^2$ then

$|\Phi|$ must become a harmonic function as $r \to \infty$, in which case we must have $|\Phi| \to c - \kappa r^{-1}$ where κ is a constant, and the problem reduces to the determination of κ. To determine it we note from (9.5) that on the z-axis we have

$$\Delta_o(z,\rho=0) = \frac{\sinh z}{H_\ell(z)}, \quad \Delta_o(z,\rho=0) = 0, \ s \neq 0, \qquad (9.11)$$

for odd ℓ (and a similar expression for even ℓ). Hence from (3.5) we have

$$\Phi(z,0) = \frac{\partial}{\partial z} \ln \Delta_o(z,0) = \frac{\cosh cz}{\sinh cz} - \frac{H'_\ell(z)}{H_\ell(z)}. \qquad (9.12)$$

Comparing (9.12) with $c - \kappa/r$ for $r = |z|$ we obtain $\kappa = \ell$. Thus the solutions (9.3) describe monopole systems of total charge $n = \ell$.

10. PROPERTIES OF THE WARD FAMILY

The first question concerning the generalized Ward solutions of the previous section is whether they describe single monopoles of charge n, or systems of separate monopoles. It turns out that they describe single monopoles. This was verified directly by Ward for n = 2 by showing that the Higgs field had only a single zero, and in general it can be shown by noting that the family of solutions is axially symmetric. Then one can either invoke the result of section 5 which states that axial symmetry implies single monopoles, or check from (9.9) that there are no zeros on the symmetry axis except at the origin.

In order to see that the Ward solutions are axi-symmetric, one notes from (9.1) that a rotation around the z-axis in x-space can be absorbed in the phase of ξ, and then, if ρ does not contain ξ explicitly, we obtain a trivial phase change of Δ_μ after the integration in (9.5). It will be seen in section 12 that separated monopole solutions can be found by letting ρ depend explicitly on ξ.

As a matter of fact the Ward solutions are not only axisymmetric but are mirror-symmetric and symmetric with respect to the reflexion $z \to -z$. Both of these discrete symmetries can be seen from eqs. (9.1) and (9.5).

Another interesting property of the solutions (9.3) for arbitrary n is that the central functions Δ_o of the Durham strings (the starting points for the Bäcklund transformations IB of section 3) may be obtained from the spherically symmetric (n = 1) solution sinh r/r by the 'translations'

$$\Delta_o(x) = (T_z)^{n-1}\left(\frac{\sinh r}{r}\right) \text{ where } T_z = e^{\frac{i\pi}{2}\frac{\partial}{\partial z}} + e^{-\frac{i\pi}{2}\frac{\partial}{\partial z}} . \tag{10.1}$$

Thus the solution for arbitrary n is generated from the n = 1 solution not by the Bäcklund transformations IB alone but by the combination of the IB transformations and the translations T_z. In fact, since T_z commutes with the Cauchy-Riemann operators (3.9) which generate the IB transformations, we see that by the successive application of T_z and IB we obtain the two-dimensional lattice of solutions depicted in Fig. 1. However, in this lattice only the vertical strings with end-points generated by $T_z^r(\mathrm{IB})^{n-1}$ for r = n-1 satisfy all the conditions (i) (ii) (iii). These strings are the ones with end-points on the lines in Fig. 1 which lie at 45° to the axes.

A number of different expressions can be found for the general elements Δ_r of the Durham string in (9.5). For example, by using the identity

$$\frac{\sin(\rho\cos\psi + iz)}{(\rho\cos\psi + iz)} = \frac{i}{2}\int_{-1}^{1} e^{-i(\rho\cos\psi + iz)\cdot q}\,dq \qquad (10.2)$$

in (9.5) we obtain

$$\Delta_s(x) = \frac{e^{ct+is\psi}}{4\pi}\int_{-1}^{1} e^{zq}\left(2\cos\frac{\pi q}{2}\right)^{n-1}\left(\frac{1+q}{1-q}\right)^{s/2} I_s(\rho\sqrt{1-q^2})\,dq \qquad (10.3)$$

where I_s is the Bessel function of the first kind. Note that in (10.3) the operator T_z^{n-1} is implemented by the function $(2\cos\frac{\pi q}{2})^{n-1}$, and that this function serves to keep the integral finite at the end points.

A differential form of the Δ_s may be obtained from either (9.5) or (10.3), namely

$$\begin{bmatrix}\Delta_s \\ \\ \Delta_{-s}\end{bmatrix} = \begin{bmatrix} e^{is\psi}\delta_s(\rho,z) \\ \\ e^{-is\psi}\delta_s(\rho,-z)\end{bmatrix} \qquad \text{where} \quad \delta_s(\rho,z) =$$

$$= (T_z)^{n-1}\left(\frac{1-\partial_z}{\rho}\right)^s r^{2s-1}\left(\frac{d}{rdr}\right)\sinh r, \qquad (10.4)$$

for $s \geq 0$. Here the $\delta_s(\rho,z)$ may be generated by the function

$$e^{-uz/\rho}\sinh\{r(1+\frac{2u}{\rho})^{1/2}\} = r\sum_{s=0}^{\infty}\frac{s^n}{n!}\,\delta_s(\rho,z)\ . \qquad (10.5)$$

Finally an intuitive feeling for the 'shape' of the monopole can be obtained from a very interesting and elegant result on the asymptotic form of the solutions due to Prasad and Rossi [27]. These authors observed that if one makes the expansion

$$\Delta_o(x) = \sum \binom{n}{r} \frac{\sin hr(k)}{r(k)} = \frac{1}{2}\sum \binom{n}{r} \frac{e^{r(k)}}{r(k)} + O(e^{-(r(k)+\bar{r}(k))}); \tag{10.6}$$

in (8.1) then the corresponding expansion for the norm of the Higgs field takes the form

$$|\Phi| = c - \sum \frac{1}{r(k)} + O(e^{-(r(k) + \bar{r}(k))}) \quad . \tag{10.7}$$

The form (10.7) enables us to identify the asymptotic form of the Higgs field as $c - \sum r(k)^{-1}$, and to identify the 'core' of the monopole as the region where $\exp(-(r(k)+\bar{r}(k)))$ is not negligible. Because of the sharp fall off in the exponential (approximately 25 units for every unit of r) the core is very sharply defined, and a short calculation shows that it consists of the nested ellipses

$$\frac{z^2}{a^2} + \frac{\rho^2}{a^2 + k^2\pi^2} = 1 \quad \text{where} \quad 1 \lesssim a \lesssim \pi \ , \quad |k| \leq \frac{n-1}{2} \quad . \tag{10.8}$$

Thus the monopole core takes the shape of a discus whose width is fixed, but whose radius increases linearly with n (see Fig. 2).

11. POSITIVITY OF THE DETERMINANT

As mentioned before, the Ansatz (7.3) yields regular gauge fields $(\vec{A},\Phi)$ provided only that the determinant D(n) in (3.10) does not vanish. The non-vanishing of D(2) and D(3) has been demonstrated explicitly, but a rigorous proof for $n \geq 4$ is still lacking. (The same problem occurs in the construction of Palla et al..) However, enough progress has been made, notably by Prasad and Rossi [27], to make it virtually certain that D(n) does not vanish for general n, and we wish to briefly sketch that progress.

The first step is to note that, since the points $r(k) = 0$ in (9.7) lie within the monopole core (at $(c,\pi k)$ for $k \leq (n-1)/2$), Φ in (10.7) is non-singular outside the core. From (3.15) this implies that det $D(n) \neq 0$ outside the core, and so the problem reduces to proving that det $D(n)$ is not zero inside the monopole core (as defined by (10.8)).

To investigate the situation inside the core one uses (9.5) to write the determinant in the form

$$\det D(n) = \left(\frac{e^{2t}}{2\pi}\right) \prod_{s=1}^{n} \int_0^{2\pi} d\psi_s \, e^{-\rho\cos\psi_s} \frac{\sin(\rho\sin\psi_s + iz)}{H_n(\rho\sin\psi_s + iz)} J(n,\psi),$$

$$\text{where } J = \begin{bmatrix} \delta(n-1) & \delta(n-2) & \cdot & \delta(0) \\ \delta(n-1) & \delta(n-3) & \cdot & \delta(-1) \\ \cdot & \cdot & \cdot & \cdot \\ \delta(1) & \delta(0) & \cdot & \delta(2-n) \\ \delta(0) & \delta(-1) & \cdot & \delta(1-n) \end{bmatrix}, \qquad (11.1)$$

and $\delta(k) = \exp ik\,\psi_k$. Then noting that

$$\frac{\sin(\rho\sin\psi_k + iz)}{H_n(\rho\sin\psi_k + iz)} = \prod_{\frac{n+1}{2}}^{\infty} \left(1 - \frac{(\rho\sin\psi + iz)^2}{m^2\pi^2}\right), \qquad (11.2)$$

and recalling [25] from the theory of unitary groups the Weyl identity,

$$J(n,\psi) = \prod_{q<s}^{n} \sin^2\left(\frac{\psi_s - \psi_q}{2}\right), \qquad (11.3)$$

we see that the determinant can be written in the form

$$|D(n)| \equiv \det D(n) = \int d\mu(\psi)\, e^{-\rho \sum_{s=1}^{n} \cos\psi_s} \prod_{s=1}^{n} \left| \frac{\sin(\rho\sin\psi_s + iz)}{H_n(\rho\sin\psi_s + iz)} \right| \cos\theta, \tag{11.4}$$

for odd n (and a similar expression for even n), where $d\mu(\psi)$ is the positive measure obtained by weighting $d\psi_1 \ldots d\psi_n$ with (11.3), and θ is the phase

$$\theta = \sum_{s=1}^{n} \sum_{m=\frac{n+1}{2}}^{\infty} \tan^{-1} \frac{2z\rho\sin\psi_s}{m^2\pi^2 + z^2 - \rho^2\sin^2\psi_s} \quad . \tag{11.5}$$

From (11.3) and (11.4) it is clear that $|D(n)|$ is positive on the coordinates axes $z = 0$ and $\rho = 0$. Furthermore, since a good approximation to θ in (11.5) is

$$\theta \simeq \frac{2z\rho}{n\pi^2} \sum_{s=1}^{n} \sin\psi_s \leq \frac{2|z|}{\pi^2} , \tag{11.6}$$

we see that it remains positive in the region

$$z\rho < \frac{\pi^3}{4} \tag{11.7}$$

surrounding the axes. Thus the determinant is not zero in the heart of the core as shown in Fig. 3.

Since the determinant does not vanish either at the heart of the core monopole core or outside the core, it is likely that it does not vanish anywhere, and, indeed, more refined arguments can be used to reduce the region of uncertainty (striped region in Fig. 3) still further [17].

Some insight into the location of the zeros may also be obtained by letting the coordinate z become complex. For complex z the formulae (10.7) (9.11) are still valid and the core generalizes to

$$\frac{(\mathrm{Re}\ z)^2}{a^2} + \frac{\rho^2}{a^2 + (k\pi)^2 + (\mathrm{Im}\ z)^2} = 1 , \quad 1 \lesssim a \lesssim \pi . \qquad (11.8)$$

Thus, even for z complex, $|\mathrm{Re}\ z| \lesssim \pi$ in the core. Now for $\rho = 0$ we see from (9.11) that the only zeros of $|D(n)|$ are at $z = im\pi$, $m \geq n+1$. Since these points are quite distant from the $(\mathrm{Re}\ z,\rho)$-plane, we then see that unless the zeros turn quickly toward the plane as ρ increases from 0 to $n\pi$ (and leave it again when $2\rho > (n+1)\pi$) the determinant will have no zeros for real z and ρ.

12. TOWARD SEPARATED MONOPOLE SOLUTIONS

Since separated monopole solutions are known to exist from Taubes'result and cannot be axially symmetric, it might be asked whether the Ward construction could be generalized to describe separated monopoles by relaxing the condition of axial symmetry. Ward has now shown that such a generalization is possible at least for two monopoles and small separations. Furthermore, the number of parameters for this case (seven) agrees with the number (4n-1) for any n predicted by the index theorem [20].

The starting point for Ward's generalization is the denominator $H_2(\gamma)$ in (9.3). Replacing the function $H_2(\gamma)$ by the most general hermitian quadratic in ξ and ξ^{-1} one obtains

$$H_2(\gamma,\xi) = \gamma^2 + \gamma\varepsilon + \delta^2 , \qquad \begin{aligned} \varepsilon &= a + \alpha\xi - \bar{\alpha}\,\xi^{-1} , \\ \delta^2 &= b + \beta\xi - \bar{\beta}\xi^{-1} + \sigma\xi^2 + \bar{\sigma}\xi^{-2} ; \end{aligned} \qquad (12.1)$$

a and b are real, and a trivial overall constant has been omitted. There are 8 real parameters in (12.1), but, as we shall see, there is a normalization condition which reduces them to seven.

Since $2\gamma = 2z + x_+\xi - x_-\xi^{-1}$ it is clear that the linear term in (12.1) can be eliminated by a space-translation, and that after it has been eliminated in this way (12.1) reduces to the sum of squares

$$H_2(\gamma,\xi) = \gamma^2 + \delta^2(\xi) \quad . \tag{12.2}$$

As a matter of fact, by using also space rotations, three of the parameters in $\delta(\xi)$ can be eliminated. Ward has chosen to eliminate them so that

$$\delta^2(\xi) = d^2[\cos 2\alpha + \frac{1}{2}\sin 2\alpha(\xi-\xi^{-1})] \tag{12.3a}$$

where d and α are real, and for reasons of symmetry it will be convenient also to eliminate them (i.e. choose the axes) in such a way that

$$\delta^2(\xi) = d^2[\cos^2\alpha - \frac{1}{4}\sin^2\alpha\ (\xi-\xi^{-1})^2] \tag{12.3b}$$

with the same d and α. Note that the axisymmetric limit is obtained when $\sin\alpha = 0$. The general relationship between the 6 parameters of the Euclidean group and the 8 parameters in (12.1), including the reduction from (12.1) to (12.3), is given in Appendix C.

The Ansatz proposed by Ward for the separated monopole solution is the generalization of (9.10) which is obtained by replacing $H_2(\gamma)$ by the $H_2(\gamma,\xi)$ defined in (12.2) and hence is

$$\tilde{g} = \begin{bmatrix} \frac{e^{f} + e^{-f}}{H_2(\gamma,\xi)} & -e^{-f}\xi^{2} \\ e^{-f}\xi^{-2} & H_2(\gamma,\xi)e^{-f} \end{bmatrix} \quad \text{where } f = \left(\frac{\pi}{2}\right)\frac{\gamma}{\delta} . \tag{12.4}$$

Here $\tilde{g}$ is in the real time-independent gauge, and to transform it to the AW gauge one uses the gauge-transformations

$$X(\xi^{-1}) = \begin{bmatrix} e^{\hat{\nu}/2} & 0 \\ 0 & e^{-\hat{\nu}/2} \end{bmatrix} , \quad Y(\xi) = \begin{bmatrix} e^{\hat{\mu}/2} & -H_2(\gamma,\xi)\xi^{2}e^{-\hat{\mu}/2} \\ 0 & e^{-\hat{\mu}/2} \end{bmatrix} \tag{12.5}$$

in analogy to (9.10), but where $\hat{\mu}$ and $\hat{\nu}$ are now the parts of f which are regular for $|\xi| \gtrless 1$. Thus

$$\begin{aligned} \hat{\mu}(\xi) &= 2\Gamma(\xi), \quad |\xi| < 1 , \\ \hat{\nu}(\xi^{-1}) &= 2\Gamma(\xi), \quad |\xi| > 1 , \end{aligned} \qquad \text{where } \Gamma(\xi) = \frac{1}{2\pi i}\int \frac{d\eta}{\eta-\xi} f + U(x) \tag{12.6}$$

and $U(x)$ is a ξ-independent function of x. Then g takes the form

$$g = X\tilde{g}Y = \begin{bmatrix} \rho & -\xi^{2} \\ \xi^{-2} & 0 \end{bmatrix} \quad \text{where } \rho = \frac{e^{\hat{\mu}}+e^{\hat{\nu}}}{H_2(\gamma,\xi)} = \frac{e^{f} + e^{-f}}{H_2(\gamma,\xi)}\, e^{\frac{\hat{\mu}+\hat{\nu}}{2}} . \tag{12.7}$$

The function ρ in (12.7) will generate solutions of the Bogomolny equation, of course, only if ρ, and hence $\hat{\mu}$ and $\hat{\nu}$, are functions of the twistor coordinates μ, ν, ξ only. The condition for this is

$$(P_{\mu}\partial_{\mu})\hat{\mu} = (P_{\mu}\partial_{\mu})\hat{\nu} = 0 \tag{12.8}$$

where the P_μ are the operators defined in (7.10). On applying (12.8) to (12.6) and using the fact that γ already satisfies (12.8), one finds that the denominator $(\eta-\xi)$ in (12.7) is cancelled, and one obtains

$$(P_\mu \partial_\mu) U(x) = \frac{-1}{2\pi i} \oint \frac{d\eta}{\eta\delta} \binom{1}{\eta} \quad . \tag{12.9}$$

It is easy to see that (12.9) will be satisfied if and only if

$$(-4\pi i) U(x) = \bar{v} \int \frac{d\eta}{\eta} \frac{1}{\delta} + u \int \frac{d\eta}{\delta}, \text{ where } \begin{array}{l} u = x + iy, \\ \bar{v} = z - ix_H \end{array} . \tag{12.10}$$

Thus the condition (12.8) is satisfied and U(x) is determined at one stroke.

Since the time coordinate enters in (12.7) only through U(x) and since, from (9.7) and the boundary condition $\Phi^2 \to c^2$, the time dependence of ρ must be of the form exp(ct), we see that Q must satisfy the normalization condition

$$\frac{1}{2\pi i} \oint \frac{d\eta}{\eta} \frac{1}{\delta(\eta)} = c \quad . \tag{12.11}$$

This is the normalization condition referred to earlier, and if we use the form (12.3) for $\delta(\xi)$, it reduces to

$$d = \frac{1}{2\pi c} \int_0^{2\pi} \frac{d\psi}{(\cos 2\alpha + i \sin 2\alpha \sin \psi)^{1/2}} = \frac{1}{2\pi c} \int_0^{2\pi} \frac{d\psi}{(\cos^2\alpha + \sin^2\alpha \sin^2\psi)^{1/2}} \tag{12.12}$$

thus reducing the 2 parameters d and α to one.

Since from (12.7) (12.12) it is clear that the Ansatz (12.4) describes a real time-independent solution of the Bogomolny equations, it remains only to verify that the solution is regular, satisfies the boundary condition and describes separated monopoles. The verification of the first two points depends essentially on the fact that Ansatz (12.4) is obtained by a continuous deformation of the parameter α from the axisymmetric Ansatz (9.2). Using this fact one easily verifies that the boundary condition is satisfied for all α. Similarly since ρ in (12.5) is regular in $\vec{x}$, and the determinant $\Delta_0^2 - \Delta_1\Delta_{-1}$ does not vanish in the axisymmetric limit, one sees that the solution will be regular for small deformations (i.e. small separations) at least. A proof for large separations is still lacking, however.

Finally, one has to establish that the Ansatz (12.4) does actually describe separated system, i.e. the Higgs field has separated zeros. For this purpose it is convenient to note that the Ansatz (9.3) is symmetric with respect to rotations through an angle π around the coordinate axes, provided that the coordinate axes are suitably chosen. To see this we use the result of Appendix C to implement these rotations by the transformations $(\xi \to \xi^{-1})$, $(\xi \to -\xi^{-1})$ and $(\xi \to -\xi)$ in ξ-space. We then see that (12.3a) is invariant with respect to the 2nd transformation $(\xi \to -\xi^{-1})$ while (12.3b) is invariant with respect to all three. Thus if we choose axes so that δ takes the form (12.3b) we have invariance with respect to π-rotations around each of the coordinate axes. (This is why we have preferred the form (12.3b) for δ to Ward's original form.)

Once it is established that the Ansatz is invariant with respect to π-rotations around the three axes, it follows from the fact that there are only two monopoles (zeros of the Higgs field $\Phi(x)$) that they must lie on one of the coordinate axes, and, if they are not separated, must lie at the origin. Hence to show that the zeros are separated it suffices to show that $\Phi(0) \neq 0$. A computation shows that indeed $\Phi(0) \neq 0$, unless $\sin\alpha = 0$. Hence for all values of the parameter $\sin\alpha$, except zero, the Ansatz (12.4) (if regular) describes separated monopoles. Note that, by continuity, the axis on which the monopoles are located must be the same for all $\sin\alpha$, and hence to determine it one need only compute the zeros for small $\sin\alpha$. Using the results of Ward for small $\sin\alpha$, it is easy to see that the axis in question is the x-axis. Thus, for any $\sin\alpha \neq 0$, the monopoles are separated and lie on the x-axis.

13. A PROPOSAL FOR AN N-SEPARATED-MONOPOLE SOLUTION

Just as these notes were being completed we received a preprint [28] proposing a generalization of the $n = 2$ separated Ansatz (12.4) to arbitrary n. The proposal admits (4n-1) independent parameters, as predicted by the index theorem [20].

The essence of the proposal is to replace the polynomials H_2 and f in (12.4) by the polynomials

$$H_n(\gamma,\xi) = \sum_{s=0}^{n} a_s(\xi)\gamma^s \quad \text{and} \quad f_n = 2\pi i \sum_{s=1}^{n} n_s \prod_{r\neq s} \left(\frac{\gamma-\gamma_r}{\gamma_s-\gamma_r}\right) , \tag{13.1}$$

respectively, where the $a_s(\xi)$ are polynomials of degree n-s in ξ and ξ^{-1} satisfying the reality condition $a_s(\xi) = \bar{a}_s(-\xi^{-1})$ the $\gamma_s(\xi)$ are the roots of γ in $H_n(\gamma,\xi)$, and the n_s are integers (or half-odd-integers for odd n). It is evident that for suitable choice of the integers n_s the zeros of H_n will be cancelled by the zeros of the numerator in (12.4), and the integers chosen are the smallest with this property (this may also be required by the boundary conditions). It is also evident that the number of real parameters in (13.1) is the number in H_n and this is just $n(n+2)$.

The condition that there should exist a gauge-transformation from (12.4) to the AW-gauge (8.1) is exactly the same as in the $n = 2$ case, namely that there should exist a $\hat{\mu}$ and a $\hat{\nu}$ satisfying (12.8), (with f replaced by f_n). Accordingly, in analogy to (12.9) the condition can be written as

$$(P_\mu \partial_\mu) U(x) = \oint \frac{d\eta}{\eta} \left(\frac{\partial f_n}{\partial \gamma}\right) \binom{1}{\eta} \quad . \tag{13.2}$$

However, in contrast to the $n = 2$ case, eq.(13.2) can be integrated to give

$$(-4\pi i) U(x) = \bar{v} \int \frac{d\eta}{\eta} \left(\frac{\partial f_n}{\partial \gamma}\right) + k(u) , \text{ where } \frac{\partial k(u)}{\partial u} = \int d\eta \left(\frac{\partial f_n}{\partial \gamma}\right) , \tag{13.3}$$

if and only if the integrability conditions

$$\oint \frac{d\eta}{\eta} \eta^r \frac{\partial^2 f}{\partial \gamma^2} = 0 , \qquad r = 0, \pm 1 , \tag{13.4}$$

are satisfied. Note that these conditions are equivalent to the condition that the coefficient of $\bar{v} = z - ix_4$ in (13.3) be constant.

By expanding f_n in powers of γ one sees that the integrability conditions may be written as

$$\oint \frac{d\eta}{\eta} \eta^q b_p(\eta) = 0 \quad , \quad \begin{matrix} |q| < p \\ \\ 2 \leq p \leq n-1 \end{matrix} \quad , \quad f_n(\gamma,\eta) = \sum_{r=0}^{n=1} b_r(\eta) \gamma^r , \tag{13.5}$$

and it is then evident that there are $n(n-2)$ conditions altogether. Together with the normalization condition from

(13.3) these conditions reduce the number of independent parameters to (4n-1).

The conditions (13.5) can also be stated as the condition that f_n must be of the form

$$f_n(\xi,\gamma) = \sum_{i=0}^{n-1} \sum_{j=-\infty}^{j=\infty} a_{ij}\, \gamma^i\, \xi^j \;,\; a_{i+2j} = 0 \;,\; |j| \leq i+1, \qquad (13.6)$$

and the compatibility of (13.6) and (13.1) yield the n(n-2) conditions on the parameters in algebraic form.

We still have to consider, of course, whether the proposed solution is regular, satisfies the boundary conditions and describes n separated monopoles. That the boundary condition will be satisfied is guaranteed by the Yukawa equation (9.7) and the exponential time-dependence in (12.7) and may be verified explicitly by noting that the integrability conditions (13.4) (which incorporate the time-independence of $f(\xi,\gamma)$) limit the exponential growth of the computed from (12.7) to be of order one. The regularity of the solution depends, as usual, on the non-vanishing of the Durham determinant and, if the determinant obtained in the axisymmetric limit really is non-zero, then the small-separation argument used for n = 2 should again be valid. Finally, the fact that there is the full complement of parameters allowed by the index theorem suggests that the monopoles can indeed be separated. Thus there is strong prima facie evidence that the proposal (13.1) is correct.

APPENDIX A

We wish to show that for static fields generated by the Durham string, we have [22]

$$\Phi^2 = c^2 = \Delta \ln (\det D) \;, \qquad (A1)$$

and that if the system is axially symmetric, we have

$$(\Phi,\omega) = \partial_z \ln(\det D) \text{ and } \omega^2 + (c^2-\Phi^2) = 2\rho\partial\rho \ln(\det D) . \tag{A2}$$

Here it is assumed that since the gauge potentials are static, the string-functions $\Delta_r(x)$ have only exponential time-dependence

$$\Delta_r(x) = \Delta_r(\vec{x})\, e^{ct} , \tag{A3}$$

and that the axial symmetry condition (5.1) may be reduced to the standard form [24]

$$\partial_\varphi \Phi = n\sigma_3 \Phi \quad \text{where} \quad \omega = n\sigma_3 + A_\varphi, \quad \sigma_3 = \begin{pmatrix} 1 & 0 \\ 0 & -1 \end{pmatrix} , \tag{A4}$$

and n is the topological charge. The reason that $\omega - A_\varphi$ can be reduced to the constant vector $n\sigma_3$ is that ω is smooth and D_φ has only integer eigenvalues. (The identification of the integer with the topological charge comes from the formula (2.5).) It will also be convenient to write the formula (3.5) for the components A_4 and A_φ of the gauge-potential as

$$A_4 = \frac{-1}{2f} \begin{bmatrix} f,_z & e,_u \\ g,_{\bar{u}} & -f,_z \end{bmatrix} , \quad A_\varphi = \frac{-\rho}{2f} \begin{bmatrix} f,_\rho & e^{i\varphi}\, e,_{\bar{v}} \\ e^{-i\varphi} g,_v & -f,_\rho \end{bmatrix} . \tag{A5}$$

To establish (A1) we note from (A5) that

$$\Phi^2 = \operatorname{tr} A_4^2 = \frac{1}{4f^2} [2f,_z^2 + e,_u\, g,_{\bar{u}}] . \tag{A6}$$

Hence under the Bäcklund transformation B in (3.7) we have

$$\Phi^2 \to \frac{1}{4f^2}\,[2f_z^2 - e_{,\bar{v}}\, g_{,v}] \;=\; \Phi^2 - \Delta \ln f\,, \tag{A7}$$

where in the second step we have used the Yang equation (3.3). Furthermore, since the fields are static, Φ^2 is invariant with respect to the gauge transformations I. Hence (A7) holds also for the combined transformation BI. Iterating the BI transformation n times we then have

$$\Phi^2 = c^2 - \Delta \ln (f_n\, f_{n-1} \ldots f_2\, f_1)\ . \tag{A8}$$

But from (3.9) and (3.10) we have, by definition,

$$f_r = \det D(r)/\det D(r-1)\ . \tag{A9}$$

Inserting (A9) into (A8) we obtain (A1) as required.

To establish (A2) we note first that the invariants ω^2 and (ω,Φ) are given by

$$\omega^2 = \mathrm{tr}\,(A_\varphi + n\sigma_3)^2 \quad \text{and} \quad (\omega,\Phi) = \mathrm{tr}\,(A_\varphi + n\sigma_3)A_4, \tag{A10}$$

where, from (A5),

$$\mathrm{tr}(A_\varphi + n\sigma_3)^2 = \frac{\rho^2}{2f^2}[f_\rho^2 + e_{\bar{v}}\, g_v] - \frac{2n}{f}\, f_\rho + 2n^2$$

and

$$\mathrm{tr}(A_\varphi + n\sigma_3)A_4 = \frac{\rho}{4f^2}\,[2f_\rho f_z + e_u g_v + e_{\bar{v}} g_{\bar{u}}] - \frac{n}{f}\, f_z\ . \tag{A11}$$

Under the Bäcklund transformations B the integer n changes to -(n-1) (the minus because e and g interchange positions). Taking the change in n and the Yang equations (3.3) into account, we find from (A11) and (A10) that

$$\omega^2 \to \omega^2 + (2\rho\partial_\rho - \Delta)\ln f \quad \text{and} \quad (\omega,\Phi) \to (\omega,\Phi) + \partial_z \ln f\ , \tag{A12}$$

under the transformations B and BI. Iterating the BI transformation n times and using (A9) we obtain (A2) as required.

APPENDIX B

To show that the planes defined by $\omega = x\pi$ are self-dual consider any displacement dx in such a plane. Since dx belongs to the $D(\frac{1}{2}\frac{1}{2})$ representation of SO(4) it can be written as the direct product $\bar{\eta} \times \lambda$ of two 2-spinors. Then the condition dx $\pi = 0$ that dx lies in the plane takes the form

$$\bar{\eta} \times (\lambda \wedge \pi) = 0 , \tag{B1}$$

where wedge denotes outer product in the 2-space. Since (B1) implies that λ is parallel to π, we see that displacements dx in the plane are just those which are of the form

$$dx = \bar{\eta} \times \pi , \tag{B2}$$

where η is free and π is fixed. Let us now consider the anti-symmetric product of two such displacements dx, dx'. Normally such products belong to the D(10) $\oplus$ D(01) representation of SO(4), but because π is fixed we obtain

$$(\bar{\eta} \wedge \bar{\eta}')(\pi\cdot\pi') + (\bar{\eta}\cdot\bar{\eta}')(\pi \wedge \pi) = (\bar{\eta} \wedge \bar{\eta}')(\pi\cdot\pi') , \tag{B3}$$

which manifestly belongs only to the D(10) representation. But belonging to the D(10) part of D(10) + D(01) is just the definition of anti-self-duality. Hence the planes $\omega = x\pi$ are anti-self-dual, as required.

APPENDIX C

In order to obtain the connection between the parameters of the Euclidean group and those of the 2-form (12.1) we first note from (7.1) that (π_1,π_2) is a translational scalar and rotational 2-spinor. It follows that the quantity

$$\vec{\Sigma} = \tilde{\pi}\, C\, \vec{\sigma}\, \pi \qquad \text{where} \qquad C = \begin{pmatrix} 0 & 1 \\ -1 & 0 \end{pmatrix} , \tag{C1}$$

tilde denotes transpose, and $\vec{\sigma}$ are the Pauli matrices, is a translational scalar and rotational 3-vector. On the other hand, if we recall that $\xi = \pi_1/\pi_2$ it is easy to verify that the 2-form (12.1) may be written in the homogeneous form

$$H_2(\gamma,\xi) = \{(x\cdot\Sigma)^2 + 2(x\cdot\Sigma)(g\cdot\Sigma) + \frac{d^2}{4}(e\cdot\Sigma)(e'\cdot\Sigma)\}/\Sigma_3^2, \tag{C2}$$

where d^2 is a positive constant, g, e, e', are real constant 3-vectors (with e, e' normalized to unity), and these 8 parameters are related to the original 8 parameters in (12.1) in an obvious manner.

By inspection of (C2) one sees that the vector g may be eliminated by a space translation and the vectors e, e' rotated to any position so long as the angle between them remains fixed. Thus the Euclidean invariant parameters in (C2) are d^2 and the inner-product $(e\cdot e')$.

When g has been eliminated and {e,e'} brought to the positions {(0 0 1), (sin 2α, 0, cos 2α)} and {(sin α, 0, cos α), (-sin α, 0, cos α)}, the expression $H_2(\gamma,\xi)$ reduces to

$$H_2(\gamma,\xi) = \{(x\cdot\Sigma)^2 + \frac{d^2}{4}\Sigma_3(\Sigma_3 \cos 2\alpha + \Sigma_1 \sin 2\alpha)\}/\Sigma'^2_3 =$$

$$= \{\gamma^2 + d^2[\cos 2\alpha + \tfrac{1}{2}\sin 2\alpha(\xi-\xi^{-1})]\}\left(\frac{\Sigma_3}{\Sigma'_3}\right)^2 \tag{C3a}$$

and

$$H_2(\gamma,\xi) = \{(x\cdot\Sigma)^2 + \frac{d^2}{4}(\Sigma_3^2 \cos^2\alpha - \Sigma_1^2 \sin^2\alpha)\}/\Sigma'^2_3 =$$

$$= \{\gamma^2 + d^2[\cos^2\alpha - \tfrac{1}{4}\sin^2\alpha(\xi-\xi^{-1})^2]\}(\frac{\Sigma_3}{\Sigma_3'})^2 \tag{C3b}$$

respectively. These are just the expressions (12.2), (12.3) except for the factors

$$\left[\frac{\Sigma_3}{\Sigma_3'}\right]^2 = \left[\frac{\pi_1}{\pi_1'}\right]^2 \left[\frac{\pi_2}{\pi_2'}\right]^2 \tag{C4}$$

where

$$\frac{\pi_1'}{\pi_1} = \cos\beta_2 e^{i(\beta_1+\beta_3)} + \xi^{-1}\sin\beta_2 e^{i(\beta_1-\beta_3)} ,$$

$$\frac{\pi_2'}{\pi_2} = \cos\beta_2 e^{-i(\beta_1+\beta_3)} - \xi\sin\beta_2 e^{i(\beta_3-\beta_1)} , \tag{C5}$$

and the β's are half the Euler angles for the required rotations. But since the terms in (C4) are analytic for $|\xi| > 1$ and $|\xi| < 1$ respectively, the factors $(\Sigma_3/\Sigma_3')^2$ are gauged to zero by the gauge transformations

$$X(\xi^{-1}) = \text{diag}\left(\frac{\pi_1}{\pi_1'}, \frac{\pi_1'}{\pi_1}\right) , \qquad Y(\xi) = \text{diag}\left(\frac{\pi_2}{\pi_2'}, \frac{\pi_2'}{\pi_2}\right) . \tag{C6}$$

Thus finally H_2 can be reduced to the simple forms

$$H_2(\gamma,\xi) = \gamma^2 + d^2(\cos 2\alpha + \tfrac{1}{2}(\xi-\xi^{-1})\sin 2\alpha) ,$$

$$H_2(\gamma,\xi) = \gamma^2 + d^2(\cos^2\alpha - \tfrac{1}{4}(\xi-\xi^{-1})^2\sin^2\alpha) ,$$

respectively, as anticipated in (12.3).

It is clear from this discussion that for any function of γ and ξ, the spatial rotations may be implemented by a suitable transformation of ξ. In particular, the rotations through an angle π about the x, y, and z axes used in section 12 may be implemented by $\xi \to \xi^{-1}$, $\xi \to -\xi^{-1}$ and $\xi \to -\xi$ respectively.

ACKNOWLEDGEMENTS

The authors wish to thank Richard Ward for many illuminating discussions and to thank Drs. A. Fordy and V. Soucek for clarifying some particular points.

REFERENCES

1. D. Politzer, Phys. Reports 14C (1974) 129.
2. H. Nielsen, P. Oleson, Nucl. Phys. B61 (1973) 45.
   G.'t Hooft, Nucl. Phys. B79 (1974) 276.
   A. Polyakov, JETP Lett. 20 (1974) 194.
   A. Belavin, A. Polyakov, A. Schwartz, Y. Yyupkin, Phys. Lett. 59B (1975) 85.
3. S. Mandelstam, Phys. Reports C67 (1980) 109.
   G.'t Hooft, Nucl. Phys. B138 (1978) 1.
4. E. Bogomolny, Sov. J. Nucl. Phys. 24 (1976) 449.
   S. Coleman, S. Parke, A. Neveu, C. Sommerfield, Phys. Rev. D15 (1976) 544.
5. C.N. Yang, Phys. Rev. Lett. 38 (1977) 1377.
6. E. Corrigan, D. Fairlie, P. Goddard, R. Yates, Comm. Math. Phys. 58 (1978) 223.
7. C. Taubes, Comm. Math. Phys. (in press).
   A. Jaffe, C. Taubes, Vortices and Monopoles (Birkhauser, Boston, 1980).
8. P. Houston, L.O'Raifeartaigh, Phys. Lett. 93B, 151, 94B (1980) 153.
   Proc. Conf. Diff. Geom. Methods in Math. Phys., Claus-thal 1980, ed. Doebner, Springer-Verlag (in press).

9. N. Manton, Nucl. Phys. B135 (1978) 319.
P. Houston, L. O'Raifeartaigh, Zeit. f. Phys. C6 (1981)1.
10. F. Ernst, Phys. Rev. 167 (1968) 1175.
11. R. Ward, Comm. Math. Phys. (in press).
12. P. Forgacs, Z. Horvath, L. Palla, Phys. Rev. Lett. 45 (1980) 505.
Hungarian Academy Preprint KFKI-122 (1980).
13. M. Atiyah, R. Ward, Comm. Math. Phys. 55 (1977) 117.
14. J. Arafune, P. Freund, C. Goebel, J. Math. Phys. 16 (1975) 433.
15. N. Monastyrski, A. Perelmov, JETP Lett. 21 (1975) 43.
16. V. Fateev, A. Shvartz, Y. Tyupkin, Teor. Mat. Fiz. 26 (1976) 270.
17. J. Rawnsley, J. Phys. A10 (1977) L139-L141.
18. M. Prasad, C. Sommerfield, Phys. Rev. Lett. 35 (1975)760.
19. A. Guth, E. Weinberg, Phys. Rev. D14 (1976) 1660.
L. O'Raifeartaigh, Nuovo Cim. Lett. 18 (1977) 205.
20. E. Weinberg, Phys. Rev. D20 (1979) 936.
21. L. O'Raifeartaigh, S-Y. Park, K.C. Wali, Phys. Rev. 20D (1979) 1941.
N. Manton, Nucl. Phys. B126 (1977) 525.
22. M. Prasad, Comm. Math. Phys. (in press).
23. P. Jang, s. Park, K.C. Wali, Phys. Rev. D17 (1978) 1641.
24. R. Jackiw, Proc. 1980 Schladming Conf. Acta Physica Austriaca.
V. Romanov, A. Schwarz, Y. Tyupkin, Nucl. Phys. B130 (1977) 209.
25. S. Adler, Proc. Intl. Conf. on High Energy Physics, Madison, Wisconsin 1980.
26. C. Rebbi, P. Rossi, Phys. Rev. 22D (1980) 2010.
27. M. Prasad, P. Rossi, MIT Preprint, December 1980.
28. E. Corrigan, P. Goddard, Cambridge University Preprint, March 1981.

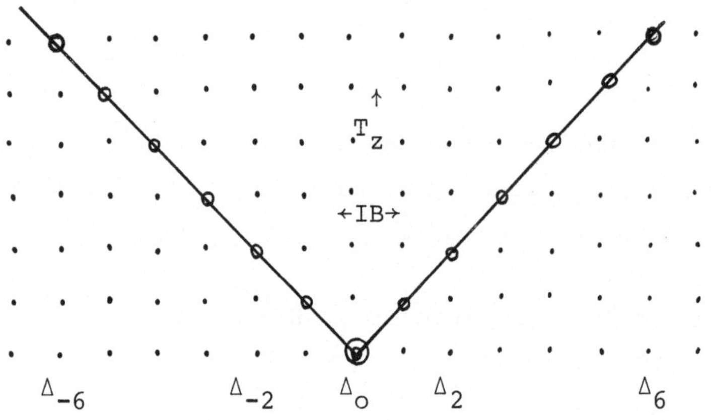

Fig. 1. Plot of the Durham strings Δ_r, $-\ell \leq r \leq \ell$, generated from the spherically symmetric solution $\Delta_o = \sinh(r)/r$ (doubly circled) by successive operations of the 'translation' T_z and the Bäcklund transformations IB. The strings are drawn horizontally, and only those whose ends are circled satisfy the reality and boundary conditions.

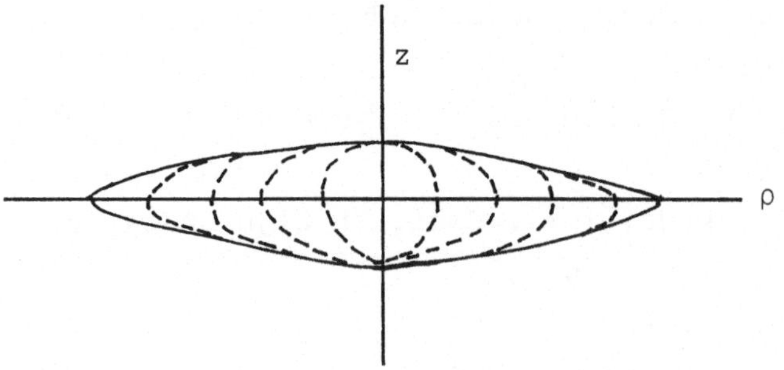

Fig. 2. Plot of the Monopole Core for Charge Q = 5, showing the nested ellipses as dotted curves.

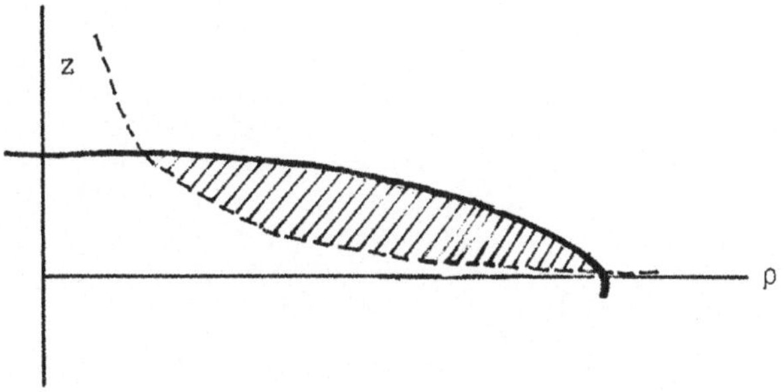

Fig. 3. Positive Quadrant of the Monopole Core with curve $z\rho = \pi^3/4$ shown as dotted line. The solutions have been shown to be non-singular outside the core and inside the dotted curve, leaving only the shadded area open to any doubt.

Acta Physica Austriaca, Suppl. XXIII, 577–585 (1981)

ON THE UNIVERSAL LOW ENERGY LIMIT IN NONRELATIVISTIC SCATTERING THEORY[+]

by

S. ALBEVERIO
Mathematisches Institut, Ruhr-Univ. Bochum
4630 Bochum 1, BRD

F. GESZTESY[x*]
Fakultät für Physik, Universität Bielefeld
4800 Bielefeld 1, BRD

R. HØEGH-KROHN
Matematisk Institutt, Universitetet i Oslo
Blindern-Oslo 3, Norge

Based on a precise formulation of the physical intuition that low momenta correspond to large distances we show that scattering quantities have a universal behaviour around the zero energy limit. Our methods rely on unitary implementation of scaling $\underline{x} \to \underline{x}/\varepsilon$, $\underline{p} \to \varepsilon\underline{p}$ in the Hilbert space $L^2(R^3)$ and on a corresponding expansion of the scattering matrix and amplitude in powers of ε around $\varepsilon = 0$. The results are largely independent of the shape of the interaction, and the leading coefficients in these ex-

---

[+] Seminar given at XX. Internationale Universitätswochen für Kernphysik, Schladming, Austria, February 17-26, 1981.
[x] Alexander von Humboldt Research Fellow
[*] On leave of absence from Institut für Theoretische Physik, Universität Graz, Austria

pansions are described in terms of suitable point interactions.

Let U_ε denote the unitary dilation group in $L^2(R^3)$,

$$(U_\varepsilon h)(\underline{x}) = \varepsilon^{-3/2} h(\underline{x}/\varepsilon), \ \varepsilon > 0, \ h \in L^2(R^3), \tag{1}$$

and assume λ: $[0,1] \to R$ to be analytic with $\lambda(0_+) = 1$. Throughout this work we assume V: $R^3 \to R$ to be measurable such that $\int d^3x d^3y e^{2a(|\underline{x}|+|\underline{y}|)} |V(\underline{x})||V(\underline{y})||\underline{x}-\underline{y}|^{-2} < \infty$ for some $a > 0$, and introduce the following Hamiltonians (defined as quadratic forms) in $L^2(R^3)$:

$$H = -\Delta + V(\underline{x}) \ , \tag{2}$$

$$H(\varepsilon) = -\Delta + \lambda(\varepsilon)V(\underline{x}), \ 0 \leqq \varepsilon \leqq 1 \ , \tag{3}$$

$$H_\varepsilon = \varepsilon^{-2} U_\varepsilon H(\varepsilon) U_\varepsilon^{-1} = -\Delta + \lambda(\varepsilon)\varepsilon^{-2} V(\underline{x}/\varepsilon), \ 0 < \varepsilon \leqq 1 \ . \tag{4}$$

In addition we abbreviate

$$G_k = (-\Delta - k^2)^{-1}, \ \mathrm{Im}k > 0, \tag{5}$$

and $v(\underline{x}) = |V(\underline{x})|^{1/2}$, $u(\underline{x}) = |V(\underline{x})|^{1/2} \operatorname{sign} V(\underline{x})$.

If $f(\varepsilon,\underline{p},\underline{q},k)$ and $f_\varepsilon(\underline{p},\underline{q},k)$ denote the off-shell scattering amplitudes of $H(\varepsilon)$ and H_ε respectively, we have

$$f(\varepsilon,\underline{p},\underline{q},k) = -(4\pi)^{-1}\lambda(\varepsilon)(ve^{i\underline{p}\cdot}, (\lambda(\varepsilon)uG_k v + 1)^{-1} ue^{i\underline{q}\cdot}), \tag{6}$$

and with the help of

$$(H(\varepsilon) - (\varepsilon k)^2)^{-1} = \varepsilon^{-2} U_\varepsilon^{-1} (H_\varepsilon - k^2)^{-1} U_\varepsilon \tag{7}$$

we obtain

$$f_\varepsilon(\underline{p},\underline{q},k) = \varepsilon f(\varepsilon\underline{p},\varepsilon\underline{q},\varepsilon k), \ |\mathrm{Im}\underline{p}| < a, \ |\mathrm{Im}\underline{q}| < a, \ \mathrm{Im}k > -a \ . \tag{8}$$

Thus for an expansion of the scattering amplitudes around $\varepsilon = 0$ we need an expansion of $\varepsilon(\lambda(\varepsilon)uG_{\varepsilon k}v + 1)^{-1}$. Since the explicit expansion of this quantity depends on whether H has a zero energy resonance resp. zero energy bound states or not we briefly discuss the definition of zero energy resonance functions for H[1-3]. Let -1 be an eigenvalue of uG_ov, i.e. $uG_ov\phi_j = -\phi_j$ for some $\phi_j \in L^2(R^3)$, $j = 1,\ldots,N$; then the functions

$$\psi_j(\underline{x}) = (G_o v\phi_j)(\underline{x}),\ j = 1,\ldots,N, \tag{9}$$

are called (zero energy) resonance functions of H and we distinguish the following cases (see also [2,4-6]):

<u>Case I):</u> There exist no resonance functions ψ_j, i.e. -1 is not an eigenvalue of uG_ov.

<u>Case II):</u> There exists precisely one resonance function ψ (i.e. -1 is a simple eigenvalue of uG_ov) and ψ is not in $L^2(R^3)$.

<u>Case III):</u> There exist $N \geq 1$ resonance functions ψ_j, $j = 1,\ldots,N$, which are all in $L^2(R^3)$.

<u>Case IV):</u> There exist $N \geq 2$ resonance functions ψ_j, $j = 1,\ldots,N$, and at least one of them is not in $L^2(R^3)$.

As a first result we present the connection between H_ε and point interactions. Let $-\Delta_\alpha$ denote the Hamiltonian with point interaction of parameter $\alpha \in R$ centered at $\underline{x} = 0$, i.e. $-\Delta_\alpha$ is the self-adjoint extension of $-\Delta\big|_{C_o^\infty(R^3-\{0\})}$ determined by the boundary condition

$$[-4\pi\alpha|\underline{x}|g(|\underline{x}|) + (\hat{\underline{x}}\cdot\nabla)(|\underline{x}|g(|\underline{x}|))]\Big|_{|\underline{x}|=0} = 0,\ \hat{\underline{x}} = \underline{x}/|\underline{x}|, \tag{10}$$

in the partial wave subspace of angular momentum zero. Then we have [1,2]

<u>Theorem 1.</u> H_ε converges to $-\Delta_\alpha$ in strong resolvent sense, where α is given by

$$\alpha = \begin{cases} \infty \text{ in case I) and case III)} , \\ -\lambda'(0_+)\,|(v,\phi)|^{-2} \text{ in case II)} , \\ -\lambda'(0_+)\,(\sum_{j=1}^{N} |(v,\phi_j)|^2)^{-1} \text{ in case IV)} , \end{cases} \tag{11}$$

and the normalization $(\phi_j, \text{ sign } V\phi_\ell) = -\delta_{j\ell}$ has been used. (If $\alpha = \infty$ one has $-\Delta_\alpha = -\Delta$.)

The proof of this and the following theorems is a direct consequence of the above-mentioned Laurent (resp. Taylor) expansion of $\varepsilon(\lambda(\varepsilon)uG_{\varepsilon k}v+1)^{-1}$ according to cases I) - IV). We refer to [2] for the explicit derivation of these expansions and now turn to the low energy behaviour of the scattering amplitude for H by putting $\lambda(\varepsilon) = 1$ [2].

Theorem 2. Let $f(\underline{p},\underline{q},k)$ denote the off-shell scattering amplitude associated with $H = -\Delta + V(\underline{x})$, i.e.

$$f(\underline{p},\underline{q},k) = -(4\pi)^{-1}(ve^{i\underline{p}\cdot}, (uG_k v + 1)^{-1} ue^{i\underline{q}\cdot}) ; \tag{12}$$

then we have the low energy expansions

$$f(\varepsilon\underline{p},\varepsilon\underline{q},\varepsilon k) = -(4\pi)^{-1}(v,Tu)+i(4\pi)^{-2}(\varepsilon k)(v,Tu)^2-(4\pi i)^{-1}(\varepsilon\underline{p}\cdot v,T\text{u}$$

$$+ (4\pi i)^{-1}(T^* v,\varepsilon\underline{q}\cdot u) + O(\varepsilon^2) \text{ in case I)} , \tag{13}$$

$$f(\varepsilon\underline{p},\varepsilon\underline{q},\varepsilon k) = \frac{i}{\varepsilon k} - 4\pi|(v,\phi)|^{-2}(\phi,C\phi) + k^{-1}(v,\phi)^{-1}(\underline{p}\cdot v,\phi)$$

$$-k^{-1}(\phi,v)^{-1}(\phi,\underline{q}\cdot v) + O(\varepsilon) \text{ in case II)} , \tag{14}$$

$$f(\varepsilon\underline{p},\varepsilon\underline{q},\varepsilon k) = -(4\pi)^{-1}(v,Tu) - (4\pi)^{-1}k^{-2}\sum_{j,\ell=1}^{N}(\underline{p}\cdot v,\phi_j)\,(\phi,C\phi)^{-1}_{j\ell}\,\cdot$$

$$\cdot(\phi_\ell,\underline{q}\cdot v) + O(\varepsilon) \text{ in case III)} , \tag{15}$$

$$f(\varepsilon\underline{p},\varepsilon\underline{q},\varepsilon k) = \frac{i}{\varepsilon k} + O(1) \text{ in case IV)} . \tag{16}$$

Here $(\phi,C\phi)^{-1}_{j\ell}$ denotes the inverse of the matrix $(\phi_j,C\phi_\ell)$, and C and T are defined as

$$(Ch)(\underline{x}) = -(8\pi)^{-1}\int d^3y v(\underline{x})|\underline{x}-\underline{y}|v(\underline{y})h(\underline{y}), \quad h \in L^2(R^3), \tag{17}$$

and

$$T = n - \lim_{\varepsilon\to 0_+} (uG_o v + 1 + \varepsilon)^{-1}(1-P) \tag{18}$$

where $P = -\sum_{j=1}^{N} (\text{sign } V\phi_j,.)\phi_j$ is the projector onto the eigenspace of $uG_o v$ to the eigenvalue -1.

If $f(k,\underline{\omega},\underline{\omega}') = f(\underline{p},\underline{q},k)\big|_{|\underline{p}|=|\underline{q}|=k}$ denotes the on-shell scattering amplitude associated with H, where $\underline{\omega} = \underline{p}/|\underline{p}|$, $\underline{\omega}' = \underline{q}/|\underline{q}|$, then the on-shell scattering operator $S(k)$ in $L^2(S^{(2)})$ ($S^{(2)}$ the unit sphere in R^3) reads

$$(S(k)h)(\underline{\omega}) = h(\underline{\omega}) - (2\pi i)^{-1}k\int d\omega' f(k,\underline{\omega},\underline{\omega}')h(\underline{\omega}'), \quad h \in L^2(S^{(2)}), \tag{19}$$

and we have [2]

Theorem 3. Let $S(k)$, $k > 0$, be the scattering operator associated with the pair $H,-\Delta$. Then $S(\varepsilon k)$ is analytic at $\varepsilon = 0$, and

$$S(\varepsilon k)=1+(2\pi i)^{-1}(\varepsilon k)(v,Tu)(Y_o,.)Y_o-(8\pi^2)^{-1}(\varepsilon k)^2(v,Tu)^2(Y_o,.)Y_o$$
$$-(\varepsilon k)^2(Y_o,.)Y_1+(\varepsilon k)^2(Y_1,.)Y_o+O((\varepsilon k)^3) \text{ in case I)}, \tag{20}$$

where

$Y_o(\underline{\omega}) = (4\pi)^{-1/2}$ and $Y_1(\underline{\omega}) = (4\pi^{3/2})^{-1}(\underline{\omega}.v,Tu)$;

$$S(\varepsilon k) = 1-2(Y_o,.)Y_o-(8\pi i)(\varepsilon k)|(v,\phi)|^{-2}(\phi,C\phi)(Y_o,.)Y_o$$
$$+i(\varepsilon k)(v,\phi)^{-1}(Y_o,.)\hat{Y}_1-i(\varepsilon k)(\phi,v)^{-1}(\hat{Y}_1,.)Y_o +$$

$$+ O((\varepsilon k)^2) \text{ in case II)} , \tag{21}$$

where

$$\hat{Y}_1(\underline{\omega}) = \pi^{-1/2}(\underline{\omega}.v,\phi) ;$$

$$S(\varepsilon k) = 1 + (2\pi i)^{-1}(\varepsilon k)(v,Tu)(Y_o,.)Y_o$$

$$+ (8\pi i)^{-1}(\varepsilon k) \sum_{j,\ell=1}^{N} (\phi,C\phi)^{-1}_{j\ell}(\hat{Y}_{1\ell},.)\hat{Y}_{1j}+O((\varepsilon k)^2)$$

$$\text{in case III)} , \tag{22}$$

where

$$\hat{Y}_{1j}(\underline{\omega}) = \pi^{-1/2}(\underline{\omega}.v,\phi_j) ;$$

$$S(\varepsilon k) = 1-2(Y_o,.)Y_o + O(\varepsilon k) \text{ in case IV)} . \tag{23}$$

Remarks

1) Within the theory of weighted Sobolev spaces, Jensen and Kato [4] derived (20), (21), and (23). (13)-(16) and (22) appear to be new in the noncentral case.

2) In case I) and III), $-(4\pi)^{-1}(v,Tu)$ represents the scattering length.

3) Due to the existence of a zero energy resonance in cases II) and IV) the scattering length is infinite and in addition $S(\varepsilon k)$ tends to -1 in the subspace of angular momentum zero as $\varepsilon \to 0_+$. (This fact occurs in s-waves if V is spherically symmetric.)

4) If the second term in (15) is non vanishing, the scattering cross section even at zero energy never becomes isotropic in case III). (For spherically symmetric potentials this effect is caused by zero energy p-wave bound states.)

Finally we discuss Puiseux resp. Taylor expansions in ε for the eigenvalues and resonances of $H(\varepsilon)$ and H_ε. Let $k(\lambda(\varepsilon))$ and $k_\varepsilon(\lambda(\varepsilon))$ be the solutions of $D_2(1+\lambda(\varepsilon)uG_kv)=0$

resp. $D_2(1 + \lambda(\varepsilon)\varepsilon^{-2}u(\underline{.}/\varepsilon)G_k v(\underline{.}/\varepsilon)) = 0$ where D_2 denotes the modified Fredholm determinant. From (7) we get

$$k_\varepsilon(\lambda(\varepsilon)) = \varepsilon^{-1}k(\lambda(\varepsilon)) \ . \qquad (24)$$

It is shown in [3] that the functions $k(\lambda(\varepsilon))$ have at most branch points of finite order as singularities and else are holomorphic in λ. Moreover we have [2] (see also [5,7])

Theorem 4. Let V have compact support and assume $k(\lambda(\varepsilon))$ resp. $k_\varepsilon(\lambda(\varepsilon))$ to be the functions giving the eigenvalues and resonances of $H(\varepsilon) = -\Delta + \lambda(\varepsilon)V(\underline{x})$ resp. $H_\varepsilon = -\Delta + \lambda(\varepsilon)\varepsilon^{-2}V(\underline{x}/\varepsilon)$. Then we have

$$k_\varepsilon(\lambda(\varepsilon)) = \varepsilon^{-1}k(\lambda(\varepsilon)) \ ,$$

$k(\lambda(\varepsilon))$ is an analytic function of ε, except for possible branch points of finite order. As $\varepsilon \to 0_+$ we have the following expansions:

1) If $k(\lambda(0_+)) \neq 0$ then

$$k_\varepsilon(\lambda(\varepsilon)) = \frac{k(\lambda(0_+))}{\varepsilon} + O(1) \ . \qquad (25)$$

2) If $k(\lambda(0_+)) = 0$ then a) in case II) $k_\varepsilon(\lambda(\varepsilon))$ is analytic at $\varepsilon = 0$ and

$$k_\varepsilon(\lambda(\varepsilon)) = 4\pi i\lambda'(0_+)|(v,\phi)|^{-2} + O(\varepsilon) \ . \qquad (26)$$

b) In case III) with $\lambda'(0_+) \neq 0$, $k_\varepsilon(\lambda(\varepsilon))$ has N branches $k_{j,\varepsilon}(\lambda(\varepsilon))$, $j = 1,\dots,N$, behaving as follows:

$$k_{j,\varepsilon}(\lambda(\varepsilon)) = \frac{c_j\sqrt{-\lambda'(0_+)}}{\varepsilon^{1/2}} + O(1), \ c_j > 0 \ . \qquad (27)$$

For N = 1 we have

$$c = (8\pi)^{1/2}|\int d^3x d^3y v(\underline{x})\overline{\phi(\underline{x})}|\underline{x}-\underline{y}|v(\underline{y})\phi(\underline{y})|^{-1/2}.$$

If $\lambda'(0_+) = 0$ we have again N branches $k_{j,\varepsilon}(\lambda(\varepsilon))$ behaving like

$$k_{j,\varepsilon}(\lambda(\varepsilon)) = \tilde{c}_j \sqrt{-\lambda''(0_+)} + O(\varepsilon),\ \tilde{c}_j > 0 . \tag{28}$$

For N = 1 we get

$$\tilde{c} = 2^{-1/2}c .$$

c) In case IV) one of the branches $k_{j_0,\varepsilon}(\lambda(\varepsilon))$ is analytic at $\varepsilon = 0$ and

$$k_{j_0,\varepsilon}(\lambda(\varepsilon)) = 4\pi i\lambda'(0_+)\left(\sum_{j=1}^{N} |(v,\phi_j)|^2\right)^{-1} + O(\varepsilon) . \tag{29}$$

The remaining branches $k_{j,\varepsilon}(\lambda(\varepsilon))$ behave in case $\lambda'(0_+) \neq 0$ as in (27) an in case $\lambda'(0_+) = 0$ as in (28).

Depending on the sign of $(\lambda(\varepsilon) - 1)$ near $\varepsilon = 0$ we therefore get bound states ($\mathrm{Im}k_\varepsilon > 0$, $\mathrm{Re}k_\varepsilon = 0$), virtual states ($\mathrm{Im}k_\varepsilon < 0$, $\mathrm{Re}k_\varepsilon = 0$), and resonance pairs in the cases I)-IV) (see [2] for a detailed discussion). According to Theorem 1, the leading coefficients in the expansions of energy eigenvalues and resonances for $H(\varepsilon)(H_\varepsilon)$ are determined by suitable point interactions (see [2]). E.g., $k_\varepsilon(\lambda(0_+)) = 4\pi i\lambda'(0_+)|(v,\phi)|^{-2}$ in (26) corresponds precisely to the bound state (if $\lambda'(0_+) > 0$) or virtual state (if $\lambda'(0_+) < 0$) of $-\Delta_\alpha$, $\alpha = -\lambda'(0_+)|(v,\phi)|^{-2}$.

In conclusion we have shown that scattering quantities behave in an universal way in the sense that only a few parameters, which are largely independent of the shape of V, are needed in order to describe the lower order coefficients in the analytic expansions around the zero energy limit.

REFERENCES

1. S. Albeverio, R. Høegh-Krohn, Point Interactions as Limits of Short Range Interactions, to appear in J. Operator Theory.
2. S. Albeverio, F. Gesztesy, R. Høegh-Krohn, The Low Energy Expansion in Nonrelativistic Scattering Theory, Preprint, Univ. Bielefeld 1981.
3. S. Albeverio, R. Høegh-Krohn, Perturbation of Resonances in Quantum Mechanics, Preprint, Univ. Bochum 1981.
4. A. Jensen, T. Kato, Duke Math. J. 46 (1979) 583.
5. M. Klaus, B. Simon, Ann. Phys. 130 (1980) 251.
6. R.G. Newton, J.Math. Phys. 18 (1977) 1348.
7. J. Rauch, J. Funct. Anal. 35 (1980) 304.

Acta Physica Austriaca, Suppl. XXIII, 587–598 (1981)

ON CLASSICAL TIME DELAY+

by

D. BOLLÉ*

Instituut voor Theoretische Fysica

Universiteit Leuven, 3030 Leuven, Belgium

and

Lab. de Physique Théorique et Hautes Energies^x

Univ. de Paris Sud, Bât. 211, 91405 Orsay, France

1. INTRODUCTION

The definition of time delay in quantum mechanics, its relation with the S-matrix and some of its properties (e.g. Levinson's theorem) have been discussed in the lectures on time delay of quantum scattering processes [1]. We have also seen how all this can be worked out for a two-body classical scattering process in terms of phase trajectories in the lectures on classical scattering theory [2].

+Seminar given at XX. Internationale Universitätswochen für Kernphysik, Schladming, Austria, February 17-26, 1981.

*Onderzoeksleider N.F.W.O., Belgium.

^xLaboratoire associé au C.N.R.S.

From these discussions it is clear that the concept of time delay is important from the point of view of general scattering theory. It provides us e.g. with a method to extend the concept of the phase shift. Therefore we were motivated to ask ourselves the question if time delay can be defined, and how it can be related with the S-matrix in a Hilbert-space approach to classical scatttering, without going into any deep discussion about the disadvantages or advantages of this approach compared with the phase space approach.

In the following we will find that, in close analogy with the corresponding quantum result, the classical reduced time delay operator for two particle scattering via short-range interactions is equal to the logarithmic derivative of the classical reduced S-operator. The methods we use to obtain this result can be extended to more complicated scattering systems. They are sketched rather briefly here, just to understand the basic mechanism. For more details we refer to Ref.[3].

2. THE HILBERT-SPACE APPROACH TO CLASSICAL SCATTERING

We consider a classical two-particle system characterized by the Hamiltonian

$$H(\vec{r},\vec{p}) = \frac{\vec{p}^{\,2}}{2\mu} + V(\vec{r}) \quad , \tag{2.1}$$

where $\vec{r}$ is the interparticle separation in the center-of-mass system, $\vec{p}$ the relative momentum, $V(\vec{r})$ the potential and μ the reduced mass. The set $\Gamma = \{z \equiv (\vec{r},\vec{p}) : H(z) < \infty\}$ is the phase space of the system.

In the following we assume that the potential satisfies

A) $V(\vec{r})$ is bounded from below by $V_{-} > -\infty$. For any $M < \infty$, $V(\vec{r})$

is continuous with bounded derivatives up to order 2 on $\{\vec{r}:V(\vec{r}) < M\}$;

B) $|\vec{\nabla}_{\vec{r}}\, V(\vec{r})| <$ const. $r^{-2-\delta}$, $\delta > 0$.

In this case, Newton's equation of motion

$$\dot{z} = (\dot{\vec{r}},\dot{\vec{p}}) = (\frac{\vec{p}}{\mu}, -\vec{\nabla}_{\vec{r}}\, V(\vec{r})) \tag{2.2}$$

have unique solutions for any initial value $z_o \in \Gamma$, and the maps

$$\phi_t : z_o \to z_t, \quad -\infty < t < +\infty \tag{2.3}$$

form a one-parameter group of canonical transformations of Γ onto Γ [2].

In the Hilbert-space approach, the states Ψ of the scattering system are taken to be elements of the Hilbert space $L^2(\Gamma)$. The dynamics is described by a strongly continuous unitary one-parameter group on $L^2(\Gamma)$ induced by the map ϕ_t defined in eq.(2.3). viz.

$$U_t = e^{-Lt} : \Psi(z) \to \Psi(\phi_t z) \quad . \tag{2.4}$$

Here, L is the Liouville operator. Explicitly we have

$$L = L_o + L_V \quad ,$$

$$L_o = - \frac{\vec{p}}{\mu} \cdot \vec{\nabla}_{\vec{r}}, \; L_V = \vec{\nabla}_{\vec{r}}\, V(\vec{r}) \cdot \vec{\nabla}_{\vec{p}} \quad . \tag{2.5}$$

L then satisfies the equation of motion

$$\frac{d\Psi}{dt} = \{H,\Psi\} = -L\Psi \quad , \tag{2.6}$$

where $\{.\}$ is the Poisson bracket.

One can now describe bound states which must be linked

to the boundedness of orbits in configuration space. One can also show the existence of scattering states and thus the existence of classical Möller operators under the conditions (A) and (B) on the potential. In close analogy to quantum mechanical scattering, these Möller operators

$$\Omega^{\pm} = \underset{t \to \mp\infty}{\text{s-lim}}\ e^{Lt}\, e^{-L_o t} \tag{2.7}$$

satisfy the following basic properties:

$$\begin{aligned} \Omega^{\pm *}\, \Omega^{\pm} &= I \ , \\ \Omega^{\pm}\ \Omega^{\pm *} + B &= I \ , \\ e^{tL}\, \Omega^{\pm} &= \pm\, e^{tL_o} \ , \end{aligned} \tag{2.8}$$

where $\Omega^{\pm *}$ is the adjoint of $\Omega^{\pm}$, I is the identity on $L^2(\Gamma)$ and B the projection operator onto the subspace of bound states. The first property expresses the fact that the $\Omega^{\pm}$ are partial isometries. The second property means that Ω^{+} and Ω^{-} form a complete set of scattering states. The third property is called intertwining.

Finally, one can define the classical scattering operator S as

$$S = \Omega^{-*}\, \Omega^{+} \ . \tag{2.9}$$

This operator is unitary, viz.

$$SS^{*} = S^{*}S = I \ , \tag{2.10}$$

and satisfies the intertwining relation

$$SL_o = L_oS \ . \tag{2.11}$$

The last property will play an important role in our further analysis.

For more details about this approach we refer to the work of Hunziker [4].

3. DEFINITION OF CLASSICAL TWO-PARTICLE TIME DELAY

Let $f_{in}(z) \in L^2(\Gamma)$ be a function specifying the incoming state of the system at $t = 0$. Then there exists a state $\phi = \Omega^+ f_{in}$ such that the exact interacting state Ψ of the system satisfies

$$\Psi(t) = e^{-Lt}\phi \rightarrow e^{-L_o t} f_{in} \equiv \Psi_{in}(t) \text{ for } t \rightarrow -\infty \quad , \tag{3.1}$$

in the sense of the norm in $L^2(\Gamma)$. Further there also exists a function $f_{out}(z) \in L^2(\Gamma)$ specifying the outgoing state of the system such that $\phi = \Omega^- f_{out}$ and

$$\Psi(t) \rightarrow e^{-L_o t} f_{out} \equiv \Psi_{out}(t) \text{ for } t \rightarrow +\infty \ . \tag{3.2}$$

Furthermore,

$$f_{out} = S f_{in} \quad . \tag{3.3}$$

We now define time delay in terms of the exact state $\Psi(t)$ and the asymptotic states $\Psi_{in}(t)$ and $\Psi_{out}(t)$. We first look at a finite region around the scattering center in configuration space and define the following projection operator $P(\Sigma)$ in $L^2(\Gamma)$:

$$\begin{aligned} P(\Sigma)\ f(z) &= f(z) \qquad \text{for} \quad \vec{r} \in \Sigma \\ &= 0 \qquad\qquad\qquad \vec{r} \notin \Sigma \quad . \end{aligned} \tag{3.4}$$

Time delay for the region Σ is then defined as the time the interacting system, described by Ψ, is spending in the region Σ minus the time the asymptotic system, described by Ψ_{in} or Ψ_{out}, is spending in that region. This can be expressed in the following formula:

$$\tau^{in}(f_{in},\Sigma) = \int_{-\infty}^{+\infty} dt\ [(\Psi(t),P(\Sigma)\Psi(t)) - (\Psi_{in}(t),P(\Sigma)\Psi_{in}(t))] \ ,$$

$$\tau^{out}(f_{in},\Sigma) = \int_{-\infty}^{+\infty} dt[(\Psi(t),P(\Sigma)\Psi(t))-(\Psi_{out}(t),P(\Sigma)\Psi_{out}(t))]. \tag{3.5}$$

The time delay we shall consider in detail is one that is fully symmetric with respect to the asymptotic states, viz.

$$\tau(f_{in},\Sigma) = \frac{1}{2}\,\tau^{in}(f_{in},\Sigma) + \frac{1}{2}\,\tau^{out}(f_{in},\Sigma) \; . \tag{3.6}$$

Of course, we are interested in a quantity that is independent of the region Σ. So, we will study the limit $\Sigma \to R^3$ of these different time delay forms by using a method similar to the one we employed in the corresponding quantum mechanical problem [5]. We will see that all these definitions (3.5)-(3.6) have limits and are equivalent. But, in analogy with the quantum case, we expect that this is no longer true in the many-particle classical time-delay problem.

In order to consider this limit, we first want to obtain an equivalent form for $\tau(f_{in},\Sigma)$ in which the exact state $\Psi(t)$ is replaced by the asymptotic states $\Psi_{in}(t)$ and $\Psi_{out}(t)$. The following result, similar to proposition 4 of Ref. 1 , allows this simplification:

Let f_{in}, $\Psi(t)$, $\Psi_{in}(t)$ and $\Psi_{out}(t)$ be defined as above. Set

$$\Delta^{+}(f_{in},\Sigma) = \int_{0}^{\infty} dt[(\Psi(t),P(\Sigma)\Psi(t))-(\Psi_{out}(t),P(\Sigma)\Psi_{out}(t))] \;,$$

$$\Delta^{-}(f_{in},\Sigma) = \int_{-\infty}^{0} dt[(\Psi(t),P(\Sigma)\Psi(t))-(\Psi_{in}(t),P(\Sigma)\Psi_{in}(t))]; \tag{3.7}$$

then we have that [1,3]

$$\lim_{\Sigma\to R^3} \Delta^{\pm}(f_{in},\Sigma) = 0 \;. \tag{3.8}$$

As a consequence, the symmetric definition of time delay (3.6) can be written as

$$\tau(f_{in},\Sigma) = \tau^{+}(f_{in},\Sigma) + \tau^{-}(f_{in},\Sigma) + \Delta^{+}(f_{in},\Sigma) + \Delta^{-}(f_{in},\Sigma) \ , \tag{3.9}$$

where

$$\tau^{\pm}(f_{in},\Sigma) = \frac{1}{2}\int_{0}^{\pm\infty} dt[(\Psi_{out}(t),P(\Sigma)\Psi_{out}(t))-(\Psi_{in}(t),P(\Sigma)\Psi_{in}(t))] \ . \tag{3.10}$$

In the form (3.9) the exact states of the system only appear in the $\Delta^{\pm}$ and these terms vanish in the limit $\Sigma \to R^3$.

Finally, we can easily introduce the S-matrix in eq.(3.10) by employing the relations (3.1)-(3.3) and the intertwining property (2.11). We obtain

$$\tau^{\pm}(f_{in},\Sigma) = \frac{1}{2} \int_{0}^{\pm\infty} dt(e^{-L_o t} f_{in}, K(\Sigma)\ e^{-L_o t} f_{in}) \ , \tag{3.11}$$

where

$$K(\Sigma) = S^{*}P(\Sigma)S - P(\Sigma) \quad . \tag{3.12}$$

From now on we omit the subscript "in".

4. RELATION BETWEEN TIME DELAY AND THE SCATTERING OPERATOR

We now complete the study of the connection between classical time delay and the classical S-operator by calculating the limit $\Sigma \to R^3$ of $\tau(f,\Sigma)$. Although the method we use resembles the one employed in the quantum mechanical case [5], we will see that the details are very different.

Writing out the scalar product in eq.(3.11) or (3.10) we get e.g. for the second term

$$- \frac{1}{2} \int_0^{\pm\infty} dt \int d\vec{r}\, d\vec{p} \,(e^{-L_o t} f)^* (\vec{r},\vec{p}) \,(P(\Sigma) e^{-L_o t} f)(\vec{r},\vec{p})$$

$$= - \frac{1}{2} \int_0^{\pm\infty} dt \int d\vec{r}\, d\vec{p}\; f^*(\vec{r} + \frac{\vec{p}}{\mu} t,\vec{p}) \chi_\Sigma(\vec{r}) f(\vec{r} + \frac{\vec{p}}{\mu} t,\vec{p}) \quad , \tag{4.1}$$

with $\chi_\Sigma(\vec{r})$ the characteristic function for the region Σ. The difficulty here is that the time dependence is implicit and not a multiplying phase factor like in quantum mechanics. To solve this difficulty we will consider a Fourier transformation with respect to the coordinate space part of phase space.

We assume that $f(\vec{r},\vec{p})$ belongs to the Schwarz functions with compact support $S(\Gamma) \subset L^2(\Gamma)$. We then define

$$(Ff)(\vec{\alpha},\vec{p}) \equiv \tilde{f}(\vec{\alpha},\vec{p}) = \frac{1}{(2\pi)^{3/2}} \int d\vec{r}\; e^{i\vec{\alpha}.\vec{r}} f(\vec{r},\vec{p}) \quad . \tag{4.2}$$

So we have that

$$(Fe^{-L_o t} f)(\vec{\alpha},\vec{p}) = e^{-i \frac{\vec{\alpha}.\vec{p}}{\mu} t} \tilde{f}(\vec{\alpha},\vec{p}) \quad . \tag{4.3}$$

From now on we drop the $\sim$ since the rest of our analysis takes place entirely in this partly Fourier transformed phase space indicated by $(\vec{\alpha},\vec{p})$. Because this Fourier transform leaves the scalar product in eq.(3.11) invariant, we can write out the right handside of this equation as

$$\frac{1}{2} \int_0^{\pm\infty} dt\, d\vec{\alpha}'\, d\vec{p}'\; e^{i \frac{\vec{\alpha}'.\vec{p}'}{\mu}} f^*(\vec{\alpha}',\vec{p}') \int d\vec{\alpha}\, d\vec{p}\; K(\Sigma;\vec{\alpha}',\vec{p}';\vec{\alpha},\vec{p}) \cdot$$

$$e^{-i \frac{\vec{\alpha}.\vec{p}}{\mu} t} f(\vec{\alpha},\vec{p}) \quad , \tag{4.4}$$

where

$$K(\Sigma;\vec{\alpha}',\vec{p}';\vec{\alpha},\vec{p}) = (S^* \mathbf{P}(\Sigma) S)\,(\vec{\alpha}',\vec{p}';\vec{\alpha},\vec{p}) - P(\Sigma;\vec{\alpha}'-\vec{\alpha})\,\delta(\vec{p}'-\vec{p}), \tag{4.5}$$

with $P(\Sigma,\vec{\alpha}'-\vec{\alpha})$ the Fourier transform of $\chi_\Sigma(\vec{r})$. These are the forms we are going to work with.

First, we want to make the following important observation. Property (2.11) states that the S-operator intertwines with L_o. If we write out the explicit form of this property in the $\vec{\alpha},\vec{p}$ variables we find that the kernel of S, $S(\vec{\alpha}',\vec{p}';\vec{\alpha},\vec{p}) \sim \delta(\vec{\alpha}'.\vec{p}'-\vec{\alpha}.\vec{p})$. We explicitly want to take out that δ-function by defining a classically reduced s^R matrix by the kernel relation

$$S(\vec{\alpha}',\vec{p}';\vec{\alpha},\vec{p})=\delta(\vec{\alpha}'.\vec{p}'-\vec{\alpha}.\vec{p})\frac{(\vec{\alpha}.\hat{p})^3}{(\vec{\alpha}'.\vec{p}')^2}\, s^R_{\vec{\alpha}'.\vec{p}'}(\vec{\alpha}',\vec{p}';\vec{\alpha},\frac{\vec{\alpha}'.\vec{p}'}{\vec{\alpha}.\hat{p}}\hat{p}). \tag{4.6}$$

The s^R kernel on the right handside of eq.(4.6) represents an operator that is the classical equivalent of the quantum mechanical reduced on-energy-shell s operator. The factors in front of it are chosen such that the operator relations S obeys (e.q. unitarity) are also valid for s^R in exactly the same form.

With this information we continue our derivation of the time-delay relation. We first introduce the Abel limit in eq.(4.4) and carry out the t-integration. The result is [3]

$$\tau(f,\Sigma) = i\mu \int d\vec{\alpha}'\, d\vec{p}'\, d\vec{\alpha}\, d\vec{p}\, f^*(\vec{\alpha}',\vec{p}')\, \frac{K(\Sigma;\vec{\alpha}'\vec{p}';\vec{\alpha},\vec{p})}{\vec{\alpha}'.\vec{p}' - \vec{\alpha}.\vec{p}}\, f(\vec{\alpha},\vec{p})$$

$$+ \Delta^+(f,\Sigma) + \Delta^-(f,\Sigma)\ , \tag{4.7}$$

where the integral is defined as a principal-value integral.

We next have to calculate this principal-value integral (4.7) in the limit $\Sigma \to R^3$. We take the region Σ to be a sphere of radius R and use the Fourier transform properties of this sphere. After some tedious manipulations [3] we arrive at

$$\tau(f,R^3) = -i\mu \int d\vec{\alpha}'\, d\vec{p}'\, d\vec{\alpha}\, d\vec{p}\; \delta(\vec{\alpha}'.\vec{p}'-\vec{\alpha}.\vec{p})\; \frac{(\vec{\alpha}.\hat{p})^3}{(\vec{\alpha}'.\vec{p}')^2}\, f^*(\vec{\alpha}',\vec{p}')$$

$$\int d\vec{\alpha}_1\, d\hat{p}_1 [s^{R^*}(\vec{\alpha}',\vec{p}';\vec{\alpha}_1, \frac{\vec{\alpha}'.\vec{p}'}{\vec{\alpha}_1.\hat{p}_1}\hat{p}_1)\, \frac{d}{d\vec{\alpha}'.\vec{p}'}$$

$$s^R(\vec{\alpha}_1, \frac{\vec{\alpha}'.\vec{p}'}{\vec{\alpha}_1.\hat{p}_1}\hat{p}_1;\, \vec{\alpha}, \frac{\vec{\alpha}'.\vec{p}'}{\vec{\alpha}.\hat{p}}\hat{p})]\, f(\vec{\alpha},\vec{p}) \quad . \qquad (4.8)$$

From this result, we see that $\tau(f,R^3)$ contains the same δ-function as the S-matrix (compare eq.(4.6)). In fact, if we look back at the definition of $\tau(f,\Sigma)$ for finite Σ, given by eqs. (3.5) and (3.6), and write out the time-integrand of these equations explicitly, it is clear that the factor $\exp[\frac{i}{\mu}(\vec{\alpha}'.\vec{p}' - \vec{\alpha}.\vec{p})t]$ also leads to this δ-function. Therefore we are justified in defining a reduced time delay operator q^R by

$$\tau(f,\Sigma) = \int d\vec{\alpha}'\, d\vec{p}'\, d\vec{\alpha}\, d\vec{p}\;\; \delta(\vec{\alpha}'.\vec{p}'-\vec{\alpha}.\vec{p})\; \frac{(\vec{\alpha}.\hat{p})^3}{(\vec{\alpha}'.\vec{p}')^2}\, f^*(\vec{\alpha}',\vec{p}')$$

$$q^R_{\vec{\alpha}'.\vec{p}'}(\Sigma;\vec{\alpha}',\vec{p}';\vec{\alpha}, \frac{\vec{\alpha}'.\vec{p}'}{\vec{\alpha}.\hat{p}}\hat{p})\, f(\vec{\alpha},\vec{p}) \quad . \qquad (4.9)$$

Comparing this equation with eq.(4.8) we see that the connection between time delay and the S-operator can be expressed in terms of the kernels q^R and s^R as follows:

$$q^R_{\vec{\alpha}'.\vec{p}'}(\vec{\alpha}'.\vec{p}';\vec{\alpha}, \frac{\vec{\alpha}'.\vec{p}'}{\vec{\alpha}.\hat{p}}\hat{p}) = -i\int d\vec{\alpha}_1 d\hat{p}_1\, s^{R^*}_{\vec{\alpha}'.\vec{p}'}(\vec{\alpha}',\vec{p}';\vec{\alpha}_1, \frac{\vec{\alpha}'.\vec{p}'}{\vec{\alpha}_1.\hat{p}_1}\hat{p}_1)$$

$$\frac{d}{d\frac{\vec{\alpha}'.\vec{p}'}{\mu}}\, s^R_{\vec{\alpha}'.\vec{p}'}(\vec{\alpha}_1, \frac{\vec{\alpha}'.\vec{p}'}{\vec{\alpha}_1.\hat{p}_1}\hat{p}_1;\vec{\alpha}, \frac{\vec{\alpha}'.\vec{p}'}{\vec{\alpha}.\hat{p}}\hat{p}) \; , \qquad (4.10)$$

where we have written again the $\vec{\alpha}.\vec{p}$ subscript on the s^R-

matrices. On the basis of this kernel relation (4.10), we can also state that the following operator relation is valid :

$$\tilde{q}^{R}_{\vec{\alpha}.\vec{p}} = \tilde{s}^{R*}_{\vec{\alpha}.\vec{p}} \frac{d}{di\frac{\vec{\alpha}.\vec{p}}{\mu}} \tilde{s}^{R}_{\vec{\alpha}.\vec{p}} , \tag{4.11}$$

where we have introduced again the $\sim$.

Doing an inverse Fourier transformation with respect to $\vec{\alpha}$ (recall eqs. (4.2)-(4.3)), the results (4.8)-(4.11) can finally be written in the form

$$\lim_{\Sigma\to R^3} \tau(f,\Sigma) = (f, S^* \frac{dS}{dL_o} f) . \tag{4.12}$$

The final result (4.12) expressing classical time delay in terms of the classical S-operator is the analogue of the corresponding quantum relation [1]

$$\lim_{\Sigma\to R^3} \tau(f,\Sigma) = (f, S^* \frac{dS}{diH_o} f) . \tag{4.13}$$

5. DISCUSSION

In the foregoing we have derived the relation between classical time delay and the classical S-operator in a two-particle system with short-range interactions, within the context of the Hilbert-space approach to classical mechanics. The method we have used is based upon the explicit properties of the projection operator $P(\Sigma)$ onto a finite region Σ, taken to be a sphere of radius R. It would certainly be possible to establish the final result (4.12) on the basis of methods analogous to other rigorous quantum mechanical treatments [1]. However, the big advantage of the method we have used here is that its quantummechanical analogue is the only one up to now that has been shown to work in the quantum mechanical many-particle scattering problem [5]. So we are confident that with this method the same type of results can be obtained for classical many-particle systems.

The result derived here and some of the results discussed in previous lectures [1,2] tell us that the time delay relation is valid for any (scattering) system irrespective of the fact that the underlying dynamics is classical or quantum mechanical. For two-body quantum [1] and classical [2] scattering in a specific angular momentum, we know that time delay is proportional to the energy derivative of the phase shift. Thus we can state that time delay theory provides a method of defining a universal phase-shift-like functional that is a characteristic of the quantum mechanical and classical scattering system. This is stressed again by the fact that one can derive a classical analogue to the quantum mechanical Levinson's theorem [2,6].

REFERENCES

1. Ph. Martin, lecture notes in these proceedings.
2. W. Thirring, lecture notes in these proceedings.
3. D. Bollé and J. D'Hondt, to be published in Journ. Phys. A, 1981.
4. W. Hunziker, Comm. Math. Phys. 8 (1968) 282; in Lectures in Theoretical Physics, ed. by A.O. Barut and W.E. Brittin (Gordon and Breach, New York, vol. X-A, 1968), p.1; in Scattering in Mathematical Physics, ed. by J.A. LaVita and J.P.Marchand (Reidel, Boston, 1974), p.79.
5. D. Bollé and T.A. Osborn, J. Math. Phys. 20 (1979) 1121.
6. D. Bollé and T.A. Osborn, to be published in J. Math. Phys., 1981.

Acta Physica Austriaca, Suppl. XXIII, 599–603 (1981)

A FINITE-ENERGY SU(3) SOLUTION WHICH DOES NOT SATISFY THE BOGOMOLNY EQUATIONS+

by

J. BURZLAFF
Fachbereich Physik, Univ. Kaiserslautern
Kaiserslautern, Germany

We choose an SU(3) ansatz which corresponds to the 't Hooft-Polyakov ansatz, however, involves a 5-plet of an SU(2) subalgebra instead of the 3-plet and only admits the trivial solution to the Bogomolny equations. It is proven that within this ansatz a non-trivial static spherically symmetric finite-energy solution with magnetic quantum numbers (0,2) exists.

All monopole solutions [1] which are known up to now satisfy the Bogomolny equations. In this note, I present a second finite-energy solution, besides the 't Hooft-Polyakov monopole [2], for non-vanishing Higgs field self-interaction. The existence proof for this solution can be extended to the Prasad-Sommerfield limit and yields the first finite-energy solution which does not satisfy the Bogomolny equations.

+Seminar given at XX. Internationale Universitätswochen für Kernphysik, Schladming, Austria, February 17-26, 1981.

The model we are going to study is the SU(3) Yang-Mills-Higgs model in Minkowski space-time given by the Lagrangian

$$L = -\frac{1}{4} G^a_{\mu\nu} G^{\mu\nu}_a + \frac{1}{2} D_\mu \phi^a D^\mu \phi_a - V(\phi) , \tag{1}$$

$\mu,\nu,\ldots = 0,1,2,3, \quad a,b,\ldots = 1,\ldots,8$.

$G_{\mu\nu}$ and $D_\mu\phi$ are the field strength and the covariant derivative, respectively, with Gell-Mann matrices λ_a and potentials W_μ. The Higgs field self-interaction is taken to be

$$V(\phi) = p(1 - 2 \operatorname{tr} \phi^2)^2 + q(\phi^a + \sqrt{3} d_{abc} \phi^b \phi^c)^2 . \tag{2}$$

To find a solution we choose a simple ansatz [3]: We set $W_o = 0$ and

$$\phi = A(r) K , \tag{3a}$$

$$W_i = i[E,\partial_i E] + iB(r)[K,\partial_i \mathbf{E}] , \tag{3b}$$

where $\vec{E} = (\lambda_7, -\lambda_5, \lambda_2)$ are the generators of the maximal SU(2) subalgebra of SU(3), and E and K are the scalars

$$\mathbf{E} := \hat{x}_i E_i , \tag{4a}$$

$$K := \hat{x}_i \hat{x}_j K_{ij} := \hat{x}_i \hat{x}_j (E_{(i}E_{j)} - \frac{2}{3} \delta_{ij}) . \tag{4b}$$

This ansatz differs from **the** 't Hooft-Polyakov ansatz only in the fact that it involves the 5-plet K_{ij} instead of the 3-plet E_i. Its first interesting property is that for finite energy, it cannot satisfy the Bogomolny equations. In fact, the potential

$$V(\phi) = p\left(\frac{4}{3} A^2 - 1\right)^2 + \frac{4}{3} q A^2 \left(\frac{2}{\sqrt{3}} A - 1\right)^2 \tag{5}$$

enforces $A \xrightarrow[r\to\infty]{} \sqrt{3}/2$ for finite energy. On the other hand,

$$B_i = \frac{1}{r^2}\,(B^2(r) - 1)\,\hat{x}_i\,E + B'(r)\,\partial_i\,K\,, \tag{6a}$$

$$D_i\phi = A'(r)\,\hat{x}_i\,K - A(r)\,B(r)\,\partial_i\,E \tag{6b}$$

hold, and the Bogomolny equations can only be satisfied by $B_i = D_i\phi = 0$, i.e., $B^2 = 1$ and $A = 0$.

Whereas left and right hand side of the Bogomolny equations involve different E and K terms and do not match, our ansatz has the second interesting property that the equations of motion involve the same E and K terms and reduce to

$$(r^2A') = 6\,AB^2 + 4\,pr^2A\,(\tfrac{4}{3}\,A^2 - 1)$$
$$+ \frac{16}{3}\,q\,r^2\,A(A - \tfrac{\sqrt{3}}{2})\,(A - \tfrac{\sqrt{3}}{4})\,, \tag{7a}$$

$$r^2\,B'' = B\,(B^2 + r^2\,A^2 - 1)\,. \tag{7b}$$

Except for a factor 3 and the different potential, these are the 't Hooft-Polyakov equations of motion.

Furthermore, the equations of motion (7) are Euler-Lagrange equations of the energy functional

$$E(A,B) = 4\pi \int_0^\infty dr\,\{\tfrac{2}{3}\,r^2\,A'^2 + 4\,B'^1 + \frac{2}{r^2}\,(B^2 - 1)^2$$
$$+ 4A^2B^2 + pr^2(\tfrac{4}{3}A^2 - 1)^2 + \tfrac{4}{3}qr^2A^2(\tfrac{2}{\sqrt{3}}A - 1)^2\}. \tag{8}$$

This enables us to prove the existence of a finite-energy solution using the technique of Tyupkin et al. [4].

This proof can be extended to the Prasad-Sommerfield limit $(p,q) = 0$ [5]. In the Prasad-Sommerfield limit we restrict our attention to sequences $A_n(r)$ with $A_n(r) \xrightarrow[r\to\infty]{} \sqrt{3}/2$, and show with the same technique that a weak limit (A_o,B_o) exists. Because furthermore

$$|A_n(r) - \frac{\sqrt{3}}{2}| = |\int_r^\infty A_n'(r)\, dr|$$

$$\leq (\int_r^\infty A_n'^2\, r^2\, dr \cdot \int_r^\infty \frac{1}{r^2}\, dr)^{1/2} \leq \frac{D}{\sqrt{r}}\ , \qquad (9)$$

with D independent of n, holds since $\{E(A_n,B_n)\}$ is bounded, we obtain $A_o(r) \xrightarrow[r\to\infty]{} \sqrt{3}/2$. I.e., E obtains its minimum in the space of functions with this boundary condition.

To show for all cases $(p,q) \geq 0$ that A_o and B_o satisfy the Euler-Lagrange equations (7), we vary A_o and B_o by adding functions with compact support. This completes the existence proof of our solution. With the asymptotic behaviour studied in Ref.[3], one can show that this solution is a monopole with magnetic quantum number 0, isomagnetic quantum number 2, and regular ϕ, $D_i\phi$, B_i and energy density. Recently, after the talk on which this work is based, Taubes [6] pointed out that the gauge transformation $U = \exp\,(i\,\frac{\pi}{2}\,(E + K))$ even puts our ansatz (3) into the form

$$\phi = AK,\ W_i = -i\,(B-1)\,[E,\partial_i E] \qquad (10)$$

which is manifestly smooth.

In Ref.[7], for Euclidean Yang-Mills theory a non-static ansatz was chosen which has similar properties as the one we studied before. Again, the first order equations involve different E and K terms and admit only the trivial solution. The equations of motion, however, reduce to equations which except for a factor 3 are identical to Witten's equations. Because an existence proof could not yet be given it is not known whether a non-self-dual finite-action solution exists within this ansatz.

REFERENCES

1. For a recent review see L. O'Raifeartaigh and S. Rouhani's lecture at this 1981 Schladming Winter School.
2. G. 't Hooft, Nucl. Phys. B79 (1974) 276; A.M.Polyakov, Zh. Eksp. Teor. Fiz. Pis'ma Red. 20 (1974) 430 (JETP Lett. 20 (1974) 197).
3. J. Burzlaff, Phys. Rev. D23 (1981) 1329.
4. Yu. Tyupkin, V.A. Fateev, and A.S. Shvarts, Theor. Math. Phys. 26 (1976) 270.
5. D. Maison, private communication.
6. C. Taubes, private communication.
7. J. Burzlaff, Univ. Kaiserslautern Preprint, 1980, Phys. Rev. D, in press.

Acta Physica Austriaca, Suppl. XXIII, 605–611 (1981)

$\widetilde{SO}_o(3.2)$-INVARIANT SCATTERING OF DIRAC SINGLETONS+

by

L. CASTELL and W. HEIDENREICH
Max-Planck-Institut, Starnberg, Germany

1. DIRAC SINGLETONS AND THEIR TENSOR PRODUCTS

The universal covering $\widetilde{SO}_o(3.2)$ of the (3.2) de Sitter group can be contracted to the Poincaré group. So although a de Sitter-invariant physical theory is closely related to special relativity, it may have more structure than the usual Poincaré-invariant physics. An example are Dirac-singletons [1] Di and Rac, which vanish in the contraction-limit; they have only two degrees of freedom - as compared to particles which have three - and can thus be localised only on a line, not in a point in space-time. Their weights (states in the vector space) in an energy-angular momentum diagram are for the spin O Rac

$$\mathrm{Rac} = \sum_{i=0}^{\infty} (1/2 + i, i) \quad ,$$

and for the spin 1/2 Di (1)

$$\mathrm{Di} = \sum_{i=0}^{\infty} (1+i,\ 1/2+i)$$

+Seminar given at XX. Internationale Universitätswochen für Kernphysik, Schladming, Austria, February 17-26, 1981.

(see Fig. 1). There are no higher spin Dirac singletons.

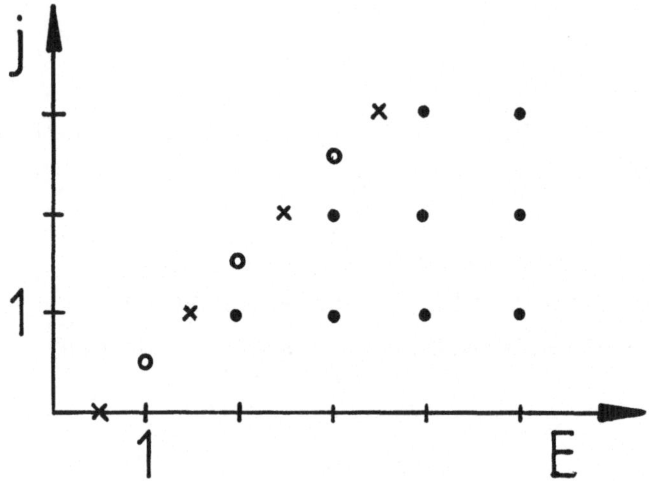

Fig. 1. The weights of the Rac(x), Di(O) and the Photon (●).

Mass zero particles with spin 0 are represented by

$$D(1,0) = \sum_{i=0}^{\infty} \sum_{n=0}^{\infty} (1 + i + 2n, i) \quad , \tag{2}$$

and those with spin s = 1/2, 1, 3/2,... have [2]

$$D(s+1,s) = \sum_{i=0}^{\infty} \sum_{n=0}^{\infty} (s+1+i+n,\ s+i) \quad . \tag{3}$$

We use the lowest weight to denote the whole representation. The independent generalised momenta for singletons are e.g. the angular momentum j and its third component j_3, for the massless representations we need the energy E in addition.

The tensor product of two Dirac singletons can be reduced graphically [3]. Using the Clebsch-Gordan series of the angular momentum,

$$(j) \otimes (j') = \sum_{J=|j-j'|}^{j+j'} (J)$$

we obtain

$$\mathrm{Di} \otimes \mathrm{Rac} = \sum_{j=0}^{\infty} (1+j,\tfrac{1}{2}+j) \otimes \sum_{j'=0}^{\infty} (\tfrac{1}{2}+j',j') =$$

$$= \sum_{j=0}^{\infty} \sum_{j'=0}^{\infty} \sum_{J=|j-j'|}^{j+j'} (\tfrac{3}{2}+j+j',\tfrac{1}{2}+J) \ . \qquad (4)$$

The first weights are shown in Fig. 2a.

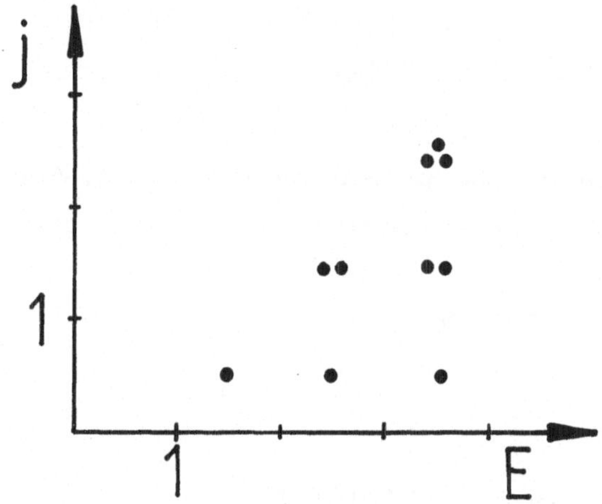

Fig. 2a. Weights of Di ⊗ Rac.

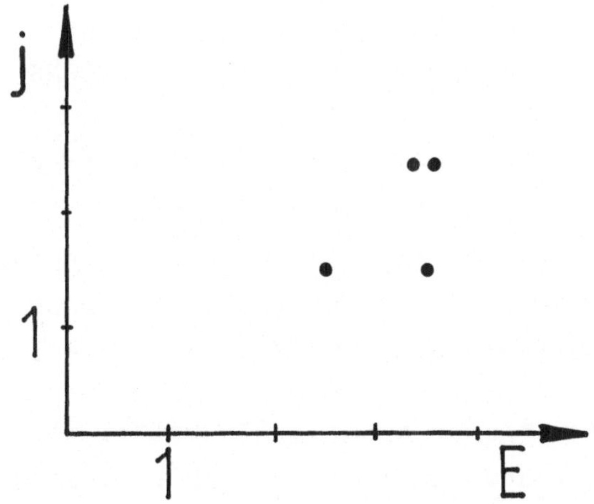

Fig. 2b. Di ⊗ Rac minus D(3/2,1/2).

The lowest weight in this fully reducible representation is

(3/2,1/2). Therefore the representation D(3/2,1/2) must occur in the Clebsch-Gordan-series of Di $\otimes$ Rac. If we remove all the weights of D(3/2,1/2) from Fig. 2a, we obtain Fig. 2b. Its lowest weight (5/2, 3/2) belongs to D(5/2,3/2). Repeating this procedure we get uniquely the first terms of the reduction of Di $\otimes$ Rac:

$$\mathrm{Di} \otimes \mathrm{Rac} = D(3/2,1/2) \oplus D(5/2,3/2) \oplus \ldots =$$

$$= \sum_{i=0}^{\infty} D(3/2+i,\ 1/2+i)\ . \tag{5}$$

The full formula can be proved by induction. The other tensor products of two Dirac singletons are given by [4]

$$\mathrm{Rac} \otimes \mathrm{Rac} = \sum_{i=0}^{\infty} D(1+i,i)$$

and

$$\mathrm{Di} \otimes \mathrm{Di} = D(2,0) \oplus \sum_{i=1}^{\infty} D(1+i,i)\ . \tag{6}$$

2. THE SCATTERING OF DIRAC-SINGLETONS

The space of states of n Dirac-singletons is, according to the rules of quantum mechanics, the tensor product space of the one-singleton-states. In the case (Di,Rac) = Di $\otimes$ Rac, we have states in the massless half-integer spin representations D(3/2 + i, 1/2 + i).

If the S-matrix is a $\tilde{S}O_o(3.2)$-scalar, it can only be nonzero for $|\mathrm{in}\rangle$ and $\langle\mathrm{out}|$ states which belong to the same representation. This strongly restricts the open channels of Dirac-singleton scattering. A Di and Rac can obviously scatter elastically, a process which is written with reduced matrix elements like

$$\sum_{j=0}^{\infty}\sum_{j'=0}^{\infty} \langle Di,Rac\|D(j+\tfrac{3}{2},j+\tfrac{1}{2})\rangle\langle D(j+\tfrac{3}{2},j+\tfrac{1}{2})\|S\|D(j'+\tfrac{3}{2},j'+\tfrac{1}{2})\rangle$$

$$\langle D(j'+\tfrac{3}{2},j'+\tfrac{1}{2})\|Di,Rac\rangle \ , \qquad (7)$$

$$\langle D(j+\tfrac{3}{2},j+\tfrac{1}{2})\|S\|D(j'+\tfrac{3}{2},j'+\tfrac{1}{2})\rangle = \delta_{jj'}\langle S_{(j+\frac{3}{2},j+\frac{1}{2})}\rangle = \delta_{jj'}\exp 2i\eta_j \ .$$

There is no other open channel into Dirac-singletons, as (2Di) or (2Rac) does not contain half-integer spin representations and the product of three or more singletons has no weights (j+1,j), and therefore cannot contain any massless representations.

With the same arguments we get

$$(2Rac) \rightarrow \begin{matrix}(2Rac)\\ (2Di),\end{matrix} \qquad \text{and} \quad (2Di) \rightarrow \begin{matrix}(2Di)\\ (2Rac)\\ (4Rac)\end{matrix} \ . \qquad (8)$$

From exact group-theoretical arguments we get the result that two Rac can scatter only (quasi)-elastically. The only inelastic channel of two-singleton scattering is

$$\langle 2Di\|D(2,0)\rangle \quad \langle S_{(2,0)}\rangle \quad \langle D(2,0)\|\,4Rac\rangle \quad . \qquad (9)$$

It is possible as the lowest weight of the (Rac) [4] product is (2,0).

Similar to the naive quark model, equations (5) and (6) suggest a constituent picture: massless particles with integer spin, like the photon, consist either of two Di or two Rac, those with half-integer spin like the neutrino of a Di and a Rac. Then the scattering of a Dirac singleton and a massless particle should be described on the constituent level: the free singleton reacts with one of the singletons which form the massless particle. The resulting singletons can again form particles.

A Rac and a neutrino can react not only elastically, but there is an additional quasielastic reaction

$$\text{Rac} \otimes \text{Neutrino} \rightarrow 3\ \text{Di} \rightarrow \text{Di} \otimes \text{Photon}\ . \tag{10}$$

The corresponding process of a Di and a neutrino is

$$\text{Di} \otimes \text{Neutrino} \rightarrow 3\ \text{Rac} \rightarrow \text{Rac} \otimes \text{Photon}\ . \tag{11}$$

There is an inelastic channel as well:

$$\text{Di} \otimes \text{Neutrino} \rightarrow 5\ \text{Rac} \rightarrow \text{Rac} \otimes 2\ \text{Photons}\ . \tag{12}$$

In this connection the photon can be replaced by any other integer spin, the neutrino by any other half-integer spin massless particle. Only these reactions between singletons and neutrinos with one single scattering are allowed by group theory.

The open channels for singleton-photon scattering are

$$\text{Rac} \otimes \text{Photon} \rightarrow \begin{array}{l} \text{Rac} \otimes \text{Photon} \\ \text{Di} \otimes \text{Neutrino}\ , \end{array} \tag{13}$$

$$\text{Di} \otimes \text{Photon} \rightarrow \begin{array}{l} \text{Di} \otimes \text{Photon} \\ \text{Rac} \otimes \text{Neutrino} \\ \text{Di} \otimes 2\ \text{Photons} \\ \text{Rac} \otimes \text{Photon} \otimes \text{Neutrino}\ . \end{array} \tag{14}$$

This simple constituent model does not allow some $\widetilde{SO}(3.2)$ invariant channels. They are given by Di $\otimes$ Neutrino $\otimes$ Photon, and Rac $\otimes$ 2 Neutrinos for reaction (12), and Di $\otimes$ 2 Neutrinos for (14).

An extension of the constituent model to massive particles is possible, as massive representations of $\tilde{SO}_0(3.2)$ lie in the tensor product of four Dirac singletons. Two massless particles in Eq. (12,14) could be replaced by one massive one. The scattering of a massive particle and a Dirac singleton can at most have one additional massless particle in the outgoing channel.

REFERENCES

1. E. Majorana, Nuovo Cimento 9 (1932) 335.
   P.A.M. Dirac, J. Math. Phys. 4 (1963) 901.
2. L. Castell, W. Heidenreich, "SO(3.2) invariant scattering and Dirac singletons", to appear Phys.Rev.D.
3. W. Heidenreich, "Tensor products of positive energy representations of $\tilde{SO}(3.2)$ and $\tilde{SO}(4.2)$", to appear J. Math. Phys..
4. M. Flato, C. Fronsdal, Lett. Math. Phys. 2 (1978) 421.

Acta Physica Austriaca, Suppl. XXIII, 613–625 (1981)

GENERATING MULTIMONOPOLES BY SOLITON THEORETIC METHOD[+x]

by

P. FORGÁCS
Central Research Inst. for Physics
H-1525 Budapest 114, P.O.B. 49, Hungary

and

Z. HORVÁTH and L. PALLA
Institute for Theoretical Physics
Roland Eötvös University
H-1088 Budapest, Hungary

ABSTRACT

The explicit form of static, axially and mirror symmetric monopoles of arbitrary topological charge is derived using Bäcklund transformations.

[+]Seminar given at the XX. Internationale Universitätswochen für Kernphysik, Schladming, Austria, February 17-26, 1981.

[x]Presented by L. Palla.

In the last year great progress was made in the theory of magnetic monopoles [1]. In addition to proving the existence [2] of static, spatially separated multi-monopoles in an SU(2) Yang-Mills-Higgs theory with vanishing Higgs potential two independent methods were proposed by Ward [3] and by the present authors [4] to construct axially symmetric (and therefore superimposed) multimonopoles.

We have shown [5] that the Bogomolny equations for the axially and mirror symmetric ansatz constructed by Manton [6] reduce to the Ernst equation [7] of general relativity and emphasized that the soliton generating techniques worked out for this equation (Bäcklund transformations [8,9] and the method of inverse scattering [10]) may be useful in the search for multimonopoles. We outlined a systematic method the use of which was illustrated by generating the one [5] (IMP) and two [4] (2MP) monopole solutions by applying a single and a double Bäcklund transformation (B) to the "vacuum" (B is a slightly modified Harrison transformation [4]). Our aim in this paper is to give the explicit formulae for monopoles of topological charge n using the n-times iterated B transformation derived in ref.[11]. These solutions are axially and mirror symmetric by construction and correspond to n superimposed monopoles located at the origin. They depend on exactly five parameters three of which correspond to the position and the two others to the direction of the symmetry axis.

The fact that multimonopoles are obtained by iterated Bäcklund transformations emphasizes their multisoliton nature, giving one more item to the list of similar properties between the Sine-Gordon and monopole theories [12], as well as illustrating how stimulating the cross-fertilization between different branches of theoretical physics may be.

The ansatz we start with is

$$\Phi^a = (0, \Phi_1, \Phi_2) \qquad A^a_\phi = -(0, \eta_1, \eta_2)$$

$$A^a_z = -(W_1, 0, 0) \qquad A^a_\rho = -(W_2, 0, 0) \qquad (1)$$

where ρ, z, ϕ are the usual polar coordinates and Φ_i, η_i, W_i are functions of ρ, z only. The Bogomolny equations for (1) can be recast in the following form [5]:

$$\text{Re}\, \varepsilon\, \Delta\, \varepsilon - (\nabla\varepsilon)^2 = 0 \qquad (2)$$

where $\varepsilon = f + i\psi$, $\Delta = \partial^2_\rho + \rho^{-1}\partial_\rho + \partial^2_z$. Eq.(2) is just the Ernst equation which describes stationary and axially symmetric spacetimes in vacuum. Φ_i, η_i, W_i are given by ε as

$$\Phi_1 = f^{-1}\psi_{,z} \qquad \eta_1 = -\rho f^{-1}\psi_{,\rho} \qquad W_1 = -f^{-1}\psi_{,z}$$

$$\Phi_2 = f^{-1} f_{,z} \qquad \eta_2 = \rho f^{-1} f_{,\rho} \qquad W_2 = -f^{-1}\psi_{,\rho} \,. \qquad (3)$$

We introduce the following new variables:

$$M_1 = \frac{1}{2} f^{-1} \varepsilon_{,1}\,; \qquad M_2 = \frac{1}{2} f^{-1} \varepsilon^*_{,1}\,; \qquad N_1 = \frac{1}{2} f^{-1} \varepsilon^*_{,2}\,;$$

$$N_2 = \frac{1}{2} f^{-1} \varepsilon_{,2}\,; \qquad \varepsilon_{,1} = \frac{\partial\varepsilon}{\partial\zeta_1}\,, \quad \zeta_1 = \rho+iz; \quad \zeta_2 = \rho-iz \,. \qquad (4)$$

In the rest of this paper we will work with M_i's, N_i's instead of ε. To carry out the Bäcklund transformations (BT's), denoted by B, the only analytic work to be done is to solve a Riccati equation for the pseudopotential $q(\zeta_1,\zeta_2)$:

$$dq = [(M_2-M_1)q + \gamma(w)(M_2-M_1q^2)]d\zeta_1 + [(N_1-N_2)q +$$

$$+ \gamma^{-1}(w)(N_1 - N_2 q^2)]d\zeta_2 \tag{5}$$

where

$$\gamma(w) = \sqrt{\frac{w - i\zeta_2}{w + i\zeta_1}} ,$$

w being an arbitrary constant. Starting with a seed solution ε_o, a single B transformation gives a new solution of eq.(2) as

$$M_1^{(1)} = BM_1^{(0)} = \frac{\gamma+q}{1+\gamma q}(q^{-1} M_2^{(0)} + \gamma(4\rho)^{-1}) ,$$

$$N_2^{(1)} = BN_2^{(0)} = \frac{\gamma+q}{1+\gamma q}(qN_2^{(0)} + (\gamma 4\rho)^{-1}) , \tag{6}$$

and similarly for $N_1^{(1)}$, $M_2^{(1)}$. From these one could determine $\varepsilon^{(1)}$ which is irrelevant for us since M_i's, N_i's are the quantities of our interest. In fact $\Phi^2 = \Phi_1^2 + \Phi_2^2$ which is most important for us is given as

$$\Phi^2 = 4(M_1 - N_2)(N_1 - M_2) . \tag{7}$$

We start with the following seed solution of eq.(2):

$$f_o = e^z , \qquad \psi_o = 0 . \tag{8}$$

(8) is a natural ground state i.e. a Higgs vacuum where $|\Phi| = 1$. Eq.(5) is readily integrated for q in this case,

$$q = -\tanh\left(\frac{R(w)}{2} - \beta\right) \tag{9}$$

where $R(w) = \sqrt{(w-z)^2 + \rho^2}$ and β is the constant of integration.

As it was already indicated in ref.[5] a possible strategy to try to generate multimonopole solutions of eq.(2) is to iterate the B transformations which is possible by purely algebraic means. Then it was demonstrated in [4] that one can generate a doubly charged monopole (2MP) by two consecutive BT's from (8). We now present our results for an arbitrary number of iterations starting with (8) as a seed solution, which enabled us to find monopole solutions of arbitrary (integer) topological charge. We found a most remarkable formula for $|\Phi|$ which is totally explicit:

$$|\Phi| = \frac{1}{2}\left| -i\left(\frac{D_2^{(2k+\varepsilon)}}{D_1^{(2k+\varepsilon)}} + \frac{D_1^{(2k+\varepsilon)}}{D_2^{(2k+\varepsilon)}}\right) + \frac{1}{\rho}\left(\frac{D_3^{(2k+\varepsilon)}}{D_1^{(2k+\varepsilon)}} - \frac{D_4^{(2k+\varepsilon)}}{D_2^{(2k+\varepsilon)}}\right)\right| \tag{10a}$$

where $D_a^{(2k+\varepsilon)}$ ($a = 1,\ldots 4$; ε is either 0 or 1) are $(2k+\varepsilon) \times (2k+\varepsilon)$ determinants and they are completely characterized by their i-th row:

if $\varepsilon = 0$,

$$\begin{aligned}
D_1^{(2k)} &= |q_i,\ \gamma_i,\ \gamma_i^2 q_i,\ \gamma_i^3,\ \gamma_i^4 q_i, \ldots,\ \gamma_i^{2k-1}| \\
D_2^{(2k)} &= |1,\ \gamma_i q_i,\ \gamma_i^2,\ \gamma_i^3 q_i,\ \gamma_i^4, \ldots,\ \gamma_i^{2k-1} q_i| \\
D_3^{(2k)} &= |1,\ \gamma_i q_i,\ \gamma_i^2,\ \gamma_i^3 q_i,\ \gamma_i^4, \ldots,\ \gamma_i^{2(k-1)},\ \gamma_i^{2k}| \\
D_4^{(2k)} &= |\gamma_i^{-1},\ \gamma_i,\ \gamma_i^2 q_i,\ \gamma_i^3,\ \gamma_i^4 q_i, \ldots,\ \gamma_i^{2k-1}|\ ;
\end{aligned} \tag{10b}$$

when $\varepsilon = 1$, the $D_a^{(2k+1)}$'s are given as

$$D_1^{(2k+1)} = |q_i, \gamma_i, \gamma_i^2 q_i, \gamma_i^3, \gamma_i^4 q_i, \ldots, \gamma_i^{2k} q_i|$$

$$D_2^{(2k+1)} = |1, \gamma_i q_i, \gamma_i^2, \gamma_i^3 q_i, \gamma_i^4, \ldots, \gamma_i^{2k}|$$

$$D_3^{(2k+1)} = |1, \gamma_i q_i, \gamma_i^2, \gamma_i^3 q_i, \gamma_i^4, \ldots, \gamma_i^{2k-1} q_i, \gamma_i^{2k+1} q_i|$$

$$D_4^{(2k+1)} = |\gamma_i^{-1}, \gamma_i, \gamma_i^2 q_i, \gamma_i^3, \gamma_i^4 q_i, \ldots, \gamma_i^{2k} q_i| \qquad (10c)$$

and $i = 1, \ldots, 2k+\varepsilon$. An inductive proof of (10) is given in [11]. Observing that q (in eq. (9)) tends to -1 exponentially for $r \to \infty$ while γ approaches to 1 here only polynomially we can prove that after n B steps the asymptotic behaviour of $|\Phi|$ is $(1-n/r)$ as expected for n monopoles. Indeed keeping in (10) the polynomial terms only we see that $D_1^{(2k+\varepsilon)}/D_2^{(2k+\varepsilon)} = (-1)^{\varepsilon}$, and using an elementary identity for the ratio of two Vandermonde-type determinants one obtains

$$\frac{D_3^{(n)}}{D_1^{(n)}} = (-1)^{n+1} \sum_{i=1}^{n} \gamma_i \ , \qquad \frac{D_4^{(n)}}{D_2^{(n)}} = (-1)^{n+1} \sum_{i=1}^{n} \frac{1}{\gamma_i} \ .$$

Now plugging these ratios into (10) we obtain

$$|\Phi| \underset{r \to \infty}{\sim} (1 - \sum_{i=1}^{n} R^{-1}(w_i)) \sim (1-\frac{n}{r}) \ .$$

Imposing the conditions

$$M_1^* = N_1 \ , \qquad M_2^* = N_2 \qquad (11)$$

will obviously guarantee the reality of the new solution. We satisfy (11) by demanding

$$q^*_{2r-1} = q^{-1}_{2r} , \qquad \gamma^*_{2r-1} = \gamma^{-1}_{2r} \tag{12a}$$

for an even number of iterations, whereas for an odd number of steps

$$q^*_1 = q_1, \ \gamma^*_1 = \gamma^{-1}_1, \ q^*_{2r} = q_{2r+1}, \ \gamma^*_{2r} = \gamma^{-1}_{2r+1} , \tag{12b}$$

$r = 1,....k$. We remark that there exist other possibilities as well; however, only (12a,b) will be interesting for us. In fact the reality conditions constrain somewhat the arbitrary parameters β_i w_i. Still, there are further constraints for these constants as we impose the condition of finite energy. This proves to be so strong as to exclude any arbitrariness in β_i, w_i. In practice, demanding regular behaviour of $|\Phi|$ on the z axis and the absence of singularities on the $z = 0$ plane singles out the only possible values of w_i, β_i that ensure finiteness of the energy. The result we obtained is

$$w_{2r-1} = w^*_{2r} = i(2r-1)\frac{\pi}{2}$$

$$q_{4\ell-3} = -\tanh\left(\frac{R[i\,\frac{4\ell-3}{2}\pi]}{2}\right) = q^{*-1}_{4\ell-2}$$

$$q_{4\ell-1} = -\coth\left(\frac{R[i\,\frac{4\ell-1}{2}\pi]}{2}\right) = q^{*-1}_{4\ell} \tag{13a}$$

for an even number of steps, while in the case of an odd number of steps: $w_1 = 0$, $w^*_{2r} = w_{2r+1} = ir\pi$, $q_1 = -\tanh\frac{R(0)}{2}$,

$$q_{4\ell-2} = -\coth\left(\frac{R[i(2\ell-1)\pi]}{2}\right) = q^*_{4\ell-1}$$

$$q_{4\ell} = - \tanh \left(\frac{R[i2\ell\pi]}{2}\right) = q^{*}_{4\ell+1} \quad . \tag{13b}$$

(We must take the first 2k or 2k+1 elements of these sequences of q_i's respectively.)

We claim that eq.(10) together with (13) yields $|\Phi|$ of $(2k+\varepsilon)$ monopoles superimposed at the origin. This is the main result of our paper.

We now illustrate the use of our general formulae with some examples. From (10) and (13) it is straightforward to calculate the functional form of $|\Phi|$ for the n monopole solution. As the resulting formulae are quite complicated we give here only $|\Phi|$ on the z axis:

$$|\Phi|^{(2k)} = \left| \text{th } z - \sum_{i=1}^{2k} \frac{2z}{z^2+|w_i|^2} \right|$$

$$|\Phi|^{(2k+1)} = \left| \coth z - \frac{1}{z} - \sum_{i=2}^{2k+1} \frac{2z}{z^2+|w_i|^2} \right| \quad . \tag{14}$$

As the structure of (10) is ideally suited for numerical investigations we calculated the moduli of the Higgs fields for various values of topological charge, n. We present here $|\Phi|$ for the case n = 3, depicted on Fig. 1. One can observe that $|\Phi|$ has a first order zero when approaching the origin from any direction including the z axis,and an n-th order zero on the z = 0 plane as suggested by topologica and symmetry arguments [13]. This reflects the somewhat surprising property of these configurations that the individual monopoles in them could be moved apart only in the z = 0 plane. Apart from this the shape of $|\Phi|$ for configurations of different topological charge is qualitatively very similar, and in all cases $|\Phi|$ tends to its asymptotic value monotonously in every direction.

Inspecting formulae (10,13) we conclude that the 2MP solution does not contain the 1MP in the sense that none of the pairs of Bäcklund parameters β_i, w_i of the 2MP ($w_1 = w_2^* = i\frac{\pi}{2}$, $\beta_1 = 0$; $\beta_2 = i\frac{\pi}{2}$) coincides with that of the 1MP ($w = \beta = 0$). This means that in generating the 2MP the second B transformation is carried out on a state entirly different from the 1MP. In fact this state cannot be interpreted in our framework, as for it the reality conditions are not satisfied.

In the general case from (10,13) one finds that monopoles with even charges are generated by iterating the double B transformation (with appropriate parameters) on the vacuum, while those with odd charges are obtained by applying the iterated double B transformation (with different parameters) to the 1MP solution. Thus monopoles with charges $n = 2k$ can be conceived as nonlinear superpositions of k 2MP's, while those with charges $n = 2k+1$ as nonlinear superpositions of the 1MP and k 2MP-like structures.

We checked if this nonlinear superposition manifests itself in "physical" quantities such as the energy density $E = \frac{1}{2}\Delta\Phi^2$ [14], and found that it does not. It turned out that even for $n > 1$ E is well localized within a doughnut shaped structure surrounding the z axis. The location of its maximum is on the $z = 0$ plane away from the origin, i.e. in these solutions - unlike in the 1MP - the energy is no longer concentrated at the location of the topological charge (i.e. at the zero of $|\Phi|$). It is interesting to note that E for $n > 4$ exhibits no trace of the presence of several "rings" in the asymptotic form of $|\Phi| = (1 - \sum_{i=1}^{n} R^{-1}(w_i))$, however the location of the maximum of E seems to be connected to the location of the outermost "ring".

In conclusion, we presented the explicit expressions for monopoles of arbitrary (integer) charges. These solutions are axially and mirror symmetric, and - given the fact that they have only one zero of the corresponding order - describe n superimposed monopoles located at the origin.

After this work was completed a different line of attack on the multimonopole problem [15] - generalizing Ward's construction of the 2MP [3] using the ansätze of Atiyah and Ward [16] - came to our attention. The connection between the two methods deserves further study.

REFERENCES

1. L.O'Raifeartaigh and S. Rouhani, Schladming Lecture Notes 1981.
2. C.H. Taubes, Harvard preprint (1980) to be published.
3. R.S. Ward, Trinity College preprint, Dublin (1980).
4. P. Forgâcs, Z. Horvâth and L. Palla, Phys. Lett. 99B (1981) 232.
5. P. Forgâcs, Z. Horvâth and L. Palla, Phys. Rev. Lett. 45 (1980) 505.
6. N.S. Manton, Nucl. Phys. B135 (1978) 319.
7. F.J. Ernst, Phys. Rev. 167 (1968) 1175.
8. B.K. Harrison, Phys. Rev. Lett. 41 (1978) 1197.
9. G. Neugebauer, J. Phys. A12 (1979) L67.
10. V.A. Belinski and V.E. Zakharov, Sov. Phys. JETP 50 (1979) 1.
11. P. Forgâcs, Z. Horvâth and L. Palla, CRIP Report Budapest 1981.
12. P. Goddard and D.I. Olive, Rep. on Progress in Physics 41 (1978) 1357.
13. C. Rebbi and P. Rossi, Phys. Rev. D22 (1980) 2010.
14. P. Forgâcs, Z. Horvâth and L. Palla, CRIP Report

Budapest (1981).

15. M.K. Prasad, A. Sinha, L.L.C. Wang, ITP-SB-80-73 preprint.
M.K. Prasad, MIT preprint 1980.

16. M.F. Atiyah and R.S. Ward, Comm. Math. Phys. 55 (1977) 117.

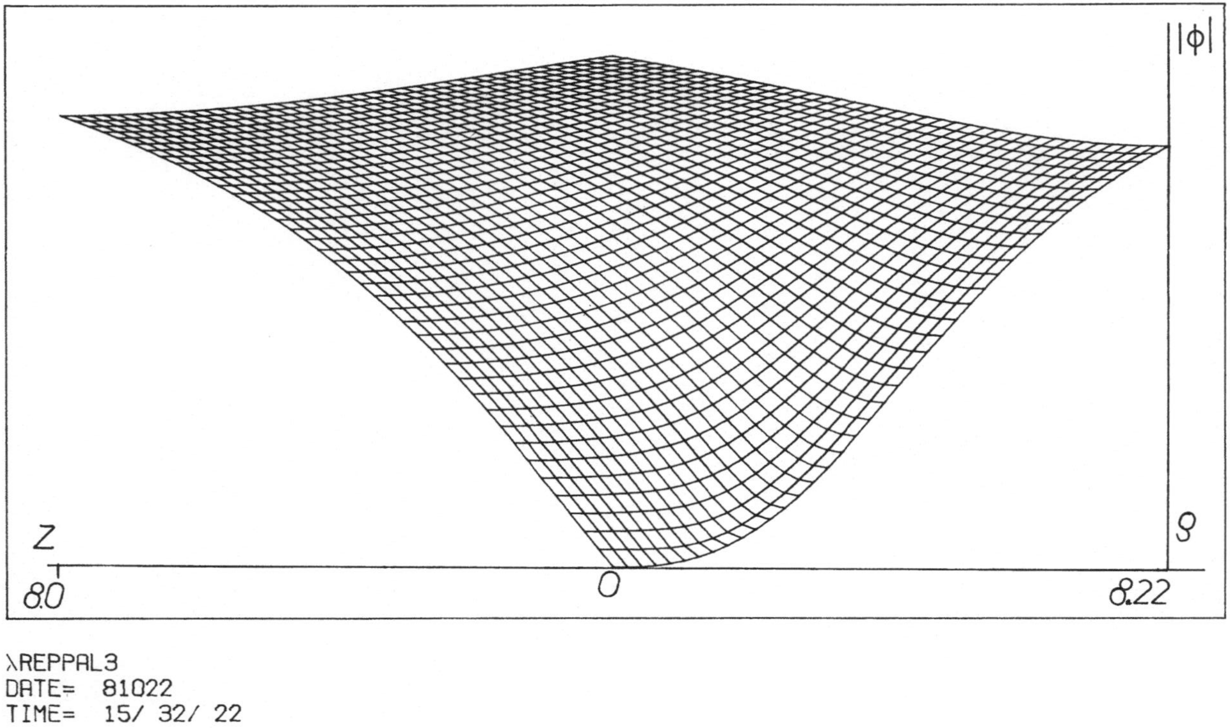

Fig.1. The axonometric view of the $|\Phi|$ of the 3 monopole configuration. The plane of projection inclines at 45° with both the ρ and z axis. The coordinates of O are $\rho = 0.22$, $z = 0$ respectively.

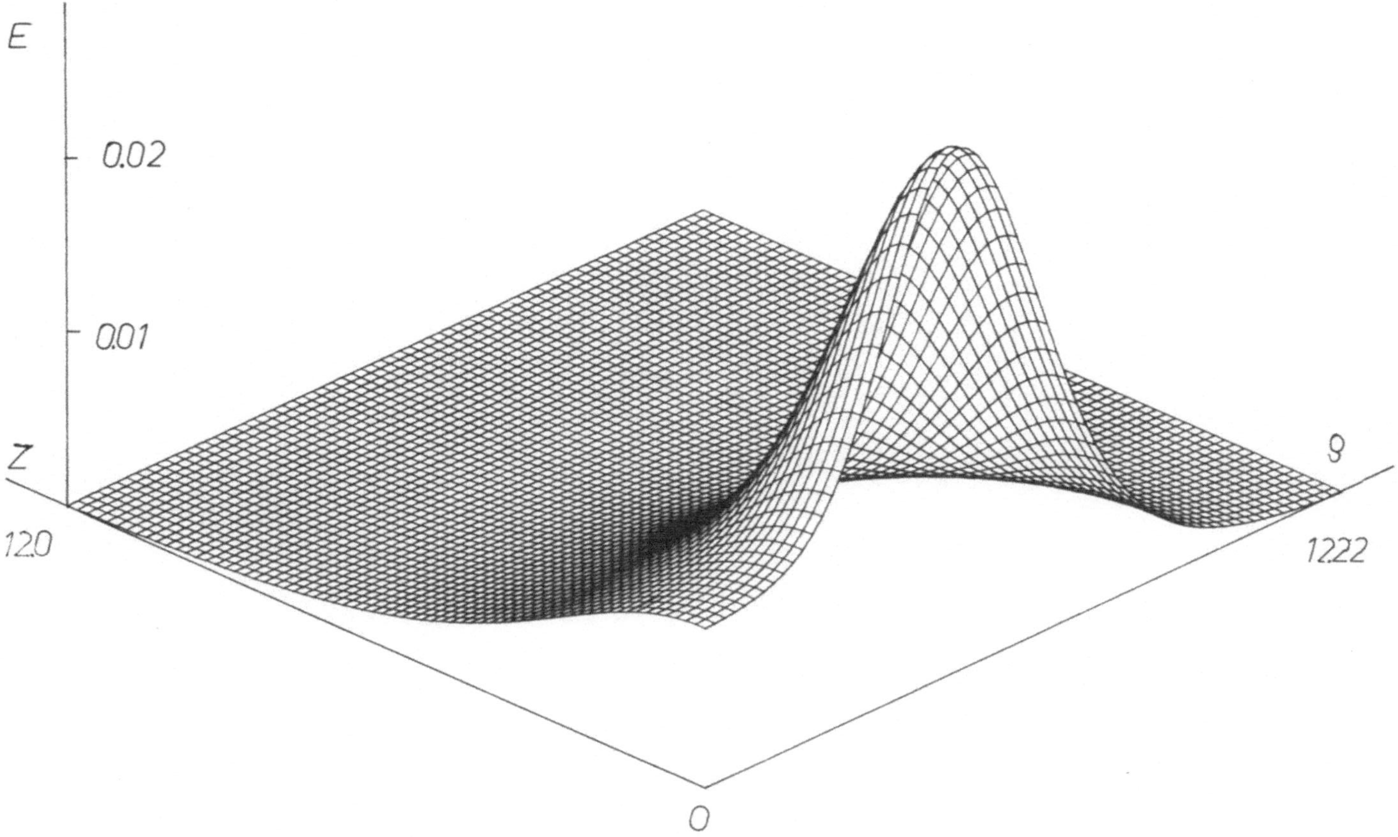

Fig. 2. The energy density of the 4 monopole configuration. The coordinates of O are $\rho = 0.23$, $z = 0$ respectively.

Acta Physica Austriaca, Suppl. XXIII, 627–639 (1981)

REMARKS ON LATTICE GAUGE MODELS[+x]

by

H. GROSSE
Institut für Theoretische Physik
Universität Wien, Austria

ABSTRACT

We report on a study of the phase structure of lattice gauge models where one takes as a gauge group a non-abelian discrete subgroup of SU(3). In addition we comment on a lattice action proposed recently by Manton and observe that it violates a positivity property.

I. INTRODUCTION

A complete study of a gauge field theory would consist first of a treatment of the classical theory, second of finding classical solutions to the field equations, and third of an attempt to develop a quantization scheme. Since any study of the quantum field theoretical aspects requires the introduction of a cut-off, and intro-

---

[+] Seminar given at XX. Internationale Universitätswochen für Kernphysik, Schladming, Austria, February 17-26, 1981.

[x] Supported in part by Fonds zur Förderung der wissenschaftlichen Forschung in Österreich, project no. 3569.

ducing a cut-off destroys certain symmetries the theory has, one intends to find a procedure which allows to maintain as many properties as possible. The only ultraviolet cut-off procedure known to be consistent with gauge invariance which allows to stay in the appropriate space-time dimensions is Wilson's [1] method of formulating the theory on a lattice.

We intend to start with a few comments on the requirements one would like to have fulfilled by a quantum field theory [2], and discuss next ideas of testing confinement. In a second chapter we give the usual Wilson lattice approach fulfilling most of these requirements but being representation dependent. In order to overcome this disease Manton [3] proposed recently an alternative action which as we shall show violates Osterwalder-Schrader positivity [4] for any gauge group containing a U(1) subgroup. In a third part we comment on known phase-properties of U(1) and SU(2) gauge theories [5] and add properties for theories having a discrete non-abelian subgroup of SU(3) as gauge group [6].

In a classical treatment of gauge theories one starts with a gauge potential A and a field strength $F = dA + [A,A]$, both being Lie algebra-valued forms. Clearly the relevant quantity is not A themselves, but the orbit of A under gauge transformations [A] which can be characterized by the holonomy group consisting of elements

$$g_c = P\,(\exp\,[i \int_c dz^\mu A_\mu(z)]) \tag{1}$$

out of the gauge group G, where c denotes a nodeless closed curve (an element out of the loop space Ω) and P means path ordering. Invariant functions on G can be characterized by the set $\{\chi_i(g_c)\,|\,c \in \Omega,\ \chi_i = \text{irreducible character of } G\}$. In trying to quantize a gauge theory, this set plays an important role; n-point functions being gauge dependent are replaced by n-loop functions $S_n(W_{i_1}(c_1)\ldots W_{i_n}(c_n))$ where the Wilson-loop observables $W_i(c)$ should result from $\chi_i(g_c)$ by a suitable normal ordering [7]. If one would be able to construct a set of n-loop functions S_n by some limiting procedure, the requirements one would like to check are:

a) local gauge invariance,
b) euclidean invariance implying Poincaré invariance in Minkowski space-time,
c) Osterwalder-Schrader positivity, implying a positive definite scalar product and the spectral condition,
d) cluster properties in order to check whether one has a dynamical mass generation or not.

On a lattice one replaces b) by

b') lattice translation invariance, and adds in addition:
e) formally correct $a \to 0$ limit, with a being a lattice constant,
f) representation independence.

Next one would intend to study further properties the theory has, especially confinement properties, where one can imagine a number of suitable definitions for a SU(n) theory [7]:

a) Fields transforming nontrivial under Z_n should not connect the vacuum to the physical one-particle state.
b) Asymptotic states transform trivial under the center.
c) Physical states are bound states of two and three fermions.

In practice one clearly likes to simplify and one tries to get information about the effective potential, which two fermions will feel if one couples them to the gauge field, from the pure gauge theory themselves: Define [1]

$$V(L) = - \lim_{T\to\infty} \frac{1}{T} \ln |S_1(W(C))| \tag{2}$$

where C denotes a rectangular curve of extension $L \times T$;

then we speak of confinement in the sense of Wilson iff

$$V(L) \underset{L\to\infty}{\simeq} \sigma \cdot L \cdot a^2 \qquad (3)$$

with $\sigma > 0$ being the string tension.

II. FORMULATION OF LATTICE GAUGE THEORIES

Here we discuss only the Lagrangian as opposed to the Hamiltonian approach. On a euclidean cubic lattice one is dealing with quantities defined on lattice sites, bonds and plaquettes. A gauge field is associated with bonds; a mapping $b \to g_b \in G$, such that the inverse group element is assigned to the reversed bond, defines a field configuration $\{g_b\}$ [4].

The most essential step of establishing a theory consists in fixing a lattice action. Given a character χ one gets for the Wilson loop variable and for a curve $C = \partial P$ being the boundary $\partial P = \{b_1, b_2, b_3, b_4\}$ of a plaquette P:

$$W(\partial P) = \chi(g_{b_1} g_{b_2} g_{b_3} g_{b_4}) =: \chi(g_{\partial P}) \; . \qquad (4)$$

Replacing A_μ in equ.(1) by $g_u A_\mu$, g_u being the unrenormalized coupling constant, and expanding formally (4) in terms of the lattice constant gives [1]:

$$\mathrm{Re}\, \chi(g_{\partial P}) \underset{a\to 0}{\simeq} d - \frac{a^4 g_u^2}{2} \mathrm{Tr}\, F^2 + O(a^6) \; , \qquad d = \chi(e) \; , \qquad (5)$$

where e denotes the unit element of the group. This justifies Wilson's definition for an action which leads to a gauge invariant generalization of Ising models first studied by Wegner [8]:

$$S_\Lambda(\{g\}) = \sum_{P \subset \Lambda} [d - \mathrm{Re}\,\chi(g_{\partial P})], \tag{6}$$

Λ denotes space-time volume. Expectation values of local observables (depending on finitely many bond variables) are then defined by integrating over all possible field configurations:

$$\langle F \rangle_\Lambda = Z_\Lambda^{-1} \int \prod_{b \in \Lambda} dg_b\, F(\{g_b\}) \exp\left[-\frac{1}{g_u^2} S_\Lambda(\{g\})\right], \tag{7}$$

dg_b denotes the Haar measure of the group and Z_Λ the partition function. Identifying $1/g_u^2 = \beta$ with an inverse temperature makes contact with the classical statistical mechanics interpretation of euclidean lattice theories; the two languages can therefore be identified [9]; so for instance a phase transition corresponds to a change of the vacuum, an exponential decrease of correlations corresponds to the existence of a mass gap; defects correspond to nontrivial topological configurations etc..

After taking the thermodynamic limit $\Lambda \to R^4$ one tries to check the above-mentioned requirements: a), b'), and e) are fulfilled, so the most essential question concerns the positivity property [4]. More explicitly one defines a t = 0 plane which is half way between lattice planes; this allows to define algebras $A_\pm$ of local observables defined on bond variables out of $\Lambda_\pm$, which are the subspaces of bonds with $t > 0$ or $t < 0$. One then asks for an antilinear mapping Θ of the field algebra A to A with

$$\Theta F(\{g_{xy}\}) = \overline{F(\{g_{rx,ry}\})} \quad , \tag{8}$$

where the bar means complex conjugation and r means the reflection on the plane $t = 0$; so $\Theta|A_\pm = A_\mp$. O.S. positivity is then the requirement that

$$\langle F \cdot \Theta F \rangle_\Lambda \geq 0 \qquad \forall F \in A_+ \quad . \tag{9}$$

Since (9) implies on the one hand that the underlying space carries a positive definite scalar product, and on the other hand one obtains a positive transfer matrix (besides chess-board estimates), requirement c) seems to be essential. It has been shown in [4] that Wilson's action fulfills (9).

Here we note that in a theory based on an alternative action proposed by Manton [3], which has been used recently in a number of calculations [10] and seems to allow for a smoother continuum limit, positivity is violated:

Theorem: Assume that the gauge group contains a U(1) subgroup and take as an action

$$S_\Lambda^M(\{g\}) = \sum_{P \subset \Lambda} D^2(e, g_{\partial P}) \tag{10}$$

where $D(e, g_{\partial P})$ denotes the length of the smallest geodesic connecting the unit element to $g_{\partial P}$; then O.S. positivity is violated.

Remarks: Clearly a violation for some subgroup implies the same for the larger group, so we may restrict ourselves to the U(1) case. The next step consists in choosing a gauge in which all bond variables for bonds crossing the t = 0 plane are set equal to unity, which implies a factorization of the integrations involved in (9). Positivity is implied by showing that $\exp\{-\beta D^2(e, g_{\partial P})\}$ is a function of positive type on the group

$$\int dg \int dh\ F^*(g)\ e^{-\beta D^2(e, gh^{-1})} F(h) \geq 0 \tag{11}$$

which means that all Fourier coefficients have to be positive. For the U(1) case it is simple to establish that the coefficients $I_n(\beta)$:

$$I_n(\beta) = \int_{-\pi}^{\pi} \frac{d\alpha}{2\pi}\, e^{in\alpha}\, e^{-\beta\alpha^2} \underset{\beta\to 0}{\simeq} (-)^{n+1}\,\frac{2\beta}{n^2} + O(\beta^2),\ n>0, \quad (12)$$

change sign depending on whether n is even or odd, so that (11) is violated.

In the next chapter we will mainly concentrate on the observable energy per plaquette:

$$E(\beta) = \langle\chi(g_{\partial P})\rangle = -\frac{1}{6}\frac{\partial F}{\partial\beta}, \quad F = \lim_{N\to\infty}\frac{\ln Z_\Lambda}{N}, \quad (13)$$

with N being the number of lattice points. For an attempt to match the strong coupling behaviour with asymptotic freedom results the string tension as defined in equ.(3) is of great importance too, although it is not a local observable.

May be it is interesting to mention two general results: It has been shown in [4] that the cluster expansion for local observables has a finite radius of convergence, implying analyticity of these quantities in the complex β plane for $|\beta|$ small. In addition the cluster property d) is verified with a generated mass $m \geq c\,(\ln g_u)/a$ for large g_u. Furthermore Wilson's area law is realized in that phase.

There is actually a general proof saying that the potential V(L) defined in (2) is bounded by the linear potential [11] $V(L) \leq cL$.

III. PHASE TRANSITIONS

A physical mass has to be a function of the form $m_{ph} = f(g_u)/a$ (in a pure gauge theory); so since $a \to 0$ corresponds to $g_u \to 0$ the continuum limit corresponds to an approach of a critical point of the theory. Clearly the phase, in which the theory will be for large $\beta = 1/g_u^2$, will determine the properties of the continuum theory. Since for large coupling constant all lattice theories show confinement, one likes to have one phase transition in the U(1) theory, while none should be present for SU(2) and SU(3). In this context it is interesting to note that a phase transition has been observed for the SU(5) theory [12].

a) Abelian gauge theories:

Some time ago A. Guth [13] obtained after a work of Glimm and Jaffe a rigorous estimate on the Wilson loop variable for the U(1) theory implying that the potential is nonconfining for large β. Together with our previous remarks this implies the existence of at least one phase transition.

In addition to rigorous results one has been able to obtain further insight into the phase structure of lattice theories by using Monte Carlo simulations. In that way one starts from an initial configuration, goes through the lattice a number of times, and tries to determine an equilibrium configuration which allows to determine expectation values approximately:

$$\langle F \rangle = \lim_{T\to\infty} \frac{1}{T} \int_0^T d\tau\, F(\{g_\tau\}) \quad . \tag{14}$$

The first result for U(1) obtained by Creutz, Jacobs and Rebbi [5] indicated exactly one phase transition. Additional support was obtained by studying discrete subgroups of U(1).

For Z_2, Z_3 and Z_4 one observes one phase transition at the inverse temperatures determined by duality. Z_2 and Z_4 are actually very special theories. Their critical points fulfill $\beta_c^{Z_4} = 2\beta_c^{Z_2}$. In addition it was known [14] that the two spin systems in two dimensions are actually equivalent up to a scale transformation. So it was natural to look for a similar relation between the gauge theories [15].

With the help of the strong coupling expansion [16] we have been able to show equality between two noninteracting Z_2 theories putting them onto the same lattice and one Z_4 theory with a scaled temperature. For the partition functions and any volume Λ we obtained equality

$$Z_{\Lambda}^{Z_4}(2\beta) = Z_{\Lambda}^{Z_2}(\beta)^2 \tag{15}$$

by identifying terms in the high temperature expansion. The actual proof is unfortunately technical and has up to now not been generalized to other theories.

Going up to Z_n with $n \geq 5$ one observes two phase transitions. One moves rapidly out to low temperature, the other approaches the critical point of the U(1) theory.

b) Non-abelian theories:

One could ask why one believes to learn something about the phase structure of a theory through the study of theories where one takes a discrete subgroup as a gauge group. One result pointing in that direction was obtained first by Mack and Petkova [17] and generalized further in Ref.[18] and says that confinement within the Z_n theories implies (up to a scale transformation) confinement in the SU(n) case.

For SU(2) one expects a roughening transition around $\beta \simeq 2$, the region where one observes a turn-over from the strong coupling behaviour to the asymptotic freedom behaviour. A study of the non-abelian subgroups (there exist finitely many) shows one transition moving out up to $\beta \simeq 6$ [19] for the icosahedral group which has 120 elements.

So for SU(2) one is in a satisfying position: Not only is it possible to parametrize the group manifold themselves easily; through the study of subgroups it is also possible to study the range of β values up to $\beta \simeq 6$ which includes the roughening point.

For SU(3) things are not in such a good shape: Besides attempts to calculate directly the averaged action per plaquette [20] only a few points for the Wilson loop variable have been obtained. We asked ourselves the question how much one can learn by taking discrete nonabelian subgroups as a gauge group.

At first that program seems to be promising. There exist infinitely many nonabelian subgroups of SU(3) [21]. Besides finitely many crystal-like groups two infinite sequences of subgroups $\Delta(3n^2)$ and $\Delta(6n^2)$ exist, which are actually semidirect products of $Z_n \times Z_n$ with Z_3 and S_3:

$$\Delta(3n^2) = Z_n \times Z_n \underset{s}{\times} Z_3 , \quad \Delta(6n^2) = Z_n \times Z_n \underset{s}{\times} S_3 . \quad (16)$$

The multiplication laws can be written down compactly,and all irreducible representations and characters are known in closed form. Certain sequences of subgroups together with their representations have been studied recently by the Bonn group [22].

We have performed the familiar thermal cycles in order to get insight into the phase structure of these theories,and have compared the results with high temperature and low temperature expansions [6]. For all investigated groups (we took first $\Delta(3n^2)$) the Monte Carlo results agree nicely with series expansions up to a certain value of β; then a phase change occurs. But by increasing n beyond 5 we observed two phase transitions, one moving out to large values of β, the other one staying around $\beta \simeq 2.2$ (see Figures 1 and 2). Unfortunately it is clear from [20] that a turn-over from the strong to the weak coupling regime occurs around $\beta \simeq 6$, so that the nice feature of the SU(2) case is not reproduced here.

In the meantime we have investigated the $\Delta(6n^2)$ groups [23] and observed a similar situation (with a transition around 2.6). Since Bhanot and Rebbi [24] obtained a phase change for the largest crystal-like subgroup S(1080) around $\beta \simeq 3.0$, we have to conclude that through the study of these subgroups it is not possible to get information about the turn-over point.

REFERENCES

1. K.G. Wilson, Phys. Rev. D10 (1974) 2445.
2. B. Simon, The $P(\Phi)_2$ Euclidean Quantum Field Theory.
3. N.S. Manton, Phys. Lett. 96B (1980) 328.
4. K. Osterwalder and E. Seiler, Ann. Phys. 110 (1978) 440.
5. M. Creutz, L. Jacobs and C. Rebbi, Phys. Rev. D20 (1979) 1915.
   M. Creutz, Phys. Rev. Lett. 43 (1979) 553.
   M. Creutz, Phys. Rev. D21 (1980) 2308.
   B. Lautrup and M. Nauenberg, Phys. Rev. Lett. 45 (1980) 1755.
6. H. Grosse and H. Kühnelt, Phys. Lett. 101B (1981) 77.
7. J. Fröhlich, Phys. Rep. 67 (1980) 137.
8. F. Wegner, Jour. Math. Phys. 12 (1971) 2259.
9. G. Parisi, Talk presented at the XX. Conference on High Energy Physics, Madison 1980.
10. C.B. Lang, P. Salomonson and B.S. Skagerstam, Phys. Lett. 101B (1981) 173.
11. E. Seiler, Phys. Rev. D18 (1978) 482.
12. M. Creutz, preprint BNL 29301 (Febr. 1981).
13. A. Guth, Phys. Rev. D21 (1980) 2291.
14. M. Suzuki, Prog. Theor. Phys. 37 (1967) 770.

15. H. Grosse, C.B. Lang and H. Nicolai, Phys. Lett. 98B (1981) 69.
16. R. Balian, J.M. Drouffe and C. Itzykson, Phys. Rev. D11 (1975) 2104.
17. G. Mack and V. Petkova, to be published in Ann. Phys. (1981).
18. J. Fröhlich, Phys. Lett. 83B (1979) 195.
C.P. Korthals Altes, Nucl. Phys. B170 FS1 (1980) 98.
19. G. Bhanot and C. Rebbi, CERN preprint TH 2979 (1980).
20. M. Creutz, BNL preprint (1980).
R.C. Edgar, L. McCrossen and K.J.M. Moriarty, Royal Holloway College preprint (1980).
E. Pietarinen, HU-TFT 80-49 preprint (1981).
21. W.M. Fairbain, T. Fulton and W.H. Klink, Journ. Math. Phys. 5 (1964) 1038.
22. A. Bovier, M. Lüling and D. Wyler, Bonn University preprint (1980).
23. H. Grosse and H. Kühnelt, to be published.
24. G. Bhanot and C. Rebbi, BNL preprint (1981).

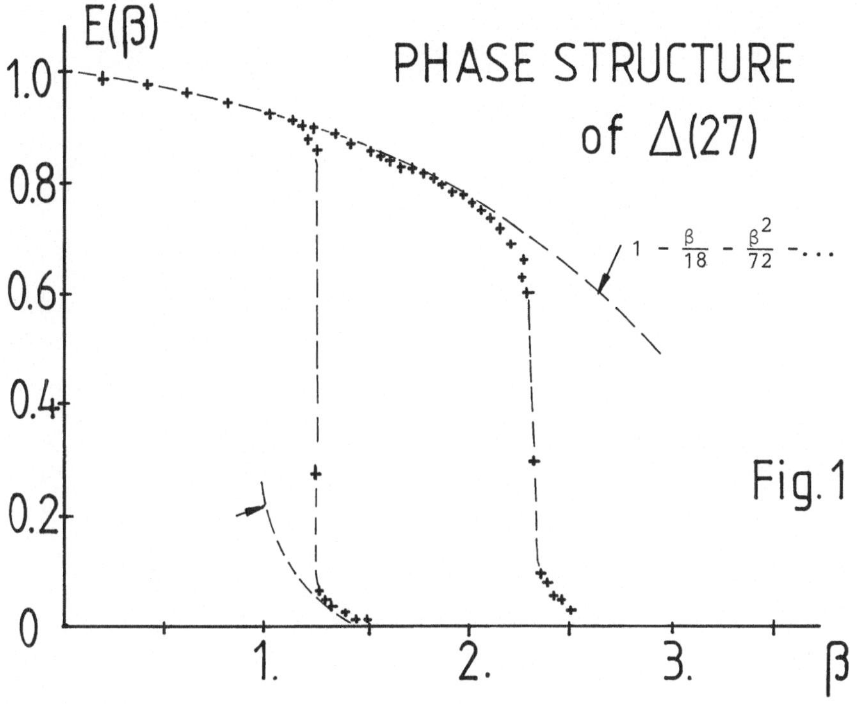
E(β)
1.0
0.8
0.6
0.4
0.2
0
PHASE STRUCTURE
of Δ(27)
1 - β/18 - β²/72 -...
1.
2.
3.
β

Fig.1

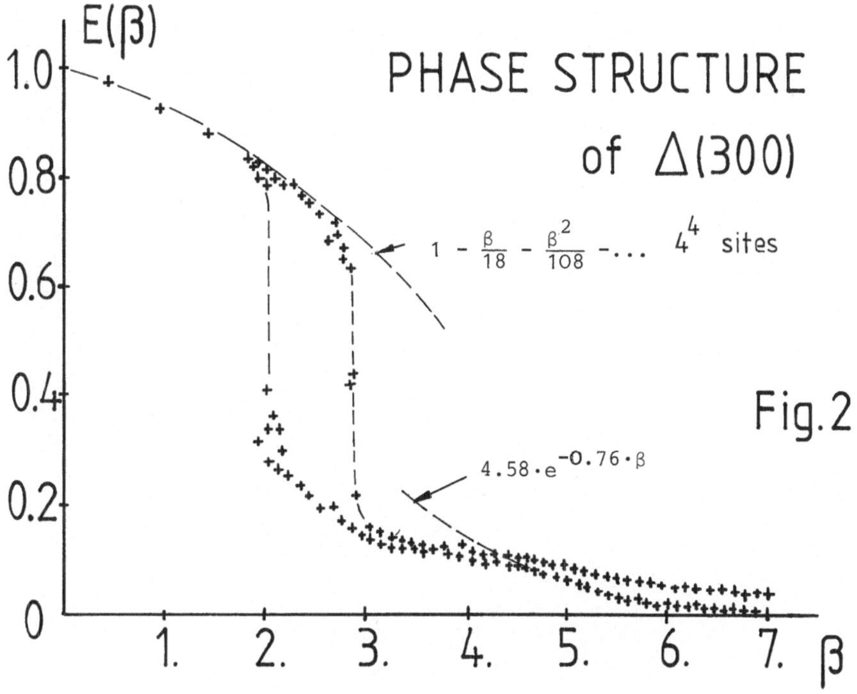
E(β)
1.0
0.8
0.6
0.4
0.2
0
PHASE STRUCTURE
of Δ(300)
1 - β/18 - β²/108 -... 4⁴ sites
4.58·e^(-0.76·β)
1.
2.
3.
4.
5.
6.
7.
β

Fig.2

Acta Physica Austriaca, Suppl. XXIII, 641–652 (1981)

THE KLEIN-KALUZA THEORY WITH A TORSION+

by

M.W. KALINOWSKI
Institute of Philosophy
00330 Warsaw, Nowy Świat 72, Poland

ABSTRACT

The Klein-Kaluza theory with a nonvanishing torsion is developed.

The torsion is associated with spin and polarization of electromagnetic field. The electromagnetic polarization is considered as a source of additional components of torsion connected with 5th dimension. It is proved that new effects are bigger 10^{36} times than the effects from Einstein-Cartan theory.

1. INTRODUCTION AND SUMMARY

The aim of this paper is to generalize the Klein-Kaluza theory [1,2] to a situation with nonvanishing torsion of the connection. The polarization of a electromagnetic field and spin will be associated to torsion. Our generalization of the Klein-Kaluza theory is analogous to the relation of the Einstein-Cartan theory to the general theory

+Seminar given at XX. Internationale Universitätswochen für Kernphysik, Schladming, Austria, February 17-26, 1981.

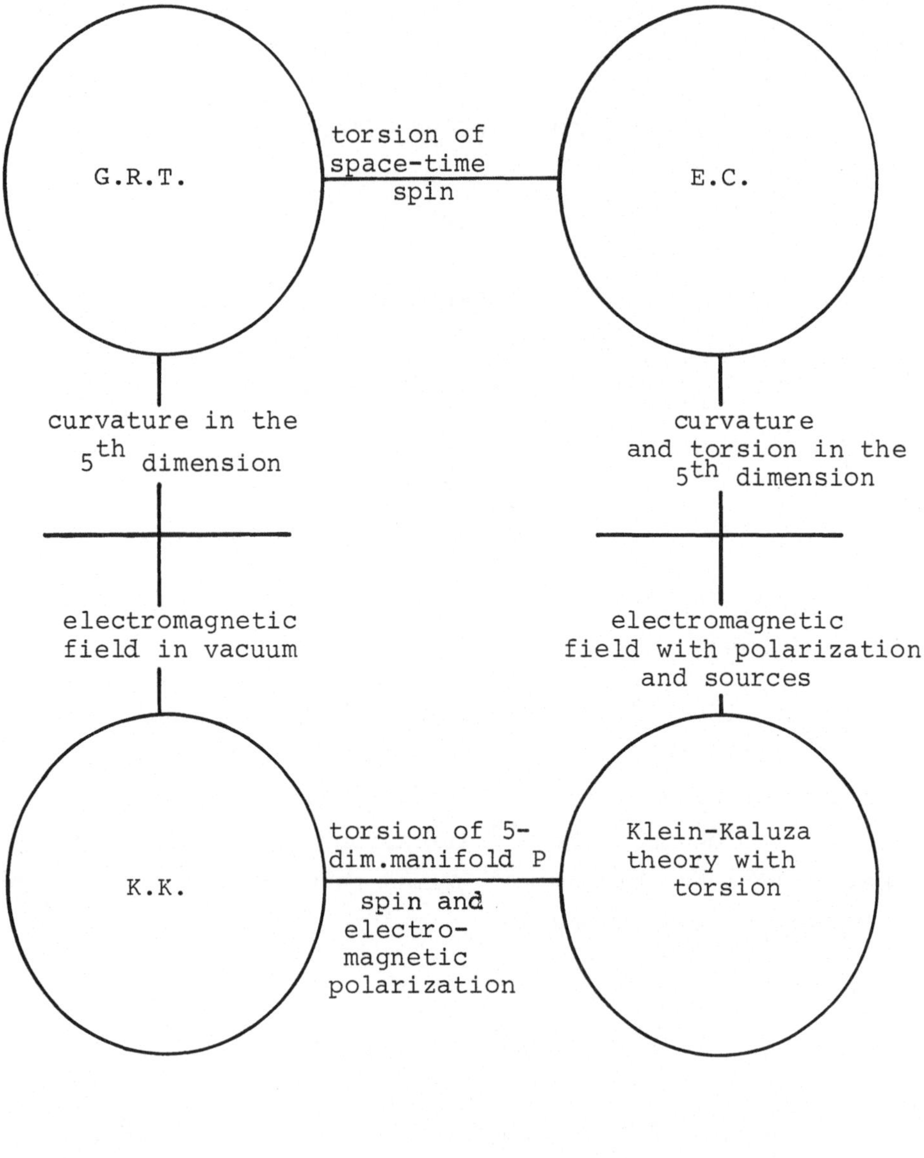

Fig. 1. The position of the Klein-Kaluza theory with torsion among the General Relativity, the Einstein-Cartan theory, and the classical Klein-Kaluza theory.

of relativity. The diagram (Fig. 1) places the Klein-Kaluza theory with torsion among the above mentioned theories. A new geometric element in our theory is the torsion in the 5th dimension the source of which is electromagnetic polarization $M_{\alpha\beta}$. Roughly speaking, if one says that "mass curves space-time", "spin twists it" and "electric charge curves the 5th dimension". Naturally the 5th dimension is understood as a dimension connected with gauging. The general plan of the paper is following. We introduce on P-metrized electromagnetic bundle the connection with nonvanishing torsion. This connection is invariant with respect to transformations of group U(1). In the paper we also assume that a torsion of the connection is horizontal. Next we construct a form of a scalar curvature for this connection and introduce sources. Then from a variational principle we obtain equations of fields and interpret them. According to the postulate of geometrization of physical quantities we shall obtain equations where on the left-hand side there will be geometric quantities and on the right-hand side matter quantities. In this way matter quantities will be sources of geometry. We shall obtain an interpretation of electromagnetic polarization as a torsion related to the 5th dimension. We get equations of gravitation in the Einstein-Cartan theory. On the right-hand side as source will be the sum of energy-momentum tensors of: electromagnetic field with polarization of matter in the form given by W. Israel [3,4,5] and of matter. Additionally there will be also a component $\pi\, g_{\alpha\beta}\, M_{\mu\nu}\, M^{\mu\nu}$ where $M_{\mu\nu}$ is tensor of electromagnetic polarization of matter. This additional component has been obtained similarly as the component with contact interaction (spin) x (spin) in Einstein-Cartan's theory. The new component may be treated as a contact interaction (electromagnetic moment) x (electromagnetic moment). The role of this component will be estimated and compared with the effects originated, from Einstein-Cartan theory. We derive the second pair of Maxwell equations in terms of derivatives with respect to connection with torsion. This will give us an additional internal current related to spin. From Bianchi's identity we get conservation laws of energy-momentum, angular momentum and charge.

2. THE KLEIN-KALUZA THEORY WITH TORSION

We introduce electromagnetic bundle P (for details see [6,7]) with natural metrization and a metrical connection ω^A_B invariant under a transformation of U(1). The connection ω^A_B is not necessarily Riemannian. We also define a connection $\bar{\omega}^{\alpha\beta}$ metrical, but unnecessarily Riemannian on E. Now we have $(E, g, \bar{\omega}^\alpha{}_\beta)$ a 4-dimensional manifold with metrical connection, metrical tensor g with the signature (---+), $(P, \gamma, \omega^A{}_B)$ a 5-dimensional manifold with the metrical connection $\omega^A{}_B$. We suppose that A,B = 1,2,...5, $\alpha, \beta, \sigma = 1,2,3,4$. γ has the signature (---+-). Thus

$$D\gamma_{AB} = 0 \, , \qquad \underset{\xi_5}{L} \, \omega^A{}_B = 0 \, , \tag{2.1}$$

where $\gamma = \pi^* g - \theta^5 \otimes \theta^5$ and $\theta^A = (\theta^\alpha, \theta^5 = \lambda\alpha)$, $\lambda = 2$ for Gauss' system, in C.G.S. $\lambda = \frac{\sqrt{G}}{c^2}$; ξ_5 is a Killing vector, and ξ_A forms a dual frame of θ^A. $\gamma(\xi_A) = \gamma_{AB}\,\theta^B$, α is a electromagnetic connection on P. We calculate forms of torsion of ω_{AB} :

$$\textcircled{H}^A = D\,\theta^A \quad , \tag{2.2}$$

and suppose

$$\textcircled{H}^A = \mathrm{hor}\ \textcircled{H}^A \quad . \tag{2.3}$$

From (2.3) we easily get

$$\textcircled{H}^\mu = \bar{\textcircled{H}}^\mu \quad , \tag{2.4}$$

$$\textcircled{H}^{5} = K_{\alpha\beta}\, \Theta^{\beta} \wedge \Theta^{\alpha} \tag{2.5}$$

where $K_{\alpha\beta}$ is a tensor on E. Using (2.4) and (2.5) we obtain

$$\omega_{\alpha\beta} = \bar{\omega}_{\alpha\beta} + H_{\alpha\beta}\, \Theta^{5},\ \omega_{\alpha 5} = -\omega_{5\alpha} = H_{\alpha\gamma}\, \Theta^{\gamma}\ \omega_{55} = 0 \tag{2.6}$$

where $H_{\alpha\beta} = F_{\alpha\beta} + K_{\alpha\beta}$.

3. VARIATIONAL PRINCIPLE. FIELD EQUATIONS

Now we formulate a variational principle for a scalar of curvature K constructed for ω_{AB},

$$K = \frac{1}{2}\, \eta_{AB} \wedge \Omega^{AB} \quad . \tag{3.1}$$

We vary K with respect to ω_{AB}, γ_{AB}, and Θ^{A}. Since independent quantities are only $\bar{\omega}_{\alpha\beta}$, $K_{\alpha\beta}$, $g_{\alpha\beta}$, Θ^{5}, Θ^{α} (due to constraints) we vary K with respect to these quantities. Thus we have:

$$\delta K = \delta\Theta^{A} \wedge e_{A} + \frac{1}{2}\delta\, g_{\alpha\beta} E^{\alpha\beta} - \frac{1}{2}\delta\bar{\omega}^{\alpha}{}_{\beta} \wedge \bar{p}_{\alpha}{}^{\beta} \wedge \Theta^{5}$$

$$+ \delta K^{\alpha\beta}\, \frac{\delta K}{\delta K^{\alpha\beta}} + \text{exact form} \quad , \tag{3.2}$$

where

$$e_{A} = \frac{1}{2}\, \eta^{B}{}_{CA} \wedge \Omega^{C}{}_{B} E^{\alpha\beta} = \frac{1}{2}(\gamma^{\alpha\beta}\eta^{D}_{C} - \gamma^{\alpha D}\eta^{\beta}_{C} - \gamma^{\beta D}\cdot\eta^{A}_{C}) \wedge \Omega^{C}_{D} \tag{3.3}$$

and

$$\frac{\delta K}{\delta K^{\alpha\beta}} = -K_{\alpha\beta}\, \bar{\eta} \wedge \Theta^{5},\ \bar{p}_{\alpha\beta} = \bar{\bar{D}}\eta_{\alpha\beta} \quad . \tag{3.4}$$

Now we introduce material sources. Let be 5-form,

$$L = L(g_{\alpha\beta}, \Theta^{\alpha}, \Theta^{5}, \bar{\omega}^{\alpha}{}_{\beta}, H^{\alpha}{}_{\beta}, \psi_{a}) .$$

We vary L with respect to independent quantities, i.e. $g_{\alpha\beta}$, θ^{α}, θ^{5}, $\bar{\omega}_{\alpha\beta}$, $K_{\alpha\beta}$, Ψ_{a}, and simply get:

$$\delta L=\delta\theta^{A}\wedge t_{A} + \frac{1}{2}\delta g_{\alpha\beta}T^{\alpha\beta}\wedge\theta^{5} + \frac{1}{2}\bar{\omega}^{\alpha}\wedge\bar{S}_{\alpha}\wedge\theta^{5} +$$

$$+ \frac{1}{2}\delta K^{\alpha}{}_{\beta}M^{\beta}{}_{\alpha}\pi\wedge\theta^{5}+L^{a}\delta\psi_{a}+\text{exact form} , \qquad (3.5)$$

where $\bar{S}_{\alpha}{}^{\beta}$ is a form of spin and $T^{\alpha\beta}$ is a tensor of energy-momentum of matter. Quantities Ψ_{a} may be either fields or macroscopic variables, i.e. density, enthalpy, pressure. From variational principle

$$\delta \int_{V} (K - 8\pi L) = 0 , \qquad V\subset P , \qquad (3.6)$$

we get

$$\bar{p}_{\alpha\beta} = -8\pi\ \bar{S}_{\alpha\beta}, \qquad K_{\alpha\beta} = -4\pi\ M_{\alpha\beta} \qquad . \qquad (3.7)$$

The first equation of (3.8) is Cartan's equations known from the Einstein-Cartan-theory [8,9,10]. Using (3.7) we see that

$$H_{\alpha\beta} = F_{\alpha\beta} - 4\pi\ M_{\alpha\beta} , \qquad (3.8)$$

and we will interprete $H_{\alpha\beta}$ as the second tensor of strength of electromagnetic field and $M_{\alpha\beta}$ as an electromagnetic polarization. The form of t_{A} is

$$t_{\alpha} = \bar{t}_{\alpha}\wedge\theta^{5} + \frac{1}{2}i_{\alpha}\bar{\eta}; \quad t_{5} = \frac{1}{2}j^{\mu}\bar{\eta}_{\mu}\wedge\theta^{5} + t\bar{\eta} \quad ; \qquad (3.9)$$

$\bar{t}_{\alpha}$ is a form of energy-momentum of matter and $j = j^{\mu}\bar{\eta}_{\mu}$ is a form of current. The quantity i_{α} is only auxiliary one and will be eliminated by the Bianchi identity, t is related to a horizontal part of t_{5} and eliminated from field equations by Lagrange multipliers. Varying K-8πL with respect to $g_{\alpha\beta}$ we obtain the following equation :

$$\bar{E}^{\alpha\beta} = 8\pi \; (\overset{em}{T}{}^{\alpha\beta} + T^{\alpha\beta} + g^{\alpha\beta} M^2 \bar{\eta}) \quad , \tag{3.10}$$

where

$$\overset{em}{T}{}^{\alpha\beta} = \frac{1}{8\pi} \; (H^{\mu\alpha}F_{\mu}{}^{\beta} + H^{\mu\beta}F_{\mu}{}^{\alpha})\bar{\eta} - \frac{1}{16\pi} g^{\alpha\beta}F^2\bar{\eta} \; , F^2 = F_{\mu\nu}F^{\mu\nu}, \; M^2 = M_{\mu\nu}M^{\mu\nu} .$$

Variations with respect to $\Theta^A = (\Theta^\alpha, \Theta^5)$ yield the following equations:

$$\bar{e}_\alpha = 8\pi \; (\overset{em}{t}_\alpha + \bar{t}_\alpha + \pi M^2 \bar{\eta}_\alpha) \quad ,$$

where

$$\overset{em}{t}_\alpha = \frac{1}{4\pi} \; (H^{\mu\gamma} \; F_{\mu\alpha} \; \bar{\eta}_\gamma - \frac{1}{4} F^2 \; \bar{\eta}_\alpha) \tag{3.11}$$

and

$$\bar{\nabla}_\gamma \; H^{\gamma\beta} = 4\pi j^\gamma \; , \; \bar{\nabla}_\gamma \; H^\gamma{}_\alpha - H^\rho{}_\beta \Theta_\rho{}^\beta{}_\alpha = 4\pi \; i_\alpha \quad . \tag{3.12}$$

Constraints eliminate the equation in which t appears. The second equation (3.12) can be reduced to the first by Bianchi's identities. For more details see [7]. $\overset{em}{t}_\alpha$ is a form of energy-momentum reported in Israel's papers [3,4,5] and $\overset{em}{T}{}^{\alpha\beta}$ a symmetrized form of Israel's tensor. The second equation (3.7) defines a relation between torsion in 5th dimension and electromagnetic polarization of matter. It is an illustration of Einstein's postulate that matter quantities are on the right-hand side of equations and geometrical ones on the left-hand side of them. Electromagnetic polarization becomes the source of torsion. Thus equations (3.10) are the Einstein-Cartan equations with the sum of the following energy-momentum tensors: of matter, of electromagnetic field (of Israel-type with polarization of matter taken into account) plus a new additional pressure-type component. The additional components with $M^2 = M_{\mu\nu}M^{\mu\nu}$ may be treated as a kind of electromagnetic interaction. Perhaps this interaction is related to a non-linear electrodynamics.

4. BIANCHI'S IDENTITIES. CONSERVATION LAWS

The forms of curvature Ω^A_B and torsion $Ⓗ^A$ obey Bianchi's identities

$$D\Omega^A_B = 0 , \qquad D Ⓗ^A = \Omega^A_B \wedge \Theta^B \quad . \tag{4.1}$$

From second of (4.1) we obtain the angular momentum conservation

$$\bar{D}\bar{S}_{\alpha\beta} = \Theta_\alpha \wedge \overset{tot}{t}_\beta - \Theta_\beta \wedge \overset{tot}{t}_\alpha \ , \quad \overset{tot}{t}_\alpha = \overset{em}{t}_\alpha + \bar{t}_\alpha + \pi M^2 \bar{\eta}_\alpha \ ; \tag{4.2}$$

$$4\pi (j_\alpha + i_\alpha) = \frac{1}{2} H_\alpha{}^\delta \bar{S}_\alpha + H^{\gamma\delta} \bar{S}_{\gamma\delta\alpha}, \text{ where } \quad \bar{S}_\delta = \bar{S}_\delta{}^\gamma{}_\gamma . \tag{4.3}$$

The last identity is not a new conservation law. It only establishes the relation between j_α and i_α. In this way it eliminates second equation (3.12) reducing it to the first. This equation is written by aid of covariant derivatives with respect to a connection with torsion. We can write them as Riemannian derivatives on E and spin. We have

$$\tilde{\bar{\nabla}}_\beta H^{\gamma\beta} = 4\pi \overset{tot}{j^\gamma} \qquad \text{where} \quad \overset{tot}{j^\gamma} = j^\gamma + j^\gamma_i = j^\gamma - \frac{1}{4} \bar{S}^\gamma{}_{\beta\alpha} H^{\beta\alpha} . \tag{4.4}$$

The internal current $j^\gamma_i = -\frac{1}{4} \bar{S}^\gamma{}_{\beta\alpha} H^{\beta\alpha}$ satisfies generalized "Ohm's law" (proportionality to strength of field). We notice that j^γ_i is of gravitational-electromagnetic nature. Thanks to this both gravitational and electromagnetic fields are more strongly interrelated than in classical Klein-Kaluza theory. Now let us turn back to the first equation of (4.1). We get from it

$$\bar{D} \overset{tot}{t}_\alpha = \bar{Q}^\beta{}_\alpha \wedge \overset{tot}{t}_\beta - \frac{1}{2} \bar{R}^{\beta\gamma}{}_\alpha \wedge \bar{S}_{\beta\gamma} \ , \tag{4.5}$$

i.e. conservation law of energy-momentum in the Einstein-Cartan theory for total energy of matter and electromagnetic and gravitational fields [7,8]. We also obtain the continuity equation

$$d\overset{tot}{j} = 0 \quad , \qquad \overset{tot}{j} = \overset{tot}{j^{\gamma}} \; \bar{\eta}_{\gamma} \quad , \tag{4.6}$$

i.e. conservation of a charge.

5. APPLICATIONS

In this paragraph we discuss new physical effects appearing in the Klein-Kaluza theory with torsion. We estimate the contribution of a new additional pressure-type component. The equation (3.11) may be written in a tensor notation (we turn back to the C.G.S. system)

$$\bar{R}_{\alpha\beta} - \frac{1}{2} g_{\alpha\beta} \bar{R} = \frac{8\pi G}{C^4} (t_{\alpha\beta} + \overset{em}{t}_{\alpha\beta} + \pi M^2 g_{\alpha\beta})$$

where

$$\overset{em}{t}_{\alpha\beta} = \frac{1}{4\pi} (H^{\mu}{}_{\alpha} F_{\mu\beta} - \frac{1}{4} F^2 g_{\alpha\beta}) \quad . \tag{5.1}$$

Now let us consider the simple model described by following formulae :

$$t^{\alpha}{}_{\beta} = \mu^{\alpha} h_{\beta} - p\delta^{\alpha}_{\beta} \quad ; \quad \bar{S}^{\alpha}{}_{\beta\gamma} = u^{\alpha} S_{\beta\gamma}, \; u^{\beta} S_{\beta\gamma} = 0 \quad ,$$

$$M_{\alpha\beta} = \rho S_{\alpha\beta}; \; j_{\alpha} = n\, q\, C\, u_{\alpha} \quad . \tag{5.2}$$

This is a some dust model, where: h_{β} - a four-vector of enthalpy, u_{α} - a four-vector of velocity, ρ - a giromagnetic ratio, n - a concentration of dust particles, q - a charge of a dust particle. Quantities (5.2) must obey the Bianchi's identities i.e. conservations laws, in particular angular momentum conservation which has the form

$$\bar{D}S_{\alpha\beta} = \frac{1}{C}\theta_{\alpha}\wedge\bar{t}_{\beta} - \frac{1}{C}\theta_{\beta}\wedge\bar{t}_{\alpha} - \frac{1}{C} (M^{\mu}{}_{\alpha}F_{\mu\beta} + M^{\mu}{}_{\beta}F_{\mu\alpha})\bar{\eta} \quad . \tag{5.3}$$

Substituting (5.2) into the formula (5.3) we obtain a formula for $t_{\alpha\beta}$:

$$t_{\alpha\beta} = (e+p)u_\alpha u_\beta + pg_{\alpha\beta} + u_\alpha u^\delta u^\gamma \tilde{\overline{\nabla}}_\gamma S_{\beta\delta} + \mathbf{q}\mathbf{u}_\alpha S^\delta{}_\beta F_{\delta\gamma} u^\gamma , \tag{5.4}$$

where $e = t_{\alpha\beta} u^\alpha u^\beta$ is density of energy. Applying the formula (5.1), (5.4) and a formula from[11]we get:

$$\tilde{R}_{\alpha\beta} - \frac{1}{2}\tilde{R}g_{\alpha\beta} = \frac{8\pi G}{c^4}[t^{\text{em.vac}}_{\alpha\beta} + (e+p - \frac{4\pi G}{c^2}s^2)u_\alpha u_\beta - \tag{5.5}$$

$$-(p-\frac{2\pi G}{c^2}s^2-2\pi q^2 s^2)g_{\alpha\beta}-C(g^{\delta\gamma}+u^\delta u^\gamma)\tilde{\overline{\nabla}}S_\gamma{}_{(\alpha}u_{\beta)}+q(u_\alpha S^\delta{}_\beta F_{\delta\gamma}u^\gamma-S^\mu{}_\alpha F_{\mu\beta})]$$

In the formula (5.5) we have explicite a term of coupling between spin and electromagnetic field. Namely it is

$$q(u_\alpha u^\gamma S^\delta{}_\beta F_{\delta\gamma} - S^\mu{}_\alpha F_{\mu\beta}) . \tag{5.6}$$

Now we conclude that (5.6) is the term that has been searched for in the Einstein-Cartan theory.

$s^2 = \frac{1}{2} S_{\alpha\beta} S^{\alpha\beta}$ and $t_{\alpha\beta}$ is a tensor of energy-momentum of electromagnetic field in vacuum. It would be very interesting to analyse this term in relation to a theory of neutron stars. In such a case q would be a nuclear magneton. The formula (5.5) really differs from the analogical formula given in [11]. First of all there is a component giving a coupling between electromagnetic field and spin on the right-hand side of equation. Secondly there have appeared additional pressure-type components.This component gives a correction to e and p. Now we estimate the contribution of this new correction in relation to that known from the Einstein-Cartan theory. To do this we assume that our fluid is a nuclear fluid and $q \rightarrow \frac{q}{m_p c}$. Taking the ratio of these two corrections we have:

$$\frac{\text{new component connected to the } 5^{\text{th}} \text{ dim.}}{\text{addit.comp.known in Einstein-Cartan theory}} = \frac{q^2}{m_p^2 G} \simeq 10^{36} .$$

An interesting fact is that ratio $\frac{q^2}{m_p^2 G}$ is simply equal to the ratio of electric interaction of two protons to their gravitational interaction. Now let us estimate a density for which this correction will be comparable to the density of energy e. We have

$$S \simeq \frac{1}{2} n\hbar, \quad 2\pi q^2 S^2 \simeq \frac{q^2 n^2 \hbar^2}{4m_p^2 C^2} \text{ and } e \simeq m_p C^2 n \quad .$$

Assuming that $e \simeq m_p C^2 n \simeq \frac{q^2 n^2 \hbar^2}{4m_p^2 C^2}$ we obtain $n_1 \simeq 10^{43} \frac{1}{cm^3}$. But it is known that the correction to e in the Einstein-Cartan theory is comparable to e for a concentration $n_2 \simeq 10^{79} \frac{1}{cm^3}$. Concentrations n_1 and n_2 correspond to matter densities:

$$q_1 = 10^{18} \frac{g}{cm^3}, \quad q_2 = 10^{54} \frac{g}{cm^3} .$$

The same order of concentration as n_1 is possible in the centre of a neutron star. So this correction should play a certain role in the evolution of neutron star, gravitational collaps, and in cosmological models [7].

REFERENCES

1. T. Kaluza, Sitzungsber. Preuss. Akdad. Wiss. 966-1002 (1921).
2. A. Lichnerowicz, "Theorie relativistes de la gravitation et de l'électromagnetisme", Masson, Paris 1955.
3. W. Bailey, W. Israel, Comm. Math. Ph. 42 (1975) 65-82.
4. W. Israel, Ph. Lett. 67 (1977) 125-128.
5. W. Israel,"Foundation of relativistic kinetic theory of spinning particles", Colloque Internationaux CNRSN° 235 Theorie Cinetiques classiques et relativists, 169-182 (1975).
6. A. Trautman, Rep. of Math. Ph. 1 (1970) 29-62.

7. M.W. Kalinowski,"Gauge fields with torsion",part I and part II, Warsaw University, preprint IFT/2/1980.
8. A. Trautman, Bull. Pol. Sci. Ser. Sci. Math. Astr. Phys. 20 (1972), 184-190, 503-506, 895-1005; 21 (1972) 345.
9. A. Trautman, Symposia Mathematica 12 (1973) 139-162.
10. F. Hehl, P. von der Heyde, G.G. Kerlich, J.M. Nester, Rev. of Modern Ph. 48 (1976) 393-416.
11. W. Arkuszewski, W. Kopczyński, V.N. Ponomariew, Ann. Inst. Henri Poincaré A21 (1974) 89-95.

Acta Physica Austriaca, Suppl. XXIII, 653–660 (1981)

ATTEMPTS AT A GEOMETRICAL UNDERSTANDING OF HIGGS FIELDS+

by

W. MECKLENBURG
Internat. Centre for Theor. Physics, Trieste

and

Scuola Internat. Superiore di Studi Anvanzati, Trieste

I describe a number of attempts for a geometrical understanding of Higgs fields in high (> 4)-dimensional theories.

The ad hoc introduction of Higgs fields in the construction of unified models [1] is widely considered as a very unappealing feature of such theories. However, as the presence of Higgs fields seems to be a necessity one might as well ask for a "good" reason for their presence.

The work to be talked about in this note is based on the assumption that among the possible "good" explanations geometrical ones rank high. In fact, the

---

+Seminar given at the XX. Internationale Universitätswochen für Kernphysik, Schladming, Austria, February 17-26, 1981.

attempts to be described below aim at a link between the appearance of Higgs fields and the properties of an underlying space-time.

Among the presently popular field theories the two most prominent ones have an intrinsic geometrical interpretation: Einstein's theory of gravity and Yang-Mills theories. Moreover, these theories may be treated on the same footing in the Kaluza-Klein approach [2].

The starting point for the models to be presented here is the observation that gravity and Yang Mills theories can be formulated quite independently of the dimension of the underlying space-time. One then has a theory containing fields ϕ depending on Minkowski (x) and N internal (y) coordinates, $\phi = \phi\,(x,y)$. In order to obtain a theory whose dynamics is that of a four-dimensional one, one then requires that the Lagrangian be independent of y apart from an overall weight factor:

$$L_{4+N}(x,y) = \mu(y)\,L_4(x) \quad . \tag{1}$$

It is not generally appreciated that Yang-Mills theories constitute a prototype for such a construction. This is the content of the celebrated Kaluza-Klein theory [2]. This theory is a special case of Einstein's gravity theory in 4+N dimensions. I will give here a brief outline of the construction, but omit gravity for simplicity.

In fact, consider a 4+N-dimensional space with flat Minkowskian part whose metric has been chosen such that a basis of orthonormal vector fields is given by (s,t,... and $\sigma,\tau,\ldots$ are internal, $\alpha,\beta,\ldots$ are Minkowskian indices)

$$\theta_s = h_s^\sigma\,\partial_\sigma \,, \qquad \partial_\sigma = \partial/\partial x^\sigma \;; \tag{2}$$

$$\theta_\alpha = \partial_\alpha + A_\alpha^\sigma\,\partial_\sigma = \partial_\alpha + A_\alpha^s\,\theta_s,\ \partial_\alpha = \partial/\partial x^\alpha \quad . \tag{3}$$

The connection coefficients of the 4+N-dimensional space can then be read off from the commutators of the vector fields (2), (3) [3]. The internal part of the space is chosen to be isomorphic to the group as a manifold, the θ_s are then the generators of the group. (This group turns out to be the gauge group.) The generator property of θ_s gives

$$[\theta_s, \theta_t] = f^v{}_{st} \; \theta_v \quad . \tag{4}$$

For θ_α to be a Minkowskian vector one requires

$$[\theta_s, \theta_\alpha] = 0 \; . \tag{5}$$

Using (4) and (5) one finds

$$\theta_s \, A^t_\alpha = f^t{}_{sv} \; A^v_\alpha \tag{6}$$

and

$$[\theta_\alpha, \theta_\beta] = F_{\alpha\beta}{}^s \; \theta_s , \; \theta_s \, F_{\alpha\beta}{}^t \;\; f^t{}_{sv} \;\; F_{\alpha\beta}{}^v \quad , \tag{7}$$

where

$$F_{\alpha\beta}{}^s = \{\partial_\alpha A^s_\beta - \partial_\beta A^s_\alpha + f^s{}_{tv} A^t_\alpha \, A^v_\beta\} \; \theta_s \quad . \tag{8}$$

Note that because of (6) and (7) both A^s_α and $F_{\alpha\beta}{}^s$ depend on the group coordinates.

As a Lagrangian one then chooses the curvature scalar of the 4+N-dimensional space; one then has, with R_G being the curvature scalar of the group,

$$R_{4+N} = -\frac{1}{4} \, (F_{\alpha\beta}{}^s F^{\alpha\beta s}) + R_G \; , \tag{9}$$

or, if gravity were to be included,

$$R_{4+N} = R_{4,\mathrm{Einstein}} - \frac{1}{4} \, F^2 + R_G \quad . \tag{10}$$

The statement (10), namely that the Yang-Mills density is part of the curvature scalar of a certain 4+N-dimensional space (N being the order of the gauge group) is the fundamental result of the Kaluza-Klein theory. Note that the Yang-Mills density F^2 is independent of the group coordinates. We thus have indeed an example of the general construction outlined above.

Attempts have been made to include Higgs fields into this framework[4,5]. This is done by relating the h^{σ}_{s} (which up to now are just the vielbeins on the group space) to the would-be Higgs fields. The most detailed analysis of such a construction is given in [5]. The Higgs potentials arrived at in this way are rather peculiar, however, and it seems difficult to see how matters can be adapted to physically relevant situations.

Another class of attempts to obtain a framework for a natural emergence of Higgs fields proceeds by exploiting the structure of high-dimensional Yang-Mills theories (compare references[6-10]). In fact, even more than gravity theories, high dimensional Yang-Mills theories lend themselves to a natural suggestion for a geometrical introduction of Higgs fields. Consider a high-dimensional Yang-Mills theory ($\alpha,\beta\ldots = 1\ldots 4$; $\sigma,\tau,\ldots = 5\ldots 4+N$),

$$L_{4+N} = \frac{1}{4}\,\{\mathrm{tr}\, F_{\alpha\beta}F^{\alpha\beta} + 2\,\mathrm{tr}\, F_{\alpha\sigma}F^{\alpha\sigma} + \mathrm{tr}\, F_{\sigma\tau}\, F^{\sigma\tau}\}\ . \qquad (11)$$

It contains gauge fields, A_{α}, and a set of would-be Higgs fields, A_{σ}. In order to recover (1), the simplest possibility is to impose the conditions

$$\partial_{\sigma}\, A_{\alpha} = \partial_{\sigma}\, A_{\tau} = 0\ , \qquad (12)$$

and to keep as an effect from the use of higher dimensions only the proliferation of the extra fields. We then get a Yang-Mills-Higgs Lagrangian with a set of Higgs fields in the adjoint representation,

$$L_{4+N} \to L_4 = \frac{1}{4} \{\mathrm{tr}\, F_{\alpha\beta}^2 + 2\, \mathrm{tr}[D_\alpha, A_\sigma]^2 + \mathrm{tr}[A_\sigma, A_\tau]^2\}, \quad (13)$$

where the covariant derivative D is defined in the usual way, $[D_\alpha, A_\sigma] = [\partial_\alpha + A_\alpha, A_\sigma]$. The Lagrangian (13) does not give rise to a spontaneous symmetry breaking of the gauge symmetry, however. In fact, even though non-zero constant ground state solutions $A_\sigma \| A_\tau$ exist for (13), it can be shown that these solutions are not classically stable [7] and that quantum corrections tend to worsen the situation [11].

An attempt to improve this situation is to re-introduce a dependence on the internal coordinates. A systematic way to do so has been given in reference [12]. The essential trick is to admit only such dependencies on the internal coordinates that could be re-expressed by local gauge transformation in 4+N dimensions. Gauge invariance then ensures the validity of (1). This requirement gives the restriction that the internal dimensions comprise a coset space of the gauge group under consideration. In practice [13], the fields are given a specified y-dependence, and gauge invariance is taken care of by algebraic constraints to be fulfilled by boundary values of the fields.

The method still poses difficulties, however. Firstly, it is not clear whether the stability problem mentioned above is avoided, as the dependence on internal coordinates is given rather ad hoc and not by solving field equations. Secondly the theory, unlike the Kaluza-Klein theory, poses an interpretation problem for the internal coordinates.

Both these difficulties would be overcome if the internal coordinates in (11) were the same as in Kaluza-Klein theory, that is, if they were the group coordinates. This then leads to a Lagrangian of the form

$$L_{4+N} = \frac{1}{4}\{F_{\alpha\beta}{}^{\sigma}F^{\alpha\beta}{}_{\sigma} + 2F_{\alpha\sigma}{}^{\tau}F^{\alpha\sigma}{}_{\tau} + F_{\sigma\tau\rho}\,F^{\sigma\tau\rho}\} \quad , \tag{14}$$

where for $M,N,\ldots = (\alpha,\sigma)$, $(\beta,\tau),\ldots$ all tensors can be written

$$F_{MNL} = A_{MN,L} - A_{ML,N} + G_{MNL} \quad , \tag{15}$$

provided we adopt the convention $A_{\alpha\beta} = 0$. The term G_{KLM} is bilinear in the fields and determined by the requirements of gauge invariance. The details of this construction have been given in reference[14]for SU_2 and in reference[15] for general gauge groups.

The model has the specific property that the would-be Higgs field $A_{\sigma\tau}$ is to be identified with the physical Higgs field. In fact in (14) the vector bosons are already massive due to a dependence on internal coordinates like in (6). By analogy with $A_{\alpha}{}^{\sigma}$,$A_{\tau}{}^{\sigma}$transforms inhomogeneously, as befits a physical Higgs field. Altogether, (14) turns out to be equivalent to a Yang-Mills-Higgs Lagrangian displaying a Goldstone-Higgs-Kibble mechanism. For general groups however, gauge invariance requires an algebraic constraint on the vacuum expectation value of the Higgs field to be fulfilled, a constraint which is easy to solve only for SU_2.

Models like (11) or (14) give a prediction for the ratio vector boson mass/Higgs boson mass, as a result of the symmetry between gauge and Higgs fields. However as this symmetry is only formal due to the specified dependence on the internal dimensions, it is not clear whether such a prediction will remain unaltered by quantum corrections. This particular problem has not been tackled for either model.

ACKNOWLEDGEMENTS

Thanks are due to Professor Abdus Salam, P. Budinich and L. Fonda, the International Atomic Energy Agency and UNESCO for the hospitality extended to me at the International Centre for Theoretical Physics, Trieste, and the International School for Advanced Studies, Trieste.

REFERENCES

1. S. Weinberg, Phys. Rev. Lett. 19 (1967) 1264; A. Salam, Nobel Symp. 8, Ed. Nils Svartholm (1968) 367; J.C. Pati and A. Salam, Phys. Rev. D8 (1973) 1240; H. Georgie and S.L. Glashow, Phys. Rev. Lett. 32 (1974) 438.
2. Th. Kaluza, Sitz. Preuss. Akad. Wiss. 966 (1921); O. Klein, Zeitschr. Phys. 37 (1926) 895; W. Mecklenburg, Phys. Rev. D21 (1980) 2149, and references therein.
3. For a Riemannian space, in an orthonormal basis, the connection coefficients may be determined either from Cartan's first identity

$$d\theta^k + \omega^k{}_{LN}\theta^N \wedge \theta^L = 0$$

or from the commutator relation

$$[\theta_k, \theta_L] = 2\omega^N{}_{KL}\theta_N \ .$$

On a group space, $\omega_{KLN} = f_{KLN}$ where f_{KLN} are the structure constants of the group.
4. Y.M. Cho and P.G.O. Freund, Phys. Rev. D12 (1975) 1711.
5. J. Scherk and J.H. Schwarz, Nucl. Phys. B153 (1979) 61.
6. G.B. Mainland and L. O'Raifeartaigh, Lett. Nuov. Cim. 10 (1974) 733.
7. W. Mecklenburg and D.P. O'Brien, Lett. Nuov. Cim. 23 (1978) 566.

8. D. Olive, Nucl. Phys. B153 (1979) 1.

9. D.B. Fairlie, J. Phys. G5 (1979) L55;
Phys. Lett. B82 (1979) 97.

10. J.G. Taylor, Phys. Lett. B83 (1979) 331.

11. W. Mecklenburg, Southampton preprint SHEP 79/80-10,
Phys. Rev. D (in press).

12. P. Forgacs and N.S. Manton, Comm. Math. Phys. 72 (1980)
15.

13. N.S. Manton, Nucl. Phys. B158 (1979) 141.

14. W. Mecklenburg, Journ. Phys. G6 (1980) 1049.

15. W. Mecklenburg, Trieste preprint IC/81/8;
Phys. Rev. D (in press).

Acta Physica Austriaca, Suppl. XXIII, 661–663 (1981)
© by Springer-Verlag 1981

MONOPOLES IN SU2-YANG-MILLS THEORY[+]

by

H. MITTER

Institut für Theoretische Physik
Universität Graz, Austria

Static solutions of the SU2-Yang-Mills theory with given charge- and current sources have been studied for spherically symmetric situations. In addition to the well-known Coulomb-Biot-Savart solution we have found a large class of nonabelian solutions, which generalize the ones found previously [1] for purely electric sources. Here I shall concentrate on the situation without electric sources, which can be described as follws. The source currents are (in a particular gauge)

$$j_a^o = 0, \quad gr_o^3 \, j_a^k = \varepsilon_{kal} \, e_{\ell} m(x) \, ,$$

where $e_k = r^k/r$, $x = r/r_o$ and r_o is an arbitrary length scale. The potentials resp. fields read

$$A_a^o = 0, \quad gA_a^k = \varepsilon_{kal} \, e_{\ell} (1-a(x))/r \, ,$$

$$E_a^k = 0, \quad gB_a^k = [e_a e_k (1-a^2) + (e_a e_k - \delta_{ak}) xa']/r^2 \, .$$

[+]Seminar given at XX. Internationale Universitätswochen für Kernphysik, Schladming, Austria, February 17-26, 1981

The static field equations amount to

$$x^2 a'' + a(1-a^2) = x^3 m \quad .$$

We assume $m(0) = 0$, $m(x) \propto x^{-4-\varepsilon}$, $\varepsilon \geq 0$ at large x. At the origin we have $a = 1 + a_2 x^2 + \ldots$ For the asymptotic behaviour at large x there are three possibilities indicating three types of solutions:

$$a_I \to +1, \quad a_{II} \to -1, \quad a_{III} \to x^{-2-\delta}, \ \delta \geq 0 \ .$$

The types I and II are found also in presence of electric sources. Type III corresponds to a monopole of fixed strength, since the leading term at large distances in B_a^k is radial and $\propto r^{-2}$. Analytical examples with exponential decrease are

$$a = x/\sinh x \ , \ m = (1-x \coth x)^2/x^2 \sinh x$$

and

$$a = P \exp(-x), m = [R-P^3 \exp(-2x)]x^{-3} \exp(-x),$$

$$P(x) = 1 + x + x^3/3 \ , \quad R = 1 + x - x^3(1-x)/3.$$

There are, of course, many more possibilities. It is noted that the magnetic field cannot always be represented as the covariant gradient of a scalar-isovector potential (which would correspond to Bogomolny's condition in a Higgs model).

We have also performed calculations with a shell distribution $m = M\delta(x-1)$.

In this case the three types are neatly separated from each other in M: type I requires $M < 0$, type II $M > 2.33$, type III $M = 1.458$. The field energy $\frac{1}{2} \int d^3r \ (B_a^k)^2$ (which is always finite) is larger than the energy of the corresponding abelian solution, for which $j_a^k \sim \delta_{a3}$ and A_3^k is given by the Biot-Savart integral.

Monopoles are therefore quite natural phenomena in classical Yang-Mills theory. It has to be observed, however, that - at least for the situations studied here - the presence of electric sources excludes monopoles: the asymptotic behaviour of type III is not possible, if $j_a^o \neq 0$. For further details cf. [2].

REFERENCES

1. R. Jackiw, L. Jacobs, C. Rebbi, Phys. Rev. D20 (1979) 474.
2. L. Mathelitsch, H. Mitter, F. Widder, Phys. Rev. D (in print).

Acta Physica Austriaca, Suppl. XXIII, 665–670 (1981)

EXTENDED KALUZA-KLEIN UNIFIED GAUGE THEORIES+

by

J.W. MOFFAT[x]
Dept. of Physics, Univ. of Toronto
Toronto, Ontario M5S 1A7, Canada

Much progress has been made in unifying strong, weak and electromagnetic interactions using gauge theories [1-4]. However the unification of these interactions with space-time still remains to be understood. One difficulty has been the ongoing problem of how to satisfactorily combine gravity and quantum theory to produce a finite perturbation theory of gravitational interactions. A complete unification of all the fields of nature should occur at the Planck energy $(\hbar c^5/G)^{1/2} \approx 10^{19}$ GeV, where gravitation becomes as important at short distances as the other forces of nature.

The possibility of constructing unified gauge theories of gravitation in a higher-dimensional extended Kaluza-Klein [5] scheme has received renewed attention [6-9]. The differences between the four dimensions of space-time and the additional microscopic ones would be generated by a spontaneous breakdown of the vacuum

---

+Seminar given at XX. Internationale Universitätswochen für Kernphysik, Schladming, Austria, February 17-26, 1981.

xSupported by the Natural Sciences and Engineering Research Council of Canada.

symmetry with an associated compactification of the extra dimensions. These extra dimensions should be considered as "real", physical dimensions at the ultimate energy level of unification $\approx 10^{19}$ GeV, corresponding to the Planck distance $(\frac{\hbar G}{c^3})^{1/2} \approx 10^{-33}$ cm [10].

We begin with a theory of gravitation in 4+N dimensions with a ground state M^{4+N}-Minkowski space-time in 4+N dimensions It is assumed that there exists a product space $M^4 \times D$ where D is a compact N-dimensional manifold and $M^4 \times D$ should be a solution of the classical field equations. The local gauge symmetries of the 4-dimensional world will be symmetries of the manifold D. The gauge symmetries "observed" at the Planck energy will be the symmetries of a manifold A of the 4+N-dimensional world and the problem of unification amounts to discovering A and D.

The coordinates of the 4+N dimensional space are $x^\Sigma = (x^\mu, y^\alpha)$, where x^μ are the coordinates of space-time and y^α the coordinates of the compact space D with $\alpha = 1,\dots,N$. Let T^i, $i = 1,\dots,k$, denote the generators of the symmetry group H of D. The action of the symmetry generators T^i on the y^α is: $y^\alpha \rightarrow y^\alpha + K^{i\alpha}(y)$, where $K^{i\alpha}(y)$ is the Killing vector associated with the T^i. The metric of the 4+N dimensional space is written in block form:

$$g_{\Sigma\Lambda}(x^\mu, y^\alpha) = \begin{bmatrix} g_{\mu\nu}(x) & \sum_i A^i_\mu(x) K^{i\alpha}(y) \\ \sum_i A^i_\mu(x) K^{i\alpha}(y) & g_{\alpha\beta}(y) \end{bmatrix}, \quad (1)$$

where the $A^i_\mu(x)$ are the massless vector gauge fields of the group H, while the $g_{\alpha\beta}(y)$ is the metric of the internal space D (scalar fields in 4-dimensional space-time).

The action in 4+N dimensions is [10]

$$S = -\frac{1}{4\kappa^2}\int d^4x \int \frac{d^N y}{V(N)} \sqrt{-g}\, [R + i\bar{\psi}_A (\gamma^\Sigma)^A_B D_\Sigma \psi^B + \lambda] , \qquad (2)$$

where $g = \mathrm{Det}(g_{\Sigma\Lambda})$, $V(N)$ is the volume of the N-dimensional space D, $\kappa = (4\pi G)^{1/2}$ and R is the scalar curvature of the 4+N dimensional space. The ψ_A are spin 1/2 fields (trivially scaled with respect to κ), where A denotes a (flat) tangent space suffix in 4+N dimensions associated with the vielbeins e^A_Σ [11]. Moreover, D_Σ is a gauge derivative operator defined by

$$D_\Sigma e^A_\Lambda = \partial_\Sigma e^A_\Lambda + (\omega_\Sigma)^A_C e^C_\Lambda \quad , \qquad (3)$$

where $(\omega_\Sigma)^A_C$ is the spin connection of the gauge fiber bundle [11]. λ is a cosmological constant.

After dimensional reduction, the action S can be shown to contain in 4 dimensions the scalar curvature R_4 of gravity and the correct (non-Abelian) kinetic energy term $\sum_i (F^i_{\mu\nu})^2$. It also contains $i\bar{\psi}_a (\gamma^\mu)^a_b D_\mu \psi^b$ and $g^{\rho\lambda} D_\rho g_{\alpha\beta} D_\lambda g^{\alpha\beta}$, a Higgs symmetry breaking potential $V(\phi)$ and a Yukawa-type coupling which can break the fermion masses [10]. The cosmological constant λ is needed to stabilize the vacuum. The coordinate invariance group G in 4+N dimensions has been spontaneously broken down to the symmetries of $D \times M^4$.

We must now find the minimal group G which contains $D \times M^4$ i.e. contains H such that $H \supset SU_C(3) \times SU(2) \times U(1)$, since known particle interactions can be described by $SU_C(3) \times SU(2) \times U(1)$. The total dimensions d of the 4+N dimensional space must be even to accomodate the correct (complex) spinor structure ($\psi_L \neq \psi_R$) [9]. There are two minimal choices:

i) $SO(13,1) \supset SO(10) \times SO(3,1)$

ii) $SU(8,1) \supset SU(5) \times U(1) \times SU(3,1)$.

In case i), the grand unified gauge scheme SO(10) [4] is

combined with the local gauge group SO(3,1) of general relativity in the minimal non-compact group SO(13,1) in a fourteen-dimensional space. In ii), the grand unified gauge scheme SU(5) × U(1) = U(5) (SU(5) is the Georgi-Glashow model [3]) is combined with SU(3,1) as the minimal product sub-group in four dimensions, where SU(3,1) is the local gauge group of a theory of gravity [12-15] based on a sesquilinear metric [11,16]. The minimum number of (real) dimensions of the manifold A with the symmetry SU(8,1) is d = 16, but we could also consider the higher-dimensional space d = 18. The groups SU(3,1) and SU(8,1) have double covering groups,e.g. for SU(3,1) it is C spin (3,1)/W = = SU(3,1). Since the topological manifold of SU(3,1) is $SU(3) \times S^1 \times R^6$ (S^1 is the circle) with d = 15, we know that SU(3,1) is not simply connected and therefore has a double covering group (because of S^1) with spin 1/2 representations. For SO(3,1), the double covering group is well-known to be SL(2,C).

The symmetry breaking sequence for case i) is:

$$SO(13,1) \to SO(10) \times SO(3,1)$$
$$\to SU(5) \times SO(3,1) \to SU_C(3) \times SU(2) \times U(1) \times SO(3,1)$$
$$\to SU_C(3) \times U(1) \times SO(3,1)$$

or

$$SO(13,1) \to SO(10) \times SO(3,1)$$
$$\to SU_C(4) \times SU(2) \times SU(2) \times SO(3,1)$$
$$\to SU_C(3) \times SU(2) \times U(1) \times SO(3,1)$$
$$\to SU_C(3) \times U(1) \times SO(3,1) \quad .$$

Here SO(13,1) has a complex 64-dimensional fermion representation, while SO(10) has a 16-dimensional anomaly-free fermion representation. The symmetry breaking sequence for the case ii) is:

$$SU(8,1) \to SU(5) \times U(1) \times U(1) \times SU(3,1)$$
$$\to SU_C(3) \times SU(2) \times U(1) \times U(1) \times SU(3,1)$$
$$\to SU_C(3) \times U(1) \times SU(3,1) \quad .$$

There is an additional superheavy neutral boson (U(1)symmetry) that could predict neutral current structure. SU(8,1) has a complex 9-dimensional fermion representation, while SU(5) has the conventional 5-dimensional anomaly-free fermion representation. SU(3,1) contains a spin 2 graviton and a massless spin 0 boson.

An interesting problem that needs further investigation is how the fundamental single couple constant, $\kappa = (4\pi G)^{1/2}$, gets broken down to the four "low energy" coupling constants observed in nature.

REFERENCES

1. A. Salam, in Proceedings of the Eighth Nobel Symposium on Elementary Particle Theory, Relativistic Groups, and Analyticity, Stockholm, Sweden, 1968, edited by N. Svartholm (Almquist and Wiksell, Stockholm, 1968).
2. S. Weinberg, Phys. Lett. 19 (1967) 1264.
3. H. Georgi and S. Glashow, Phys. Rev. Lett. 32 (1974) 438.
4. H. Fritzsch and P. Minkowski, Ann. Phys. (N.Y.) 93 (1975) 193.
   H. Georgi, in Particles and Fields, G.E. Carlson (ed.) (AIP-N.Y. 1975).
   H. Georgi and D.V. Nanopoulos, Phys. Lett. B82 (1979) 392. Nucl. Phys. B155 (1979) 52.
5. Th. Kaluza, Sitzungsber. Preuss. Akad. Wiss. Berlin, Math.-Phys. Kl., (1921) 966.
   O. Klein, Z. Phys. 37 (1926) 895; Arkiv. Mat. Astron. Fysik B 34A (1946); Helv. Phys. Acta Suppl. IV (1956) 58.
6. Y.M. Cho and P.G.O. Freund, Phys. Rev. D12 (1975) 1711.

7. J. Scherk and J.H. Schwarz, Nucl. Phys. B135 (1979) 61.
8. E. Cremmer and B. Julia, Phys. Lett. 57B (1975) 463.
9. E. Witten, University of Princeton Report (1981).
10. J.W. Moffat, University of Toronto Reports, May 1981.
11. J.W. Moffat, Ann. Inst. Henri Poincaré 34 (1981) 85.
12. J.W. Moffat, Phys. Rev. D19 (1979) 3554; D19 (1979) 3562.
13. J.W. Moffat, J. Math. Phys. 21 (1980) 1798.
14. J.W. Moffat, Can. J. Phys. 59 (1981) 283.
15. R.B. Mann, J.W. Moffat and J.G. Taylor, Phys. Lett. 97B (1980) 73.
16. G. Kunstatter and R. Yates, J. Phys. A, Math. Gen. 14 (1981) 847.

Acta Physica Austriaca, Suppl. XXIII, 671–678 (1981)

RIGOROUS ESTIMATES OF THE ELASTIC $e^-(\mu^- p)$ SCATTERING AMPLITUDE[+]

by

L. PITTNER[+x*]
Institut für Theoretische Physik
Universität Graz, Austria

The following efforts shall be concerned with rigorous estimates of the scattering amplitude for charged particles off muonic hydrogen atoms below the excitation threshold. In this low energy region non-relativistic quantum mechanics may be used. Since the Coulomb potential of the proton is screened by the ground state muon, time-independent scattering theory can be applied without any long range troubles. In order to avoid the difficulties arising from the so-called Hughes-Eckart term [1], the proton is kept fixed throughout our investigations.

The well-known projection method [1] can be used to prove that no discrete bound state of the $(e^-\mu^- p)$ system exists, which enables one to construct rigorous bounds of the $e^-(\mu^- p)$ scattering amplitude. An appropriate method starts from some suitable trial potential instead of the actual scattering potential, which allows for an easy cal-

---

[+] Seminar given at XX. Internationale Universitätswochen für Kernphysik, Schladming, Austria, February 17-26, 1981.

[x] Work supported in part by the Fonds zur Förderung der wissenschaftlichen Forschung in Österreich, Projekt Nr.3569.

[*] This work has been performed during the authors stage at the Institut für Theoretische Physik, University of Vienna.

culation of the corresponding trial transition amplitude, then aiming at an estimate of the resulting variational error. Physical intuition leads to the trial potential which is obtained by projecting the scattering potential onto the $(\mu^- p)$ ground state. An operator formulation of Kohn's variational principle [2] due to H. Narnhofer and W. Thirring [1] turns then out to be convenient to control the variational error. This error amplitude is then expanded into an appropriate multiple-stage Born series, which in turn can be majorized by some rather complicated combination of geometrical series, essentially an expansion into powers of $(m_1/m_2)^{1/2}$ where m_1 and m_2 denote the masses of e^- and μ^-, respectively.

The above mentioned projection method can be used to estimate the variational error just at the scattering threshold [3], but it fails to succeed for non-zero scattering energies.

SCATTERING AMPLITUDE

The transition operator for $e^-(\mu^- p)$ scattering is defined by

$$T(\lambda) := V_s - V_s R(\lambda) V_s = (V_s^{-1} + R_{as}(\lambda))^{-1}$$

with the resolvents

$$R(\lambda) := (H-\lambda)^{-1}, \quad R_{as}(\lambda) := (H_{as}-\lambda)^{-1}, \quad \lambda \text{ non-real,}$$

of the non-relativistic Hamiltonian operators

$$H := H_{as} + V_s, \quad V_s := \frac{\alpha}{r_{12}} - \frac{\alpha}{r_1}, \quad r_{12} := |\vec{x}_1 - \vec{x}_2| ,$$

$$H_{as} := K_1 + C_2, \quad C_2 := K_2 - \frac{\alpha}{r_2} ,$$

where the kinetic energies of electron and muon are denoted by

$$K_1 : = -\Delta_1/2m_1 ,\ K_2 : = -\Delta_2/2m_2 ,\ m_2/m_1 = 206.8 ,$$

and α denotes the fine-structure constant. These definitions describe $e^-(\mu^- p)$ scattering in the fixed-proton limit.

This transition operator has to be sandwiched between the asymptotic states

$$\Phi_1(\vec{x}_1,\vec{x}_2) : = (2\pi)^{-3/2} \exp(i\vec{p}_1\cdot\vec{x}_1) g(r_2) ,$$

$$\Phi_1'(\vec{x}_1,\vec{x}_2) : = (2\pi)^{-3/2} \exp(i\vec{p}_1'\cdot\vec{x}_1) g(r_2) ,$$

with the normalized muon ground state

$$g(r_2) : = \pi^{-1/2}(m_2\alpha)^{3/2}\exp(-m_2\alpha r_2),\quad (C_2 + m_2\alpha^2/2)g(r_2) = 0.$$

The conserved scattering energy

$$E : = \vec{p}_1^{\,2}/2m_1 - m_2\alpha^2/2 = \vec{p}_1'^{\,2}/2m_1 - m_2\alpha^2/2$$

is restricted to the interval between the scattering threshold and the muon excitation threshold, i.e.

$$-m_2\alpha^2/2 \leq E < -m_2\alpha^2/8 .$$

The scattering angle Θ is related to the momentum transfer $\vec{q}$ by

$$\vec{q}^{\,2} = 2\vec{p}_1^{\,2}(1-\cos\Theta),\quad \Theta : = \arccos(\vec{p}_1\cdot\vec{p}_1'/\vec{p}_1^{\,2});\ \vec{q} : = \vec{p}_1' - \vec{p}_1 .$$

In order tp perform an accurate estimate of the scattering amplitude $\lim_{\varepsilon\to 0} \langle\Phi_1'|T(E+i\varepsilon)\Phi_1\rangle$, we shall compare this amplitude with an appropriate trial transition amplitude.

VARIATIONAL PRINCIPLE

As indicated in the introduction, we start from some suitable trial potential V_t instead of the scattering potential V_s such that the corresponding trial transition amplitude can be calculated easily. Following the physical considerations in the introduction we choose

$$H_t := H_{as} + V_t ,$$

$$V_t(r_1) := \int d^3x_2\, g^2(r_2) V_s(\vec{x}_1,\vec{x}_2) = -\alpha(r_1^{-1}+\alpha m_2)\exp(-2m_2\alpha r_1),$$

such that the corresponding trial transition operator

$$T_t(\lambda) := V_t - V_t R_t(\lambda) V_t = (V_t^{-1} + R_{as}(\lambda))^{-1}$$

with the trial resolvent

$$R_t(\lambda) := (H_t - \lambda)^{-1}$$

shall be compared to the transition operator $T(\lambda)$ between the above defined asymptotic states, for complex $\lambda := E+i\varepsilon$, $\varepsilon > 0$, $\varepsilon \to 0$.

The harmonic mean equations

$$T^{-1}(\lambda) = V_s^{-1} + R_{as}(\lambda), \qquad T_t^{-1}(\lambda) = V_t^{-1} + R_{as}(\lambda)$$

yield the relations

$$T^{-1}(\lambda) - T_t^{-1}(\lambda) = V_s^{-1} - V_t^{-1}, \quad T(\lambda) - T_t(\lambda) = T_t(\lambda)(V_t^{-1} - V_s^{-1})T(\lambda),$$

hence the variational principle [1]

$$T(\lambda) - T_t(\lambda) = T_t(\lambda) V_t^{-1} (W - WR(\lambda)W) V_t^{-1} T_t(\lambda) ,$$

where the potential difference W is defined by

$$W: = V_s - V_t = \frac{\alpha}{r_{12}} + V_b(r_1)$$

with the bounded negative potential

$$V_b(r_1): = -\frac{\alpha}{r_1}(1-\exp(-2m_2\alpha r_1)) + \alpha^2 m_2 \exp(-2m_2\alpha r_1).$$

Since obviously

$$V_t^{-1}T_t(\lambda)\Phi_1(\vec{x}_1,\vec{x}_2) = (2\pi)^{-3/2}(1-(K_1+V_t-m_2\alpha^2/2-\lambda)^{-1}V_t) \cdot \exp(i\vec{p}_1\cdot\vec{x}_1)g(r_2)$$

and analogously for $\Phi_1'(\vec{x}_1,\vec{x}_2)$, we enjoy

$$\langle\Phi_1'|T_t(\lambda)V_t^{-1}WV_t^{-1}T_t(\lambda)\Phi_1\rangle = 0 ;$$

therefore the variational error is of second order in terms of the potential difference W.

BORN EXPANSIONS

The trial transition amplitude $T_t(\lambda)$ can be expanded easily with respect to the effective potential V_t, using the spectral decomposition of the muon Coulomb operator and the Rollnik norm $\|V_t\|_R$ [4]; one obtains the estimate

$$|\langle\Phi_1'|(T_t(\lambda) - V_t)\Phi_1\rangle| \leq \frac{\rho}{1-\rho} v,$$

$$v: = -\langle\Phi_1|V_t\Phi_1\rangle = \frac{4\pi}{(2\pi)^3 2m_2^2\alpha},$$

$$\rho: = \frac{2m_1}{4\pi}\|V_t\|_R = \sqrt{7/6}\,\frac{m_1}{m_2} .$$

The main part of our work consists in an appropriate

multiple-stage Born expansion of the variational error amplitude

$$<\Phi_1'|(T(\lambda)-T_t(\lambda))\Phi_1> = -<\Phi_1'|T_t(\lambda)V_t^{-1}WR(\lambda)WV_t^{-1}T_t(\lambda)\Phi_1> \quad .$$

Whereas the products $T_t(\lambda)V_t^{-1}$ and $V_t^{-1}T_t(\lambda)$ can be estimated easily via Born expansions as mentioned just above, convergence of the Born series

$$WR(\lambda)W = - \sum_{k=1}^{\infty} W(-R_t(\lambda)W)^k$$

can be proved only by rather complicated methods. Estimates involving the trial resolvent $R_t(\lambda)$ are always performed with the aid of the spectral resolution of the muon Coulomb operator.

One then needs bounds on the operator norm $\|WR_t(\lambda)W\|_{op}$, i.e. estimates of the kind

$$<\Psi|r_{12}^{-1}(K_1-\lambda)^{-1}r_{12}^{-1}\Psi> \leq 4m_1\|\Psi\|^2 \quad ,$$

which can be derived in coordinate space by spherical rearrangement [5] and using Hilbert's inequality [6]; one ends up with the estimate [7]

$$\|WR_t(\lambda)W\|_{op} \leq 16\ m_1\ \alpha^2(1-\rho)^{-1} \quad .$$

Since one succeeds only with operator norm estimates of $WR_t(\lambda)W$, but not of $\sqrt{W}R_t(\lambda)\sqrt{W}$, rather complicated combinatorial considerations are necessary to majorize the Born expansion of $WR(\lambda)W$. One uses the fact $<g|Wg> = 0$ which follows trivially from the very definition of W, and the obvious estimate

$$\|R_t^{(tr)}(E)\|_{op} \leq (\mu_2 - E)^{-1} \quad ,$$

where $R_t^{(tr)}$ denotes the "truncated" trial resolvent obtained by excluding the muon ground state, and μ_2 denotes the lowest level of hydrogen excitation.

RESULTS AND OUTLOOK

The geometrical-type majorant of the Born expansion of the variational error amplitude converges in the energy region

$$-m_2\ \alpha^2/2 \leq E < -(0.657)\ m_2\ \alpha^2/2\ .$$

The relative error, i.e. variational error amplitude compared to $v = -\langle\Phi_1|V_t\Phi_1\rangle$, amounts to 65.3% at the threshold and reaches 100% at the energy $-(0.88)\ m_2\ \alpha^2/2$.

Although the numerical methods based on Schwinger's variational principle [8] yield results of astonishing accuracy, the foregoing operator analysis furnishes an insight into the mathematical structure of the Coulomb three-particle system in the energy region of elastic scattering, and allows especially for an estimate of the error resulting from projection of the scattering potential onto the hydrogen ground state.

Our next effort shall be devoted to extend the domain of convergence of the Born expansion of the variational error amplitude beyond the first muon excitation level by use of an appropriate trial potential $V_t^{(N)}$ which is constructed by projecting the scattering potential onto the lowest N energy levels of the muon, i.e.

$$V_t^{(N)} := P^{(N)} V_s P^{(N)}\ , \qquad P^{(N)} := \sum_{n=1}^{N} P_n\ .$$

The highest object of one's desire should be estimates

of the electron-hydrogen scattering amplitude for comparison with experimental data (compare data and ref. in the survey over "The variational method in atomic scattering" by J. Callaway [9]), but apart from the mass ratio which is then equal to one, statistics gives rise to awful troubles here, as is known at least since R.N. Hill performed his ingenious proof of the existence of exactly one bound state of the H^- ion [10].

ACKNOWLEDGEMENTS

Thanks are due to W. Thirring for initiating this work, and to B. Baumgartner, H. Grosse, H. Narnhofer, A. Pflug, and W. Thirring for stimulating discussions.

REFERENCES

1. W. Thirring, Lehrbuch der Mathematischen Physik 3, Springer (1979).
2. M. Reed and B. Simon, Methods of Modern Mathematical Physics 3, Academic Press (1979).
3. H. Grosse, H. Narnhofer, and W. Thirring, Journ.Phys. B12 (1979) L189.
4. B. Simon, Quantum Mechanics for Hamiltonians defined as Quadratic Forms, Princeton Univ. Press (1971).
5. H.J. Brascamp, E.H. Lieb, and J.M. Luttinger, Journ. Funct. Anal. 17 (1974) 227.
6. G.H. Hardy, J.E. Littlewood, G. Pólya, Inequalities, Cambridge Univ. Press (1967), equ. 9.9.1.
7. An extensive version of the author's work will be published in the Reports on Math. Phys..
8. K. Takatsuka and V. McKoy, Phys.Rev.Lett. 45 (1980) 1396, and references therein.
9. J. Callaway, Physics Reports 45 (1978) 89.
10. R.N. Hill, Journ. Math. Phys. 18 (1977) 2316.

Acta Physica Austriaca, Suppl. XXIII, 679–682 (1981)

ON THE RELATION BETWEEN BARE AND DRESSED CHARGE IN QED+

by

J. RAYSKI
Institute of Physics
Jagellonian University, Krakow, Poland

Our aim is to show that the exact relation between the bare charge e_o and the dressed e is (in a properly handled QED) of the same type as in the Lee model [1], i.e., together with the cutoff being removed, the bare charge approaches the value zero from the domain of imaginary values whereas the dressed charge remains finite, different from zero, and of course real.

This finding sharply contradicts the usual view that the dressed elementary charge must be smaller than the bare charge [2] and that the dressed charge tends to zero so that QED is, in fact, an interaction-free theory [3] leading to a trivial S-matrix S = 1.

Let us introduce first a regularization of the D-functions and photon propagators as suggested first by Stueckelberg and constituting a part of the procedure called Pauli-Villars regularization. The denominator of

+Seminar given at XX. Internationale Universitätswochen für Kernphysik, Schladming, Austria, February 17-26, 1981.

the regularized photon propagator is

$$\frac{M^2}{k^2(M^2-k^2)} \quad \text{or better} \quad \frac{2M^4}{k^2(M^2-k^2)(2M^2-k^2)} \qquad (1)$$

instead of $1/k^2$, with a suitable prescription for bypassing the poles on the real k_o-axis. A regularization of the fields describing the sources of the electromagnetic field is not necessary for our purposes.

It is easily seen that the theory regularized according to (1), as long as M is finite, possesses a status of a superrenormalizable formalism! A simple power counting shows that the only divergent integrals are those represented by the lowest order graphs, viz. closed loops describing vacuum polarization, whereas higher order graphs mean corrections that yield already finite contributions. Therefore a discussion of the results obtained in the lowest order of the perturbation calculus is already significant and sufficient as regards the peculiarities of the charge renormalization due to the appearance of divergent integrals.

As is well known, the relation between the bare and the dressed charge of the electron (it is better to use a mixture of charged fields with spins 0 and 1/2 forming the simplest supermultiplet with s = 0) obtained in the one-loop approximation is [2,4]

$$e = \frac{e_o}{1 + Be_o^2} \qquad (2)$$

where

$$B = \frac{1}{6\pi^2}\left(\frac{1}{3} + \frac{1}{2}\int_0^\infty \frac{d\alpha}{\alpha}\ \text{sgn}\ \alpha \quad \cos m^2\alpha\right) \quad . \qquad (3)$$

The coefficient B is infinite because the integral involved

in (3) is divergent. Therefore let us introduce a cutoff by replacing zero by ε/m^2 in the lower limit of integration. A superficial look confirms apparently the usual view that $e < e_o$ ($B(\varepsilon)$ is positive unless the cutoff ε is unreasonably large) as well as the fact that the limit is $e = 0$ if the cutoff is removed ($\varepsilon \to 0$). However, if the relation (2) is significant and already characteristic for the peculiarities arising from the divergent expressions connected with the charge renormalization, one should proceed as follows: First of all solve (2) with respect to e_o,

$$e_o = (1 \pm (1 - 4B(\varepsilon)e^2)^{1/2})/2B(\varepsilon)e \quad , \tag{4}$$

and choose the solution with the minus sign since in the limit $\varepsilon \to \infty$ or $B(\varepsilon) \to 0$ one should get $e = e_o$. An interesting and rather unexpected result is obtained when diminishing the cutoff ε gradually: if $B(\varepsilon)$ becomes larger than $1/4e^2$ the bare charge (4) becomes complex which is interpretable in the following terms: The sequence of formalisms characterized by a variable degree of concentration of the elementary charge exhibits an effect similar to a transition of the phase along with the change of concentration of the charged matter. As long as $B(\varepsilon)$ is finite, there is a unique relation between e and e_o although unexpectedly e_o appears to be complex. In the limit of a full removal of the cutoff the unique relation between e and e_o disappears: e_o tends to zero from the domain of imaginary values

$$e_o \to -iB^{-1/2} \tag{5}$$

while e keeps an arbitrary finite value.

From the above analysis there emerges the following procedure: (a) First of all, introduce the regularization (1). (b) Introduce the cutoff ε for the integrals involved in the expressions for the vacuum polarization, and perform the charge renormalization. (c) Remove the cutoff in the integrals involved in (3), and (d) perform the limit transition with the auxiliary masses $M \to \infty$ at the very end. This sequence of procedures and limit transitions seems to be quite essential!

The circumstance that the bare charge ultimately vanishes is satisfactory. In view of the presence of infinities in the local QED it is plausible that a finite result $0 < e < \infty$ may arise only because of a vanishing e_o. On the other hand, the effect of a "phase transition" as well as the fact that e_o approaches zero from the domain of imaginary values is rather unexpected.

Our result contradicts the "exact" one of Källén [2]. The explanation of this discrepancy may be seen in the fact that one of Källén's assumptions (viz. a regular character of the energy spectrum and absence of ghosts) is not satisfied in QED as well as in other local quantum field theories.

Similar results hold true in superrenormalizable versions of the theory, e.g. in QED in two dimensions, or in a non-local but gauge invariant electrodynamics of a Peierls-McManus type [5].

REFERENCES

1. T.D. Lee, Phys. Rev. 95 (1954) 1329.
2. G. Källén, Quantum Electrodynamics, Springer Verlag 1972.
3. L. Landau, Y. Pomeranchuk, Dokl. Akad. N. SSSR 102 (1955) 48.
4. R. Jost, J. Rayski, Helv. Phys. Acta 22 (1949) 457.
5. J. Rayski, Intern. Journal of Theor. Phys., in press.

Acta Physica Austriaca, Suppl. XXIII, 683–687 (1981)

DECOMPOSITION OF THE COHERENCE RELATION IN C^*–ALGEBRAIC MANY BODY PHYSICS+

by

A. RIECKERS
Institut für Theoretische Physik
Univ. Tübingen, 7400 Tübingen, Germany

In quantum optics it is well known that interference effects of coherent light scattering do not depend on a monochromatic pure state preparation. A similar generalization of coherence is especially desirable in many body physics where the scattering of local perturbations and quasi-particle excitations of global equilibrium states are investigated, and where it is known that these are never pure states for finite temperatures.

We use here the C^*-algebraic description of (infinite) many body systems, which is based firstly on a quasi-local C^*-algebra $A = \overline{U_\Lambda A_\Lambda}^{norm}$ ($A_\Lambda = B(H_\Lambda)$, Λ bounded region in R^3 or Z^3, H_Λ separable Hilbert space), and secondly on a full folium S_o of locally normal states on A [1,2]. With the pair (A, S_o) there is canonically associated the von Neumann algebra $M_o = (\Sigma_{\phi \in S_o} \pi_\phi(A))''$, $(\pi_\phi, H_\phi, \Omega_\phi)$ being the GNS triple of ϕ. There is an affine bijection $\phi \to \tilde{\phi}$ of S_o onto the normal states $S_n(M_o)$ [2]. We then define for $\phi \in S_o$ the support (central support) as that of $\tilde{\phi}$ in M_o and denote

+Seminar given at XX. Internationale Universitätswochen für Kernphysik, Schladming, Austria, February 17-26, 1981.

it by $S_\phi (C_\phi)$. This enables us to transfer the definitions and reasonings of [3] to the present situation.

Definition: Three states $\phi,\psi,\chi \in S_o$ are said to satisfy the coherence relation $K(\phi,\psi,\chi)$, if $S_\phi \wedge S_\psi = S_\phi \wedge S_\chi = S_\psi \wedge S_\chi = 0$ and $S_\phi \vee S_\psi = S_\phi \vee S_\chi = S_\psi \vee S_\chi$.

We also write $K(\phi,\psi)$, if there is a $\chi \in S_o$ with $K(\phi,\psi,\chi)$. It is shown in [3] that $K(\phi,\psi)$ is valid, iff $S_\phi \wedge S_\chi = 0$ and $S_\phi \sim S_\psi$ ($\sim$ is equivalence of projections). This implies $C_\phi = C_\psi$. A simultaneous decomposition of ϕ and ψ into factor states on A can be achieved if one finds a faithful state ρ, with $S_\phi, S_\psi \leq S_\rho$ and H_ρ separable. For, then $A \subset M_\rho = \pi_\rho(A)''$ and ϕ,ψ are extendable to normal states on M_ρ. The central measure μ of ρ decomposes the quantities $M_\rho = \int M_\rho^\omega d\mu(\omega)$ as well as $\phi = \int \phi^\omega d\mu(\omega)$ and $\psi = \int \psi^\omega d\mu(\omega)$, where ϕ^ω, ψ^ω are normal states on M_ρ^ω [4]. But M_ρ^ω can be identified with the GNS-representation of M_ρ with respect to $\omega \in S(M_\rho)$ ([1], Ch.4.2.2). Because of $A \subset M_\rho$ the ω, ϕ^ω, and ψ^ω can be also interpretated as states on A.

Theorem: Employing the notation introduced above we have for $\phi,\psi \in S_o$, $K(\phi,\psi)$, iff there is a faithful $\rho \in S_o$ with central measure μ, such that $S_\phi, S_\psi \leq S_\rho$ and $K(\phi^\omega,\psi^\omega)$ is valid μ - a.e..

Proof: (α) $K(\phi,\psi) \rightarrow C_\phi = C_\psi =: C$. C is σ-finite (i.e. dominates an at most countable family of pairwise orthogonal projections in M_o), and thus there is a $\rho \in S_o$ with $S_\rho = C = C_\rho$. Since by construction A is simple, π_ρ is faithful. Thus $A \subset M_\rho = M_o C_\rho$ and ρ is not only faithful on M_ρ but also on A. (β) Assume now $\rho \in S_o$ is faithful, $S_\phi, S_\psi \leq S_\rho$, H_ρ is separable, and use the central measure μ to decompose quantities Then one can show that

$$\begin{array}{llll} S_\phi^\omega = S_{\phi^\omega} & \mu \text{ - a.e.}, & & \\ S_\phi \wedge S_\psi = 0 & \text{iff } S_\phi^\omega \; S_\psi^\omega = 0 & & \mu \text{ - a.e.}, \\ S_\phi \sim S_\psi & \text{iff } S_\phi^\omega \sim S_\psi^\omega & & \mu \text{ - a.e.} \end{array}$$

(for the last two statements cf.[5]). With this,

$$K(\phi,\psi) \quad \text{iff} \quad K(\phi^{\omega},\psi^{\omega}) \quad \mu - \text{a.e.} \quad .$$

For illustration of our result consider the limiting Gibbs state ρ_{β}, at natural temperature β, of an infinite many body system. If a phase transition occurs at β_o, ρ_{β} has a non-trivial central decomposition $\int \rho_{\beta}^{\omega}\, d\mu(\omega)$ for $\beta > \beta_o$. In fact does the state preparation "Increase β slowly above β_o!" correspond to ρ_{β} itself and not to a pure phase ρ_{β}^{ω}, since the values of the classical observables which serve to identify the pure phases are not completely determined by this prescription. For S_o one may take the smallest folium which contains ρ_{β}, since this is indeed full in virtue of the faithfulness of ρ_{β}.

Assume now that during the macroscopic state preparation there occur local disturbances of the equilibrium state such as the excitation of local phonons or impurity centers by light. The prepared states $\phi,\psi,\ldots$ are then in S_o and decompose by means of μ into local perturbations of the pure phases. Under which conditions is it possible that $K(\phi,\psi)$ is valid? According to our Theorem we investigate the relations $K(\phi^{\omega},\psi^{\omega})$ where ω is in the support of μ. But $\phi^{\omega},\psi^{\omega}$ are then normal states on the type III-factor M_{ω}. If they have non-intersecting supports $S_{\phi^{\omega}}, S_{\psi^{\omega}}$ then $S_{\phi^{\omega}} \sim S_{\psi^{\omega}}$ is automatically satisfied [3]. Thus $K(\phi,\psi)$ is valid iff $S_{\phi} \wedge S_{\psi} = 0$, a condition which can experimentally be realized without difficulty.

Especially instructive in this respect is the BCS-model, the local Hamiltonian of which reads in the quasi-spin formulation

$$H_{\Lambda} = \sum_{i \in \Lambda} \varepsilon(1-\sigma_i^z) - \frac{g}{2|\Lambda|} \sum_{i,j \in \Lambda} \sigma_i^- \sigma_j^+ \quad .$$

Here the first term describes the reduced kinetic energy of the electrons and the second gives the pair-interaction.

(Λ is a finite subset of the lattice and $|\Lambda|$ the number of points in Λ.) For $A \in A_\Lambda$ fixed one calculates the expectation value in the limiting Gibbs state as ([6,7]):

$$<\rho_\beta;A> = \int_0^{2\pi} <\rho_\beta^\Theta;\ A > d\Theta/2\pi$$

$$= \int_0^{2\pi} tr_{H_\Lambda} \{c \exp[\beta \sum_{i\in\Lambda} (\sigma_i^x r \cos\Theta + \sigma_i^y r \sin\Theta + \varepsilon\sigma_i^z)]A\}\frac{d\Theta}{2\pi}$$

where c is a normalization constant and r depends on β. The very calculation of ρ_β leads automatically to the central decomposition of this state, where the phase angle Θ provides a natural parametrization of the supporting set of the central measure. As mentioned before, the same measure induces also a decomposition of local perturbations ϕ,ψ of ρ_β. The phase transition leads thus to a new classical observable $\hat{\Theta} := \int_0^{2\pi}\Theta 1^\Theta\, d\Theta/2\pi$ in the center of $M_{\rho\beta}$. The values of the macroscopic phase operator $\hat{\Theta}$ create new super-selection rules and decompose the coherence relation. Even if one allows for additional manipulations whatsoever in the super-conducting state preparation it is not possible to fix $\hat{\Theta}$ at a prescribed value. Thus to the isolated super-conductor applies the coherence relation only in the integrated form $K(\phi,\psi)$ and not in the pure phase version $K(\phi^\Theta,\psi^\Theta)$.

REFERENCES

1. O. Bratteli and D.W. Robinson, Operator Algebras and Quantum Statistical Mechanics I (1979) and II (1981), Springer, Berlin.
2. H. Roos, Physica 100 A (1980) 183.
3. G.A. Raggio and A. Rieckers, Coherence and Incompatibility in W^*-Algebraic Quantum Theory, to be published.

4. M. Takesaki, Theory of Operator Algebras I, Van Nortrand, Princeton, 1979.
5. J. von Neumann, Ann. Math. 50 (1949) 401.
6. W. Thirring, Commun. Math. Phys. 7 (1968) 181.
7. W. Fleig, Diplom thesis, Tübingen 1981.

Acta Physica Austriaca, Suppl. XXIII, 689–693 (1981)

OPTICAL THEOREM FOR THREE-THREE SCATTERING+

by

S. SERVADIO
Istituto di Fisica dell'Università
Pisa, I.N.F.N. Sezione di Pisa, Italy

The elastic 3-3 scattering is known to have [1] a non-uniform asymptotic behaviour in the relative coordinate space R_6. Most difficulties stem from double-scatterings. Double scattering with intermediate state on mass shell, i.e. singularities of connected graphs, provide an anomalous $O(\rho^{-2})$ wave in those directions that can be reached by classical double collision. Elsewhere double scattered waves appear as spherical waves with the same fall-off $O(\rho^{-5/2})$ as the R_6 Green's function. At any finite ρ the matching is realized by Fresnel's integral of a complicated phase argument.

The 3-body wave function however has additional peculiar features. No matter how far one goes from the origin of R_6 there are portions of the spherical surface where some pair is close; if that pair has scattered last its wave is not yet asymptotic. Although these regions have small weight compared to the whole spherical surface the wave function is substantially larger than elsewhere and care is needed to assess their importance.

+Seminar given at XX. Internationale Universitätswochen für Kernphysik, Schladming, Austria, February 17-26, 1981.

Apart from unconnected graphs and the anomalous double scatterings the elastic 3-3 scattering has an $O(\rho^{-5/2})$ outgoing spherical wave contributed by off-shell double scattering and the higher order collisions. This is usually referred to as the "truly 3-body scattering" since it cannot be reduced to on-shell 2-body t-matrices and is regarded as the interesting part. What condition does unitarity impose on this amplitude? Do we know an optical theorem for 3-3 scattering?, i.e.something like a relation between the forward amplitude and the total cross section?

The 2-body case suggests the following. Take the stationary solution arising out of an incident plane wave; calculate the flux of representative points across a spherical surface of radius ρ in R_6. That flux is zero, so each coefficient of its asymptotic expansion as $\rho \to \infty$ must vanish. The first non-trivial term will give the "truly 3-body optical theorem".

The flux is $\Phi(\rho) = \oint d\vec{\sigma}\cdot\mathrm{Re}(i\Psi \;\; \vec{\nabla}\Psi)$ where $d\vec{\sigma}$ is the 5-dimensional surface element of radius ρ. The wave function is analysed in a multiple scattering approach as

$$\Psi = \chi + \sum_i \Phi^i + \sum_{i\neq j} \Phi^{i,j} + \Phi^{\text{higher}} ,$$

where the indices run from 1 to 3 labelling which particle has not taken part in the collision. For example, $\Phi^{1,3}$ is the double scattering in which first pair (1,2), then pair (2,3) scatter. Hyperspherical coordinates can be introduced for points of the surface (i.e. ray directions in R_6) in several orthogonally equivalent manners. One can choose $(\omega,\hat{X},\hat{Y})$ where $\omega = \mathrm{arctg}\,(Y/X)$, $\vec{X} = \frac{\sqrt{3}}{2}\,\vec{r}_1$ and $\vec{Y} = \frac{1}{2}\,(\vec{r}_2 - \vec{r}_3)$. Chosing some $R = O(\rho)$, a small spherical cap near $\omega = 0$ will be distorted into a hypercylindrical surface of constant X. Integration over the hypercylinder reduces to a density integration over the Y coordinates of the close pair. If $(R/\rho) \to 1$ as $\rho \to \infty$ that covers the whole relative coordinate space.

By order of magnitude estimate, the whole sphere being $O(\rho^5)$ and Φ^{13} being $O(\rho^{-2})$ there must be $O(\rho)$ fluxes. Some $O(\rho)$ fluxes involve χ or Φ^i and one must then apply the principle of stationary phase to a multidimensional integral. The result is that there are indeed several $O(\rho)$ fluxes which relate to on-shell scattering and sum up to zero by virtue of the usual 2-body optical theorem. The cancellation pattern is the following:

$$\{\chi,\Phi^{13}\} + \{\Phi^1,\Phi^{13}\} + \{\Phi^1,\Phi^3\} + \{\chi,\Phi^{31}\} + \{\Phi^3,\Phi^{31}\} =$$

$$= \text{zero } O(\rho) + \ldots, \{\Phi^{13},\Phi^{13}\} + \{\Phi^3,\Phi^{13}\} = \text{zero } O(\rho) + \ldots$$

The expectation is that the truly 3-body optical theorem should occur at the $O(1)$ level since the spherical waves interfere to that order. Of the same order turn out to be the contributions by the hyper-cylinders. For example, when pair (2,3) are "close" one gets:

$$\{\chi,\Phi^{13}\} + \{\Phi^1,\Phi^3\} + \{\Phi^1,\Phi^{13}\} \sim 2^{-13/2} 3^{-1} \pi^{-6} \mathrm{Im}\langle\vec{Q}'|t_3(Q'^2)|\vec{Q}'\rangle \cdot$$

$$\cdot \left(\frac{P'}{Q'}\hat{P}'\cdot\hat{Q}' - \sqrt{3}\right) \int d^3y \left(|\Psi^1_{\vec{q}}(\vec{y})|^2 - |e^{i\vec{q}\cdot\vec{y}}|^2\right) ,$$

$$\{\Phi^{13},\Phi^{13}\} + \{\Phi^3,\Phi^{13}\} \sim 2^{-8} 3^{-1} \pi^{-5} \int d\Omega_{\hat{x}} \left[|\langle\hat{x}\vec{Q}'|t_3(Q'^2)|\vec{Q}'\rangle|^2 \cdot \right.$$

$$\left. \cdot \left(\tfrac{1}{2}\vec{P}'\cdot\hat{x} - \frac{\sqrt{3}}{2} Q'\right) \int d^3y \left(|\Psi^!_{\vec{q}}(\vec{y})|^2 - |e^{i\vec{q}\cdot\vec{y}}|^2\right)\right] ,$$

where the excess normalization integral is the Wigner-Eisenbud time-delay and can be related to the energy derivative of the 2-body S-matrix. This result is appealing since no information on the details of the potential has come in other than through the S-matrix, and the contribution gets very large when the subsystem is nearly binding.

Φ^{13} can be written as

$$\Phi^{13} = \Phi_O^{13} + \Phi_F^{13} + U^{13}$$

$$= \frac{e^{i\tau(\bar{P})}}{\rho^2} G(\omega,\hat{x}) I_{\hat{y}}(\bar{P})\Theta(\beta)$$

$$- \frac{e^{i\pi/4}}{\sqrt{\pi}} \frac{e^{i\tau(\bar{P})}}{\rho^2} G(\omega,\hat{x}) I_{\hat{y}}(\bar{P}) \operatorname{sgn}(\beta) F(|\tau(P_O)-\tau(\bar{P})|^{1/2})$$

$$- 2^{-3}\pi^{-2} 3^{-3/2} e^{-i\pi/4} \frac{\gamma_1 e^{\frac{i}{2}\vec{P}\cdot\vec{x}}}{Y\rho^{3/2}} e^{i\tau(P_O)} .$$

Φ_O^{13} is the $O(\rho^{-2})$ wave that went into the $O(\rho)$ flux; Φ_F^{13} is Fresnel's and U^{13} is the spherical wave. Interference fluxes involving Φ_F^{13} must be investigated. Fresnel's wave is $O(\rho^{-2})$ when its argument is small but it is $O(\rho^{-5/2})$ when it is large. At any finite ρ there is a continuous matching over a width $O(\rho^{-1/2})$. Such fluxes lead to evaluation of integrals with integrals of non-uniform asymptotic behaviour. The asymptotic series has contributions $O(1)$ from where the argument is large (replace $F \to F_{asympt.}$) but also special contributions from the matching. Taken literally, the substitution $F \to F_{asympt.}$ is meaningless [2] since it leads to divergent integrals. In fact correct treatment also provides unambiguous regularization [3] of otherwise divergent integrals. The importance of the matching should not be overlooked since its largest terms are $O(\rho^{1/2})$. For these terms one can prove the following cancellations:

$$\{\chi,\Phi^{13}\} = \text{zero } O(\rho^{1/2}) + \ldots ,$$

$$\{\Phi^{1},\Phi^{13}\} = \text{zero } O(\rho^{1/2}) + \ldots ,$$

$$\{\Phi_O^{13},\Phi^{13}\} + \{\Phi^{13},\Phi^{13}\} = \text{zero } O(\rho^{1/2}) + \ldots .$$

The truly 3-body optical theorem requires knowledge

of all O(1) fluxes. Previous [4] $O(\rho)$ and $O(\rho^{1/2})$ calculations must be pushed to higher order and cancellations do not carry on to the O(1) level. Moreover, since the $O(\rho^{-2})$ waves yeald $O(\rho)$ fluxes one must also investigate the $O(\rho^{-3})$ pieces and their contributions. All these O(1) terms involve the on-shell 2-body t-matrices and their energy derivatives. These derivatives come in since the energy of the latter scattering depends on the angle of the former scattering, which is a function of the direction in R_6. The method of stationary phase pushed to higher order involves also first and second order derivatives of the integrand at the stationary point [5] . Although some cancellations obtain these terms do not cancel out so that energy derivatives must be involved in the truly 3-body optical theorem.

Spelling out the O(1) terms presents now no more difficulty than correct bookkeeping and application of theorems of asymptotics. The ultimate result appears to be quite complicated as for the details of the coefficients. That should perhaps be of no great surprise since we are effectively looking for the explicit coordinate representation of a likely complicated operator relation.

REFERENCES

1. J. Nuttall, J. Math. Phys. 12 (1971) 197.
   S.P. Merkuriev, Teor. Mat. Fiz. 8 (1971) 235.
2. R.G. Newton, Ann. of Phys. 74 (1972) 324.
   R.G. Newton et al., Phys. Rev. A14 (1975) 642.
3. S. Servadio, preprint 1980.
4. S. Servadio, preprint 1980.
5. F.W.J. Olver, Asymptotics and Special Functions, Academic Press, 1974.

Acta Physica Austriaca, Suppl. XXIII, 695–701 (1981)

ON THE DETERMINATION OF πN PHASE SHIFTS FROM ISOSPIN CONSTRAINTS AND FIXED t ANALYTICITY[+]

by

I. S. STEFANESCU
Institut für Theoretische Kernphysik
Universität Karlsruhe, Germany

This seminar deals with the problem of the determination of the phase of the πN scattering amplitudes from observable quantities (differential cross-sections and polarizations). Its aim is to present, following Ref.[1], conditions on the amplitudes under which a unique construction of the phase can be achieved in some neighbourhood of the forward direction, using constraints coming from analyticity at fixed momentum transfer (t), isospin invariance and unitarity.

The interest in this question is motivated by the progress that has been achieved recently in the incorporation of analyticity at fixed t (Ref.[2]) (or along curves in the s-t space [3]) in phase shift analysis. As a result, one can claim nowadays that (to a good approximation), in πN scattering, amplitudes A(s,t), B(s,t) (the notation is that of Ref.[4]) can be constructed, which reproduce (within errors) the data ($d\sigma/d\Omega$, $P(\pi^+ p \to \pi^+ p, \pi^- p \to \pi^- p, \pi^- p \to \pi^0 n)$, $\sigma_{tot}(\pi^{\pm}-p)$),

---

[+]Seminar given at XX. Internationale Universitätswochen für Kernphysik, Schladming, Austria, February 17-26, 1981.

satisfy unitarity in all partial waves, are holomorphic in the s plane with cuts $s,u \geq (m+\mu)^2$ $(s+t+u = 2m^2 + 2\mu^2$, $m = m_N$, $\mu = m_\pi)$, for each fixed real t, $-t_o \leq t \leq 0$, and are consistent with analyticity in the "small" Lehmann-Martin ellipse, at fixed $s,u \geq (m+\mu)^2$. These amplitudes are consistent with crossing and isospin invariance, i.e. using $\nu = (s-u)/4m$, $A_+ = A(\pi^+p)$, etc., the observables for π^-p scattering are reproduced by $A_-(\nu,t) = A_+(-\nu,t)$, $B_-(\nu,t) = -B_+(-\nu,t)$, and those for $\pi^-p \to \pi^0 n$ by

$$A_o(\nu,t) = \frac{1}{\sqrt{2}} (A_+(\nu,t) - A_-(\nu,t)) \qquad (1)$$

and the same for B. Equations like (1) are known to imply [5] that the quantities $f^{\pm}_{\pm,o} \equiv (\frac{d\sigma}{d\Omega}(1 \pm P))^{1/2}_{\pm,o}(\nu,t)$ are constrained such that they make up an "isospin triangle": $|f^{\pm}_+ - f^{\pm}_-| \leq f^{\pm}_o \leq f^{\pm}_+ + f^{\pm}_-$.

There exist reasons to doubt that the construction above leads to unique amplitudes, without additional assumptions. E.g., if one tries to construct the phase at one fixed energy from measurements of $\frac{d\sigma}{d\Omega}$, P on the whole angular range, one meets discrete and continuum ambiguities: the discrete ones [6] are related to reflections of the zeros [7] of the extrapolated combinations $\frac{d\sigma}{d\Omega}(1 \pm P)(t)$; the continuum one is related to the nonpolynomial character of the amplitude [8,9] and its extent can be precisely investigated [10]. These ambiguities persist even if isospin invariance is used [8].

In discussing the construction of amplitudes, when analyticity in t is taken into account, one must in principle avoid using arguments based on analytic continuation, since the latter cannot be performed in general with controllable errors [11]. However, the use of fixed t analyticity requires information on the amplitudes in the "unphysical region" [4] present at $t < 0$, for $s,u \geq (m+\mu)^2$. A compromise, following

Refs.[12,13] is to assume that the observables $(\frac{d\sigma}{d\Omega}(1\pm P))_{\pm,0}$ can be continued with finite errors off the physical region to an interval $-t_o \leq t \leq 4\mu^2$ and then avoid, apart from this, analytic continuation completely.

The (very difficult) problem of principle is therefore to find the set of pairs of functions A(s,t), B(s,t) for $-t_o \leq t \leq 4\mu^2$, with the properties stated above and reproducing, via (1), the set of six data available over the three reactions at each point of $D \equiv \{s,u \geq (m+\mu^2),\ -t_o \leq t \leq 4\mu^2\} \cup \{s,t,u$ in the physical regions of the s,u channels$\}$. Further, one assumes the combination [14]

$$D(s,t) = A(s,t) + \nu B(s,t) \qquad (2)$$

is known at t = 0, for all s [13], via $(\frac{d\sigma}{dt})_{t=0}$ and the optical theorem, e.g. $\text{Im}\, D_+(s,0) \sim \sigma_{tot}(\pi^+p)$. We follow Ref.[13] in extending the study of uniqueness to the unphysical region $0 \leq t \leq 4\mu^2$ where use can be made of the positivity properties of D(s,t) [14]. One also wishes that, if a unique solution can be achieved in some way, it should be insensitive to small variations of the data.

In this form the problem is unsolved, because of the complicated form of the unitarity constraints. However, it appears it is possible to make progress in the following way [1]: one can find at each t, $-t_o \leq t \leq 4\mu^2$, the complete set of functions leading to prescribed values of the observables, holomorphic in the corresponding cut s plane (and polynomially bounded) and consistent with isospin invariance and (some form) of fixed s analyticity; further, we can show that the positivity properties of D together with some "experimental facts" and (apparently weak) assumptions can fix the amplitude completely, in some neighbourhood of t = 0. In Ref.[1], it is shown that: if the true amplitudes A(s,t), B(s,t) are such that (a1) they are continuous functions of s,t in D; (a2) there exist constants $\nu_o, c, \alpha', a' > 0$ so that, for $\nu \geq \nu_o$, $|A^+(\nu,t)|/|D^+(\nu,t)| < c$ in $-\alpha' \leq t \leq a'$ ($A^+ = A_+(\nu) + A_+(-\nu)$); this assumption can be checked for $t < 0$ by spin rotation measurements [15]; (a3) there exist constants C_R, ν'_o, so that, for $\nu \geq \nu'_o$, $-\alpha' \leq t \leq 0$, $|\mathrm{Re}\, C^+(\nu,t)| \leq C_R\, \mathrm{Im}\, C^+(\nu,t)$; (a4) the phase of a certain combination of amplitudes (h^+, Ref.[4]) is Hölder continuous and finite along the cuts, and if the following "facts" are true: (f1) there is no point in the physical region where the isospin triangles reduce to points; (f2) there exist constants c_6, s_6, so that $\sigma_{tot}(\pi^+p) + \sigma_{tot}(\pi^-p) > c_6$, $s \geq s_6$; (f'2) there exist constants c^+, ν_o, so that

$$(\frac{d\sigma}{d\Omega})_+(1\pm P)_+ + (\frac{d\sigma}{d\Omega})_-(1\pm P)_- - \frac{1}{2}(\frac{d\sigma}{d\Omega})_o(1\pm P)_o \geq c_+,$$

$\nu \geq \nu_o$, $-\alpha' \leq t \leq 0$; (f3) the function $D_+(\nu,0)$ does not vanish at symmetrical points of the ν plane, then: there exists an interval $-\alpha < t < a$, $\alpha, a > 0$, on which A,B (s,t) are uniquely determined by the set of six observables in D.

To achieve uniqueness in $0 \leq t \leq a$ for some a, one can give up (a3), (f'2). By uniqueness we mean that the determination of amplitudes is equivalent (under the assumptions above) to the construction of a meromorphic function in some unbounded domain (the ratio of two polynomially bounded holomorphic functions) from its values (modulus and phase) along the boundary. There remains a question of stability, since the number of zeros and poles of the meromorphic function must be controlled, and the problem of actually determining the interval $[-\alpha, a]$, given the real data. It is shwon in Ref.[1] that a (slight) strengthening of the assumptions allows one to settle these questions.

An outline is now given, on an artificial example (spinless scattering), of the main points of the argument.

Assume the following information on $a(\nu,t)$ is available, $0 \leq t \leq 4\mu^2$:

$$|a(\nu,t)| = f(\nu,t) \ , \tag{3a}$$

$$|a(\nu,t) - a(-\nu,t)| = f_o(\nu,t) \ , \tag{3b}$$

and $a^+(\nu,t) = a(\nu,t) + a(-\nu,t)$ has positivity properties in $0 \leq t \leq 4\mu^2$. Then, dividing (3b) by (3a) and squaring, we obtain the values of $r(\nu,t) = a(-\nu,t)/a(\nu,t)$ for $|\nu| \geq \nu_{th}$, up to a twofold ambiguity, since the sign of $\mathrm{Im}\, r(\nu,t)$ cannot be determined. However

$$\mathrm{Im}\, \frac{a(-\nu,t)}{a(\nu,t)} = \frac{\mathrm{Im}\, a^*(-\nu,t)a(\nu,t)}{|a(\nu,t)|^2} = \frac{2 \text{ area of "isospin triangle"}}{f^2(\nu,t)} \tag{4}$$

(condition (3b) replaces (1)). The area of the triangle constructed from $f(\nu,t)$, $f(-\nu,t)$, $f_o(\nu,t)$ is a holomorphic function of t at fixed s, and so the sign can be determined from its knowledge in one point, e.g. at $t = 0$ (this is no analytic continuation). Since $a(\nu,t)$ is completely determined by its zeros in the complex s plane and modulus on the cuts, knowledge of $r(\nu,t)$ for $|\nu| \geq \nu_{th}$ fixes it completely, unless $a(-\nu,t)$, $a(\nu,t)$ or, equivalently, $a^+(\nu,t)$, $a^-(\nu,t) \equiv a(\nu) - a(-\nu)$, have coincident zeros in the ν plane. However, if this happened at some $t > 0$, we follow such a pair to decreasing t it cannot reach $t = 0$, because of (f3); it cannot disappear through the cut, for sufficiently small a, because of (a1) and (f1) (see Ref.[1]); further, the coincident zeros cannot move to infinity since, as a consequence of (f2) and the positivity property of $a^+(s,t)$: $\mathrm{Im}\, a^+(s,t) \geq \mathrm{Im}\, a^+(s,0)$, $a^+(\nu,t)$ cannot vanish for $|\nu|$ sufficiently high in the complex plane: e.g. let $x = \nu^2(\cos\Theta + i \sin\Theta)$, $0 \leq \Theta \leq \frac{3\pi}{4}$, and use (f2):

$$|a^+(x,t)| \geq \frac{2|x|}{\pi}\left|\mathrm{Im}\int_{\nu_{th}^2}^{\infty}\frac{\mathrm{Im}\, a^+(x',t)}{x'(x'-x)}\,dx'\right| - \mathrm{const} \geq$$

$$\geq \frac{2|x|}{\pi}\int_{\bar{\nu}_6^2}^{\infty}\frac{dx'\,|x|\sin\Theta}{\sqrt{x'}[(x'-|x|\cos\Theta)^2+|x|^2\sin^2\Theta]} - \mathrm{const} \tag{5}$$

$$\geq 2\,|x|\bar{c}_6\ \mathrm{Im}\frac{1}{\sqrt{-x}} - \mathrm{const}\quad .$$

For $\pi \geq \Theta \geq \frac{3\pi}{4}$ one uses the lower bound given by $|\mathrm{Re}\, a^+(\nu,t)|$ with a similar result. So, there cannot exist coincident zeros of $a^+(\nu,t), a^-(\nu,t)$, for a small enough, and the amplitude is completely determined from knowledge of $r(\nu,t)$.

The argument for πN scattering is more elaborate, but follows the same lines. There is one difference: there exists only one amplitude with crossing symmetry and positivity, $D^+(s,t)$; this obliges us to use the supplementary assumption (a2) for $t > 0$. The role of (a3),(f'2) for $t < 0$ is to supplement the missing information coming from the positivity of $\mathrm{Im}\, D^+$ at $t > 0$.

REFERENCES

1. I.S. Stefanescu, TKP preprint 80/3, to appear in J. Math. Phys..
2. E. Pietarinen, Nucl. Phys. B107 (1976) 21.
3. R.E. Cutkosky et al., preprint COO-3066-117, LBL-8553 (1979), D.P. Hodgkinson et al., Proceedings of the 1976 Conference on Baryonic Resonances, p. 41.
4. G. Höhler et al., Handbook of Pion Nucleon Scattering, ZAED report 12-1 (1979).
5. L. Michel, Nuovo Cim. 22 (1961) 203.
6. J. H. Crichton, Nuovo Cim. 45A (1966) 256.

7. E. Barrelet, Nuovo Cim. 8A (1972) 331, F.A. Berends, Nucl. Phys. B64 (1973) 236.
8. D. Atkinson, G. Mahoux, F.J. Yndurain, Nucl. Phys. B66 (1973) 429.
9. G.R. Bart, P.W. Johnson, R.L. Warnock, J. Math. Phys.14 (1973) 1558, Nucl. Phys. B72 (1974) 329.
10. D. Atkinson, A.C. Heemskerk, S.D. Swiestra, Nucl. Phys. B109 (1976) 322.
11. S. Ciulli, C. Pomponiu, I.S. Stefanescu, Phys. Rep. 17C (1975) 135.
12. E. Pietarinen, Phys. Scripta 14 (1976) 11.
13. H. Burkhardt, A. Martin, Nuovo Cim. 29A (1975) 141.
14. G. Sommer, Nuovo Cim. 52A (1967) 373.
15. J. Pierrard et al., Phys. Lett. 57B (1975) 393.

Acta Physica Austriaca

Supplementum XXII

Field Theory and Strong Interactions

Edited by **P. Urban**

1980. 245 figures. V, 815 pages.
Cloth DM 166,–, S 1190,–. ISBN 3-211-81615-1

Supplementum XXI

Quarks and Leptons as Fundamental Particles

Edited by **P. Urban**

1979. 184 figures. V, 716 pages.
Cloth DM 149,–, S 1070,–. ISBN 3-211-81564-3

Supplementum XX

Laserspektroskopie

Neue Entwicklungen und Anwendungen

Herausgegeben von **F. Aussenegg**

1979. 103 Abbildungen. VI, 299 Seiten.
Gebunden DM 65,–, S 448,–. ISBN 3-211-81515-5

Supplementum XIX

Facts and Prospects of Gauge Theories

Edited by **P. Urban**

1978. 181 figures. VI, 889 pages.
Cloth DM 208,–, S 1488,–. ISBN 3-211-81514-7

Acta Physica Austriaca

Supplementum XVIII

Contacts Between High Energy Physics and Other Fields of Physics

Edited by **P. Urban**

1977. 170 figures. VI, 897 pages.
Cloth DM 208,–, S 1488,–. ISBN 3-211-81454-X

Supplementum XVII

The Schrödinger Equation

Edited by **W. Thirring** and **P. Urban**

1977. 13 figures and 1 portrait. VII, 224 pages.
Cloth DM 66,–, S 472,–. ISBN 3-211-81437-X

Supplementum XVI

Quantum Dynamics: Models and Mathematics

Edited by **L. Streit**

1976. 13 figures. X, 239 pages.
Cloth DM 66,–, S 472,–. ISBN 3-211-81414-0

Supplementum XV

Current Problems in Elementary Particle and Mathematical Physics

Edited by **P. Urban**

1976. 59 figures. VI, 638 pages.
Cloth DM 148,–, S 1020,–. ISBN 3-211-81401-9

Supplementum XIV

Electromagnetic Interactions and Field Theory

Edited by **P. Urban**

1975. 107 figures. V, 681 pages.
Cloth DM 148,–, S 1020,–. ISBN 3-211-81333-0